13

Springer Series on Fluorescence

Methods and Applications

Series Editor: M. Hof

For further volumes:
http://www.springer.com/series/4243

Springer Series on Fluorescence

Series Editor: M. Hof

Recently Published and Forthcoming Volumes

Fluorescent Methods to Study Biological Membranes
Volume Editors: Y. Mély and G. Duportail
Vol. 13, 2013

Fluorescent Proteins II
Application of Fluorescent Protein Technology
Volume Editor: G. Jung
Vol. 12, 2012

Fluorescent Proteins I
From Understanding to Design
Volume Editor: G. Jung
Vol. 11, 2012

Advanced Fluorescence Reporters in Chemistry and Biology III
Applications in Sensing and Imaging
Volume Editor: A.P. Demchenko
Vol. 10, 2011

Advanced Fluorescence Reporters in Chemistry and Biology II
Molecular Constructions, Polymers and Nanoparticles
Volume Editor: A.P. Demchenko
Vol. 9, 2010

Advanced Fluorescence Reporters in Chemistry and Biology I
Fundamentals and Molecular Design
Volume Editor: A.P. Demchenko
Vol. 8, 2010

Lanthanide Luminescence
Photophysical, Analytical and Biological Aspects
Volume Editors: P. Hänninen and H. Härmä
Vol. 7, 2011

Standardization and Quality Assurance in Fluorescence Measurements II
Bioanalytical and Biomedical Applications
Volume Editor: Resch-Genger, U.
Vol. 6, 2008

Standardization and Quality Assurance in Fluorescence Measurements I
Techniques
Volume Editor: U. Resch-Genger
Vol. 5, 2008

Fluorescence of Supermolecules, Polymeres, and Nanosystems
Volume Editor: M.N. Berberan-Santos
Vol. 4, 2007

Fluorescence Spectroscopy in Biology
Volume Editor: M. Hof
Vol. 3, 2004

Fluorescence Spectroscopy, Imaging and Probes
Volume Editor: R. Kraayenhof
Vol. 2, 2002

New Trends in Fluorescence Spectroscopy
Volume Editor: B. Valeur
Vol. 1, 2001

Fluorescent Methods to Study Biological Membranes

Volume Editors:
Yves Mély
Guy Duportail

With contributions by

L.A. Bagatolli · C. Billaudeau · R. Bizzarri · S. Brasselet · C.E. Butler · A. Chattopadhyay · M. Cottet · J.-L. Darlix · P. Didier · G. Duportail · T. Durroux · C. Eggeling · O. Faklaris · P. Ferrand · H. Gacem · M.F. Garcia-Parajo · A.J. García-Sáez · A. Gasecka · P. Gonnord · G. Gorbenko · J. Graham · S. Haldar · Y. Hamon · H.-T. He · M. Heilemann · M. Hof · M. Jafurulla · P. Jurkiewicz · B. Kaltschmidt · C. Kaltschmidt · P.K.J. Kinnunen · A.S. Klymchenko · A. Kress · L.M.S. Loura · F.W. Lund · S. Mailfert · C. Manzo · D. Marguet · Y. Mély · F. Neugart · A. Olżyńska · J.-P. Pin · S. Pokorna · M. Prieto · H. Ranchon · H. de Rocquigny · V. Rouger · L.M. Solanko · M.T. Stöckl · Y. Subburaj · V. Subramaniam · E. Trinquet · T. Trombik · K.M. Tyler · R. Varma · X. Wang · G. Wheeler · D. Widera · D. Wüstner · T.S. van Zanten · M. Zelman-Femiak

Springer

Volume Editors
Professor Dr. Yves Mély
Dr. Guy Duportail

Université de Strasbourg
Faculté de Pharmacie
UMR 7213 du CNRS
Laboratoire de Biophotonique et Pharmacologie
Illkirch, France

ISSN 1617-1306 ISSN 1865-1313 (electronic)
ISBN 978-3-642-33127-5 ISBN 978-3-642-33128-2 (eBook)
DOI 10.1007/978-3-642-33128-2
Springer Heidelberg New York Dordrecht London

Library of Congress Control Number: 2012948982

Printed on acid-free paper

Springer is part of Springer Science+Business Media (www.springer.com)

Series Editor

Prof. Dr. Martin Hof

Academy of Sciences of the Czech Republic
J. Heyrovsky Institute of Physical Chemistry
Department of Biophysical Chemistry
Dolejskova 3
16223 Prague 8
Czech Republic
martin.hof@jh-inst.cas.cz

Aims and Scope

Fluorescence spectroscopy, fluorescence imaging and fluorescent probes are indispensible tools in numerous fields of modern medicine and science, including molecular biology, biophysics, biochemistry, clinical diagnosis and analytical and environmental chemistry. Applications stretch from spectroscopy and sensor technology to microscopy and imaging, to single molecule detection, to the development of novel fluorescent probes, and to proteomics and genomics. The *Springer Series on Fluorescence* aims at publishing state-of-the-art articles that can serve as invaluable tools for both practitioners and researchers being active in this highly interdisciplinary field. The carefully edited collection of papers in each volume will give continuous inspiration for new research and will point to exciting new trends.

Preface

Biological membranes play a central role in cell structure, shape, and functions. While the plasma membrane surrounds the cell and separates its interior from the extracellular environment, a number of intracellular membranes compartmentalize the cell interior into organelles having different functions. Being selectively permeable to ions and organic molecules, cell membranes primarily control the exchange of substances and nutriments in and out of cells. They are also involved in a variety of cellular processes, such as cell adhesion and signaling. Biological membranes of eukaryotic cells are essentially built up by a lipid bilayer, constituted of a large number of different lipids, where a large variety of proteins are inserted. Glycerophospholipids are the main structural components of cellular membranes, either the plasma membrane or membranes of organelles. Together with the other types of lipids, including sphingolipids and cholesterol, these amphipathic molecules spontaneously self-assemble into lipid bilayers so that their hydrophobic tails are isolated from the surrounding polar medium and their more hydrophilic head groups are exposed to the extracellular medium and the cytosol. Initially, according to the "fluid mosaic model," biological membranes and, notably, the plasma membrane were simply perceived as a lipidic "sea," in which membrane proteins were freely diffusing. Nowadays, it is well recognized that this simplistic picture is inappropriate and that biological membranes are highly heterogeneous systems in which embedded molecules show complex diffusion patterns. In fact, due to their high structural and chemical diversity and their highly heterogeneous spatiotemporal distribution, lipids play a key role in membrane functions by regulating the conformational state and functions of proteins inserted into or associated with the membranes. Lipids also directly participate in signaling pathways and constitute direct molecular receptors of a number of external agents. An important consequence of the diversity and dynamics of membrane lipids is the formation of transient but highly ordered lipid domains, often referred to as membrane "rafts" that allow tight spatial packing of important proteins. These domains are thought to function as signaling platforms in a variety of important biological functions. Although the membrane properties and functions are of key importance, their investigation proved to be difficult due to their intrinsic characteristics.

Indeed, the depth of the membrane bilayer is only 4–5 nm, which is two orders of magnitude below the resolution of optical microscopes, precluding any direct visualization of membranes in the "z" direction. Similarly, the highly dynamic nature as well as the small size of lipid "rafts" (below 100 nm) precluded, up to now, their direct observation. Another key issue is the highly anisotropic structure of the bilayer, which generates steep gradients at the nanometer scale. For instance, in phospholipids characterized by polar head groups and highly apolar aliphatic chains, there are steep gradients of polarity and hydration between the membrane surface and its interior. Moreover, the asymmetric distribution of charges at the membrane level generates a number of static electric fields with different localizations, often classified as surface (Ψs), dipole (Ψd), and transmembrane (Ψt) potentials. Another issue of key importance in biomembranes is their asymmetric lipidic distribution between the inner and outer leaflets, especially in the plasma membrane, since negatively charged phospholipids, like phosphatidylserine, are mainly distributed in the inner leaflet while sphingolipids are distributed in the outer leaflet. In addition, to ensure the efficient absorption of external molecules needed for cell metabolism, the plasma membrane continuously undergoes endocytosis, a process in which the plasma membrane engulfs the extracellular content through small vesicles that detach into the cytoplasm. As a consequence, it is considered that the whole membrane surface of fibroblasts is internalized every 16 minutes. Taking into account these various aspects, it is obvious that the design of membrane probes and instrumentation for characterizing membrane properties and functions is particularly challenging. Nevertheless, due to the consistent efforts of a large scientific community and the decisive impact of fluorescence-based techniques, remarkable progress has been made in the last decades in the understanding of membrane characteristics and functions.

In this context, this book illustrates some of these major advances through a collection of review articles written by invited authors who are experts in the subjects of membranes and fluorescence. A number of these authors have presented their work at the 12th International Conference on Methods and Applications of Fluorescence (MAF-12), held in Strasbourg, France, in September 2011. Since their inception in 1989, the biennial MAF conferences have become the largest meetings dedicated to studies concerning fluorescence, being highly interdisciplinary and covering areas including physics, chemistry, nanotechnology, biology, and medicine. The MAF-12 conference has attracted more than 400 attendees from 40 countries worldwide.

This book is organized into three parts. The first part deals with membrane probes and model membranes. Two "historical" membrane probes, namely, Laurdan- and NBD-labeled lipids, are presented, with a discussion on both their early and present applications. The two-color 3-hydroxychromone probes are also described, with a focus on the importance of their precise location and orientation in the membrane for their applications in sensing membrane potentials, lipid domains, and apoptosis. Three review articles on membrane models, such as lipid vesicles and supported bilayers, are also presented. Model membranes with controlled composition and environment continue to play a central role in understanding the basic physicochemical properties of biological membranes. These review articles focus specifically on the use of FRET to probe the lateral heterogeneity of membranes and lipid–protein interactions and the use of time-resolved fluorescence techniques to monitor hydration and mobility changes across the bilayer. The second part describes the use of fluorescent-based microscopy techniques, and notably of advanced quantitative and high-resolution techniques, to explore the properties of biological membranes. This part illustrates the key advances that have been made in the understanding of membrane organization, dynamics, and interactions using techniques such as wide-field microscopy; multiphoton microscopy; fluorescence lifetime imaging microscopy (FLIM); fluorescence correlation spectroscopy (FCS), spot variation FCS, and FCS combined with stimulated emission depletion (STED) microscopy; polarization-resolved fluorescence microscopy; and near-field optical nanoscopy. The last part focuses on the investigation by these techniques of membrane proteins and, notably, membrane receptors that play a central role in a number of signaling pathways and are, thus, largely targeted in therapeutic strategies. Two review articles discuss the strategies that can be used to explore either the biophysical properties of a hypothetical membrane receptor or the oligomerization of G-protein-coupled receptors (GPCR). Two additional review articles focus on the characterization of serotonin 1A and TNF receptors. Finally, the last review article deals with the assembly of retroviral HIV-1 particles at the plasma membrane level.

Strasbourg, France Yves Mély and Guy Duportail

Contents

Part I
Fluorescence Probes and Model Membranes

LAURDAN Fluorescence Properties in Membranes: A Journey from the Fluorometer to the Microscope

L.A. Bagatolli

Abstract After 32 years since its introduction, the particular fluorescence properties of 6-lauroyl-2-(dimethylamino)-naphtalene (LAURDAN) in model and biological membranes are revisited. This review includes a historical perspective about the design, synthesis and initial description of the probe's fluorescent properties, a discussion about the proposed mechanism of LAURDAN sensitivity to membrane lateral packing, and a detailed description of the definition of the Generalized Polarization function. This article includes as well examples of the different experimental strategies involving LAURDAN in model and biological membranes, using both bulk fluorescence spectroscopy measurements and spatially resolved information from fluorescence microscopy. The value of this probe in the study of membrane structure and dynamics is reflected in more than 330 papers reported in the existing literature.

Keywords Biological membranes · Dipolar relaxation process · Fluorescence microscopy/spectroscopy · Generalized polarization function · Membrane lateral structure

Contents

L.A. Bagatolli (✉)
Membrane Biophysics and Biophotonics Group/MEMPHYS-Center for Biomembrane Physics, Department of Biochemistry and Molecular Biology, University of Southern Denmark, Campusvej 55, DK-5230 Odense M, Denmark
e-mail: bagatolli@memphys.sdu.dk

Y. Mély and G. Duportail (eds.), *Fluorescent Methods to Study Biological Membranes*,
Springer Ser Fluoresc (2013) 13: 3–36, DOI 10.1007/4243_2012_42,

Abbreviations

ACDAN	6-Acethyl-2-dimethylamine-naphthalene
ACRYLODAN	6-Acryloyl-2-dimethylaminonaphtalene
C-LAURDAN	6-Dodecanoyl-2-[N-methyl-N-(carboxymethyl)amino] naphthalene
Cer	(ceramide) *N*-acylsphingosine
Cholesterol	(3β)-Cholest-5-en-3-ol
D_2O	Deuterium oxide
DANCA	2′-(*N,N*-dimethyl)amino-6-naphtophyl-4-trans-cyclohexanoic acid
DHPC	1,2-Di-O-hexadecyl-*sn*-glycero-3-phosphocholine
DLPC	1,2-Dilauroyl-*sn*-glycero-3-phosphocholine
DMPC	1,2-Dimiristoyl-*sn*-glycero-3-phosphocholine
DOPC	1-Oleoyl-2-palmitoyl-*sn*-glycero-3-phosphocholine
DPPC	1,2-Dipalmitoyl-*sn*-glycero-3-phosphocholine
Gal	Galactose
G_{D1a}	NeuAc2α → 3Galβ1 → 3Gal-NAcβ1 → 4Gal (3 ← 2αNeuAc)β1 → 4Glcβ1 → 1'Cer
GalCer	Galβ1 → 1'Cer
Glc	Glucose
Gg_3Cer	(Asialo GM2) Gal-NAcβ1 → 4Galβ1 → 4Glcβ1 → 1'Cer
Gg_4Cer	(Asialo GM1) Galβ1 → 3Gal-NAcβ1 → 4Galβ1 → 4Glcβ1 → 1'Cer
G_{M3}	NeuAc2α → 3Galβ1 → 4Glcβ1 → 1'Cer
G_{M2}	Gal-NAcβ1 → 4Gal(3 ← 2αNeuAc)β1 → 4Glcβ1 → 1'Cer
G_{M1}	Galβ1 → 3Gal-NAcβ1 → 4Gal(3 ← 2αNeuAc)β1 → 4Glcβ1 → 1'Cer
GP	Generalized Polarization function
GP_{ex}	Generalized Polarization, excitation
GP_{em}	Generalized Polarization, emission
G_{T1b}	NeuAc2α → 3Galβ1 → 3GalNAcβ1 → 4Gal(3 ← 2αNeuAc8 ← 2αNeuAc)β1 → 4Glcβ1 → 1'Cer
LAURDAN	6-Dodecanoyl-2-dimethylamine-naphthalene
LAURISAN	2-Diisopropylamino-6-lauroylnaphthalene
LAURMEN	2-Methoxy-6-lauroylnapthalene

LAURNA	2-Hydroxy-6-lauroylnapthalene
ld	Liquid disordered phase
lo	Liquid ordered phase
NeuAc	Neuraminic acid
PATMAN	6-Palmitoyl-2-[[(2-trimethylammonium)ethyl]methyl]amino] naphthalene
Phre	(phrenosine) Galβ1 → 1'(*N*-[α-OH]acylsphingosine)
PRODAN	6-Propionyl-2-dimethylamine-naphthalene
so	Solid ordered (gel) phase
Sphingomyelin	*N*-acyl-D-*erythro*-sphingosylphosphorylcholine
Sulf	(Sulfatide) HSO3 → 3Galβ1 → 1'Cer
TPEFM	Two photon excitation fluorescence microscopy

1 Introduction

The displacement of the fluorescence spectrum to longer wavelengths (red shift) with increasing polarity of solvents has been the object of both theory and experiment since the 50s. This type of red shift is largely dependent upon a large increase in the molecule's dipole moment in the fluorescence state over that of the ground state [1]. Particular substituted aromatic molecules can satisfy this condition. In 1979, Gregorio Weber introduced a family of environmentally sensitive dyes based on a naphthalene structure substituted with a electron donor (alkylamino group) and acceptor (acyl substituted carbonyl group) groups in position 2 and 6 [1]. This structure possesses a maximum distance between the electron donor and acceptor groups, resulting in a lowest excited state with an important charge transfer character. In his seminal publication, Weber presented a general method for the synthesis of 2-(dimethylamino)-6-acylnaphtalenes, i.e., 6-propionyl (PRODAN), 6-acetyl (ACDAN), and 6-lauroyl (LAURDAN), together with a careful study of the response of PRODAN's absorption and fluorescence properties to solvents of different polarity. In this description the dipole interaction theory of Lippert [2] was considered and generalized to the other synthesized naphtalene derivatives [1]. For example, a 130 nm red shift (one of the largest reported so far) was described for PRODAN's emission maximum from cyclohexane to water with an important change in the magnitude of its transition dipole. Additionally, a significant (larger) Stoke's shift was reported in solvents that can form hydrogen bonds (i.e., that can act as proton donors) with respect to aprotic solvents [1]. Later on, further studies of PRODAN's response to solvent polarity were conducted in Weber's laboratory, where the Langevin distribution of electrostatic interactions was considered [3].

Weber's contribution went beyond the rational design of environmentally sensitive fluorescent molecules and the characterization of their response in solvents with different polarity. An essential contribution was to apply this knowledge to the study of biological material. More precisely, in 1979 Weber presented a study of the interaction of PRODAN with bovine serum albumin, proposing the use of this 2-(dimethylamino)-6-acylnaphtalene derivative as "a relaxation probe of various biological environments" [1]. In a later work, Weber exploited the nanosecond relaxation phenomenon in a protein matrix (addressed earlier in his laboratory [4]) to determine the polarity of the myoglobin haem pocket [5]. Specifically, Weber designed and used another 2,6 substituted naphthalene derivative, i.e., 2′-(*N*, *N*-dimethyl)amino-6-naphtophyl-4-*trans*-cyclohexanoic acid (DANCA) [5], a probe with more affinity than PRODAN for apomyo-globin. In a similar way, Prendergast et al. synthesized 6-Acryloyl-2-dimethylaminonaphtalene (ACRYLODAN), a 2,6 substituted naphthalene derivative that selectively labels thiol moieties in proteins [6]. In the Prendergast et al. paper, the usefulness of ACRYLODAN in the study of "hydrophobic" domains, conformational changes, and dipolar relaxation processes in proteins was demonstrated by measurements of fluorescence spectra and lifetimes of a mercaptoethanol adduct in different solvents and of adducts of this agent with parvalbumin, troponin C, papain, and carbonic anhydrase [6].

The interest of environmental relaxation processes in biological systems was not restricted only to proteins. Similar ideas were used to explore the this phenomenon in membranes using different 2-(dimethylamino)-6-acylnaphtalene derivatives introduced by Weber, i.e., LAURDAN, PRODAN, ACDAN, and 6-palmitoyl-2-[[(2-trimethylammonium)ethyl]methyl]amino] naphthalene (PATMAN), the last one introduced by Lakowicz et al. in 1983 [7]. Other derivatives had been also synthesized in Weber's group, i.e., 2-diisopropylamino-6-lauroylnaphthalene (LAURISAN), 2-methoxy-6-lauroylnapthalene (LAURMEN), and 2-hydroxy-6-lauroylnapthalene (LAURNA), and further characterized by Parasassi et al. [8]. Very recently, a new 2,6 naphthalene derivative has been introduced 6-dodecanoyl-2-[*N*-methyl-*N*-(carboxymethyl)amino]naphthalene (C-LAURDAN), to explore membranes mainly using fluorescence microscopy [9, 10]. The molecular structures of some of these compounds are illustrated in Fig. 1.

To the best of my knowledge, the first membrane study using LAURDAN (along with other membrane probes) was published by Lakowicz and Sheppard in 1981 [11]. This paper focused on the membrane properties of intact fibroblast from both normal subjects and patients with Huntington's disease. In a similar type of study, Sumbilla and Lakowicz [12] reported a year later another comparative membrane study using LAURDAN-labeled red blood cells from normal subjects and Huntington's disease patients. This last paper comments on the high sensitivity of LAURDAN to the polarity and relaxation time of its membrane environment. However, it was not until 1986 that the first extensive characterization of LAURDAN fluorescence properties was performed in membrane model systems (glycerophospholipid vesicles) by Parasassi et al. [13]. In this work, the effect of solvent relaxation, reflected as a pronounced red shift of the steady state and time-resolved emission spectra of LAURDAN, was observed at (and above) the solid ordered (s_o) to liquid disordered (l_d) phase transition occurring in the membrane. In order to explain the particular response of the probe the authors invoked the existence of an environment that can orient with the LAURDAN excited state dipole [13]. Furthermore, from the time-resolved fluorescence experiments these authors computed the relaxation rate of the probe during the lipid phase transition, that occurs in a similar time window (nanoseconds) to that of the probe's fluorescence lifetime. A two state model was used to describe some of the features

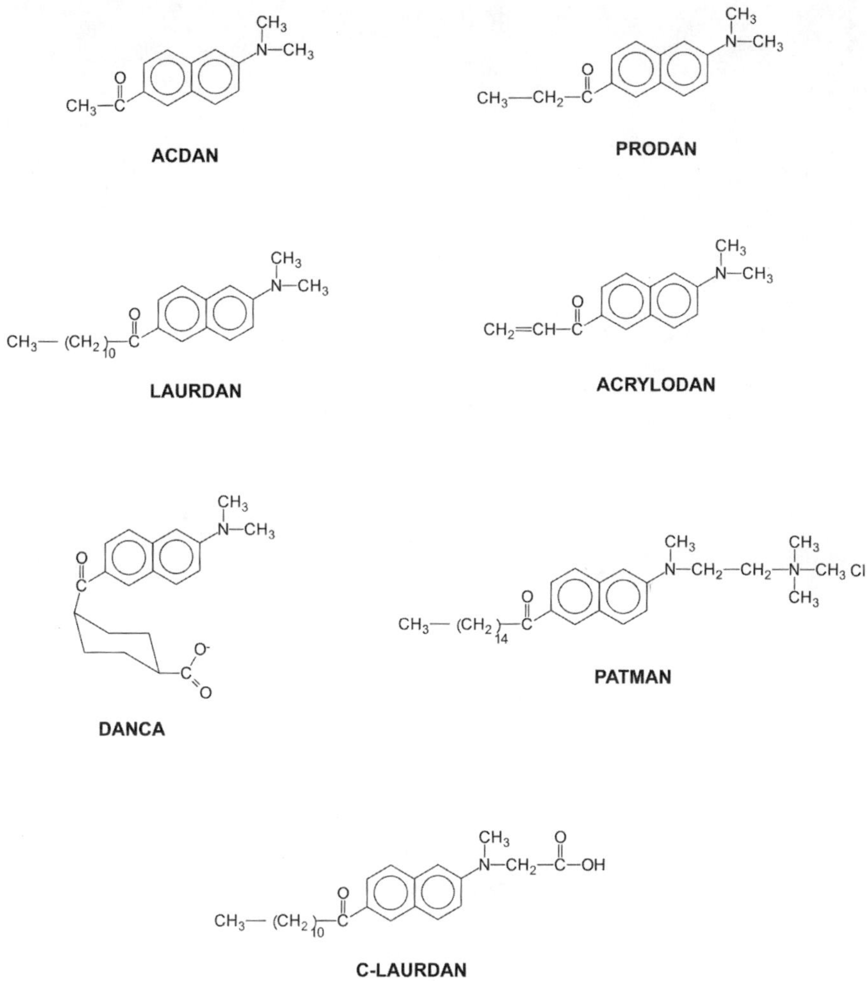

Fig. 1 Chemical structures of different 2,6 substituted naphthalene derivatives

observed during the relaxation process [13, 14], providing the basis for the concept of the Generalized Polarization (GP) function introduced by these authors in subsequent publications [15, 16]. The GP function was defined to exploit a simple steady state parameter (the probe's emission spectra) to study structural and dynamical processes in model and biological membranes (see Sect. 4 below).

Although LAURDAN is currently the most used 2-(dimethylamino)-6-acylnaphtalene derivative for membrane studies, PRODAN, ACDAN, and PATMAN were also employed to study relaxation processes in model membranes. Similar to LAURDAN, experiments with PRODAN were performed to explore dipolar relaxation in membranes undergoing phase transitions [17, 18] as well as to study the effects of pressure in model lipid membranes [19, 20]. Unlike LAURDAN, PRODAN

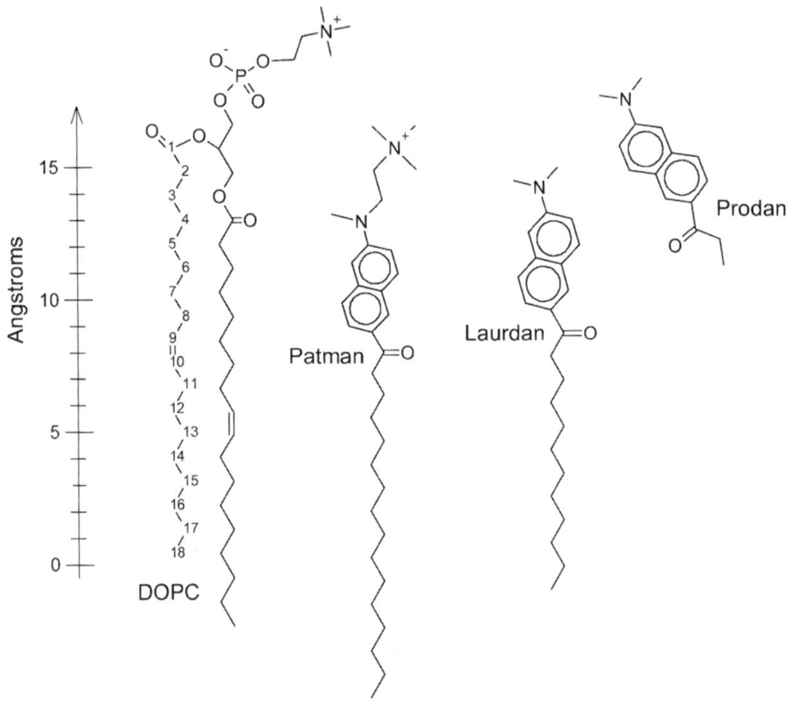

Fig. 2 Schematic diagram of the location of PATMAN, LAURDAN, and PRODAN in a membrane composed of dioleoylphosphatidylcholine (DOPC). The carbon atoms of the sn_2 chain are numbered. The scale on the left shows the approximate distance from the bilayer center. Adapted from [23] with permission

possesses a propionyl tail that is more loosely anchored in the lipid bilayer and the localization of this probe was shown to be closer to the membrane interface [20]. Additionally, the partition of PRODAN for phospholipid membranes is ~2 orders of magnitude lower than LAURDAN and, therefore, a fraction of PRODAN is contributing from the aqueous medium (a situation that is reflected in a characteristic shoulder at 520 nm in its emission spectrum) [21]. LAURDAN's fluorescent group was reported to be located ~10Å from the center of the bilayer [22], quite similar to PATMAN but considerably deeper than PRODAN. PATMAN is located ~1Å deeper than LAURDAN in the membrane [23] (see Fig. 2). PATMAN has been mainly used to study dipolar relaxation processes in membranes composed of different lipids [23–25] using a solvent relaxation technique that is based on simple time-resolved fluorescence measurements [26]. Finally, ACDAN (together with LAURDAN and PRODAN) has been used to study lipid interdigitation caused by ethanol in phospholipid bilayers [21]. This probe has an even shorter alkyl substitution in position 6 of the naphthalene ring with respect to LAURDAN (an acetyl tail, see Fig. 1) showing a very low partition to the membrane interface (three orders of magnitude lower than LAURDAN). In the presence of membranes, the dominant contribution in the ACDAN emission spectrum comes from the aqueous environment (maximum at ~520 nm) [21]. Since this review deals mainly with LAURDAN, I strongly recommend the readers to explore the aforementioned references for more information about the other naphthalene derivatives.

2 A Model for LAURDAN Relaxation in Membranes

The very important features of LAURDAN are: (i) the negligible contribution of the probe from water (its partition to the membrane is highly favored) and (ii) the even partition of LAURDAN in membranes displaying phase coexistence [27–29]. When LAURDAN is located in glycerophospholipid membranes an emission red shift of ~50 nm is observed when the membrane undergoes a s_o to l_d phase transition (Fig. 3a) [13, 15, 16]. As seen also in Fig. 3a, when s_o and l_d phases coexist in the membrane the probe shows an emission spectrum with intermediate properties [29]. This membrane-phase-dependent emission shift was reported to be independent of the nature of the glycerophospholipid polar head group (pH between 4 and 10) [16], and was also observed for some sphingolipid species forming bilayers [30, 31]. Since changes in the "static" dielectric constant between the two membrane phases is not sufficient to explain the observed fluorescence emission shift, a model to interpret the changes in LAURDAN's emission properties was originally provided by Parasassi et al. [16]. These authors proposed that the nanosecond relaxation process observed in the l_d phase [13, 15, 32] is caused by the presence of water molecules with restricted mobility in the region where LAURDAN is located (nearby the glycerol backbone of the glycerophospholipids, see Fig. 2) [16, 22, 23]. Part of LAURDAN's excited state energy is utilized for the reorientation of the water dipoles diminishing the single excited state (S_1) energy (see Fig. 3b) and consequently shifting the emission spectrum of the probe to longer wavelengths. Importantly, the relaxation caused by these water molecules is different to the water molecules existing in bulk that have an orientational relaxation time below 1 ps [16]. The relaxation time have been measured for the l_d and s_o phases using an equivalent expression of the classical Perrin equation for the GP (for details of the GP definition see below, Sect. 4), assuming a two state process [32]. These values were reported to be $2.5 \times 10^9\ s^{-1}$ for the l_d phase and $4 \times 10^7\ s^{-1}$ for the s_o phase. Notice that the relaxation time measured in the l_d phase is similar to the lifetime of the probe (~3 ns [30]).

There is considerable experimental evidence that supports the idea that structured water molecules in the vicinity of the probe are the cause of LAURDAN's emission shift in model membranes:

(a) The relaxation process is independent of the nature of glycerophospholipid polar head group and pH, i.e., the emission spectrum position depends on the membrane phase [16]. Additionally, this behavior is also observed in ether derivatives of glycerophospholipids [33] and sphingolipid-containing bilayers [30].

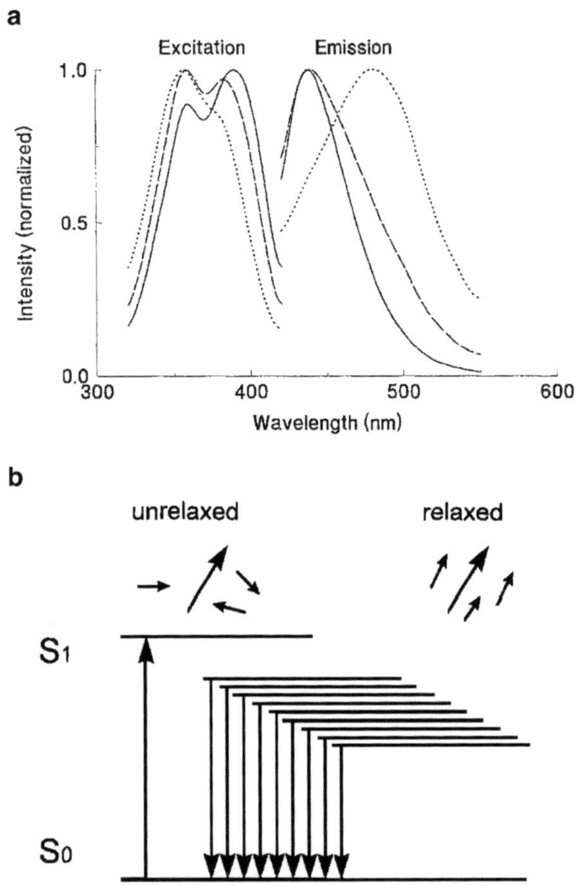

Fig. 3 (**a**) Normalized LAURDAN excitation and emission spectra in phospholipid multilamellar vesicles displaying a s_o phase (*continuous line*), l_d phase (*dotted line*), and s_o/l_d phase coexistence (*dashed line*). Phospholipid composition of the vesicles and the temperature of measurements are DLPC at 40°C (*dotted line*), DPPC at 5°C (*continuous line*), and an equimolar mixtures of the two phospholipids at 20°C; (**b**) Schematics of the ground (S_0) and excited state (S_1) energy levels in the presence of the solvent dipolar relaxation. S_1 decreases in energy as solvent dipolar relaxation proceeds. Adapted from [8] and [37] with permission

(b) Experiments in membranes composed of DMPC in D_2O show slower relaxation dynamics at and above the phase transition (higher GP value in D_2O), although the main phase transition temperature is the same with respect to that observed for the same lipid membranes in H_2O [8].
(c) Information obtained from classical polarization spectra excludes probe reorientation along its molecular axis as the main cause of the observed emission shift [8].
(d) Results obtained using LAURDAN derivatives suggest that the partial charge separation is necessary to observe the emission shift (experiments with

LAURNA and LAURMEN); also, steric hindrance to intramolecular reorientation expected for LAURISAN (which has a di-isopropilamino group in the position 2 of the naphtalene ring) has no effect on the red shift. These experiments point out that the molecular entity responsible for the dipolar relaxation cannot be the fluorophore itself [8].

(e) Based on experiments performed in alcohols at very low temperature, Viard et al. [34] proposed that depending on the relative orientation of the dimethylamino and carbonyl groups, two states contribute to the emission of LAURDAN: the locally excited state and charge transfer state. This model was applied with some modifications to explore the nanosecond dynamics of membrane-water interface in reverse micelles of AOT-water in isooctane [35]. Although the locally excited state can be confirmed in apolar solvents (where the lifetime and quantum yield is low), there is no clear evidence that this state contributes to LAURDAN emission in glycerophospholipid membranes. For example, the high quantum yield observed in membranes together with a clear lifetime dependence on the membrane phase (~3 and 6 ns in the s_o and l_d phases, respectively [30]) and the lack of isosbestic point during the phase transition demonstrate that the charge transfer state is dominant (see also point (d) above) [8], i.e., there is one state that relaxes upon the lipid main phase transition causing the emission spectrum shift, not an intramolecular relaxation of the probe.

Based on novel observations obtained by two photon excitation fluorescence microscopy (TPEFM, see Sect. 6), Parasassi et al. refined the model of water relaxation [36]. This seminal article reported for the first time spatially resolved information of the LAURDAN GP function in micrometer-sized (multilamellar) vesicles, showing that the GP images obtained in l_d phase membranes show a much broader GP distribution relative to the s_o phase. This observation indicates the existence of a large dynamical heterogeneity in the l_d phase, unexpected to the authors for a membrane displaying a single phase. To explain these observations Parasassi et al. proposed a distribution of different sites (or cavities) in which the LAURDAN molecule can reside [36]. These sites are characterized by a different number of dynamically restricted water molecules, since the GP value is sensitive to the water content of the membrane [37]. The average number of water molecules at the location of the LAURDAN fluorescent moiety was estimated to be not more than two or three [36]. Considering a Poisson distribution of these few water molecules, the authors concluded that there is a distribution of LAURDAN environments with no, one, two, three, etc. molecules of water. For example, for an average of two molecules of water per cavity around the LAURDAN fluorescent moiety, the Poisson distribution of water molecules at the different sites is $0 \rightarrow 0.135$, $1 \rightarrow 0.270$, $2 \rightarrow 0.270$, $3 \rightarrow 0.203$, $4 \rightarrow 0.090$, $5 \rightarrow 0.031$, and more than $5 \rightarrow 0.020$, concluding that the larger the number of water molecules, the lower is the GP, and the larger is the cavity around the probe. This model has been recently supported by scanning-fluctuation correlation spectroscopy measurements of the LAURDAN GP function on the scale of few pixels in single glycerophospholipid membranes in the l_d phase [38].

The model of LAURDAN relaxation in membranes has been recently revisited by Jurkiewicz et al. [39]. Based on time-dependent fluorescence shift measurements complemented with molecular dynamic simulations, these authors mention that the nanosecond relaxation times reported by LAURDAN in membranes carry information exclusively on the mobility of hydrated lipid moieties at the level where the probes are located and not about the dynamic of water molecules, which interchange or even exchange with those from bulk water on a picosecond timescale. The author's conclusion is based on the fact that in glycerophospholipid bilayers, and particularly at the glycerol backbone level where the dimethylaminonaphtalene group of the probe is located, water molecules are sparse and fully bound to lipids (mainly to the lipid carbonyl groups). However, the D_2O experiments (point (c) above) together with the fact that a similar response of the probe emission is observed in bilayers where the chemical environment is different at the probe's position (point (a), i.e., sphingomyelin, Gg_3Cer, Phrenosine, sulfatide, galactosyl ceramide [30], and DPPC and its ether derivative [33] undergoing a s_o to l_d phase transition), reinforce the idea that water relaxation is the main cause of the observed probe's emission shift.

3 LAURDAN Ground State Information

In addition to the emission spectral features of the naphthalene probes, their absorption and excitation spectra also show the existence of different excitation states. In polar (protic) solvents, the excitation spectra of LAURDAN (and PRODAN) are composed of at least two bands, centered at about 350 and 390 nm [8, 16]. The excitation band centered at 390 nm has been called the red excitation band. The excitation spectra of LAURMEN and LAURNA in ethanol do not show the red excitation band, suggesting that the presence of a partial charge separation due to the carbonyl and the dimethylamino residue is a necessary condition for its observation [8]. This excitation band at 390 nm has been interpreted as the absorption of LAURDAN molecules stabilized in a ground state Lα conformation by surrounding dipoles [16]. In glycerophospholipid vesicles (independent of the nature of the polar head group), the excitation spectra of LAURDAN are also composed of at least two bands [15, 16]. With respect to polar solvents, the red excitation band is particularly intense in membranes displaying a s_o phase constituting the excitation spectrum maximum (Fig. 3a). In the glycerophospholipid l_d phase, the red excitation band is less intense than in the s_o phase and it is no longer the excitation spectrum maximum (see Fig. 3a) [8]. Although this behavior is general for glycerophospholipid membranes, an ether derivative of DPPC (DHPC) and the sphingolipid sphingomyelin behave differently [33]. Specifically, the dominant red excitation band observed in glycerophospholipids in the s_o phase gradually decreases in the order DPPC > DHPC > sphingomyelin, suggesting that the stabilization of ground state Lα conformation responds differently to changes in the chemical groups present in the vicinity the probe. In other words, this band is responsive to the presence of ester (DPPC), ether (DHPC), or amide linkages (sphingomyelin), consistently with the location of the probe in the bilayer (~ 10 Å from the center) [33]. For other sphingolipids forming bilayers (Gg_3Cer, Phrenosine, sulfatide, and galactosyl ceramide) the same trend is observed as in sphingomyelin-containing membranes, i.e., the red band is not the excitation maximum. However, in those sphingolipids forming micelles (gangliosides), the red excitation band is somehow stabilized and constitutes the excitation maximum below their phase transition [30, 31]. Interestingly, this stabilization of the LAURDAN red excitation band in micelles below their main phase transition is accompanied with a highly relaxed emission spectrum, completely the opposite to that observed in glycerophospholipids [30, 31]. This behavior may be due to the presence of a different environment in the vicinity of the probe with respect to bilayers, influencing the stabilization of the probe ground state [33].

4 LAURDAN Generalized Polarization Function

The Generalized Polarization (GP) function was proposed by Parasassi et al. in 1990 [15] to introduce an analytical method to quantitatively determine the relative amount of s_o and l_d phases when they coexist in a model membrane and to study their temporal fluctuations. The GP function was originally defined as:

$$GP = \frac{I_B - I_R}{I_B + I_R} \tag{1}$$

where I_B and I_R are the measured fluorescence intensities under conditions in which a wavelength (or a band of wavelengths) B and R are both observed using a given excitation wavelength. This definition corresponds to the classical fluorescence polarization definition [40] if B and R represent two different orientations of the observation polarizers. The advantage of the GP for the analysis of the spectral properties of LAURDAN is related to the well-known properties of the classical polarization function, which contains information on the interconversion between different "states." In the classical polarization definition the "states" correspond to different orientations of the emitting dipole with respect to the laboratory axis. In the GP function the condition of interconversion between two states *a* and *b*, i.e., unrelaxed and relaxed, has been linked respectively to the extent of solvent relaxation observed in the s_o and l_d phases in glycerophospholipid membranes, respectively [15, 16]. The general expression for the decay of *a two state system* can be written in the following way [15]:

$$a(t) = a_1 e^{-m_1 t} - a_2 e^{-m_2 t} \tag{2}$$

$$b(t) = b_1 e^{-m_1 t} - b_2 e^{-m_2 t} \tag{3}$$

$$a_1 = \frac{a_0(k_a - m_2) - b_0 k_{ab}}{m_1 - m_2} \quad (4)$$

$$a_2 = \frac{a_0(k_a - m_1) - b_0 k_{ab}}{m_1 - m_2} \quad (5)$$

$$b_1 = \frac{b_0(k_b - m_2) - a_0 k_{ba}}{m_1 - m_2} \quad (6)$$

$$b_2 = \frac{b_0(k_b - m_1) - a_0 k_{ba}}{m_1 - m_2} \quad (7)$$

$$m_1, m_2 = \frac{1}{2}\left[k_a + k_b \pm \sqrt{(k_a - k_b)^2 + 4k_{ab}k_{ba}}\right] \quad (8)$$

where *a(t)* and *b(t)* represent the fluorescence emission from state *a* (unrelaxed) and *b* (relaxed), k_a and k_b are the decay rates of the decays *a* and *b*, a_0 and b_0 are the initial excitation state for the *a* and *b* states, respectively, and k_{ab} and k_{ba} the interconversion rates from state *a* to *b* and vice-versa (the forward and backward relaxation rates, respectively). The model assumes that the total emission is a linear combination of the decays *a(t)* and *b(t)* given by the superposition of two characteristic spectra.

These authors considered $B(t)$ and $R(t)$ to be the total intensity decay observed under the observation conditions B and R. In the case of LAURDAN, these two conditions correspond to observations in the blue and red parts of the emission spectra. By analogy to the classical fluorescence polarization function, these conditions correspond to the use of vertical and horizontal polarizers (however, in this case two emission filters are used):

$$B(t) = Ba(t) + (1 - B)b(t) \quad (9)$$

$$R(t) = Ra(t) + (1 - R)b(t) \quad (10)$$

Here, B and R correspond to the relative observation of state *a(t)* in the blue and red part of the emission spectrum, respectively. This assumes that B and R have values between 0 and 1 because they represent relative fractional intensities. Three expressions for the GP have been defined according to Eq. 1, i.e., steady state GP, time-resolved GP, and time zero GP [15], respectively, as:

$$\langle GP \rangle = \frac{\langle B \rangle - \langle R \rangle}{\langle B \rangle + \langle R \rangle} \quad (11)$$

$$GP(t) = \frac{B(t) - R(t)}{B(t) + R(t)} \quad (12)$$

$$GP_0 = \frac{B_0 - R_0}{B_0 + R_0} \tag{13}$$

By replacing Eqs. 9 and 10 in the definition of GP an expression can be obtained where the details of how the system is observed (left side of Eq. 14) are separated with parameters that represent the behavior of the physical system [15, 16]:

$$\frac{B - R}{GP} - (B + R) = \frac{2b(t)}{a(t) - b(t)} \tag{14}$$

By considering B and R to be the parallel and perpendicular observation conditions and considering *a(t)* and *b(t)*, respectively, the intensity decays in these two conditions, the equation for the classical polarization function can be retrieved (see [15] for more details). Therefore, the GP function satisfies several important properties of the classical polarization function, in particular Weber's law of addition [41] and the Perrin equation [40].

By replacing Eqs. 2–8 in Eq. 14 [15, 16] an expression for the GP directly in terms of the physical constants (Eq. 15 below) is obtained. The expression for the steady state GP is then:

$$\frac{B - R}{\langle GP \rangle} - (B + R) = \frac{2(b_0 + \frac{k_{ab}}{k_a})}{a_0 \frac{k_b}{k_a} - b_0 + \frac{k_{ba}}{k_a} - \frac{k_{ab}}{k_a}} \tag{15}$$

It is reasonable to assume that $k_a = k_b$ (the intrinsic decay rates are equal) and that $k_{ba} = 0$ (the backward relaxation rate), i.e., the decay rate is independent of the relaxation process and the back reaction is slow. After these assumptions Eq. 15 reduces to:

$$\frac{B - R}{GP} - (B + R) = \frac{2(b_0 + \frac{k_{ab}}{k_a})}{a_0 - b_0 - \frac{k_{ab}}{k_a}} \tag{16}$$

From this expression it is interesting to see that it is the ratio k_{ab}/k_a that determines the modalities of the relaxation process and the GP value. Parasassi et al. analyzed three particular limiting cases in the context of this definition [16]:

1. $k_{ab} >> k_a$, fast relaxation or long lifetime value of the probe. In this case the GP value depends only on the spectral emission properties of the unrelaxed state at the wavelength of excitation, i.e., on B and R and not on the relaxation process of the solvent.
2. $k_{ab} << k_a$, slow relaxation or short lifetime value of the probe. The GP depends also on the relative absorption of both the relaxed and unrelaxed state at the wavelength of excitation. If no relaxed state is excited ($b_0 = 0$), then there should be no dependence of the GP value on the excitation wavelength because $a_0 = 1$.

3. $k_{ab} = k_a$, the GP value also depends on the dynamic properties of the solvent that relaxes the probe's excited state.

Condition 2 satisfies what is observed in membranes displaying s_o phase whereas condition 3 reflects what is observed in a l_d phase [16].

5 Fluorescence Spectroscopic Measurements in Bulk

Originally the GP function was defined to introduce an analytical method to determine both the relative amount of s_o and l_d phases and their temporal fluctuations when they coexist in a model membrane [15, 16]. This strategy has been applied, however, to systems other than exclusively lipid membranes. For example, many studies have made use of the LAURDAN GP approach to dissect membrane interactions in bulk with peptides, proteins, or other ligands, as well to investigate natural membranes (e.g., bacteria, mammalian cells), see, e.g., [22, 42–45]. Since space restrictions do not allow a review of the over 330 papers reported, this section will describe two general strategies involving bulk GP experiments: (i) measurements of membrane phase transitions and (ii) detection of phase coexistence in membranes. Particularly, the characteristics of the (emission and excitation) LAURDAN GP spectra to detect phase coexistence will be discussed for membranous aggregates containing glycerophospholipids, sphingolipids, and their mixtures with cholesterol. This will provide a general idea of the potentialities and limitations of the technique to obtain relevant information about membrane structure and dynamics.

5.1 *Measurements of Phase Transitions*

Several studies by Parasassi and collaborators focused on membranes composed of single glycerophospholipids species, their mixtures as well as mixtures containing cholesterol [15, 16, 29, 32, 37, 43, 46–48]. From these studies, and especially for single glycerophospholipid components, the lipid main phase transition temperature (s_o to l_d) can be easily detected by measuring the GP function versus temperature [15] (see DPPC in Fig. 4a). Also, by adding cholesterol to the system the very well-known decrease in the cooperativity and abolishment of the glycerophospholipid main phase transition can also be observed [47].

The characterization of the spectroscopic properties of LAURDAN, including the ability of the GP function to detect phase transitions in a series of sphingolipids and their mixtures with glycerophospholipids, were reported by Bagatolli et al. [30, 31]. These studies showed that the extent of relaxation detected by the probe was strongly influenced by the nature of the sphingolipid polar head group. For some of these lipids, particular those forming bilayers, e.g., Gg_4Cer, Gg_3Cer, Phrenosine, sulfatide,

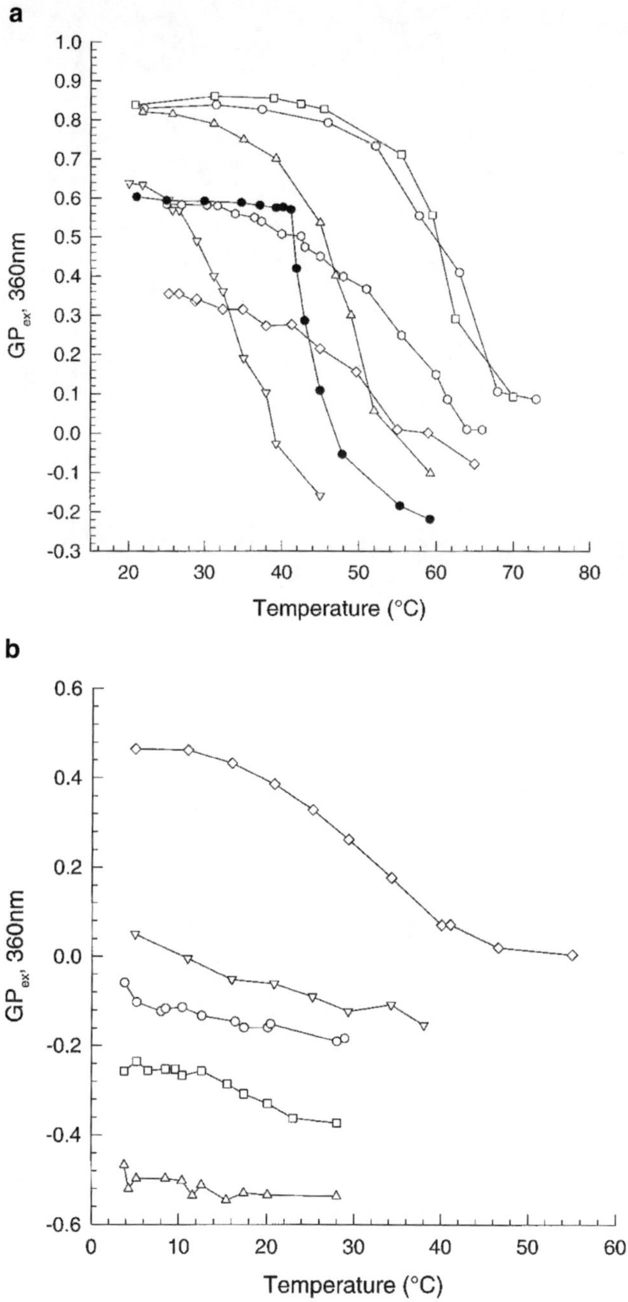

Fig. 4 LAURDAN GP dependence with the temperature in membranes. (**a**) DPPC (*black circles*); Sulfatide (*triangles*), Phrenosine (*squares*), GalCer (*open circles*), Shingomyelin (*inverted triangles*) Gg_4Cer (*diamonds*), and Gg_3Cer (*hexagons*); (**b**) G_{T1b} (*triangles*), G_{D1a} (*squares*), G_{M1} (*circles*), G_{M2} (*inverted triangles*), and G_{M3} (*diamond*). The main phase transition temperature was calculated using $\partial^2 GP/\partial T^2$. Adapted from [30] with permission

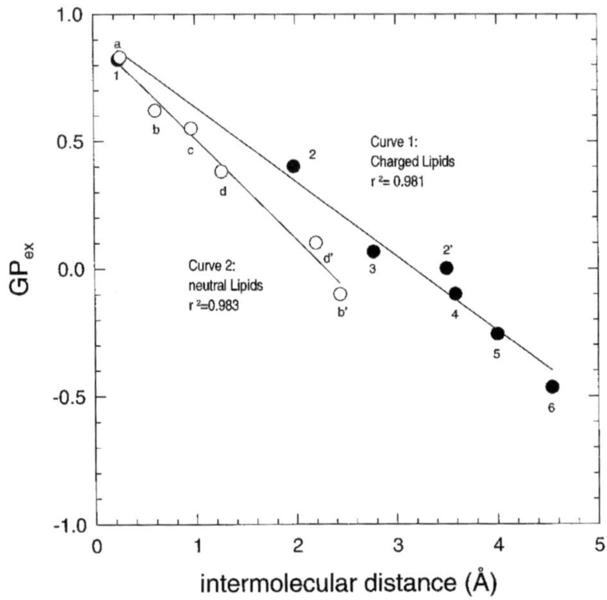

Fig. 5 Variation of LAURDAN GP_{ex} with the calculated intermolecular distance (measured in monolayers) of anionic GSLs (*filled symbols*) and neutral GSLs and phospholipids (*empty symbols*). Neutral lipids: GalCer (a); DPPC below (b), and above (b') the transition temperature; Gg_3Cer (c); Gg_4Cer below (d) and at (d') the transition temperature. Anionic lipids: Sulf (1); G_{M3} below (2), and above (2') the transition temperature, G_{M2} (3); G_{M1} (4); G_{D1a} (5); G_{T1b} (6). Adapted from [30] with permission

GalCer, and sphingomyelin, the GP function shows a characteristic s_o to l_d phase transition at particular temperatures that depend on the lipid studied (Fig. 4a). However, in glycosphingolipids that have a bulky sugar-based polar head group (e.g., the gangliosides G_{M1}, G_{D1a}, G_{T1b}) the response of LAURDAN is different. Gangliosides form micelles and in this type of structure the GP values measured below and above the lipid main phase transition temperature (detected by differential scanning calorimetry) are very low and invariant with respect to that observed in bilayers (with the exception of G_{M2}), see Fig. 4b [30, 31]. This lack of sensitivity is consistent with the probe experiencing a highly relaxed environment below and above the micelle phase transition, i.e., a highly curved and hydrated interface [49] compared to that observed in bilayers. In the study of this sphingolipid series, a linear dependence between the GP function and the intermolecular spacing among these lipids (measured at 30 mN/m in monolayers) was reported for both neutral and anionic species, Fig. 5 [30]. This finding strongly supports the correlation between water content/dynamics with the response of the GP, i.e., LAURDAN resides in cavities containing different amounts of associated water in the membrane [30, 31] (Sect. 2).

5.2 Detection of Phase Coexistence in Membranes

By measurements of time-resolved emission spectra, Parasassi et al. demonstrated that in DLPC/DPPC mixtures phase separated domains (s_o and l_d) exist in the membrane [29]. The authors estimated domain dimensions of ~20–50 Å and measured domain fluctuations with kinetics of ~25 ns [37]. By applying the additive property of the GP function [16, 29], the fractions of the two coexisting phases were obtained, concluding that the characteristic of the s_o and l_d phases for pure components is somehow modified when these phases coexist (i.e., total lipid immiscibility is not present in the mixture) [29].

Parasassi et al. [16] presents a strategy to ascertain phase coexistence in membranes from bulk steady state fluorescent measurements. This was based on the particular response of LAURDAN emission and excitation spectra to the phase state (s_o, l_d or s_o/l_d phase coexistence) of *glycerophospholipid*-containing membranes. These authors showed that the (excitation or emission) wavelength response of the GP function provides a fingerprint to discriminate among different membrane phase state scenarios [16, 37, 47]. Two GP spectra are considered in this strategy, the excitation GP (GP_{ex}) and emission GP (GP_{em}).

The GP_{ex} spectrum is sensitive to processes occurring in the probe's excited state [16, 47]. In order to obtain a GP_{ex} spectrum the excitation spectra of LAURDAN are measured at two fixed emission wavelengths (440 and 490 nm) and the GP calculated at each excitation wavelength according to Eq. 1 (I_B is the intensity at 440 nm and I_R the intensity at 490 nm). In glycerophospholipid membranes in the s_o phase the GP_{ex} spectrum is independent of the excitation wavelength, since no relaxation occurs. However, in the l_d phase the response is different. One of the characteristics of the dipolar relaxation process is that by exciting in the blue part of the excitation spectrum, molecules with an energetically unfavorable ground state (surrounded by randomly oriented dipoles) are photoselected. As a consequence, by moving the excitation toward the blue, more blue emitting molecules are excited. Since the GP_{ex} value depends on the difference between the emission intensities at 440 and 490 nm, when relaxation occurs (as it happens in the l_d phase), higher GP values are expected at shorter excitation values, i.e., the slope of the GP_{ex} spectrum is negative, Fig. 6 [15, 16, 47]. When the membrane displays s_o/l_d phase coexistence the GP_{ex} spectrum shows a positive slope [16], since there are relaxed and unrelaxed states coexisting in the membrane. Particularly, the unrelaxed state contributes to the red region of the excitation spectrum (see Sect. 2) and in the blue side of the emission spectrum, increasing the GP value at longer excitation wavelengths, Fig. 6 [37].

A similar analysis can be performed by measuring the response of LAURDAN along the emission wavelength when two excitation wavelengths are used, i.e., using the GP_{em}. This parameter is calculated using two wavelengths of excitation, 410 nm and 340 nm that correspond to I_B and I_R, respectively, in Eq. 1 [16, 37, 47]. The GP_{em} spectrum responds to processes occurring in the probe's ground state. Similar to the GP_{ex} spectrum, its behavior in the s_o phase is invariant with

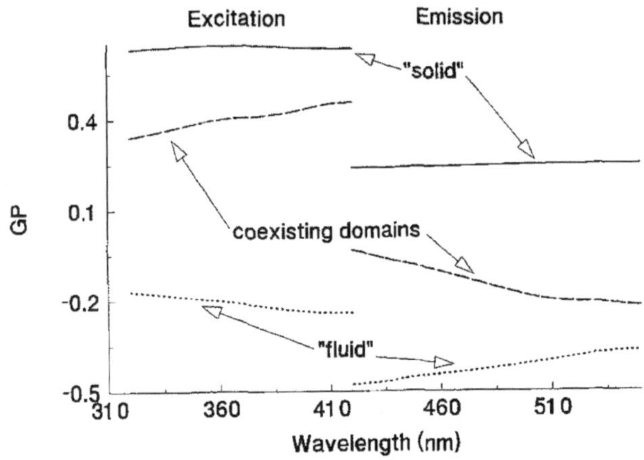

Fig. 6 Laurdan GP_{ex} and GP_{em} spectra obtained in phospholipids in different phase scenarios. GP_{ex} spectra were calculated by $GP = (I_{440} - I_{490})/(I_{440} + I_{490})$, using excitation wavelengths from 320 to 420 nm. GP_{em} spectra were calculated by $GP = (I_{410} - I_{340})/(I_{410} + I_{340})$, using emission wavelengths from 420 to 550 nm. Adapted from [37] with permission

wavelength since no relaxation occurs in the membrane (see Fig. 6). However, the GP_{em} spectrum responds opposite to the GP_{ex} spectrum both in the l_d phase and when s_o/l_d phases coexist. If the membrane is in the l_d phase, excitation at 340 nm slightly selects LAURDAN molecules with energetically unfavorable orientation of the surrounding dipoles, i.e. only a small fraction of the probe's population relax [15, 16, 37]. On the contrary, excitation at 410 nm preferentially selects LAURDAN molecules mostly surrounded by favorably oriented dipoles, which in the l_d phase will show a relaxed emission. Consequently, by observing the red emission at 490 nm, higher intensity will originate from the 410 nm's excitation band. By observing the emission toward the blue, the intensity arising from 410 nm excitation will decrease and so will the GP_{em} values; i.e., the slope is positive (Fig. 6) [16, 37, 47]. When the membrane displays s_o/l_d phase coexistence the GP_{em} spectrum shows a negative slope, since the contribution of the unrelaxed state dominates at shorter emission wavelengths (see Fig. 6).

Different to glycerophospholipids, the LAURDAN excitation red band is not a maximum in sphingolipids forming bilayers in the s_o phase (Sect. 2) [30, 33]. This has important consequences in the behavior of the GP_{ex} and GP_{em} spectra, i.e., under these conditions the wavelength dependence of both GP_{ex} and GP_{em} spectra is affected and becomes insensitive to phase coexistence [30, 33]. In experiments with gangliosides a similar insensitivity to phase coexistence in the GP_{ex} and GP_{em} spectra is observed although the reasons are different. In this last case the excitation red band is a maximum below the phase transition temperature but the probe's *emission spectrum* observed at the same conditions is already red shifted (the probe is relaxed) [30, 31]. Taking into account the GP spectra response in sphingolipids, caution is recommended if the LAURDAN GP spectra are used to ascertain membrane phase coexistence, particularly for lipids other than glycerophospholipids or membranes with complex composition. It is highly recommended in these cases to perform a detailed characterization of the probe's excitation and emission spectra. For example, bulk measurements of LAURDAN GP spectra failed to detect domain coexistence in different mammalian cell membranes [43]. Assuming that membrane heterogeneity exists, this failure may very well be due to the inherent response of the probe to the complex membrane composition existing in the cells. Alternatively, this insensitivity may also be connected to the particular spectroscopic response of LAURDAN to lipid membranes containing cholesterol discussed below.

5.3 Cholesterol Effects

It is of particular interest to discuss the effect of the incorporation of cholesterol into model membranes on the fluorescent properties of LAURDAN. This effect has been extensively characterized by Parasassi et al., particularly in glycerophospholipid membranes [46–48]. As mentioned above (Sect. 5.1), the evolution of the GP function with temperature reflects the very well-known effect of cholesterol on the glycerophospholipid main phase transition, i.e., the transition is gradually abolished as the cholesterol molar fraction is increased.

Depending on the initial phase state of the membrane cholesterol distinct effects are produced on the LAURDAN fluorescence excitation and emission spectra. First, as cholesterol concentration is increased in membranes displaying l_d phase, a blue shift is observed in the probe's emission spectrum together with an increase in the intensity of the red excitation band. This effect causes an increase in the GP values (both emission and excitation), interpreted as a decrease in the extent of relaxation processes in the membrane, which is consistent with the formation of a liquid ordered (l_o) phase [46, 48, 50]. Second, upon cholesterol addition to membranes displaying a s_o phase, the LAURDAN emission spectrum experiences a small blue shift and the excitation spectrum experiences a slight decrease in the intensity of the red excitation band. These observations are indicative of a less polar (dehydrated) environment [37, 47, 48]. Finally, when cholesterol is incorporated in a membrane displaying s_o/l_d phase coexistence (e.g., DLPC/DPPC) a similar effect to that described above for the l_d phase was reported [37, 47]. Particularly, above 10 mol % cholesterol the coexistence of s_o/l_d domains observed for DLPC/DPPC is no longer detected in the GP_{ex} and GP_{em} spectra [47], i.e., no separate LAURDAN emission signals characteristic of s_o and l_d domains are apparent (Fig. 7). Under these conditions the behavior of the GP spectra correspond to a homogeneous l_d phase, despite the absolute GP values being relatively high and closer to that observed in the s_o phase [43]. This intermediate feature can be associated to the gradual formation of a l_o phase. Parasassi et al. proposed that cholesterol causes the decrease in both polarity and dipolar relaxation in the neighborhood of LAURDAN's naphthalene moiety [48]. Additionally, these authors reported on particular cholesterol concentrations (at about 5, 10, 15, 30, and 45 mol % with respect to glycerophospholipids) at which abrupt variation of the

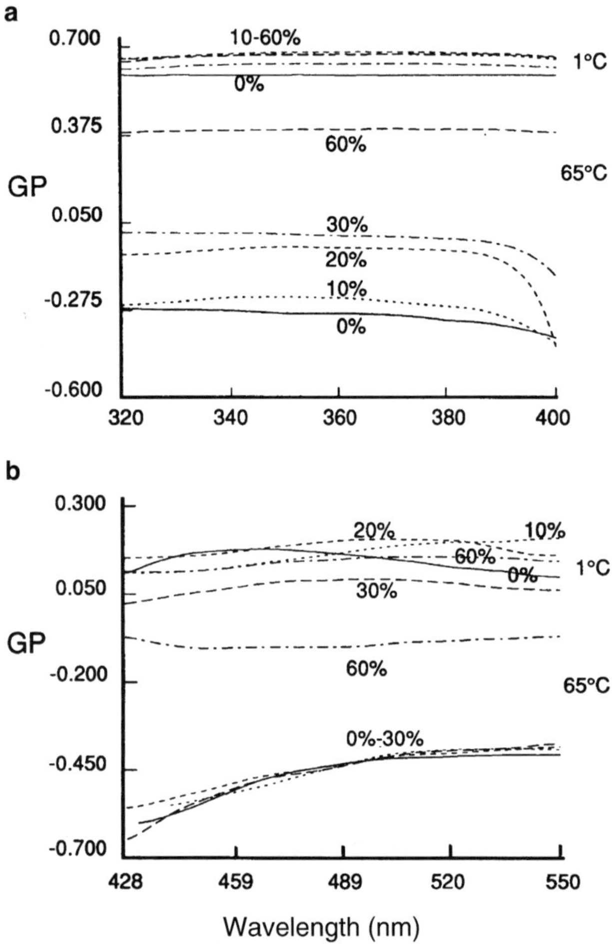

Fig. 7 LAURDAN GP_{ex} (**a**) and GP_{em} (**b**) spectra obtained in the equimolar DLPC-DPPC mixture at various cholesterol concentrations, at 1°C and at 65°C. The percent values represent mol% of cholesterol in phospholipids. Adapted from [47] with permission

membrane properties can be observed using the GP function. They proposed that the formation of ordered molecular microdomains at critical cholesterol concentrations can explain the occurrence of this observed discontinuities. These concentrations are in partial agreement with experimental measurements in similar systems using other probes [51] and theoretical calculations of a hexagonal superlattice structure formed by phospholipids and a host molecule [52].

Comparative studies in the presence and absence of cholesterol were also reported by Bagatolli et al. using complex glycosphingolipid aggregates [30]. Upon addition of cholesterol to the glycosphingolipid asialo G_{M1} (Gg_4Cer), the fluorescence behavior of LAURDAN was similar to that of pure cerebrosides and sphingomyelin bilayers suggesting a decrease in the water content of Gg_4Cer membranes. Gangliosides (G_{M1}, G_{D1a}, and G_{T1b}) micelles – that show a high extent of relaxation compared to the sphingolipids forming bilayers – also showed a very significant augment in their GP values upon addition of cholesterol [30, 31]. These observations were attributed to a cholesterol-dependent interfacial dehydration due to changes in the shape and size of the micelles [30].

6 Spatially Resolved Information in Membranes

One of the limitations of the classical "cuvette" studies is the measurement of an "average" or bulk parameter. In LAURDAN experiments, the GP values measured in bulk correspond to the sum of the different relative contributions existing in the sample (because of the additive property of the GP discussed in Section 4). In 1996, Yu et al. presented the first measurement of the GP function in cellular systems using TPEFM [53]. This strategy permits mapping the spatial distribution of the GP function (limited of course by the resolution of the microscope), allowing to disentangle the different contributions of the GP in the specimen. A year later Parasassi et al. report on application of the GP images in multilamellar vesicles composed of glycerophospholipids and cells [36]. Two years later Bagatolli and Gratton start the characterization of the GP in giant unilamellar vesicles composed of different lipids and their mixtures [27, 28, 54, 55]. All these experiments were key to further characterize the behavior of LAURDAN in membranes.

Acquisition of GP images in the microscope requires a dichroic mirror coupled to a two channel detection setup equipped with adequate band-pass filters. This hardware is necessary to split and select the two different fluorescent intensities (I_B and I_R, see Eq. 1) necessary to compute the GP. Additionally, since two channels are used a calibration is required in the system using a G factor (described in [56]). The use of laser scanning multiphoton excitation flourescene microscopy showed high improvements in the acquisition of GP images comparing with one photon excitation wide field fluorescence microscopy systems, i.e., the overall high photobleaching effect observed in the latter case (using UV excitation sources) is largely reduced by the use of multiphoton excitation microscopy [36, 53, 54]. Recently, however, some reports indicates that it is possible to perform LAURDAN GP measurements using one photon excitation wide field illumination [57, 58], opening a more accessible (economically) alternative to acquire LAURDAN GP images.

6.1 LAURDAN GP Imaging in Free Standing Bilayers

Fluorescence microscopy experiments using giant unilamellar vesicles (GUVs) composed of distinct lipid mixtures further advanced understanding of lateral phase separation in membranes. Using LAURDAN-labeled GUVs and TPEFM, direct information on local membrane packing can be easily obtained for membranes displaying coexisting domains at the level of single vesicles [59, 60]. This is in part possible because of the sensitivity of LAURDAN to solvent relaxation and the even partition of the probe in membranes displaying phase coexistence.

The LAURDAN GP imaging approach is superior to other similar confocal fluorescence microscopy strategies, which use pairs of lipophilic probes that preferentially partition to each of the coexisting membrane regions. Particularly, it has been demonstrated that partition of these probes depends on the chemical nature of the domains, i.e., probe partition *is not* necessarily phase specific [59–61]. In this case additional experiments must be performed besides imaging to obtain local information on domain packing [62] (e.g., diffusion coefficient measurements of the probe using FCS). These additional experiments are not needed for LAURDAN [59].

The effects of temperature on the lateral structure of LAURDAN-labeled GUVs composed of pure phospholipids and their different mixtures were first reported in 1999 and 2000 [27, 28, 63]. Micrometer-sized domains with s_o character, surrounded by membrane regions corresponding to a l_d phase, where observed at particular temperatures depending on the lipid mixture. These domains spanned the bilayer, demonstrating inter leaflet coupling in the membrane [27, 28]. Additionally, the experiments showed that generation of micrometer-sized domains in bilayers was possible, disproving the general assumption of nanometer-sized domains in those lipid mixtures. In one of these studies a correlation between domain shape and lipid miscibility was reported for binary mixtures of different phospholipids displaying s_o/l_d phase coexistence [27].

The same strategy was implemented to explore the effect of cholesterol in lipid mixtures. Particularly, the lateral structure of canonical raft mixtures (DOPC/sphingomyelin/cholesterol) was first visualized in free standing bilayers using LAURDAN and TPEM [55]. For this mixture, the coexistence of two liquid phases (l_o/l_d) was described, characterized by the presence of round micrometer-sized domains (that supports the idea of fluid immiscibility in the membrane) and LAURDAN GP values intermediate to those observed for the s_o/l_d case.

The information gathered using LAURDAN and TPEFM in several lipid mixtures (including lipopolysaccharides, other sphingolipids and their mixtures with glycerophospholipids and cholesterol [64–67]) has aided in the characterization of distinct phase coexistence regimes, providing particular structural signatures for lateral phase separation phenomena in membranes. Figure 8 compares the s_o/l_d and l_o/l_d phase coexistence scenarios as revealed by LAURDAN. At a first glance it can be observed that the shapes of the domains are quite different between these two phase coexistence regimes. Particularly for the s_o/l_d phase coexistence a number of options are possible for the shape of the s_o domains [27, 28, 54, 64, 65]. Second, the coexistence of l_o/l_d phases show a clear reduction in the GP differences between the coexisting domains compared to that found in the s_o/l_d phase coexistence [28, 55]. Finally the probe's photoselection effect (see Sect. 6.2 below) is much more prominent in the s_o phase compared to that observed in the l_o phase, confirming the fluid character of the latter. The LAURDAN GP imaging approach has been also successfully

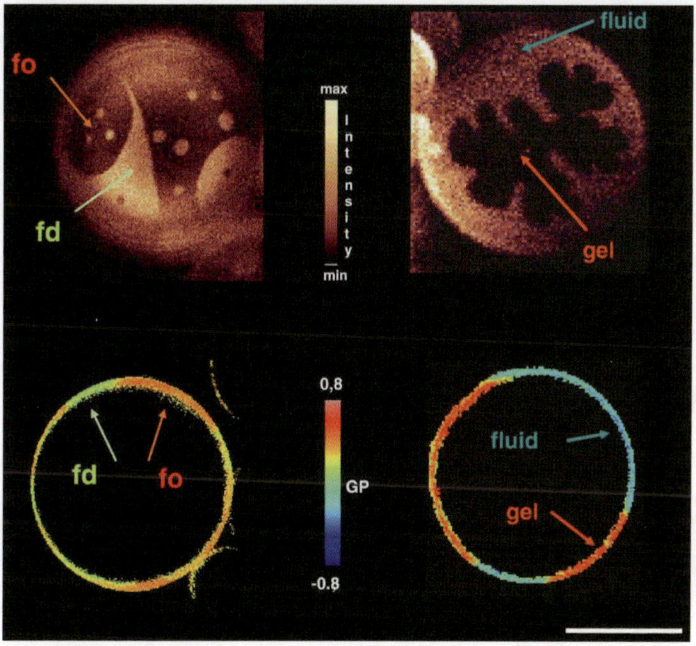

Fig. 8 Two photon excitation LAURDAN fluorescent images (taken at the polar region of the vesicle) and LAURDAN GP images (taken at the equatorial region of the vesicle) of DOPC/Cholesterol/Sphingomyelin 1:1:1 mol displaying l_o/l_d phase coexistence (*left panel*, indicated in the figure as fo and fd, respectively) and DPPC/DPPE 3:7 mol displaying s_o/l_d phase coexistence (*right panel*, indicated in the figure as gel and fluid, respectively). The bar corresponds to 20 μm. Adapted from [59] with permission

applied to the study of lateral structure in compositionally complex membranes such as skin stratum corneum lipid membranes [68, 69], lung surfactant [70, 71], and red blood cells [72].

Beside the structural information described above, a combination of imaging and fluctuation techniques using LAURDAN has been used recently to investigate the *temporal evolution* of domains in membranes [73]. These experiments are based on the very particular spectroscopic properties of LAURDAN and the even distribution of the probe in membranes displaying phase coexistence. Specifically, the temporal evolution of s_o phase domains was reported at the onset of phase separation using GUVs composed of a binary mixture of DPPC/DLPC 1:1 [73]. The data acquired at temperatures immediately above the transition temperature of these two lipids showed fluctuations in the lipid organization with a lifetime <0.1 s and a characteristic length of 1.2 μm [73]. This study also found that just below the l_d to l_d/s_o phase transition of the DPPC/DLPC mixture, coupling between the two leaflets of the bilayer was observed to begin within the first 5 min after the onset of phase separation. However, this study also found that domains confined to only one leaflet can be found during the first 45–50 min after the onset of phase separation [73].

Finally, studies using scanning-fluctuation correlation spectroscopy were also used to detect subresolution organizational fluctuations in membranes composed of different single lipid species displaying l_d phase. This was achieved by measuring fluctuations of the LAURDAN GP function on the scale of few pixels (50 nm pixel) [38]. The calculated GP pixel variance showed an abrupt change at the membrane phase transition that decreases slowly at higher temperatures. The relatively large variance of the GP indicates that the l_d phase is quite heterogeneous, even at several degrees higher than the phase transition temperature. This result was interpreted as evidence of an underlying microscale structure of the membrane in which water is not uniformly distributed at the micron scale [38]. Interestingly, the statistical and fluctuation analysis of the GP data shows that even such simple lipid systems are capable of generating and maintaining stable structural and organizational heterogeneities [38].

6.2 LAURDAN Photoselection Effect

The photoselection effect arises from the fact that only those fluorophores which have electronic (absorption) transition moments aligned parallel or nearly so to the plane of polarization of the excitation light are excited, i.e., the excitation efficiency is proportional to the cosine to the fourth power (because of the two photon excitation absorption) of the angle between the transition moment of the probe and the polarization plane of the excitation light. If circularly polarized excitation light (confined to the x-y plane) is used, different excitation efficiencies (in a given vertical section) are observed in a spherical vesicle. This effect depends on the position of the probe's transition moment relative to the plane of polarization of the excitation light (Fig. 9a). At *the equatorial region* of the vesicle the circularly polarized excitation light allows excitation of all LAURDAN molecules with the same efficiency, independent of the lipid phase scenario present in the vesicle (one single phase or phase coexistence), i.e., the probe's electronic transition moment is always parallel to the polarization plane. This allows calculation of the GP without the influence of the photoselection effect (as seen in Fig. 9b for coexistence of s_o/l_d and l_o/l_d phases). Instead, at *the polar region* of the vesicle the dominant signal is that coming from the more fluid part of the bilayer (Fig. 9c). This effect is dictated by the fact that a component of LAURDAN's electronic transition moment is always parallel to the excitation polarization plane (the because of the relatively low lipid order). This last phenomenon does not take place in the s_o phase because the high lipid lateral order precludes the wobbling movement of LAURDAN, Fig. 9c.

In the case of l_o/l_d phase coexistence, a component of LAURDAN's transition moment (parallel to the excitation light polarization plane) will be present in both phases at the polar region of the GUV, allowing collection of the fluorescence signal from these two regions. Particularly, the l_o phase shows a lower impact of the photoselection effect compared to s_o phase regions (compare Fig. 9 panels c and d).

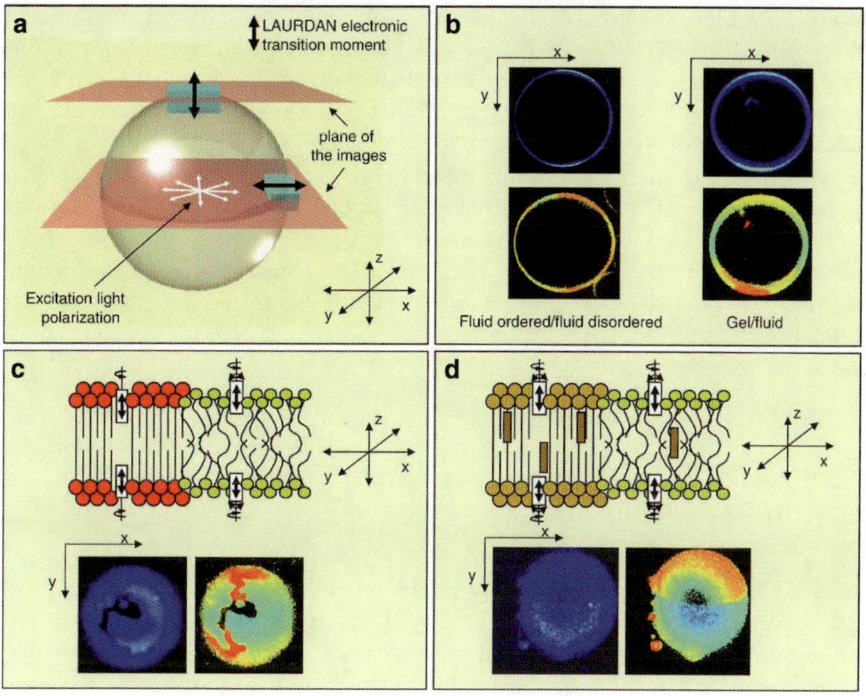

Fig. 9 The photoselection effect is dictated by the relative orientation of the LAURDAN electronic transition moment with respect to the polarization plane of the excitation light. The last is also influenced by the lipid bilayer phase state. (**a**) Sketch of the GUV and the position of the LAURDAN electronic transition moment; (**b**) the photoselection effect do not operate in the equatorial region of the GUV independent of the phase coexistence scenario (notice that l_o/l_d and s_o/l_d are indicated in the figure as fluid ordered/fluid disordered and gel/fluid, respectively); (**c**) s_o/l_d phase coexistence observed at the polar region of the GUV composed of Ceramide/POPC 1:5 mol (fluorescence intensity and GP); notice the strong photoselection effect observed in the s_o phase with respect to the l_d phase; (**d**) l_o/l_d phase coexistence observed at the polar region of the GUV (fluorescence intensity and GP) composed of DOPC/SM/cholesterol 1:1:1 mol. The photoselection effect in the l_o phase is less pronounced to that observed in the s_o phase (compare **c** and **d**). Adapted from [59] with permission

Interestingly, the photoselection effect per se allows qualitative information about lipid phases directly from the intensity images [59] and the spatial orientation of the fluorophores in the membrane [28].

LAURDAN GP images obtained at the GUV's equatorial region using *linearly polarized excitation light* can be exploited to obtain information about the coexistence of lipid domains with sizes below the resolution of the microscope (the radial resolution of the microscope is ~300 nm) [36, 63, 74, 75]. When the lipid domain size is smaller than the image pixel the size of these domains cannot be resolved in the image. However, each pixel of the image contains this information. Since linearly polarized light, which photoselects appropriately oriented LAURDAN molecules, also selects LAURDAN molecules associated with high GP values, pixels with high and low GP values can be discriminated. In other words, if the membrane displays phase coexistence with domains below the microscope resolution, the image will contain separate domains (pixels) of different GP value, i.e., the higher GP value domains will appear parallel to the orientation of the polarized excitation light and not in the perpendicular direction [36, 63, 74, 75]. This effect was used to ascertain lipid domain coexistence in GUVs composed of single phospholipid species [63] and multilamellar vesicles composed of binary mixtures of glycerophospholipids, including some intact cell membranes [36]. For a detailed description of the use of LAURDAN GP images to ascertain lipid domains below the microscope resolution the reader is encouraged to explore the following references [36, 63, 74, 75].

6.3 LAURDAN GP Imaging in Planar Supported Membranes and Langmuir Films

Most of the fluorescence microscopy studies reported on planar membranes focus on measurements of fluorescence intensity of a particular selection of probes with almost no practical combinations of spatially resolved fluorescence spectroscopic parameters (such as lifetime, polarization, or emission spectral shift). Recently, LAURDAN GP measurements have been done in planar supported bilayers composed of DOPC/DPPC [76]. The results show that lipid packing structure within s_o domains is heterogeneous and the domains display orientational texture [76]. Although the LAURDAN GP is uniform within the s_o domains, multiphoton excitation fluorescence polarization microscopy measurements using LAURDAN (whose emission intensity is sensitive to the tilt of the probe with respect to the plane of the membrane) are able to detect the lipid domain texture (see Fig. 10). The phenomenon of orientational texture adds a new level of complexity to the structure of condensed membrane areas. Correlations with similar textures in Langmuir monolayers and liquid crystals indicate that the so-called hexatic phase might be the proper designation for the phase state of certain condensed s_o domains observed in lipid bilayers. Whether or not this texture is biologically relevant is still an open question. Although very speculative, this situation may be relevant, e.g., in the case of skin stratum corneum lipid membranes, where coexistence of solid-like domains has been reported to exist at skin physiological temperatures [69].

This type of approach has been recently applied to other planar membrane models, i.e., phospholipid monolayers at the air/water interface, i.e., Langmuir films [56]. Unlike other probes LAURDAN partitions evenly in monolayers and bilayers and the coexisting domains are distinguished by different emission properties of the probe that in turn depends on local lipid packing (see Fig. 11) [59]. The fact that LAURDAN exhibits a homogeneous distribution in monolayers, particularly in systems displaying domain coexistence, overcame a general problem observed when "classical" fluorescence probes are used to label Langmuir films,

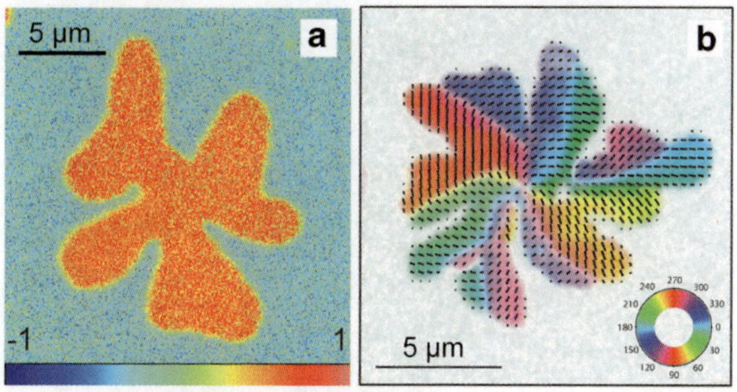

Fig. 10 (**a**) s_o/l_d phase coexistence in a binary supported membrane (DOPC/DPPC 1:1 mol on a mica substrate) observed using LAURDAN GP imaging [76]. The flower shape domain shows a higher GP value compare to the surroundings which indicates a higher membrane lateral packing. (**b**) Orientational texture of a s_o domain revealed by two photon excitation fluorescence polarization microscopy. The colors in (**b**) represent the direction of the molecular tilt of the lipid acyl chains and are obtained by image processing of the image sequence in a 360° polarization scan [76]. Adapted from [83] with permission.

i.e., the inability to obtain simultaneous information from the two coexisting membrane regions. Also, the photoselection effect discussed above (Sect. 6.2) allows: (i) to qualitatively infer tilting information of the monolayer when liquid-condensed phases are present and (ii) to provide high contrast to visualize 3D membranous structures at the film's collapse pressure. In the latter case, computation of the LAURDAN GP function provides information about lipid packing in these 3D structures [56]. Additionally, LAURDAN GP values upon compression in monolayers were compared to those obtained in compositionally similar planar bilayer systems. A correspondence between the GP was observed between monolayers and bilayers when the monolayers exist at ~28 mN/m [56], in agreement with theoretical predictions [77].

6.4 LAURDAN Measurements in Cells and Tissues

From the pioneering work of Yu et al. and Parasassi et al. [36, 53], LAURDAN was proposed to be a promising tool in exploring cell membranes. In these studies, domains of sizes below, in the same range, and above the microscope resolution limit (0.3 µm radial) were observed in OK cells, red blood cells, and brush border native membranes, respectively [36, 53]. The LAURDAN GP differences observed in compositionally complex mixtures and artificial lipid ternary mixtures containing phospholipids, sphingomyelin, and cholesterol were used to interpret LAURDAN GP images in cellular membranes [78]. In this work the LAURDAN GP function was used to directly observe transient micron size high GP regions surrounded by low GP areas in living macrophages. This result supports the presence of membrane lateral heterogeneity in cellular membranes under in vivo conditions. Similarly, LAURDAN GP imaging was performed in cells exposed to

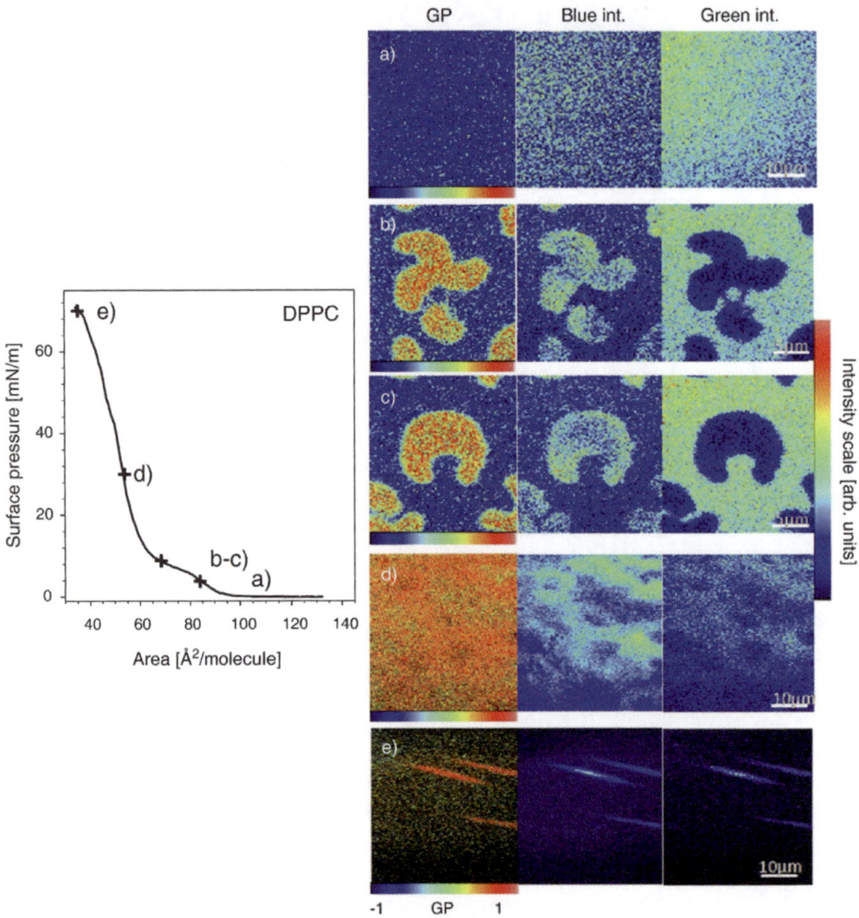

Fig. 11 Surface pressure/area isotherm for the compression of a DPPC monolayer at room temperature (21 °C) (*left*). Representative LAURDAN fluorescence intensity (*blue* and *green* channels) and GP images of a DPPC monolayer at different surface pressures (**a**–**e**) (*right*). The blue and green intensity images have the same fluorescence intensity scales and are represented in false colors. The pressures indicated in the images are: (**a**) 3.8 mN/m (average GP value is −0.58), (**b**) and (**c**) 8.8 mN/m (average GP value in liquid-expanded and liquid-condensed phases are −0.32 and 0.44, respectively), (**d**) 30 mN/m (average GP value is 0.55), (**e**) 70 mN/m (average GP value is 0.62). At the collapse pressure the gain of the detectors has been reduced to avoid saturation on the high fluorescence intensity areas, i.e., the collapsed structures. For such reason the GP values shown in (**e**), except those obtained in the collapse areas, are meaningless. Adapted from [56] with permission

Shiga toxin [79], which induces tubular membrane invaginations for its uptake into cells. LAURDAN measurements were interpreted to support lipid reorganization (reflected in an increase in the abundance of ordered domains in the cell, i.e., higher GP) an effect claimed necessary for the formation of inward membrane tubules [79].

Finally, LAURDAN GP imaging has been also applied to tissues. Sun et al. [80] have demonstrated that both LAURDAN multiphoton polarization and GP can be combined using a TPEFM to characterize the structural changes of intercellular lipids in skin tissue. This work showed how treatment with oleic acid increases water dipolar relaxation in the skin stratum corneum membranes [80]. A similar strategy using LAURDAN GP has been reported in order to characterize membranous structures in pig skin [81] and perform a comparative evaluation between lipid stratum corneum membranes of normal human skin and acquired cholesteatoma [82].

7 Concluding Remarks

The particular fluorescent properties of LAURDAN allow the study of structural and dynamical aspects of membranes, providing both valuable bulk and spatially resolved information. Particular membrane model systems (such as multilamellar vesicles, large unilamellar vesicles, giant unilamellar vesicles, and planar membrane systems) can be exploited to mimic processes that occur in the complex framework of biological membranes. The journey from the cuvette to the microscope has been an important step to better understand supramolecular properties of lipid membranes using this probe. It has allowed to link knowledge gained from compositionally very simple situations, where bulk information is easy to interpret, to compositionally more complex lipid mixtures where spatially resolved information is essential to disentangle (supramolecular) processes occurring in the membrane. As discussed in the last part of this review, the spatially resolved information (both structural and dynamical) obtained using LAURDAN provides a powerful tool to study membrane phenomena in even more complex systems such as cellular membranes and tissues.

Acknowledgments The author is greatly indebted to the following: Jonathan Brewer, Gerardo Fidelio, Enrico Gratton, Theodore Hazlett, David Jameson, Parkson Lee-Gau Chong, Tiziana Parasassi, Susana Sanchez, Patricio Sotomayor, Roberto Stock, and Gregorio Weber for the stimulating scientific discussions concerning the 2-(dimethylamino)-6-acylnaphtalene derivatives. The author wants to thank his colleagues at the Center for Biomembrane Physics (MEMPHYS), particularly Ole Mouritsen for his support during the last 10 years in Denmark. Part of the research presented in this article was supported by the Danish Research Council (FNU and FSS) and the Danish National Research Foundation (which supports MEMPHYS).

References

1. Weber G, Farris FJ (1979) Synthesis and spectral properties of a hydrophobic fluorescent probe: 6-propionyl-2-(dimethylamino)naphthalene. Biochemistry 18(14):3075–3078
2. Lippert E (1957) Spektroskopische bestimmung des dipolmomentes aromatischer verbindungen im ersten angeregten singulettzustand. Z Elektrochem 61:962–975
3. Macgregor RB, Weber G (1981) Fluorophores in polar media: spectral effects of the Langevin distribution of electrostatic interactions. Ann NY Acad Sci 366:140–154

4. Lakowicz JR, Weber G (1973) Quenching of protein fluorescence by oxygen. Detection of structural fluctuations in proteins on the nanosecond time scale Biochemistry 12(21):4171–4179
5. Macgregor RB, Weber G (1986) Estimation of the polarity of the protein interior by optical spectroscopy. Nature 319(6048):70–73
6. Prendergast FG, Meyer M, Carlson GL, Iida S, Potter JD (1983) Synthesis, spectral properties, and use of 6-acryloyl-2-dimethylaminonaphthalene (acrylodan). A thiol-selective, polarity-sensitive fluorescent probe. J Biol Chem 258 (12):7541–7544
7. Lakowicz JR, Bevan DR, Maliwal BP, Cherek H, Balter A (1983) Synthesis and characterization of a fluorescence probe of the phase transition and dynamic properties of membranes. Biochemistry 22:5714–5722
8. Parasassi T, Krasnowska EK, Bagatolli LA, Gratton E (1998) LAURDAN and PRODAN as polarity sensitive fluorescent membrane probes. J Fluorescence 8(4):365–373
9. Kim HM, Choo HJ, Jung SY, Ko YG, Park WH, Jeon SJ, Kim CH, Joo T, Cho BR (2007) A two-photon fluorescent probe for lipid raft imaging: C-LAURDAN. ChemBioChem 8(5):553–559
10. Dodes Traian MM, Gonzalez Flecha L, Levi V (2011) Imaging lipid lateral organization in membranes with C-LAURDAN in a confocal microscope. J Lipid Res. doi:jlr.D021311 [pii] 10.1194/jlr.D021311
11. Lakowicz JR, Sheppard JR (1981) Fluorescence spectroscopic studies of Huntington fibroblast membranes. Am J Human Genet 33(2):155–165
12. Sumbilla C, Lakowicz JR (1982) Fluorescence studies of red blood cell membranes from individuals with Huntington's disease. J Neurochem 38(6):1699–1708
13. Parasassi T, Conti F, Gratton E (1986) Time-resolved fluorescence emission spectra of LAURDAN in phospholipid vesicles by multifrequency phase and modulation fluorometry. Cell Mol Biol 32(1):103–108
14. Parasassi T, Conti F, Gratton E (1986) Fluorophores in a polar medium: time dependence of emission spectra detected by multifrequency phase and modulation fluorometry. Cell Mol Biol 32(1):99–102
15. Parasassi T, De Stasio G, d'Ubaldo A, Gratton E (1990) Phase fluctuation in phospholipid membranes revealed by LAURDAN fluorescence. Biophys J 57(6):1179–1186
16. Parasassi T, De Stasio G, Ravagnan G, Rusch RM, Gratton E (1991) Quantitation of lipid phases in phospholipid vesicles by the generalized polarization of LAURDAN fluorescence. Biophys J 60(1):179–189
17. Krasnowska EK, Bagatolli LA, Gratton E, Parasassi T (2001) Surface properties of cholesterol-containing membranes detected by PRODAN fluorescence. Biochim Biophys Acta 1511(2):330–340
18. Krasnowska EK, Gratton E, Parasassi T (1998) PRODAN as a membrane surface fluorescence probe: partitioning between water and phospholipid phases. Biophys J 74(4):1984–1993
19. Chong PL (1988) Effects of hydrostatic pressure on the location of PRODAN in lipid bilayers and cellular membranes. Biochemistry 27(1):399–404
20. Chong PL-G (1990) Interactions of LAURDAN and PRODAN with membranes at high pressure. High Pressure Research 5:761–763
21. Zeng J, Chong PL (1995) Effect of ethanol-induced lipid interdigitation on the membrane solubility of PRODAN, acdan, and LAURDAN. Biophys J 68(2):567–573
22. Antollini SS, Barrantes FJ (1998) Disclosure of discrete sites for phospholipid and sterols at the protein-lipid interface in native acetylcholine receptor-rich membrane. Biochemistry 37(47):16653–16662
23. Jurkiewicz P, Olzynska A, Langner M, Hof M (2006) Headgroup hydration and mobility of dotap/dopc bilayers: A fluorescence solvent relaxation study. Langmuir 22(21):8741–8749
24. Olzynska A, Zan A, Jurkiewicz P, Sykora J, Grobner G, Langner M, Hof M (2007) Molecular interpretation of fluorescence solvent relaxation of PATMAN and 2 h NMR experiments in phosphatidylcholine bilayers. Chem Phys Lipids 147(2):69–77
25. Hutterer R, Schneider FW, Sprinz H, Hof M (1996) Binding and relaxation behaviour of PRODAN and PATMAN in phospholipid vesicles: A fluorescence and 1 h NMR study. Biophys Chem 61(2–3):151–160

26. Jurkiewicz P, Sykora J, Olzynska A, Humpolickova J, Hof M (2005) Solvent relaxation in phospholipid bilayers: principles and recent applications. J Fluorescence 15(6):883–894
27. Bagatolli LA, Gratton E (2000) A correlation between lipid domain shape and binary phospholipid mixture composition in free standing bilayers: A two-photon fluorescence microscopy study. Biophys J 79(1):434–447
28. Bagatolli LA, Gratton E (2000) Two photon fluorescence microscopy of coexisting lipid domains in giant unilamellar vesicles of binary phospholipid mixtures. Biophys J 78(1):290–305
29. Parasassi T, Ravagnan G, Rusch RM, Gratton E (1993) Modulation and dynamics of phase properties in phospholipid mixtures detected by LAURDAN fluorescence. Photochem Photobiol 57(3):403–410
30. Bagatolli LA, Gratton E, Fidelio GD (1998) Water dynamics in glycosphingolipid aggregates studied by LAURDAN fluorescence. Biophys J 75(1):331–341
31. Bagatolli LA, Maggio B, Aguilar F, Sotomayor CP, Fidelio GD (1997) LAURDAN properties in glycosphingolipid-phospholipid mixtures: a comparative fluorescence and calorimetric study. Biochim Biophys Acta 1325(1):80–90
32. Parasassi T, Gratton E (1992) Packing of phospholipid vesicles studied by oxygen quenching of LAURDAN fluorescence. J Fluorescence 2(3):167–174
33. Bagatolli LA, Parasassi T, Fidelio GD, Gratton E (1999) A model for the interaction of 6-lauroyl-2-(n, n-dimethylamino)naphthalene with lipid environments: Implications for spectral properties. Photochem Photobiol 70(4):557–564
34. Viard M, Gallay J, Vincent M, Meyer O, Robert B, Paternostre M (1997) LAURDAN solvatochromism: solvent dielectric relaxation and intramolecular excited-state reaction. Biophys J 73(4):2221–2234
35. Vincent M, de Foresta B, Gallay J (2005) Nanosecond dynamics of a mimicked membrane-water interface observed by time-resolved stokes shift of LAURDAN. Biophys J 88(6):4337–4350
36. Parasassi T, Gratton E, Yu WM, Wilson P, Levi M (1997) Two-photon fluorescence microscopy of LAURDAN generalized polarization domains in model and natural membranes. Biophys J 72(6):2413–2429
37. Parasassi T, Gratton E (1995) Membrane lipid domains and dynamics as detected by LAURDAN fluorescence. J Fluorescence 5(1):59–69
38. Celli A, Gratton E (2010) Dynamics of lipid domain formation: fluctuation analysis. Biochim Biophys Acta 1798(7):1368–1376
39. Jurkiewicz P, Cwiklik L, Jungwirth P, Hof M (2012) Lipid hydration and mobility: an interplay between fluorescence solvent relaxation experiments and molecular dynamics simulations. Biochimie 94(1):26–32
40. Jameson DM, Croney JC, Moens PD (2003) Fluorescence: basic concepts, practical aspects, and some anecdotes. Methods Enzymol 360:1–43
41. Weber G (1952) Polarization of the fluorescence of macromolecules. I. Theory and experimental method. Biochem J 51(2):145–155
42. Henshaw JB, Olsen CA, Farnbach AR, Nielson KH, Bell JD (1998) Definition of the specific roles of lysolecithin and palmitic acid in altering the susceptibility of dipalmitoylphosphatidylcholine bilayers to phospholipase a2. Biochemistry 37(30):10709–10721
43. Parasassi T, Loiero M, Raimondi M, Ravagnan G, Gratton E (1993) Absence of lipid gel-phase domains in seven mammalian cell lines and in four primary cell types. Biochim Biophys Acta 1153(2):143–154
44. Vanounou S, Pines D, Pines E, Parola AH, Fishov I (2002) Coexistence of domains with distinct order and polarity in fluid bacterial membranes. Photochem Photobiol 76(1):1–11
45. Nielsen SB, Otzen DE (2010) Impact of the antimicrobial peptide novicidin on membrane structure and integrity. J Colloid Interface Sci 345:248–256
46. Parasassi T, Di Stefano M, Loiero M, Ravagnan G, Gratton E (1994) Cholesterol modifies water concentration and dynamics in phospholipid bilayers: a fluorescence study using LAURDAN probe. Biophys J 66(3 Pt 1):763–768

47. Parasassi T, Di Stefano M, Loiero M, Ravagnan G, Gratton E (1994) Influence of cholesterol on phospholipid bilayers phase domains as detected by LAURDAN fluorescence. Biophys J 66(1):120–132
48. Parasassi T, Giusti AM, Raimondi M, Gratton E (1995) Abrupt modifications of phospholipid bilayer properties at critical cholesterol concentrations. Biophys J 68(5):1895–1902
49. Arnulphi C, Levstein PR, Ramia ME, Martin CA, Fidelio GD (1997) Ganglioside hydration study by 2h-NMR: dependence on temperature and water/lipid ratio. J Lipid Res 38(7):1412–1420
50. Ipsen JH, Karlstrom G, Mouritsen OG, Wennerstrom H, Zuckermann MJ (1987) Phase equilibria in the phosphatidylcholine-cholesterol system. Biochim Biophys Acta 905(1):162–172
51. Tang D, Chong PL (1992) E/m dips. Evidence for lipids regularly distributed into hexagonal super-lattices in pyrene-pc/dmpc binary mixtures at specific concentrations. Biophys J 63(4):903–910
52. Virtanen JA, Somerharju PJ, KPK J (1988) Prediction of patterns for the regular distribution of soluted guest molecules in liquid crystalline phospholipid membranes. J Mol Electron 4:233–236
53. Yu W, So PT, French T, Gratton E (1996) Fluorescence generalized polarization of cell membranes: a two-photon scanning microscopy approach. Biophys J 70(2):626–636
54. Bagatolli L, Gratton E, Khan TK, Chong PL (2000) Two-photon fluorescence microscopy studies of bipolar tetraether giant liposomes from thermoacidophilic archaebacteria sulfolobus acidocaldarius. Biophys J 79(1):416–425
55. Dietrich C, Bagatolli LA, Volovyk ZN, Thompson NL, Levi M, Jacobson K, Gratton E (2001) Lipid rafts reconstituted in model membranes. Biophys J 80(3):1417–1428
56. Brewer J, Bernardino de la Serna J, Wagner K, Bagatolli LA (2010) Multiphoton excitation fluorescence microscopy in planar membrane systems. Biochim Biophys Acta 1798(7): 1301–1308
57. Wheeler G, Tyler KM (2011) Widefield microscopy for live imaging of lipid domains and membrane dynamics. Biochim Biophys Acta 1808(3):634–641
58. Weber PMW, Schneckenburger H (2010) Fluorescence imaging of membrane dynamics in living cells. J Biomed Opt 15(4):046017
59. Bagatolli LA (2006) To see or not to see: Lateral organization of biological membranes and fluorescence microscopy. Biochim Biophys Acta 1758(10):1541–1556
60. Bagatolli LA, Gratton E (2001) Direct observation of lipid domains in free-standing bilayers using two-photon excitation fluorescence microscopy. J Fluorescence 11(3):141–160
61. Juhasz J, Davis JH, Sharom FJ (2010) Fluorescent probe partitioning in giant unilamellar vesicles of "lipid raft" mixtures. Biochem J 430(3):415–423
62. Korlach J, Schwille P, Webb WW, Feigenson GW (1999) Characterization of lipid bilayer phases by confocal microscopy and fluorescence correlation spectroscopy. Proc Natl Acad Sci USA 96(15):8461–8466
63. Bagatolli LA, Gratton E (1999) Two-photon fluorescence microscopy observation of shape changes at the phase transition in phospholipid giant unilamellar vesicles. Biophys J 77(4):2090–2101
64. Fidorra M, Duelund L, Leidy C, Simonsen AC, Bagatolli LA (2006) Absence of fluid-ordered/ fluid-disordered phase coexistence in ceramide/popc mixtures containing cholesterol. Biophys J 90(12):4437–4451
65. Fidorra M, Heimburg T, Bagatolli LA (2009) Direct visualization of the lateral structure of porcine brain cerebrosides/popc mixtures in presence and absence of cholesterol. Biophys J 97(1):142–154
66. Kubiak J, Brewer J, Hansen S, Bagatolli LA (2011) Lipid lateral organization on giant unilamellar vesicles containing lipopolysaccharides. Biophys J 100(4):978–986
67. Sot J, Bagatolli LA, Goni FM, Alonso A (2006) Detergent-resistant, ceramide-enriched domains in sphingomyelin/ceramide bilayers. Biophys J 90(3):903–914
68. Norlen L, Plasencia I, Bagatolli L (2008) Stratum corneum lipid organization as observed by atomic force, confocal and two-photon excitation fluorescence microscopy. Int J Cosmet Sci 30(6):391–411

69. Plasencia I, Norlen L, Bagatolli LA (2007) Direct visualization of lipid domains in human skin stratum corneum's lipid membranes: Effect of ph and temperature. Biophys J 93(9):3142–3155
70. Bernardino de la Serna J, Oradd G, Bagatolli LA, Simonsen AC, Marsh D, Lindblom G, Perez-Gil J (2009) Segregated phases in pulmonary surfactant membranes do not show coexistence of lipid populations with differentiated dynamic properties. Biophys J 97 (5):1381–1389
71. Nag K, Pao JS, Harbottle RR, Possmayer F, Petersen NO, Bagatolli LA (2002) Segregation of saturated chain lipids in pulmonary surfactant films and bilayers. Biophys J 82(4):2041–2051
72. Montes LR, Alonso A, Goni FM, Bagatolli LA (2007) Giant unilamellar vesicles electroformed from native membranes and organic lipid mixtures under physiological conditions. Biophys J 93(10):3548–3554
73. Celli A, Beretta S, Gratton E (2008) Phase fluctuations on the micron-submicron scale in guvs composed of a binary lipid mixture. Biophys J 94(1):104–116
74. Bagatolli LA, Sanchez SA, Hazlett T, Gratton E (2003) Giant vesicles, LAURDAN, and two-photon fluorescence microscopy: evidence of lipid lateral separation in bilayers. Methods Enzymol 360:481–500
75. Parasassi T, Gratton E, Zajicek H, Levi M, Yu W (1999) Detecting membrane lipid microdomains by two-photon fluorescence microscopy. IEEE Eng Med Biol Mag 18(5):92–99
76. Bernchou U, Brewer J, Midtiby HS, Ipsen JH, Bagatolli LA, Simonsen AC (2009) Texture of lipid bilayer domains. J Am Chem Soc 131(40):14130–14131
77. Marsh D (1996) Lateral pressure in membranes. Biochim Biophys Acta 1286(3):183–223
78. Gaus K, Gratton E, Kable EP, Jones AS, Gelissen I, Kritharides L, Jessup W (2003) Visualizing lipid structure and raft domains in living cells with two-photon microscopy. Proc Natl Acad Sci USA 100(26):15554–15559
79. Romer W, Berland L, Chambon V, Gaus K, Windschiegl B, Tenza D, Aly MR, Fraisier V, Florent JC, Perrais D, Lamaze C, Raposo G, Steinem C, Sens P, Bassereau P, Johannes L (2007) Shiga toxin induces tubular membrane invaginations for its uptake into cells. Nature 450(7170):670–675
80. Sun Y, Lo W, Lin SJ, Jee SH, Dong CY (2004) Multiphoton polarization and generalized polarization microscopy reveal oleic-acid-induced structural changes in intercellular lipid layers of the skin. Opt Lett 29(17):2013–2015
81. Carrer DC, Vermehren C, Bagatolli LA (2008) Pig skin structure and transdermal delivery of liposomes: a two photon microscopy study. J Control Release 132(1):12–20
82. Bloksgaard M, Svane-Knudsen V, Sorensen JA, Bagatolli L, Brewer J (2012) Structural characterization and lipid composition of acquired cholesteatoma: a comparative study with normal skin. Otol Neurotol 33(2):177–183
83. Bagatolli LA, Ipsen JH, Simonsen AC, Mouritsen OG (2010) An outlook on organization of lipids in membranes: searching for a realistic connection with the organization of biological membranes. Prog Lipid Res 49(4):378–389

Application of NBD-Labeled Lipids in Membrane and Cell Biology

Sourav Haldar and Amitabha Chattopadhyay

Abstract The fluorescent NBD group has come a long way in terms of biological applications since its discovery a few decades back. Although the field of fluorescently labeled lipids has grown over the years with the introduction of new fluorescent labels, NBD-labeled lipids continue to be a popular choice in membrane and cell biological studies due to desirable fluorescence characteristics of the NBD group. In this chapter, we discuss the application of NBD-labeled lipids in membrane and cell biology taking representative examples with specific focus on the biophysical basis underlying such applications.

Keywords FRAP · Looping up · Membrane probes · REES

Contents

S. Haldar and A. Chattopadhyay (✉)
Centre for Cellular and Molecular Biology, Council of Scientific and Industrial Research, Uppal Road, Hyderabad 500 007, India
e-mail: amit@ccmb.res.in

Y. Mély and G. Duportail (eds.), *Fluorescent Methods to Study Biological Membranes*, Springer Ser Fluoresc (2013) 13: 37–50, DOI 10.1007/4243_2012_43,

Abbreviations

6-NBD-PC	1-Palmitoyl-2-(6-[*N*-(7-nitrobenz-2-oxa-1,3-diazol-4-yl) amino]hexanoyl)-*sn*-glycero-3-phosphocholine
12-NBD-PC	1-Palmitoyl-2-(12-[*N*-(7-nitrobenz-2-oxa-1,3-diazol-yl) amino]dodecanoyl)-*sn*-glycero-3-phosphocholine
6-NBD-CM	6-([*N*-(7-nitrobenz-2-oxa-1,3-diazol-4-yl)amino]hexanoyl) sphingosine
6-NBD-SM	6-([*N*-(7-nitrobenz-2-oxa-1,3-diazol-4-yl)amino]hexanoyl) sphingosylphosphocholine
25-NBD-cholesterol	25-[*N*-[(7-nitrobenz-2-oxa-1,3-diazol-4-yl)-methyl] amino]-27-norcholesterol
DOPC	Dioleoyl-*sn*-glycero-3-phosphocholine
DPPC	1,2-Dipalmitoyl-*sn*-glycero-3-phosphocholine
FRAP	Fluorescence recovery after photobleaching
NBD	7-Nitrobenz-2-oxa-1,3-diazol-4-yl
NBD-PE	*N*-(7-nitrobenz-2-oxa-1,3-diazol-4-yl)-1,2-dipalmitoyl-*sn*-glycero-3-phosphoethanolamine
NBD-PS	1,2-Dioleoyl-*sn*-glycero-3-phospho-l-serine-*N*-(7-nitrobenz-2-oxa-1,3-diazol-4-yl)
POPC	1-Palmitoyl-2-oleoyl-*sn*-glycero-3-phosphocholine
REES	Red edge excitation shift

1 Introduction

Cellular membranes represent two-dimensional, non-covalent, anisotropic, and cooperative assemblies consisting of lipids and proteins. Membranes allow cellular compartmentalization and act as the interface through which cells sense the environment and communicate with each other. They confer an identity to cells (and their organelles) and represent an appropriate milieu for the proper function of membrane proteins. In addition, membranes constitute the site of important cellular functions such as signal transduction [1] and pathogen entry [2, 3]. It has been estimated that ~50% of all biological processes occur at the cell membrane [4].

The mammalian cell is made up of a large variety of lipids [5] which orchestrate diverse cellular functions with the help of membrane proteins. Tracking individual lipids in a crowded cellular milieu poses considerable challenge. It is in this context that lipid probes assume significance (see [6] for a comprehensive account of lipid probes). Various lipid probes have proved to be useful in membrane and cell biology due to their ability to monitor lipid molecules by a variety of physicochemical approaches at increasing spatiotemporal resolution [7]. Spectroscopic and microscopic techniques using fluorescent lipid analogs represent a powerful set of approaches for monitoring membrane organization and dynamics due to their high sensitivity, suitable time resolution, and multiplicity of measurable parameters.

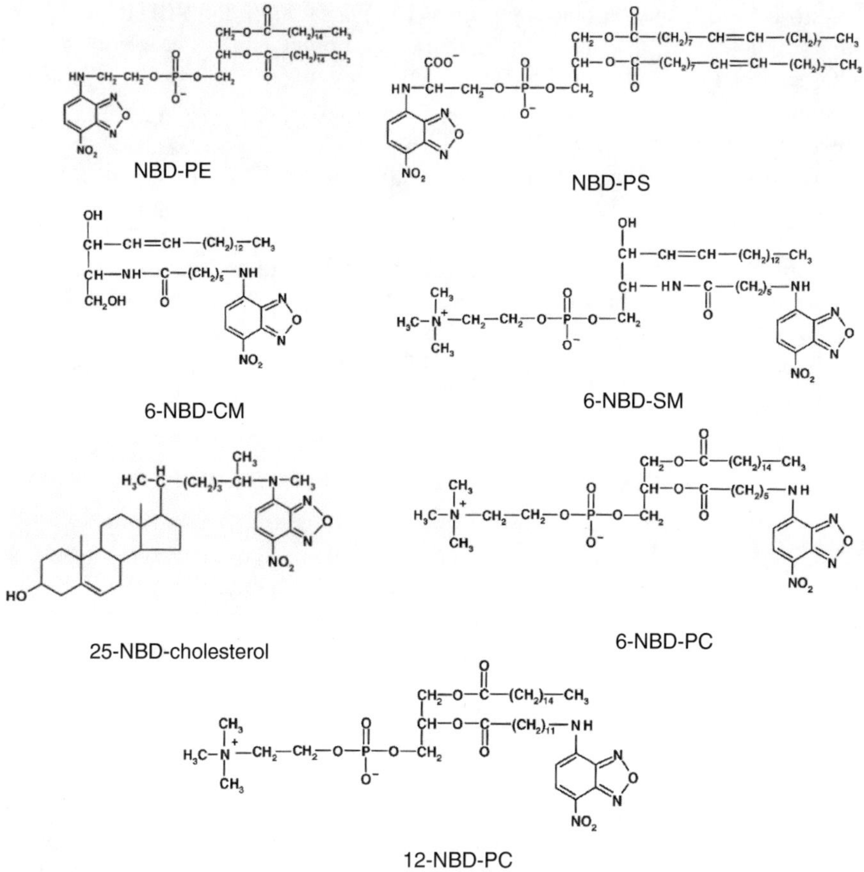

Fig. 1 Chemical structures of representative NBD-labeled lipids. NBD-labeled lipids are extensively used as fluorescent analogs of natural lipids in membrane and cell biological studies. Depending on the specific lipid type, the NBD group could be covalently attached to the polar lipid headgroup (as in NBD-PE and NBD-PS) or to the *sn*-2 fatty acyl chain of the lipid (as in NBD-SM, NBD-CM, and NBD-PC). In case of 25-NBD-cholesterol, the NBD moiety is attached to the flexible acyl chain of the sterol

Lipids covalently linked to extrinsic fluorophores are commonly used for such studies. The advantage with this approach is that one has a choice of the fluorescent label to be used, and therefore, specific probes with appropriate characteristics can be designed for specific applications.

A widely used fluorophore in biophysical, biochemical, and cell biological studies of membranes is the NBD (7-nitrobenz-2-oxa-1,3-diazol-4-yl) group (for an earlier review on NBD-labeled lipids, see [8]). NBD-labeled lipids are extensively used as fluorescent analogs of native lipids in biological and model membranes to monitor a variety of processes (see Fig. 1). This is due to the fact that the NBD group possesses some of the most desirable properties to serve as an excellent probe for both spectroscopic and microscopic applications [9]. For example, the NBD group is very weakly fluorescent in water. Yet, it fluoresces brightly in the visible range and exhibits a high degree of environmental sensitivity upon transfer to a hydrophobic medium [9–13]. Fluorescence lifetime of the NBD group exhibits sensitivity to environmental polarity [12, 14]. Lipids labeled with the NBD group have been shown to mimic endogenous lipids in a number of studies [15–18] although this appears to be not always true [19, 20].

In this chapter, we will focus on the application of NBD-labeled lipids in membrane and cell biology with representative examples. This chapter is not meant to be an exhaustive account of the literature on NBD-labeled lipids. Rather, we intend to provide the biophysical basis underlying specific applications. The reader is referred to a previous review [8] for earlier references and applications.

2 NBD Group Senses Slow Solvent Relaxation in Membranes

It has long been recognized that organized molecular assemblies (such as membranes) may be considered as large cooperative units with properties very different from the individual structural components that constitute them. An obvious consequence of such type of organization is the restriction imposed on the dynamics of their constituent structural components. Interestingly, this kind of restriction (confinement) results in coupling the motion of solvent molecules with the slow-moving molecules in the host assembly [21]. In this scenario, red edge excitation shift (REES) represents an interesting approach that relies on slow solvent reorientation in the excited state of a fluorophore which can be used to monitor the environment and dynamics around it in an organized molecular assembly [22–25]. A shift in the wavelength of maximum fluorescence emission toward higher wavelengths, caused by a shift in the excitation wavelength toward the red edge of absorption band, is termed red edge excitation shift (REES). REES arises from relatively slow rates (relative to fluorescence lifetime) of solvent relaxation (reorientation) around an excited-state fluorophore. REES therefore depends on the environment-induced motional restriction imposed on solvent molecules in the immediate proximity of the fluorophore. It allows to assess the rotational mobility of the environment itself (which is represented by the relaxing solvent molecules) utilizing the fluorophore merely as a reporter group (the definition of solvent in this context is rather pragmatic; solvent relaxation dynamics includes dynamics of restricted solvent [water] as well as the dynamics of the host dipolar matrix such as the peptide backbone in proteins [26]).

As mentioned above, an obvious consequence of high degree of organization in supramolecular assemblies such as membranes is the restriction imposed on the mobility of the constituent structural components. The biological membrane, with its viscous interior and distinct motional gradient along its vertical axis, therefore represents an ideal system for the application of REES to explore membrane phenomena ([22, 24]; see Fig. 2). The interfacial region in membranes is

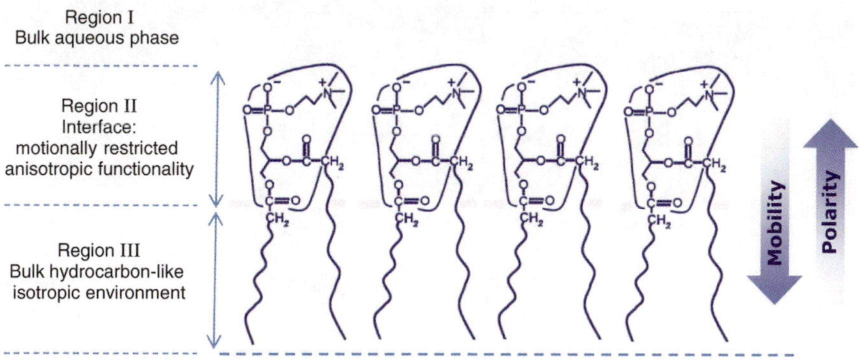

Fig. 2 Schematic representation of half of the membrane bilayer showing the asymmetric nature of membranes in terms of anisotropy in polarity and dynamics along the monolayer. The dotted line at the bottom indicates the center of the bilayer. The membrane anisotropy along the z-axis (perpendicular to the plane of the membrane) compartmentalizes the membrane leaflet into three regions exhibiting differential dynamics. Region I comprises of bulk aqueous phase characterized by fast solvent relaxation; region II is the membrane interface, characterized by slow (restricted) solvent relaxation, and water penetration (interfacial water). This region is highly heterogeneous in chemical composition; region III represents the bulk hydrocarbon-like environment, isotropic in nature, and characterized by fast solvent relaxation. In addition, a polarity (dielectric) gradient along the z-axis is also an integral feature of membranes (see [27]). Fluorescent probes and peptides localized in the membrane interface (region II) are sensitive to REES measurements (Adapted and modified from Haldar et al. [24])

characterized by unique motional and dielectric characteristics different from the bulk aqueous phase and the more isotropic hydrocarbon-like deeper regions of the membrane. The membrane interfacial region exhibits slow rates of solvent relaxation and is therefore most likely to display the REES effect. In order to explore such effect, it is necessary to choose an appropriate probe that displays suitable properties in terms of localization, polarity, and appreciable change in dipole moment upon excitation [22, 24]. The NBD group in membrane-bound NBD-PE was found to satisfy these criteria [28]. The fluorescent NBD label is covalently attached to the headgroup of a phosphatidylethanolamine molecule in NBD-PE (see Fig. 1). The orientation and location of the NBD group in membrane-bound NBD-PE has been worked out [10, 29–33]. The NBD group in NBD-PE was found to be localized at the membrane interface characterized by unique motional and dielectric properties and therefore represents an ideal probe for monitoring REES and related effects. Interestingly, the NBD group exhibits a relatively large change in dipole moment upon excitation (~4 D; [13]), a necessary condition for a fluorophore to exhibit REES [24]. The change in emission maximum with changing excitation wavelength (REES) of NBD-PE in dioleoyl-*sn*-glycero-3-phosphocholine (DOPC) membranes is shown in Fig. 3 [28, 32, 34]. Since the localization of the fluorescent NBD group in membrane-bound NBD-PE is interfacial [10, 29–33], these REES results imply that the interfacial region of the membrane offers considerable restriction to the reorientational motion of the solvent dipoles around the excited-state NBD group. It was later shown that

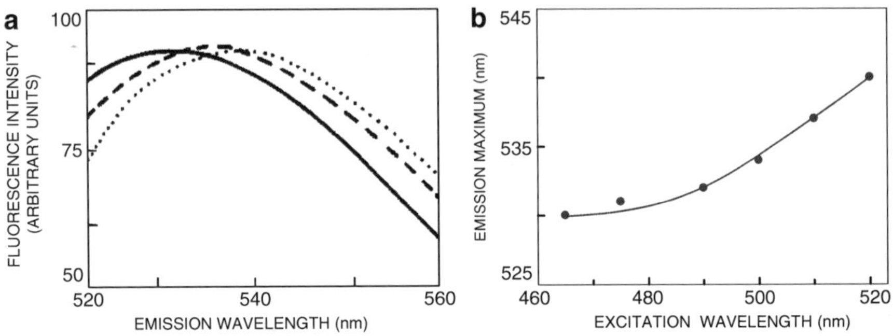

Fig. 3 NBD-PE displays REES in membranes: (**a**) typical intensity-normalized fluorescence emission spectra of NBD-PE at increasing excitation wavelengths. Excitation wavelengths used were 465 (——), 500 (— —), and 510 (------) nm. (**b**) The effect of changing excitation wavelength on the wavelength of maximum emission (REES) of NBD-PE (Adapted and modified from Chattopadhyay and Mukherjee [28])

NBD-PE exhibits REES in membrane-mimetic assemblies such as micelles and reverse micelles [9, 14, 35]. In addition, REES exhibited by the NBD group labeled in a site-specific manner in the membrane-active peptide melittin provided novel information regarding the orientation of the peptide in the membrane [36].

3 Membrane Phase Dependence of Probe Looping in Acyl Chain-Labeled NBD Lipids

An important aspect of fluorescent membrane probes is their location in the membrane [37]. In the case of NBD-labeled lipids, it has been previously shown that the NBD group of acyl chain-labeled NBD lipids such as 6- and 12-NBD-PC (Fig. 1) loops up to the membrane interface in fluid-phase membranes due to the polarity of the NBD group ([10, 30, 32, 38–40]; see Table 1 and Fig. 4). This is also consistent with the observation that the NBD group in 6- and 12-NBD-PC exhibits considerable REES in fluid-phase membrane bilayers [32] and in monolayers at the air/water interface [42] since display of REES is characteristic of interfacial probe localization. The looping up of acyl chain-labeled NBD group could be due to hydrogen bonding of the NBD group at the membrane interface. The polar imino group and the oxygen atoms of the NBD group may form hydrogen bonds with the lipid carbonyls, interfacial water molecules, and the lipid headgroup. An important consequence of the looping up of the NBD group is an increase in the headgroup area. For example, it has been estimated that in POPC membranes, looping up of the NBD group results in a ~3% increase in the headgroup area [38]. It is for this reason that the looping up tendency of the NBD group in NBD-labeled lipids has been implicated in their preferred endocytic sorting [43]. The looping up of the NBD group in acyl chain-labeled NBD lipids has been utilized to monitor lipid-protein interactions in membranes [44].

Table 1 Membrane penetration depths of the NBD group in NBD-labeled lipids by the parallax method[a]

NBD-labeled lipids	Distance from the center of the bilayer z_{cF} (Å)
NBD-PE	20.3
6-NBD-PC	20.7
12-NBD-PC	20.7
6-NBD-CM	20.8
6-NBD-SM	20.5
NBD-PS (pH 7.2)	18.8
NBD-PS (pH 5.0)	14.1
25-NBD-cholesterol[b]	5.7

[a]From Mukherjee et al. [32]
[b]From Chattopadhyay and London [30]

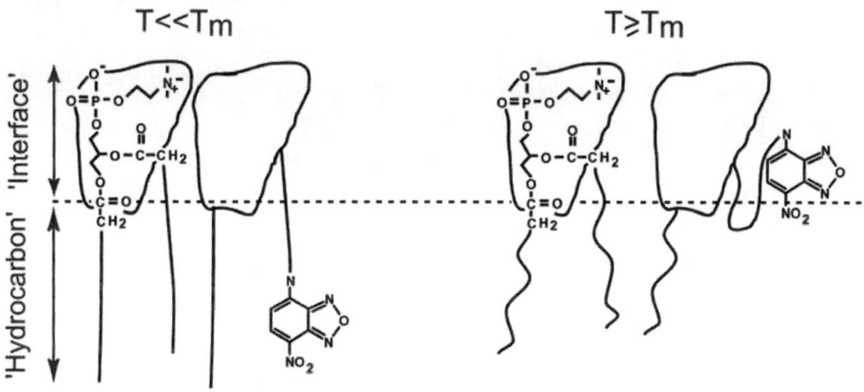

Fig. 4 To loop up or not? A schematic representation of the acyl chain conformation in acyl chain-labeled NBD lipids below (*left*) and above (*right*) phase transition temperature of the membrane. The NBD group loops up to the membrane interface in fluid-phase ($T \geq T_m$) membranes [10, 30]. Interestingly, the looping up of the NBD group is found to be absent in gel-phase ($T << T_m$) membranes [41] (Reproduced from Raghuraman et al. [41])

Interestingly, looping up of the NBD group is critically dependent on the phase state of the membrane. In contrast to the looping up of the NBD group observed in fluid-phase (i.e., above the phase transition temperature) membranes, there appears to be a vertical distribution of the NBD group in acyl chain-labeled NBD lipids in gel (ordered) phase membranes, thereby showing that looping up of the probe is not observed under these conditions ([41]; see Fig. 4). This has been attributed to change in membrane packing induced by phase transition which could influence probe localization in the membrane.

4 Application of NBD Fluorescence Sensitivity in Cell Biology

An interesting feature of NBD fluorescence is its sensitivity in response to the environment in which the fluorophore is placed. The NBD group exhibits a high degree of environmental sensitivity [9–13], and fluorescence lifetime of the NBD group displays remarkable sensitivity to environmental polarity [12, 14]. For example, NBD lifetime in hydrophobic media such as membranes is high (~7 ns; [28, 32]), while NBD lifetime is considerably reduced in presence of water [9, 11, 12]. NBD lifetime reduces to ~ 1 ns in water which has been attributed to hydrogen bonding interactions between the fluorophore and the solvent [12] that is accompanied by an increase in the rate of nonradiative decay [45]. This aspect of NBD fluorescence has been effectively utilized in a number of cell biological applications. Environmental (polarity) sensitivity of NBD lifetimes was elegantly used to address the issue of movement of the signal sequence through the ribosomal tunnel during translocation of a nascent secretory protein across the endoplasmic reticulum membrane [46]. A careful analysis of lifetimes of NBD probes attached to the signal sequence of fully assembled ribosome-nascent chain-membrane complex showed that the probes displayed lifetimes corresponding to an aqueous environment (short lifetime ~ 1 ns). Based on these results, it was concluded that the signal sequence does not insert into the nonpolar core of the endoplasmic reticulum membrane. Instead, the signal sequence is localized in an aqueous environment during the early stages of the translocation process. A similar study, utilizing polarity dependence of NBD lifetimes, revealed a novel mechanism of membrane insertion for cholesterol-dependent cytolysins [47]. A rather interesting application of temperature sensitivity of NBD lifetime is the measurement of temperature in living cells as an 'optical thermometer' [48]. Another important and widely used application of NBD-labeled lipids is to monitor membrane asymmetry by chemically modifying (reducing) the NBD group with the water-soluble reducing agent dithionite [49].

5 Transbilayer Organization of Cholesterol Monitored Using NBD Fluorescence

Although a large body of literature exists on the organization of cholesterol in plasma membranes (with high cholesterol content, typically ~30–50 mol%), very little is known about its organization in the membrane where cholesterol content is very low (< 5 mol%). Membranes from the endoplasmic reticulum (where cholesterol is synthesized) and mitochondria are characterized with low cholesterol content. Interestingly, evidence for specific organization of cholesterol molecules in membranes at low concentrations came from studies carried out using 25-NBD-cholesterol (see Fig. 1; [50–53]). The aggregation-sensitive fluorescence of the NBD group in 25-NBD-cholesterol was elegantly utilized by Mukherjee and Chattopadhyay [51] to explore the local organization of cholesterol at low

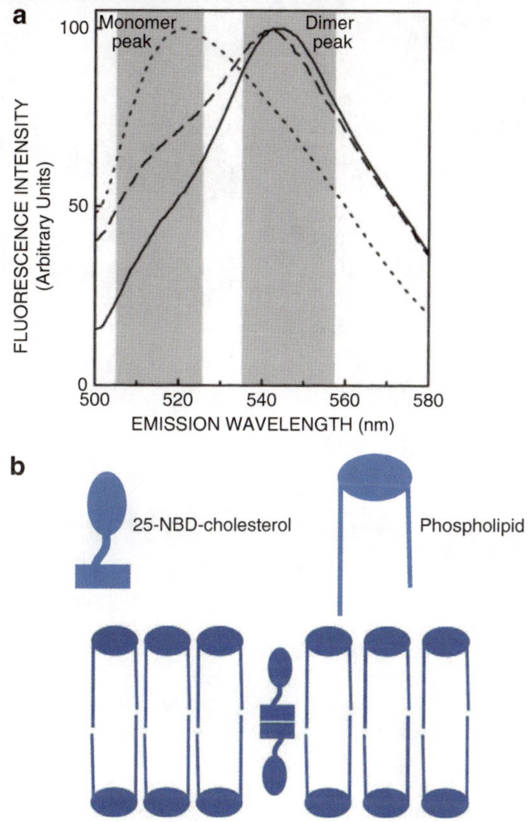

Fig. 5 (**a**) Concentration-dependent emission spectral features of 25-NBD-cholesterol: a red shift in fluorescence emission maximum is observed with increasing concentration. Fluorescence emission spectra of 25-NBD-cholesterol in gel-phase DPPC vesicles are shown. The concentration of 25-NBD-cholesterol was 0.1 mol (------), 0.5 mol (— —), and 1 mol% (———). The shaded portions of the spectra represent the two wavelength ranges (505–526 and 537–558 nm) from which fluorescence emission of 25-NBD-cholesterol was collected for FRAP measurements described in Fig. 6. More details are in [54]. (**b**) Schematic diagram of the membrane bilayer depicting the transbilayer tail-to-tail dimers of cholesterol in membranes at low concentrations (Adapted and modified from Pucadyil et al. [54])

concentrations in membranes. By careful analysis of the emission spectral features of 25-NBD-cholesterol in DPPC membranes at the concentration range of 0.1–5 mol%, the possible presence of transbilayer tail-to-tail dimers of cholesterol in such membranes was detected both in gel- and fluid-phase membranes (see Fig. 5; [51]). It was further shown by monitoring corresponding changes in the absorption spectrum that the cholesterol dimers represented the formation of a ground-state complex (rather than an excited-state interaction). The possibility that the unique spectral feature was due to nonspecific aggregation of the NBD group was ruled out by careful control experiments. This implies that these results

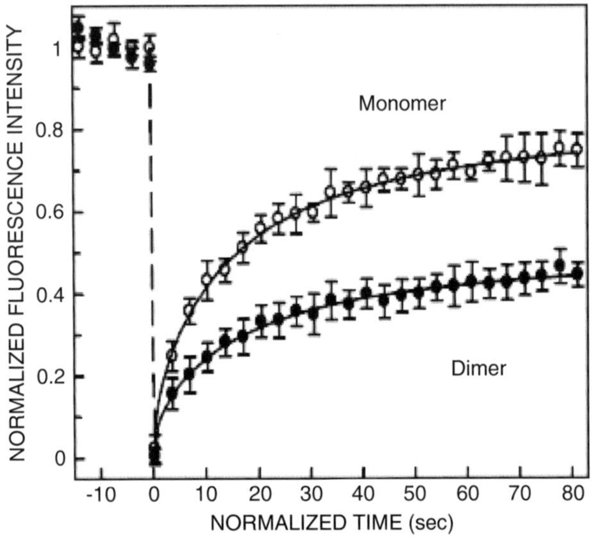

Fig. 6 Simultaneous fluorescence recovery after photobleaching (FRAP) measurement of monomeric and dimeric populations of 25-NBD-cholesterol by the wavelength-selective FRAP approach. The figure shows the wavelength-selective FRAP of 25-NBD-cholesterol in DPPC vesicles where fluorescence emission was collected from 505 to 526 nm (○), corresponding to the monomeric population of 25-NBD-cholesterol, and 537–558 (•) nm corresponding to the dimeric 25-NBD-cholesterol population (see shaded portions in Fig. 5). Note that dimeric 25-NBD-cholesterol exhibits slow diffusion relative to monomeric 25-NBD-cholesterol. Other details are in [54] (Adapted and modified from Pucadyil et al. [54])

provide novel information about cholesterol dimerization in membranes at low concentrations, rather than providing information on NBD-NBD interactions. These results were further supported by observations from other laboratories [55]. In addition, from the distinct spectral feature of 25-NBD-cholesterol in membranes of varying curvature, it was shown that the transbilayer dimer arrangement is sensitive to membrane curvature, and dimerization is not favored in highly curved membranes [53]. The organization and dynamics of cholesterol monomers and dimers were explored by REES [52]. The environment around the cholesterol dimer appears to be rigid (relative to the monomer environment) and offer more restriction to solvent reorientation.

By the application of a novel version of fluorescence recovery after photobleaching (FRAP) measurements 'wavelength-selective FRAP', lateral diffusion coefficients of dimeric and monomeric populations of cholesterol were estimated using 25-NBD-cholesterol (see Fig. 6; [54]). In these experiments, wavelength-selective FRAP measurements were carried out in DPPC membranes containing 25-NBD-cholesterol. The diffusion characteristics of the transbilayer dimer and monomer of 25-NBD-cholesterol (evident from spectral features; see Fig. 5a) were derived by analysis of FRAP results after photoselecting a given population by use of specific wavelength-range characteristic of that population.

The results showed that the organization of 25-NBD-cholesterol in DPPC membranes is heterogeneous, with the presence of fast- and slow-diffusing species. The presence of fast- and slow-diffusing populations of 25-NBD-cholesterol was interpreted to correspond to predominant populations of cholesterol monomers and dimers.

6 Conclusion and Future Perspectives

The fluorescent NBD group has come a long way in terms of biological applications since its discovery a few decades back [56]. NBD-labeled lipids were first synthesized, and their sensitive fluorescence was noted in the late 1970s [57, 58]. Since then, these lipids have been used in a number of biophysical and cell biological studies to gain a variety of information. The field of fluorescently labeled lipids has grown over the years with the introduction of new fluorescent labels with desirable properties [59]. Although NBD-labeled lipids have been shown to mimic endogenous lipids in a number of studies [15–18], concerns have been raised in some cases [19, 20]. Photostability of the NBD group could also be a concern although this can be handled by using low light intensity level and other techniques [60, 61]. Nonetheless, NBD-labeled lipids continue to be widely used for various biological applications. Future exciting applications could include simultaneous attachment of the NBD group and nitroxide group in an amphiphilic molecule to explore membrane heterogeneity [62] and monitoring amyloid fibril formation utilizing NBD-labeled lipids [63].

Acknowledgments Work in A.C.'s laboratory was supported by the Council of Scientific and Industrial Research and Department of Science and Technology, Government of India. S.H. thanks the Council of Scientific and Industrial Research for the award of a Senior Research Fellowship. A.C. is an Adjunct Professor at the Special Centre for Molecular Medicine of Jawaharlal Nehru University (New Delhi, India) and Indian Institute of Science Education and Research (Mohali, India) and Honorary Professor of the Jawaharlal Nehru Centre for Advanced Scientific Research (Bangalore, India). A.C. gratefully acknowledges J.C. Bose Fellowship (Dept. Science and Technology, Govt. of India). Some of the work described in this chapter was carried out by former members of A.C.'s research group whose contributions are gratefully acknowledged. We thank members of our laboratory for critically reading the manuscript. We dedicate this chapter to the memory of Prof. Richard E. Pagano for his seminal contribution in the development and application of NBD-labeled lipids in cell biology.

References

1. Simons K, Toomre D (2000) Lipid rafts and signal transduction. Nat Rev Mol Cell Biol 1:31–39
2. Pucadyil TJ, Chattopadhyay A (2007) Cholesterol: a potential therapeutic target in *Leishmania* infection? Trends Parasitol 23:49–53

3. Riethmüller J, Riehle A, Grassmé H, Gulbins E (2006) Membrane rafts in host-pathogen interactions. Biochim Biophys Acta 1758:2139–2147
4. Zimmerberg J (2006) Membrane biophysics. Curr Biol 16:R272–R276
5. van Meer G, de Kroon AIPM (2011) Lipid map of the mammalian cell. J Cell Sci 124:5–8
6. Chattopadhyay A (ed.) (2002) Lipid probes in membrane biology. Chem Phys Lipids 116:1–188
7. Eggeling C, Ringemann C, Medda R, Schwarzmann G, Sandhoff K, Polyakova S, Belov VN, Hein B, von Middendorff C, Schönle A, Hell SW (2009) Direct observation of the nanoscale dynamics of membrane lipids in a living cell. Nature 457:1159–1163
8. Chattopadhyay A (1990) Chemistry and biology of N-(7-nitrobenz-2-oxa-1,3-diazol-4-yl)-labeled lipids: fluorescent probes of biological and model membranes. Chem Phys Lipids 53:1–15
9. Chattopadhyay A, Mukherjee S, Raghuraman H (2002) Reverse micellar organization and dynamics: a wavelength-selective fluorescence approach. J Phys Chem B 106:13002–13009
10. Chattopadhyay A, London E (1988) Spectroscopic and ionization properties of N-(7-nitrobenz-2-oxa-1,3-diazol-4-yl)-labeled lipids in model membranes. Biochim Biophys Acta 938:24–34
11. Fery-Forgues S, Fayet JP, Lopez A (1993) Drastic changes in the fluorescence properties of NBD probes with the polarity of the medium: involvement of a TICT state? J Photochem Photobiol A 70:229–243
12. Lin S, Struve WS (1991) Time-resolved fluorescence of nitrobenzoxadiazole-aminohexanoic acid: effect of intermolecular hydrogen-bonding on non-radiative decay. Photochem Photobiol 54:361–365
13. Mukherjee S, Chattopadhyay A, Samanta A, Soujanya T (1994) Dipole moment change of NBD group upon excitation studied using solvatochromic and quantum chemical approaches: implications in membrane research. J Phys Chem 98:2809–2812
14. Rawat SS, Chattopadhyay A (1999) Structural transition in the micellar assembly: a fluorescence study. J Fluoresc 9:233–244
15. Koval M, Pagano RE (1990) Sorting of an internalized plasma membrane lipid between recycling and degradative pathways in normal and Niemann-Pick, type A fibroblasts. J Cell Biol 111:429–442
16. Pagano RE, Sleight RG (1985) Defining lipid transport pathways in animal cells. Science 229:1051–1057
17. Sparrow CP, Patel S, Baffic J, Chao Y-S, Hernandez M, Lam M-H, Montenegro J, Wright SD, Detmers PA (1999) A fluorescent cholesterol analog traces cholesterol absorption in hamsters and is esterified in vivo and in vitro. J Lipid Res 40:1747–1757
18. van Meer G, Stelzer EHK, Wijnaendts-van-Resandt RW, Simons K (1987) Sorting of sphingolipids in epithelial (Madin-Darby canine kidney) cells. J Cell Biol 105:1623–1635
19. Mukherjee S, Zha X, Tabas I, Maxfield FR (1998) Cholesterol distribution in living cells: fluorescence imaging using dehydroergosterol as a fluorescent cholesterol analog. Biophys J 75:1915–1925
20. Scheidt HA, Müller P, Herrmann A, Huster D (2003) The potential of fluorescent and spin-labeled steroid analogs to mimic natural cholesterol. J Biol Chem 278:45563–45569
21. Bhattacharyya K, Bagchi B (2000) Slow dynamics of constrained water in complex geometries. J Phys Chem A 104:10603–10613
22. Chattopadhyay A (2003) Exploring membrane organization and dynamics by the wavelength-selective fluorescence approach. Chem Phys Lipids 122:3–17
23. Demchenko AP (2008) Site-selective red-edge effects. Methods Enzymol 450:59–78
24. Haldar S, Chaudhuri A, Chattopadhyay A (2011) Organization and dynamics of membrane probes and proteins utilizing the red edge excitation shift. J Phys Chem B 115:5693–5706
25. Mukherjee S, Chattopadhyay A (1995) Wavelength-selective fluorescence as a novel tool to study organization and dynamics in complex biological systems. J Fluoresc 5:237–246

26. Haldar S, Chattopadhyay A (2007) Dipolar relaxation within the protein matrix of the green fluorescent protein: a red edge excitation shift study. J Phys Chem B 111:14436–14439
27. Stubbs CD, Ho C, Slater SJ (1995) Fluorescence techniques for probing water penetration into lipid bilayers. J Fluoresc 5:19–28
28. Chattopadhyay A, Mukherjee S (1993) Fluorophore environments in membrane-bound probes: a red edge excitation shift study. Biochemistry 32:3804–3811
29. Abrams FS, London E (1993) Extension of the parallax analysis of membrane penetration depth to the polar region of model membranes: use of fluorescence quenching by a spin-label attached to the phospholipid polar headgroup. Biochemistry 32:10826–10831
30. Chattopadhyay A, London E (1987) Parallax method for direct measurement of membrane penetration depth utilizing fluorescence quenching by spin-labeled phospholipids. Biochemistry 26:39–45
31. Mitra B, Hammes GG (1990) Membrane-protein structural mapping of chloroplast coupling factor in asolectin vesicles. Biochemistry 29:9879–9884
32. Mukherjee S, Raghuraman H, Dasgupta S, Chattopadhyay A (2004) Organization and dynamics of N-(7-nitrobenz-2-oxa-1,3-diazol-4-yl)-labeled lipids: a fluorescence approach. Chem Phys Lipids 127:91–101
33. Wolf DE, Winiski AP, Ting AE, Bocian KM, Pagano RE (1992) Determination of the transbilayer distribution of fluorescent lipid analogues by nonradiative fluorescence energy transfer. Biochemistry 31:2865–2873
34. Chattopadhyay A, Mukherjee S (1999) Red edge excitation shift of a deeply embedded membrane probe: implications in water penetration in the bilayer. J Phys Chem B 103:8180–8185
35. Rawat SS, Mukherjee S, Chattopadhyay A (1997) Micellar organization and dynamics: a wavelength-selective fluorescence approach. J Phys Chem B 101:1922–1929
36. Raghuraman H, Chattopadhyay A (2007) Orientation and dynamics of melittin in membranes of varying composition utilizing NBD fluorescence. Biophys J 92:1271–1283
37. Chattopadhyay A, Mukherjee S (1999) Depth-dependent solvent relaxation in membranes: wavelength-selective fluorescence as a membrane dipstick. Langmuir 15:2142–2148
38. Huster D, Müller P, Arnold K, Herrmann A (2001) Dynamics of membrane penetration of the fluorescent 7-nitrobenz-2-oxa-1,3-diazol-4-yl (NBD) group attached to an acyl chain of phosphatidylcholine. Biophys J 80:822–831
39. Huster D, Müller P, Arnold K, Herrmann A (2003) Dynamics of lipid chain attached fluorophore 7-nitrobenz-2-oxa-1,3-diazol-4-yl (NBD) in negatively charged membranes determined by NMR spectroscopy. Eur Biophys J 32:47–54
40. Loura LMS, Ramalho JPP (2007) Location and dynamics of acyl chain NBD-labeled phosphatidylcholine (NBD-PC) in DPPC bilayers. A molecular dynamics and time-resolved fluorescence anisotropy study. Biochim Biophys Acta 1768:467–478
41. Raghuraman H, Shrivastava S, Chattopadhyay A (2007) Monitoring the looping up of acyl chain labeled NBD lipids in membranes as a function of membrane phase state. Biochim Biophys Acta 1768:1258–1267
42. Tsukanova V, Grainger DW, Salesse C (2002) Monolayer behavior of NBD-labeled phospholipids at the air/water interface. Langmuir 18:5539–5550
43. Mukherjee S, Soe TT, Maxfield FR (1999) Endocytic sorting of lipid analogues differing solely in the chemistry of their hydrophobic tails. J Cell Biol 144:1271–1284
44. Fernandes F, Loura LMS, Koehorst R, Spruijt RB, Hemminga MA, Fedorov A, Prieto M (2004) Quantification of protein-lipid selectivity using FRET: application to the M13 major coat protein. Biophys J 87:344–352
45. Mazères S, Schram V, Tocanne J-F, Lopez A (1996) 7-Nitrobenz-2-oxa-1,3-diazole-4-yl-labeled phospholipids in lipid membranes: differences in fluorescence behavior. Biophys J 71:327–335

46. Crowley KS, Reinhart GD, Johnson AE (1993) The signal sequence moves through a ribosomal tunnel into a noncytoplasmic aqueous environment at the ER membrane early in translocation. Cell 73:1101–1115
47. Shatursky O, Heuck AP, Shepard LA, Rossjohn J, Parker MW, Johnson AE, Tweten RK (1999) The mechanism of membrane insertion for a cholesterol-dependent cytolysin: a novel paradigm for pore-forming toxins. Cell 99:293–299
48. Chapman CF, Liu Y, Sonek GJ, Tromberg BJ (1995) The use of exogenous fluorescent probes for temperature measurements in single living cells. Photochem Photobiol 62:416–425
49. McIntyre JC, Sleight RG (1991) Fluorescence assay for phospholipid membrane asymmetry. Biochemistry 30:11819–11827
50. Chaudhuri A, Chattopadhyay A (2011) Transbilayer organization of membrane cholesterol at low concentrations: implications in health and disease. Biochim Biophys Acta 1808:19–25
51. Mukherjee S, Chattopadhyay A (1996) Membrane organization at low cholesterol concentrations: a study using 7-nitrobenz-2-oxa-1,3-diazol-4-yl-labeled cholesterol. Biochemistry 35:1311–1322
52. Mukherjee S, Chattopadhyay A (2005) Monitoring cholesterol organization in membranes at low concentrations utilizing the wavelength-selective fluorescence approach. Chem Phys Lipids 134:79–84
53. Rukmini R, Rawat SS, Biswas SC, Chattopadhyay A (2001) Cholesterol organization in membranes at low concentrations: effects of curvature stress and membrane thickness. Biophys J 81:2122–2134
54. Pucadyil TJ, Mukherjee S, Chattopadhyay A (2007) Organization and dynamics of NBD-labeled lipids in membranes analyzed by fluorescence recovery after photobleaching. J Phys Chem B 111:1975–1983
55. Loura LMS, Prieto M (1997) Dehydroergosterol structural organization in aqueous medium and in a model system of membranes. Biophys J 72:2226–2236
56. Ghosh PB, Whitehouse MW (1968) 7-Chloro-4-nitrobenzo-2-oxa-1,3-diazole: a new fluorigenic reagent for amino acids and other amines. Biochem J 108:155–156
57. Monti JA, Christian ST, Shaw WA (1978) Synthesis and properties of a highly fluorescent derivative of phosphatidylethanolamine. J Lipid Res 19:222–228
58. Monti JA, Christian ST, Shaw WA, Finley WH (1977) Synthesis and properties of a fluorescent derivative of phosphatidylcholine. Life Sci 21:345–355
59. Cairo CW, Key JA, Sadek CM (2010) Fluorescent small-molecule probes of biochemistry at the plasma membrane. Curr Opin Chem Biol 14:57–63
60. Polyakova SM, Belov VN, Yan SF, Eggeling C, Ringemann C, Schwarzmann G, de Meijere A, Hell SW (2009) New GM1 ganglioside derivatives for selective single and double labelling of the natural glycosphingolipid skeleton. Eur J Org Chem 2009:5162–5177
61. Uster PS, Pagano RE (1986) Resonance energy transfer microscopy: observations of membrane-bound fluorescent probes in model membranes and in living cells. J Cell Biol 103:1221–1234
62. Pajk S, Garvas M, Štrancar J, Pečar S (2011) Nitroxide-fluorophore double probes: a potential tool for studying membrane heterogeneity by ESR and fluorescence. Org Biomol Chem 9:4150–4159
63. Ryan TM, Griffin MDW, Bailey MF, Schuck P, Howlett GJ (2011) NBD-labeled phospholipid accelerates apolipoprotein C-II amyloid fibril formation but is not incorporated into mature fibrils. Biochemistry 50:9579–9586

3-Hydroxychromone Probes Precisely Located and Oriented in Lipid Bilayers: A Toolkit for Biomembrane Research

Andrey S. Klymchenko, Guy Duportail, and Yves Mély

Abstract Environment-sensitive dyes due to the sensitivity of their spectra to the physicochemical properties of their environment are unique tools for probing model and biological membranes. Here, we describe a particular class of environment-sensitive dyes based on 3-hydroxychromones. These dyes exhibit excited-state intramolecular proton transfer resulting in dual emission, highly sensitive to environment polarity and hydration. Appropriate molecular design of the new probes allows precise localization and orientation of their fluorophore in the lipid bilayers, which confer high specificity to particular membrane properties. In this respect, interface localization of the probes allows monitoring lipid order, while vertical orientation is required to achieve sensitivity to dipole and transmembrane potentials. Finally, biological applications of these probes for sensing lipid domains (rafts) and apoptosis are shown.

Keywords Apoptosis · Fluorescence imaging · Fluorescent membrane probes · 3-Hydroxychromones · Lipid domains · Membrane electrostatics · Polarity and hydration

Contents

A.S. Klymchenko (✉), G. Duportail, and Y. Mély
Laboratoire de Biophotonique et Pharmacologie, UMR 7213 CNRS, Université de Strasbourg, Faculté de Pharmacie, 74, Route du Rhin, 67401 ILLKIRCH Cedex, France
e-mail: andrey.klymchenko@unistra.fr

Y. Mély and G. Duportail (eds.), *Fluorescent Methods to Study Biological Membranes*, Springer Ser Fluoresc (2013) 13: 51–70, DOI 10.1007/4243_2012_44,

1 Environment-Sensitive Dyes and 3-Hydroxychromones

Fluorescent probes are powerful tools for monitoring the biophysical properties of biological membranes. Particularly important in this respect are environment-sensitive dyes, which change their fluorescence intensity or color in response to changes of the physicochemical properties of their molecular environment [1–3]. The response of environment-sensitive dyes is driven by excited-state reactions (conformational change, charge, electron and proton transfer, etc.) and noncovalent interactions with the surrounding, such as universal interactions (van der Waals, dipole-dipole, dipole-external electric field, etc.), and specific H-bonding interactions.

Several types of environment-sensitive fluorophores are of particular interest for biomembrane research: molecular rotors and solvatochromic fluorescent dyes. *Molecular rotors* are dyes featuring high rotational flexibility of their conjugated system. These rotors are poorly emissive in nonviscous environments, such as water or organic solvents, while in viscous environments, such as biological membranes, their rotation mobility is restricted, which dramatically increases their fluorescence quantum yield. Therefore, these dyes respond to environment viscosity by changes in their emission intensity [4] and therefore are useful for monitoring lipid order [5]. As their response is associated with changes in their fluorescence lifetimes, they are frequently applied in fluorescence lifetime imaging (FLIM) of model and cellular membranes [6, 7]. On the other hand, *solvatochromic fluorescent dyes* exhibit shifts in their emission maxima and sometimes changes in their fluorescence quantum yield as a function of the polarity and hydration of their environment [8, 9]. The dipole moment of these dyes increases dramatically upon electronic excitation ($S_0 \rightarrow S_1$ transition) due to an intramolecular charge transfer from the electron donor to the electron acceptor group. Moreover, an increase in the solvent polarity decreases the energy of their excited state, resulting in a red shift of their emission spectra. In addition, protic solvents can also decrease the energy of the excited state due to H-bonding interactions. Therefore, the red shifts observed in polar solvents originate from superposition of dipole-dipole and H-bonding interactions. Water, as a highly dipolar and H-bond donor molecule, provides the strongest red-shifting effects on solvatochromic dyes. An additional important property of most environment-sensitive dyes is their poor fluorescence quantum yield in water. Therefore, incorporation of these dyes into lipid membranes usually increases strongly their fluorescence intensity due to efficient screening of these molecules from bulk water [1, 3, 10]. Typical examples of these dyes are Prodan, Nile Red, 4-aminophthalimide, NBD, etc. However, though numerous environment-sensitive dyes are known, none of them is comparable to classical environment-insensitive dyes (rhodamines, cyanines, BODIPY, etc.) in terms of brightness and photostability.

Within the solvatochromic dyes, *two-band solvatochromic fluorescent dyes* based on 3-hydroxychromone (3HC) derivatives are of particular interest. Due to excited-state intramolecular proton transfer (ESIPT), these dyes show two emission bands, which change their relative intensities in response to solvent polarity [1, 2]. This chapter will briefly present design and application of single-band environment-sensitive dyes, but the main focus will be made on 3HC-based dyes. For more details on single-band solvatochromic dyes and their biological applications, the reader should see the excellent review by Imperiali et al. [3].

2 Fluorophore Location in Lipid Bilayers

Lipid bilayers are characterized by steep gradients of all physical parameters. As a consequence, precise probe location in the bilayer is crucial for designing probes with desired response. The most trivial approach consists in applying rather polar fluorescent probes to study the polar membrane interface and more apolar probes to study its hydrophobic interior. However, this approach often fails as the probe molecule may exhibit several locations and orientations in the bilayer. Particularly, small fluorophores of medium polarity can distribute both in polar and apolar regions. Moreover, ground-state hydration of a probe capable to form H-bonds is an additional feature which causes this distribution. Steady-state and time-resolved fluorescence data on Nile Red [11], coumarins [12], and Prodan [13–15] demonstrate that these probes could be located simultaneously in both polar and apolar regions of the membrane. Similar conclusions were made from molecular dynamics simulations [16]. Two ground-state forms were also found for the uncharged and nonsubstituted 3HC dye 4′-(dimethylamino)-3-hydroxyflavone (probe F) in lipid bilayers, one corresponding to the H-bonded form with a surface location and the other to the non-H-bonded form located deeper [17]. Both forms showed different emission profiles, reporting on the different environments of the two locations.

Another problem is that structural changes in membranes may cause probe relocation and reorientation. Recent data on giant unilamellar vesicles (GUVs) directly showed that Prodan and Laurdan do not present any preferential orientation in fluid-phase membranes, while in liquid-ordered and gel-phase membranes, they present a constrained vertical orientation (non-H-bonded form) [18]. Moreover, hydrostatic pressure [13] and cholesterol [19] can cause Prodan relocation. Fluorophore relocation on modification of membrane properties was also proposed for NBD derivatives [20] and the 3HC probe F [17]. Moreover, relocation of a probe along the polarity gradient may also occur during the fluorescence lifetime, when the excited state is much more polar than the ground state [16, 21].

On the other hand, lipid membranes present two leaflets. In cell plasma membranes, these leaflets are composed of different lipids and proteins so that their properties can be very different. For instance, sphingomyelin is present mainly at the outer leaflet of cell membranes, while the negatively charged phosphatidylserine is mainly localized at the inner leaflet [22]. Therefore, it is important that probes bind specifically at one membrane leaflet and undergo very slow flip-flop from one leaflet to another. Neutral apolar fluorophores can readily cross cell membrane and thus are not appropriate for specific investigation of one of the two leaflets [23, 24]. Therefore, special anchor groups are needed to slow down this process.

Finally, the polarity of the fluorophore itself is important for its ground-state location. For instance, covalent attachment of NBD, a fluorophore of medium polarity, to flexible acyl chains does not allow its localization close to the bilayer center, because it loops back to the interface [25, 26]. This and other examples of fluorophore localization suggested that instead of using neutral fluorophores or modified lipids, amphiphilic constructions are more appropriate to fix the fluorophore at a desired orientation and depth [27]. This can be provided by choosing a relatively low-polar fluorophore and applying a combination of charged and apolar substituents. Typical examples are Laurdan and Patman [28], which exhibit a more precise location in lipid bilayers than the parent Prodan [29]. Recently developed fatty acid analogues of Prodan also showed improved localization at the cell membranes compared to the parent Prodan [30]. Styrylpyridinium dyes presenting a charged zwitterionic group on one side and hydrophobic chains on the other are vertically orientated dyes, showing limited flip-flop [31–33]. Similar design was applied for Nile Red derivatives, which gave an interface-localized Nile Red presenting almost negligible flip-flop and poor internalization inside living cells [23]. Moreover, a similar strategy was also applied to develop a series of 3-hydroxychromone probes, in which the conjugation of the fluorophore to both an ionic head group and a long alkyl chain(s) provides a well-defined depth and orientation to the fluorophore in the lipid bilayer.

3 Location of 3HC Fluorophores in Lipid Membranes

Application of 3HC dyes in lipid bilayers requires primarily understanding the binding and localization of the fluorophore in the membranes. For this purpose, a detailed investigation of the fluorescence excitation and emission spectra of 4′-(dimethylamino)-3-hydroxyflavone (probe F) (Fig. 1) in lipid bilayer vesicles was performed as a function of the fatty acid composition, polar heads, temperature, and cholesterol content. Due to the multiparametric nature of the spectroscopic response of 3HC dyes, complete information about the dye localization by distinguishing universal (dipole-dipole) from specific (H-bonding) interactions within the bilayer environment could be obtained.

Our spectroscopy, anisotropy, and quenching data [17] suggested the presence of two forms of probe F at different locations in the bilayer: a non-H-bonded form located deeper than the sn_1-carbonyls of phospholipids and a H-bonded form located at the polar membrane interface (Fig. 1). Using the parallax quenching method, an estimated distance between the fluorophore and the bilayer center of 9.5 and 16 Å was found for non-H-bonded and H-bonded forms, respectively. The H-bonded form is probably a complex of the dye with water, which due to its much higher polarity as compared to the non-H-bonded form locates shallower in the bilayer, close to the water-membrane interface (Fig. 1). The distribution of the probe F between these two locations may also contribute to its response to changes in the membrane properties. Moreover, analysis of the literature data on other environment-sensitive probes (Prodan, Laurdan, Nile Red, NBD-lipids, etc.) in lipid bilayers allowed us to suggest that a bimodal distribution in the lipid bilayer

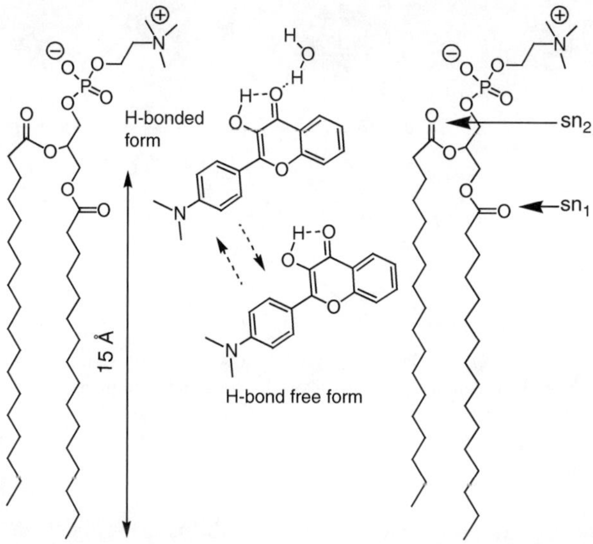

Fig. 1 Estimated location of probe F in PC lipid bilayer. Location of phospholipid functional groups is based on the study of Wiener and White [34]. Positions of sn1 and sn2 carbonyls are indicated by *horizontal arrows*. *Dashed arrows* indicate the possible distribution of probe F between the two locations

is probably a general feature of low-polar molecules with polar groups capable of H-bonding interactions [1, 17]. This indefinite probe location complicates the analysis of experimental data. To resolve this problem, we developed a number of 3HC derivatives bearing anchor groups, which fix the fluorophore at a more precise location in the lipid bilayer [27]. Moreover, by appropriate positioning of the charged groups and hydrophobic chains, it was also possible to control the orientation of the fluorophore in the lipid bilayer, which is important to address its vectorial (orientation-dependent) properties, such as electrostatics. Finally, the introduction of a zwitterionic anchor group decreased the flip-flop processes [35], allowing positioning of the fluorophore selectively at the outer leaflet of cell membranes [24]. Through systematic variation of the location and orientation of the 3HC fluorophore, we developed a series of membrane probes with different selectivities to the membrane properties.

4 Multiparametric Probing of Lipid Bilayers: Hydration and Polarity

Environment-sensitive dyes generally respond to several membrane properties simultaneously, so that it is difficult to estimate these properties independently. For instance, the commonly used probe of membrane electrostatics, di-4-ANEPPS,

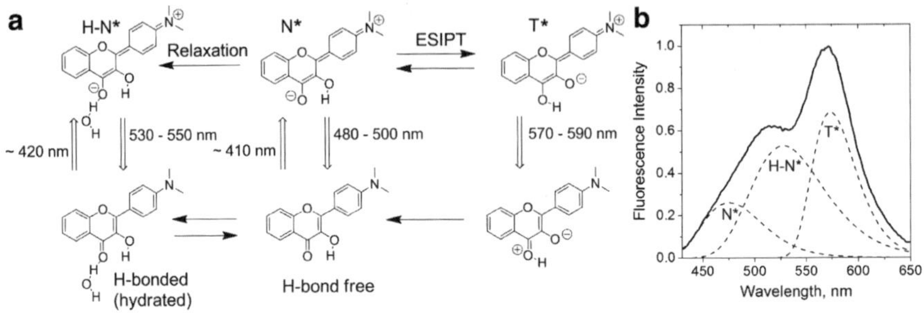

Fig. 2 (**a**) Ground (*below*) and excited (*above*) states of 4′-(dimethylamino)-3-hydroxyflavone in lipid membranes. *Upward arrows* represent excitation to the Frank-Condon excited state, while *downward arrows* represent emission. The correspondent band maxima (in nm) are indicated. In lipid membranes, the N* → H-N* conversion results from a slow relaxation process. (**b**) Fluorescence spectrum of this dye in large unilamellar vesicles composed of egg yolk phosphatidylcholine and its deconvolution into three emission bands N*, H-N*, and T*

is also sensitive to water relaxation in the bilayer [36, 37]. This side effect of water relaxation is also known for DPH-based probes, which are primarily used for estimation of membrane microviscosity through fluorescence anisotropy [38]. The ideal would be to extract several types of information from a single probe and, thus, access several membrane properties at the same site. This can be achieved by utilizing multiparametric fluorescent dyes, such as those presenting an equilibrium between several emissive states. Among all known excited-state reactions generating several emissive species, excited-state intramolecular proton transfer (ESIPT) [39] appears as the most promising, since this reaction can generate emissive species with very different sensitivity to parameters of their environment [40].

4′-(Dialkylamino)-3-hydroxyflavone dyes from the 3HC family are particularly attractive for this purpose. Due to an ESIPT reaction, they exhibit two emissive states (N* and T*) of very different properties (Fig. 2). Due to intramolecular charge transfer (ICT) from the 4′-dialkylamino to the 4-carbonyl group, the N* state attains a large dipole moment [40, 41] similar to that of classical solvatochromic dyes such as Prodan [42]. In contrast, the ESIPT product T* state exhibits a much smaller dipole moment. As the ESIPT reaction in these 3HC dyes is reversible [43], intermolecular interactions stabilize the N* state as compared to the T* state, resulting in a shift of the ESIPT equilibrium and a redistribution of the fluorescence intensities between the two bands. As a consequence, their intensity ratio (N*/T*) describes the environment polarity [40] and electrostatics [44]. In our early works based on these two-band dyes, we have developed a methodology for multiparametric probing of environment in organic solvents [40]. However, lipid bilayers are much more complicated systems due their microheterogeneous nature.

As it was mentioned above, probe F in lipid bilayers (in contrast to organic solvents) presents two different ground-state forms: the H-bonded form with water (hydrated) and the non-H-bonded (nonhydrated) form (Fig. 1). Since this dye is not

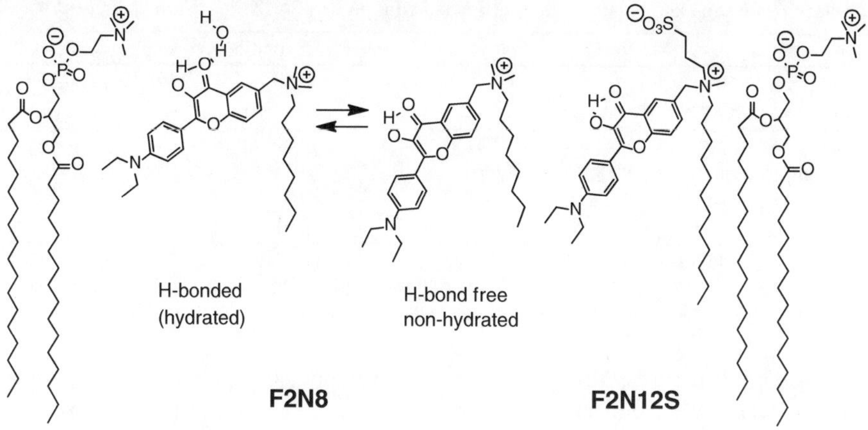

Fig. 3 Probes of the 3HC family localized at the bilayer interface. For probe F2N8, the two ground-state forms are shown. To indicate the probe locations, two phosphatidylcholine molecules constituting the outer monolayer are shown

fixed at the bilayer, these two forms distribute at different bilayer depths [17]. By fixing the fluorophore with an anchor group, a more controlled localization of these two forms in the membranes can be achieved, and the two forms can be used to obtain an additional parameter describing the membrane hydration [45]. Localization of the fluorophore at the bilayer interface was realized using a charged anchor group and a long hydrophobic chain, as shown in Fig. 3 for probes F2N8 and F2N12S. According to the parallax quenching studies, the H-bonded and non-H-bonded forms of F2N8 probe are located at 16 and 13 Å, respectively, from the bilayer center [45]. This 3-Å depth difference between the two forms is much smaller than that observed for uncharged probe F (6.5 Å), thus confirming successful localization of the fluorophore at the bilayer interface using our probe design methodology. Importantly, these two ground-state forms exhibit differences in their excitation and emission spectra. While the nonhydrated form (blue-shifted in excitation) exhibits a dual emission (N* and T* bands) as in aprotic organic solvents, the H-bonded form exhibits a single emission band (H-N* band) similar to the spectra obtained in protic solvents. To estimate the relative intensity of these bands, we deconvoluted the emission spectra of 3HC probes in lipid membranes into three bands: N*, T*, and H-N* (Fig. 2b), which allowed us to extract two independent parameters: hydration and polarity. The former was defined as the relative contribution of the hydrated form of the probe, deduced from the ratio of intensities between the H-N* band and the sum of the N* and T* bands. The polarity parameter was defined as the N*/T* ratio, similar to that initially introduced for organic solvents [40].

To better understand the photophysical behavior of the probes in lipid membranes, time-resolved studies of F2N12S in lipid bilayers were performed. These studies indicated that the ESIPT equilibrium between N* and T* forms is established on a time scale 40–70 ps, while the relaxation kinetics involving the H-N* form are completed at the nanosecond time scale [46]. These data

Table 1 Hydration and polarity values estimated with probe F2N12S in lipid membranes[a]

Sample	Hydration	Polarity
Fluid phase (EYPC)	0.56	1.27
Fluid phase (DOPC)	0.61	1.63
Fluid phase, negative surface charge (BBPS)	0.36	2.25
Fluid phase, negative surface charge (DOPS)	0.38	3.24
Liquid-disordered (fluid) phase (DOPC/Chol)	0.39	1.20
Liquid-ordered phase (SM/Chol)	0.01	0.88

[a]*DOPC* dioleoylphosphatidylcholine, *EYPC* egg yolk phosphatidylcholine, *DOPS* dioleoylphosphatidylserine, *BBPS* bovine brain phosphatidylserine, *Chol* cholesterol, *SM* bovine brain sphingomyelin. In DOPC/Chol and SM/Chol, the lipid/Chol ratio was 2/1 mole/mole

suggested that the hydration parameter describes both the ground-state hydration and the water relaxation in the lipid environment, in analogy to the generalized polarization parameter previously introduced for Prodan and Laurdan dyes [29, 47–49]. In contrast, the N*/T* ratio is controlled by the rapidly established N*-T* equilibrium in the excited state and, therefore, describes mainly the interactions with the static electric fields in the membrane. Thus, it is clear that the hydration parameter and the N*/T* parameter are intrinsically independent and describe different membrane properties.

Using this multiparametric methodology, model lipid membranes of different lipid compositions were characterized. The polarity and hydration of the bilayers were found to decrease simultaneously when the lipid order was increased (phase transition from fluid to gel phase, addition of cholesterol to the fluid phase or decrease in the lipid unsaturation) [45]. A particularly strong decrease in hydration was observed on transition from the liquid-disordered phase, presented by unsaturated lipids, to the liquid-ordered phase, presented by saturated lipids and cholesterol (Table 1). Moreover, within the liquid-disordered phase, an increase in temperature or bilayer curvature led to a higher bilayer hydration with no effect on the polarity. In contrast, when the electrostatic properties of the bilayer, such as the dipole and surface potential, were varied, a strong variation of the polarity parameter (N*/T*) was observed without considerable changes in the hydration parameter (Table 1). Thus, these studies have established a new multiparametric methodology, allowing simultaneous monitoring of two independent properties of the membranes: water content (hydration parameter) and membrane electrostatics (polarity parameter) [1, 45]. Below, some applications of this methodology will be presented in more detail.

5 Probes of Membrane Electrostatics: Dipole and Transmembrane Potentials

Since dipole (Ψ_d) and transmembrane (Ψ_t) potentials are localized inside the lipid bilayer, their detection needs an electrochromic fluorophore presenting a vertical orientation in the membrane. As these two potentials apply at different levels in the

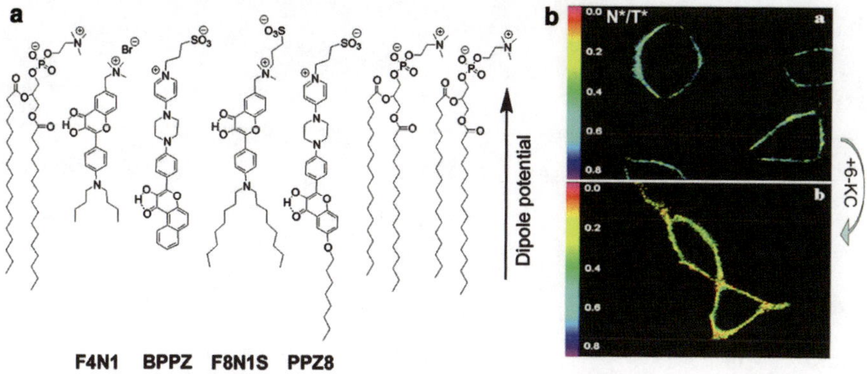

Fig. 4 Fluorescent probes of dipole potential. (**a**) Estimated location of the probes in lipid membranes. (**b**) Ratiometric imaging of dipole potential variation in CEM cells stained with probe F8N1S: (*a*) – control cells, (*b*) – cells treated with 6-ketocholestanol (6-KC) (Data from Ref. [50])

lipid bilayer, probing Ψ_d needs a location of the fluorophore between the apolar bilayer center and the membrane surface, while probing Ψ_t requires its location in the apolar region close to the bilayer center.

Critical analysis of already existing probes of dipole potential, represented mainly by 4-styrylpyrydinium dyes [31, 33, 36, 37, 51], revealed that further progresses require the development of probes that (a) do not possess a large ground-state dipole, (b) allow quantitative ratiometric recording in fluorescence emission with major contribution from the electrostatic effects in order to prevent side effects from relaxation processes, and (c) allow desired location and orientation of the fluorophore moiety. 3HC dyes and particularly 4′-(dialkylamino)-3-hydroxyflavones fit well to all these requirements. By using a similar design as for 4-styrylpyrydinium dyes, we have developed two probes (F4N1 and BBPZ), in which the fluorescent moieties and consequently the dipole moments are oriented vertically in lipid bilayers but inversely with respect to each other (Fig. 4). In both cases, the probes strongly responded to variation of Ψ_d, generated by phloretin and 6-ketocholestanol [52]. The spectroscopic response of the probes consisted of changes in the intensity ratio of their two emission bands, and it was opposite for the two dyes in accordance with their inverse orientation. Moreover, a multiparametric analysis based on band deconvolution showed that these changes are directly connected with the variation of the N*/T* ratio and are independent from the hydration parameter [53]. However, these probes were not applicable for cellular studies due to their fast internalization and/or poor staining of the plasma membrane. Therefore, we have synthesized a second generation of dipole potential probes, F8N1S and PPZ8 (Fig. 4) that are analogues of F4N1 and BBPZ, bearing zwitterionic groups and long hydrophobic chains at the opposite sides of the 3HF fluorophore. Cellular studies have shown that the improved probes bind selectively to the cell plasma membranes and exhibit strong variations in the intensity ratio of

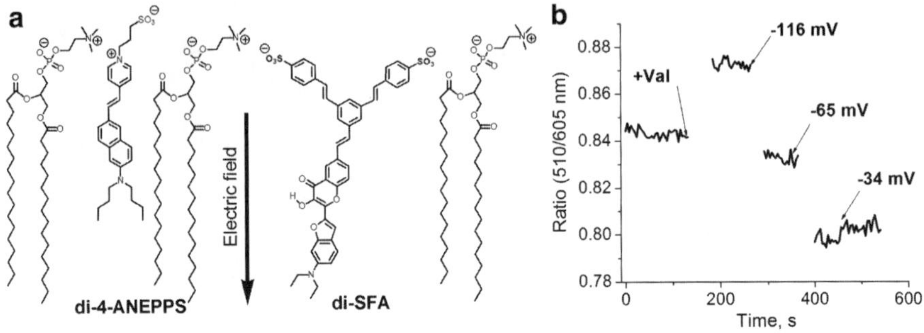

Fig. 5 New transmembrane potential probe di-SFA and the commercial probe di-4-ANEPPS. (**a**) Structure and location of the probes. (**b**) Ratiometric response of probe di-SFA to variation of transmembrane potential in suspensions of CEM cells (Data from Ref. [54])

their two emission bands as a function of Ψ_d in living cells (Fig. 4) [50], in accordance with the data on the 1st-generation probes in model membranes [52].

To monitor the transmembrane potential Ψ_t, the fluorophore has to be much deeper embedded. However, the electrochromic fluorophores used so far, which are based on 4-styrylpyrydinium, are charged and do not allow deep embedding of the fluorophore into the highly hydrophobic environment of the bilayer. For this reason, we attempted to apply the 3HC fluorophore, which is relatively low polar and noncharged. An extended 3HC fluorophore was selected [55], which showed two-color environment sensitivity at low polarities [56]. This fluorophore was coupled with two rigid arms bearing charged groups to impose a vertical orientation and a deep insertion of this fluorophore in the lipid bilayer (probe di-SFA, Fig. 5). The obtained probe showed fluorescence spectra corresponding to a very low polarity environment ($\varepsilon \sim 3$–4) in model and cellular membranes, in line with a deep insertion [54]. Cellular studies showed that the sensitivity of the probe to Ψ_t was 12% per 100 mV, which is larger than the corresponding response of the 4-styrylpyrydinium probes [54], but lesser than that of the new generation of ANNINE probes [57]. Moreover, combined patch-clamp and fluorescence microscopy measurements showed that the ratiometric response of di-SFA is faster than 1 ms, in agreement with its electrochromic mechanism. The key advantage of the present 3HC probe compared with all other probes developed so far resides in its ratiometric response in emission, which enables easy application in fluorescence microscopy by using a single excitation source and a two-color fluorescence detection setup.

6 Monitoring Hydration and Lipid Order: Lipid Domains

Monitoring lipid order in membranes is of particular importance due to the probable key role of lipid domains in membrane functions [58]. In model membranes, these domains or "rafts" form a liquid-ordered (Lo) phase that "swims" in the pool of liquid-disordered (Ld) phase [59, 60]. In cells, these hypothetic domains are believed to play a crucial role in the regulation of numerous functions of membrane proteins and membrane transport [58, 61]. Attractive tools to study lipid order are environment-sensitive dyes, which can distinguish liquid-ordered from liquid-disordered phases by changes in their emission color. The proof of concept was originally shown using Laurdan and Prodan probes, which exhibit clear-cut differences in their emission color between the two phases [18, 59]. In liquid-ordered phase, these dyes experience less polar environment and thus, show blue-shifted emission compared to liquid-disordered phase. Later on, this concept was applied with the 4-styrylpyrydinium probe 4-di-ANEPPQ, having the advantage of longer-wavelength absorption and emission as well as specific membrane binding [31, 62]. Alternatively, since molecular rotors bound to liquid-ordered phase show significant enhancement of their fluorescence, liquid-ordered phases were visualized using a molecular rotor based on a Prodan analogue [5].

3HC probes are of particular interest due to their high sensitivity to membrane properties, such as polarity and hydration. Initially, we studied our most characterized 3HC probe F2N8 in different model membrane vesicles, mimicking different membrane phases. It was found that the hydration parameter monitored by F2N8 increases in the following order: liquid ordered $<<$ gel $\approx$ liquid disordered (with cholesterol) $<$ liquid crystalline (without cholesterol). Thus, the Lo phase is characterized by an exceptionally low hydration, a feature that could be used for detection and imaging of this phase. These data on F2N8 were compared with changes of membrane fluidity which were monitored by fluorescence anisotropy of the diphenylhexatriene-based probe TMA-DPH probe [63, 64]. Remarkably, the membrane fluidity shows a somewhat different trend, namely liquid ordered $\approx$ gel $<$ liquid disordered $<$ liquid crystalline. Thus, gel- and liquid-ordered phases exhibit similar fluidity, while the last phase is significantly less hydrated. We hypothesized that cholesterol, due to its specific H-bonding interactions with lipids and its ability to fill the voids in lipid bilayers, expels efficiently water molecules from the highly ordered gel phase in order to form the liquid-ordered phase. In this work, the liquid-ordered and gel phases were for the first time clearly distinguished by their strong difference in hydration.

The strong spectroscopic response of our 3HC probes to Lo phase could be applied for visualization of lipid domains. The best model systems for visualization of lipid domains are giant unilamellar vesicles (GUVs), which being of 5–100 μm of diameter could be directly used in fluorescence microscopy. GUVs displaying Ld phase, Lo phase, or both phases were used to characterize 3HC probes F2N8 and F2N12S. It was found that F2N8 preferentially binds the Ld phase, so it is impossible to label the Lo phase when it coexists with the Ld phase. Therefore, we have tested its analogue F2N12S which bears a longer hydrophobic chain and a zwitterionic anchor group (Fig. 3). Steady-state fluorescence studies in large unilamellar vesicles showed that similarly to F2N8, the dual emission of F2N12S drastically changes with the lipid bilayer phase (Fig. 6a), which can be correlated with changes in hydration [66]. Using two-photon excitation microscopy on giant unilamellar vesicles, the F2N12S probe was found to bind both Ld and Lo phases,

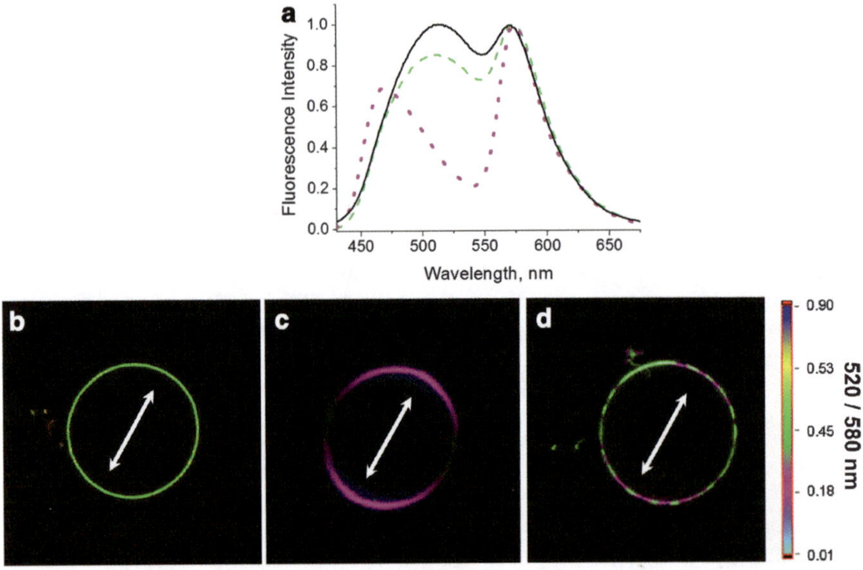

Fig. 6 Probing liquid-ordered domains using probe F2N12S. (**a**) Fluorescence spectra of F2N12S in lipid vesicles composed of DOPC (*black solid curve*) DOPC + 35 % Chol (*green dashed curve*) and SM + 35 % Chol (*magenta dotted curve*). Fluorescence ratiometric images (520/580 nm) of GUVs composed of DOPC (**b**) and SM + 35 % Chol (**c**) and the ternary mixture (DOPC/SM/Chol in the molar ratios 1/1/0.7) (**d**). Two-photon excitation (830 nm) was used. *Arrows* indicate the orientation of light polarization. Sizes of the images were 50 × 50 μm (**b**), 15 × 15 μm (**c**), and 45 × 45 μm (**d**) (Data from Ref. [65])

allowing visualization of the individual phases from the fluorescence intensity ratio of its two emission bands (Fig. 6b–d) [65]. Moreover, by using a linearly polarized excitation laser light, a strong photoselection was observed for F2N12S in the Lo phase (Fig. 6c), indicating that its fluorophore is nearly parallel to the lipid chains of the bilayer. Thus, the Lo phase imposes a vertical orientation to the 3HC fluorophore, as it was already shown with Laurdan and di-ANEPPQ [18, 31]. In contrast, the marginal photoselection observed with the Ld phase indicated no predominant orientation of the probe in the Ld phase (Fig. 6b). As it will be shown below, F2N12S designed to stain selectively the cell plasma membrane shows strong response to changes in the phase state of the cell membrane on cholesterol extraction as well as on apoptosis.

7 Monitoring Apoptosis and Its Connection with Lipid Order

Normal cells exhibit a remarkable asymmetry of the lipid distribution in the inner and outer leaflets of their plasma membranes which is lost during the early steps of apoptosis [22]. These changes in the plasma membrane can be used to detect apoptosis. In this respect, the most common detection method relies on the use of the fluorescently labeled annexin V, which interacts with PS exposed on the membrane surface [67].

Due to the high sensitivity of 3HC dyes to lipid composition, and notably on the surface charge and phase state, we attempted to use them to monitor apoptosis. Indeed, all 3HC derivatives showed strong changes in their dual emission when lipid membranes composed of anionic lipids were compared with those of neutral lipids [68]. This sensitivity to the negative surface charge could be used to detect exposure of PS during apoptosis. Moreover, the sensitivity of 3HC dyes to the lipid phase could also be important for detection of apoptosis since the loss of transmembrane asymmetry may also lead in changes of the lipid order. To detect the apoptotic loss of transmembrane asymmetry, a 3HC probe that binds selectively to the outer leaflet of the cell plasma membrane is needed. The well-characterized probe F2N8 was unsuitable for this purpose, because it readily internalizes into cells and probably redistributes rapidly between the two leaflets of the cell plasma membrane. Therefore, as already mentioned before, we have developed the probe F2N12S having a modified anchor group: the positively charged ammonium group was substituted with a zwitterionic group, while the hydrophobic chain was extended to 12 carbon atoms (Fig. 3). Fluorescence spectroscopy, flow cytometry, and microscopy measurements showed that the ratio of the two emission bands of the probe changes dramatically in response to apoptosis (Fig. 7) [24]. This response reflects the changes in the lipid composition of the outer leaflet of the cell plasma occurring on the early steps of apoptosis. Being ratiometric, the response of the new probe can be easily quantified on an absolute scale. This allows monitoring by laser scanning confocal microscopy the degree and spatial distribution of the apoptotic changes at the cell plasma membranes, a feature that can be hardly achieved with the commonly used fluorescently labeled annexin V assay.

However, the spectroscopic response of F2N12S could not be explained by the sole increase in the exposure of PS at the outer leaflet of the cell. Since F2N12S can bind both Ld and Lo phases, showing a different dual emission [65], we hypothesized that the observed response of the probe to apoptosis could also be driven by changes in the phase state (Lo/Ld) of the outer leaflet. Therefore, a comparative study of this probe in plasma membranes of living cells and lipid vesicles was performed. Using our multiparametric analysis based on band deconvolution of the fluorescence spectra of F2N12S, we characterized membranes in terms of hydration and polarity (electrostatics). It was found that plasma membranes of intact glioblastoma cells exhibit a significant fraction of liquid-ordered phase in their outer leaflet since both hydration and polarity parameters were intermediate between those observed in lipid vesicles presenting Lo and Ld phases [66]. Moreover, two-photon fluorescence microscopy experiments showed that intact living cell membranes labeled with the probe exhibited a homogeneous lipid distribution, suggesting that liquid-ordered domains are likely distributed all over the membrane and are highly dynamic. Cholesterol depletion affected dramatically the dual emission of the probe, suggesting that the Lo-like phase in cell membranes is converted into Ld-like phase (Fig. 8). These conclusions were

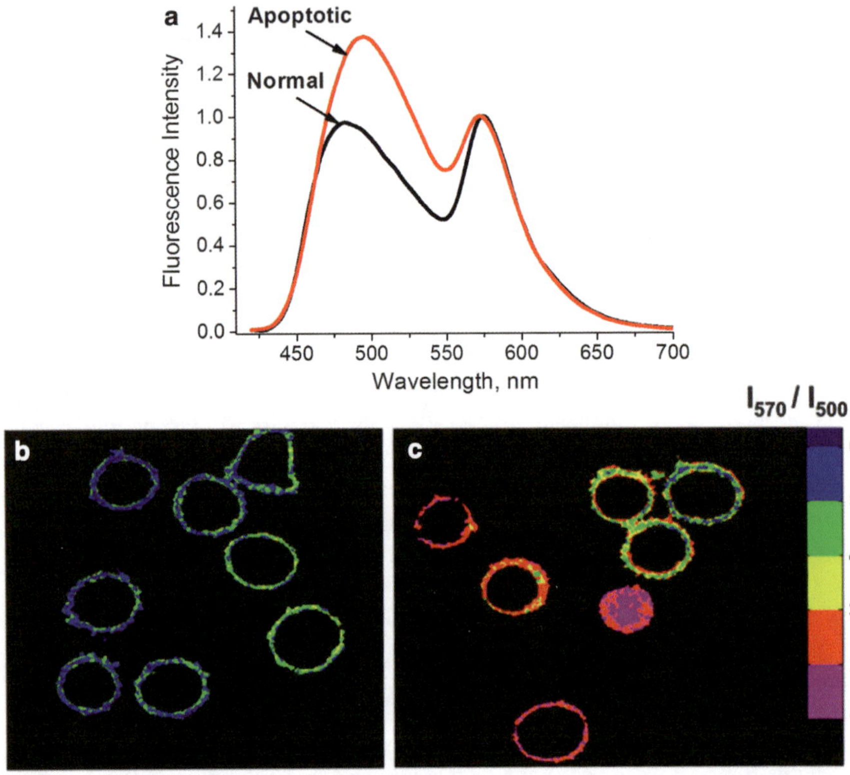

Fig. 7 Apoptosis detection using probe F2N12S. (**a**) Fluorescence spectra of F2N12S in normal and apoptotic (treated with actinomycin D) CEM cells. (**b** and **c**) Ratiometric images of normal (**b**) and actinomycin D-treated (**c**) cells stained with F2N12S. The size of the images is 60 × 73 μm (Data from Ref. [24])

corroborated by fluorescence anisotropy experiments performed with the viscosity-sensitive probe TMA-DPH. Indeed, the estimated membrane fluidity of intact cells was intermediate between those measured in the Lo and Ld phases of model membranes, while on cholesterol depletion, the fluidity increased dramatically reaching the values observed in the Ld phase.

Remarkably, analysis of the spectroscopic response of F2N12S to apoptosis revealed a strong increase in the hydration parameter, which is similar to the effect of cholesterol extraction. This observation led us to the conclusion that cell apoptosis results in a similar loss of the Lo phase, which could be attributed to a flip of sphingomyelin from the outer to the inner leaflet of the plasma membrane due to apoptosis-driven lipid scrambling [66]. In addition, we observed that apoptosis led to higher values of the polarity parameter, which correlates well with the increase in the negative surface charge resulting from the exposure of negatively charged PS lipids. Thus, these comparative studies revealed the mechanism of response of F2N12S to apoptosis and showed a link between apoptosis and changes

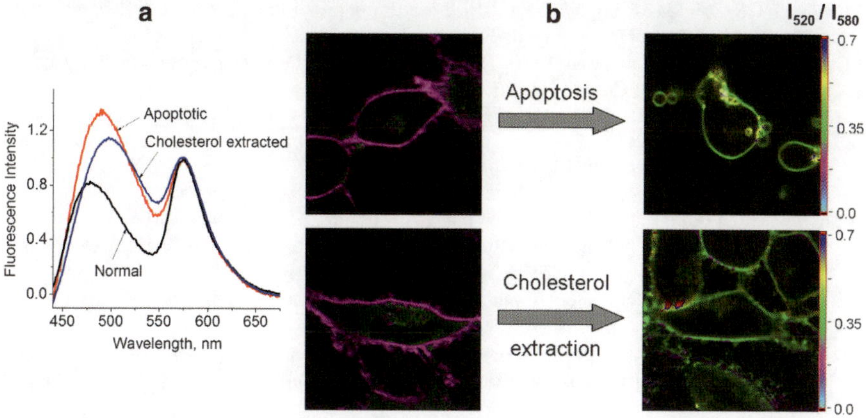

Fig. 8 Monitoring apoptosis and phase changes with F2N12S probe in U87MG cells. (**a**) Fluorescence spectra of F2N12S probe in nonapoptotic and apoptotic cells compared to the spectra of the probe in LUVs of Lo phase and to that in cells after cholesterol extraction. (**b**) Fluorescence ratiometric images of nonapoptotic, apoptotic and cholesterol-extracted cells stained with F2N12S (Data from Ref. [66])

in the liquid order at the outer leaflet of the cell plasma membranes. Finally, this work suggested a new methodology for studying the liquid-ordered phase in cell membranes, particularly in connection with their transmembrane asymmetry.

8 Conclusions

Design and application of fluorescent membrane probes based on 3-hydroxychromones led to several key achievements and conclusions. (1) ESIPT fluorophores based on 3-hydroxychromone present a unique ability to detect simultaneously polarity and hydration in the membrane environment. (2) As environment-sensitive probes present indefinite distribution in the lipid bilayers and fast flip-flop between the leaflets, it is of key importance to position the fluorophore at precise depth and orientation in the bilayer and minimize the flip-flop by introducing an anchor group. (3) Probes featuring localization at the interface are suitable for detection of lipid order and surface potential, while vertical fluorophore orientation is required for monitoring dipole and transmembrane potentials. (4) Biophysical properties monitored by these probes are strongly linked to key biological structures and functions, such as membrane lipid domains and apoptosis.

References

1. Demchenko AP, Mely Y, Duportail G, Klymchenko AS (2009) Monitoring biophysical properties of lipid membranes by environment-sensitive fluorescent probes. Biophys J 96(9):3461
2. Klymchenko AS (2012) Solvatochromic fluorescent dyes as universal tools for biological research. Actual Chim 359:20–26

3. Loving GS, Sainlos M, Imperiali B (2010) Monitoring protein interactions and dynamics with solvatochromic fluorophores. Trends Biotechnol 28(2):73–83
4. Haidekker MA, Theodorakis EA (2007) Molecular rotors – fluorescent biosensors for viscosity and flow. Org Biomol Chem 5(11):1669–1678
5. Hwan MK, Byeong HJ, Hyon JY, Myoung JA, Mun SS, Jin HH, Lee KJ, Chul HK, Joo T, Hong SC, Bong RC (2008) Two-photon fluorescent turn-on probe for lipid rafts in live cell and tissue. J Am Chem Soc 130(13):4246–4247
6. Kuimova MK, Yahioglu G, Levitt JA, Suhling K (2008) Molecular rotor measures viscosity of live cells via fluorescence lifetime imaging. J Am Chem Soc 130(21):6672–6673
7. Suhling K, Levitt JA, Chung PH, Kuimova MK, Yahioglu G (2012) Fluorescence lifetime imaging of molecular rotors in living cells. J Visual Exper (60): e2925
8. Reichardt C (1994) Solvatochromic dyes as solvent polarity indicators. Chem Rev 94 (8):2319–2358
9. Lippert EL (1975) Laser-spectroscopic studies of reorientation and other relaxation processes in solution. In: Birks JB (ed) Organic and biomolecular chemistry, vol 2. Wiley, New York, pp 1–31
10. Slavik J (1982) Anilinonaphthalene sulfonate as a probe of membrane-composition and function. Biochim Biophys Acta 694(1):1–25
11. Krishna MMG (1999) Excited-state kinetics of the hydrophobic probe Nile red in membranes and micelles. J Phys Chem A 103(19):x1
12. Sykora J, Jurkiewicz P, Epand RM, Kraayenhof R, Langner M, Hof M (2005) Influence of the curvature on the water structure in the headgroup region of phospholipid bilayer studied by the solvent relaxation technique. Chem Phys Lipids 135(2):213–221
13. Chong PL (1988) Effects of hydrostatic pressure on the location of Prodan in lipid bilayers and cellular membranes. Biochemistry 27(1):399
14. Jurkiewicz P, Olzynska A, Langner M, Hof M (2006) Headgroup hydration and mobility of DOTAP/DOPC bilayers: a fluorescence solvent relaxation study. Langmuir 22(21):8741–8749
15. Rowe BA, Neal SL (2006) Photokinetic analysis of prodan and laurdan in large unilamellar vesicles from multivariate frequency-domain fluorescence. J Phys Chem B 110 (30):15021–15028
16. Barucha-Kraszewska J, Kraszewski S, Jurkiewicz P, Ramseyer C, Hof M (2010) Numerical studies of the membrane fluorescent dyes dynamics in ground and excited states. Biochim Biophys Acta 1798(9):1724–1734
17. Klymchenko AS, Duportail G, Demchenko AP, Mely Y (2004) Bimodal distribution and fluorescence response of environment-sensitive probes in lipid bilayers. Biophys J 86 (5):2929–2941
18. Bagatolli LA (2006) To see or not to see: lateral organization of biological membranes and fluorescence microscopy. Biochim Biophys Acta 1758(10):1541
19. Bondar OP, Pivovarenko VG, Rowe ES (1998) Flavonols–new fluorescent membrane probes for studying the interdigitation of lipid bilayers. Biochim Biophys Acta 1369(1):119–130
20. Alakoskela JMI, Kinnunen PKJ (2001) Probing phospholipid main phase transition by fluorescence spectroscopy and a surface redox reaction. J Phys Chem B 105(45):11294
21. Demchenko AP, Shcherbatska NV (1985) Nanosecond dynamics of charged fluorescent probes at the polar interface of a membrane phospholipid bilayer. Biophys Chem 22 (3):131–143
22. Zwaal RFA, Schroit AJ (1997) Pathophysiologic implications of membrane phospholipid asymmetry in blood cells. Blood 89(4):1121
23. Kucherak OA, Oncul S, Darwich Z, Yushchenko DA, Arntz Y, Didier P, Mely Y, Klymchenko AS (2010) Switchable Nile red-based probe for cholesterol and lipid order at the outer leaflet of biomembranes. J Am Chem Soc 132(13):4907
24. Shynkar VV, Klymchenko AS, Kunzelmann C, Duportail G, Muller CD, Demchenko AP, Freyssinet JM, Mely Y (2007) Fluorescent biomembrane probe for ratiometric detection of apoptosis. J Am Chem Soc 129(7):2187–2193

25. Loura LM, Ramalho JP (2007) Location and dynamics of acyl chain NBD-labeled phosphatidylcholine (NBD-PC) in DPPC bilayers. A molecular dynamics and time-resolved fluorescence anisotropy study. Biochim Biophys Acta 1768(3):467–478
26. Huster D, Muller P, Arnold K, Herrmann A (2001) Dynamics of membrane penetration of the fluorescent 7-nitrobenz-2-oxa-1,3-diazol-4-yl (NBD) group attached to an acyl chain of phosphatidylcholine. Biophys J 80(2):822–831
27. Klymchenko AS, Duportail G, Ozturk T, Pivovarenko VG, Mely Y, Demchenko AP (2002) Novel two-band ratiometric fluorescence probes with different location and orientation in phospholipid membranes. Chem Biol 9(11):1199–1208
28. Lakowicz JR, Bevan DR, Maliwal BP, Cherek H, Balter A (1983) Synthesis and characterization of a fluorescence probe of the phase transition and dynamic properties of membranes. Biochemistry 22(25):5714–5722
29. Jurkiewicz P, Sykora J, Olzynska A, Humpolickova J, Hof M (2005) Solvent relaxation in phospholipid bilayers: principles and recent applications. J Fluoresc 15(6):883
30. Kim HM, Choo HJ, Jung SY, Ko YG, Park WH, Jeon SJ, Kim CH, Joo T, Cho BR (2007) A two-photon fluorescent probe for lipid raft imaging: C-laurdan. Chembiochem 8(5):553
31. Jin L, Millard AC, Wuskell JP, Dong X, Wu D, Clark HA, Loew LM (2006) Characterization and application of a new optical probe for membrane lipid domains. Biophys J 90(7):2563
32. Millard AC, Jin L, Wei MD, Wuskell JP, Lewis A, Loew LM (2004) Sensitivity of second harmonic generation from styryl dyes to transmembrane potential. Biophys J 86(2):1169
33. Wuskell JP, Boudreau D, Wei MD, Jin L, Engl R, Chebolu R, Bullen A, Hoffacker KD, Kerimo J, Cohen LB, Zochowski MR, Loew LM (2006) Synthesis, spectra, delivery and potentiometric responses of new styryl dyes with extended spectral ranges. J Neurosci Methods 151(2):200
34. Wiener MC, White SH (1992) Structure of a fluid dioleoylphosphatidylcholine bilayer determined by joint refinement of x-ray and neutron diffraction data: III. Complete structure. Biophys J 61(2I):434–447
35. Kułakowska A, Jurkiewicz P, Sýkora J, Benda A, Mely Y, Hof M (2010) Fluorescence lifetime tuning-a novel approach to study flip-flop kinetics in supported phospholipid bilayers. J Fluoresc 20(2):563–569
36. Clarke RJ, Kane DJ (1997) Optical detection of membrane dipole potential: avoidance of fluidity and dye-induced effects. Biochim Biophys Acta 1323(2):223
37. Vitha MF, Clarke RJ (2007) Comparison of excitation and emission ratiometric fluorescence methods for quantifying the membrane dipole potential. Biochim Biophys Acta 1768 (1):107–114
38. Ho C, Slater SJ, Stubbs CD (1995) Hydration and order in lipid bilayers. Biochemistry 34 (18):6188
39. Formosinho SJ, Arnaut LG (1993) Excited-state proton transfer reactions II. Intramolecular reactions. J Photochem Photobiol A 75(1):21–48
40. Klymchenko AS, Demchenko AP (2003) Multiparametric probing of intermolecular interactions with fluorescent dye exhibiting excited state intramolecular proton transfer. Phys Chem Chem Phys 5:461–468
41. Yesylevskyy SO, Klymchenko AS, Demchenko AP (2005) Semi-empirical study of two-color fluorescent dyes based on 3-hydroxychromone. J Mol Struct 755(1–3):229
42. Weber G, Farris FJ (1979) Synthesis and spectral properties of a hydrophobic fluorescent probe: 6-propionyl-2-(dimethylamino)naphthalene. Biochemistry 18(14):3075–3078
43. Shynkar V, Mely Y, Duportail G, Piemont E, Klymchenko A, Demchenko A (2003) Picosecond time-resolved fluorescence studies are consistent with reversible excited-state intramolecular proton transfer in 4′-(dialkylamino)-3-hydroxyflavones. J Phys Chem A 107 (45):9522–9529
44. Klymchenko AS, Demchenko AP (2002) Electrochromic modulation of excited-state intramolecular proton transfer: the new principle in design of fluorescence sensors. J Am Chem Soc 124(41):12372

45. Klymchenko AS, Mely Y, Demchenko AP, Duportail G (2004) Simultaneous probing of hydration and polarity of lipid bilayers with 3-hydroxyflavone fluorescent dyes. Biochim Biophys Acta 1665(1–2):6–19
46. Das R, Klymchenko AS, Duportail G, Mely Y (2008) Excited state proton transfer and solvent relaxation of a 3-hydroxyflavone probe in lipid bilayers. J Phys Chem B 112(38):11929–11935
47. Parasassi T, Di Stefano M, Loiero M, Ravagnan G, Gratton E (1994) Cholesterol modifies water concentration and dynamics in phospholipid bilayers: a fluorescence study using laurdan probe. Biophys J 66(3):763
48. Parasassi T, Krasnowska EK, Bagatolli L, Gratton E (1998) Laurdan and prodan as polarity-sensitive fluorescent membrane probes. J Fluoresc 8(4):365
49. Sykora J, Slavicek P, Jungwirth P, Barucha J, Hof M (2007) Time-dependent stokes shifts of fluorescent dyes in the hydrophobic backbone region of a phospholipid bilayer: combination of fluorescence spectroscopy and ab initio calculations. J Phys Chem B 111(21):5869
50. Shynkar VV, Klymchenko AS, Duportail G, Demchenko AP, Mely Y (2005) Two-color fluorescent probes for imaging the dipole potential of cell plasma membranes. Biochim Biophys Acta 1712(2):128–136
51. Montana V, Farkas DL, Loew LM (1989) Dual-wavelength ratiometric fluorescence measurements of membrane potential. Biochemistry 28(11):4536
52. Klymchenko AS, Duportail G, Mely Y, Demchenko AP (2003) Ultrasensitive two-color fluorescence probes for dipole potential in phospholipid membranes. Proc Natl Acad Sci USA 100(20):11219–11224
53. M'Baye G, Shynkar VV, Klymchenko AS, Mely Y, Duportail G (2006) Membrane dipole potential as measured by ratiometric 3-hydroxyflavone fluorescence probes: accounting for hydration effects. J Fluoresc 16(1):35
54. Klymchenko AS, Stoeckel H, Takeda K, Mely Y (2006) Fluorescent probe based on intramolecular proton transfer for fast ratiometric measurement of cellular transmembrane potential. J Phys Chem B 110(27):13624–13632
55. Klymchenko A, Ozturk T, Pivovarenko VG, Demchenko A (2002) Synthesis of furanochromones: a new step in improvement of fluorescence properties. Tetrahedron Lett 43:7079–7082
56. Klymchenko AS, Pivovarenko VG, Ozturk T, Demchenko AP (2003) Modulation of the solvent-dependent dual emission in 3-hydroxychromones by substituents. New J Chem 27:1336–1343
57. Fromherz P, Hubener G, Kuhn B, Hinner MJ (2008) Annine-6plus, a voltage-sensitive dye with good solubility, strong membrane binding and high sensitivity. Eur Biophys J 37(4):509
58. Simons K, Ikonen E (1997) Functional rafts in cell membranes. Nature 387(6633):569
59. Dietrich C, Bagatolli LA, Volovyk ZN, Thompson NL, Levi M, Jacobson K, Gratton E (2001) Lipid rafts reconstituted in model membranes. Biophys J 80(3):1417
60. Scherfeld D, Kahya N, Schwille P (2003) Lipid dynamics and domain formation in model membranes composed of ternary mixtures of unsaturated and saturated phosphatidylcholines and cholesterol. Biophys J 85(6):3758–3768
61. Brown DA, London E (2000) Structure and function of sphingolipid- and cholesterol-rich membrane rafts. J Biol Chem 275(23):17221
62. Owen DM, Rentero C, Magenau A, Abu-Siniyeh A, Gaus K (2012) Quantitative imaging of membrane lipid order in cells and organisms. Nat Protoc 7(1):24–35
63. Kuhry JG, Fonteneau P, Duportail G, Maechling C, Laustriat G (1983) TMA-DPH: a suitable fluorescence polarization probe for specific plasma membrane fluidity studies in intact living cells. Cell Biophys 5(2):129–140
64. Lentz BR (1989) Membrane 'fluidity' as detected by diphenylhexatriene probes. Chem Phys Lipids 50(3–4):171
65. Klymchenko AS, Oncul S, Didier P, Schaub E, Bagatolli L, Duportail G, Mely Y (2008) Visualization of lipid domains in giant unilamellar vesicles using an environment-sensitive membrane probe based on 3-hydroxyflavone. Biochim Biophys Acta 1788:495–499

66. Oncul S, Klymchenko AS, Kucherak OA, Demchenko AP, Martin S, Dontenwill M, Arntz Y, Didier P, Duportail G, Mely Y (2010) Liquid ordered phase in cell membranes evidenced by a hydration-sensitive probe: effects of cholesterol depletion and apoptosis. Biochim Biophys Acta 1798:1436–1443
67. Vermes I, Haanen C, Steffens-Nakken H, Reutelingsperger C (1995) A novel assay for apoptosis. Flow cytometric detection of phosphatidylserine expression on early apoptotic cells using fluorescein labelled annexin V. J Immunol Methods 184(1):39
68. Duportail G, Klymchenko A, Mely Y, Demchenko A (2001) Neutral fluorescence probe with strong ratiometric response to surface charge of phospholipid membranes. FEBS Lett 508 (2):196–200

Lateral Membrane Heterogeneity Probed by FRET Spectroscopy and Microscopy

Luís M.S. Loura and Manuel Prieto

Abstract Förster resonance energy transfer (FRET) is a photophysical process highly dependent on interchromophore distance. Due to this feature, it is very sensitive to membrane lateral heterogeneity, as the donor and acceptor fluorophores involved in FRET tend to have different preference for distinct types of lipid bilayer domains. In this chapter, the basic formalisms of FRET in situations of increasing complexity (from a single donor-acceptor pair at a fixed distance to non-random probe distribution) are presented and illustrated with selected examples from the literature. The importance of time-resolved fluorescence data is emphasized. It is shown that FRET can be used to study the occurrence of domain formation, allowing their detection as well as size estimation. Lateral lipid distribution heterogeneity may also result from peptide- or protein-lipid interaction. Formalisms that apply to these situations are also presented, as well as selected examples of their use. Applications of FRET under the microscope have recently come to the fore, and representative studies are mentioned.

Keywords Energy Transfer · Fluorescence · Lipid Domains · Membrane Phase Separation · Rafts

L.M.S. Loura (✉)
Faculdade de Farmácia, Universidade de Coimbra, Pólo das Ciências da Saúde, Azinhaga de Santa Comba, 3000-548 Coimbra, Portugal

Centro de Química de Coimbra, Universidade de Coimbra, Rua Larga, 3004-535 Coimbra, Portugal
e-mail: lloura@ff.uc.pt

M. Prieto
Centro de Química-Física Molecular and Institute of Nanosciences and Nanotechnology, Complexo I, IST, UTL, Av. Rovisco Pais, 1049-001 Lisboa, Portugal

Y. Mély and G. Duportail (eds.), *Fluorescent Methods to Study Biological Membranes*, Springer Ser Fluoresc (2013) 13: 71–114, DOI 10.1007/4243_2012_59,

Contents

1 Lipid Bilayer Phase Separation and Membrane Domains

The fluid mosaic model of biological membranes [122] emphasizes membrane fluidity and free lateral diffusion of membrane components. This led to the generalized idea of biomembranes as solutions of proteins embedded in bilayers of randomly distributed phospholipids. However, over the past few decades, it has become accepted in the field of membrane biophysics that lipid lateral heterogeneous distribution exists both in natural and model membranes. Because lipids in general do not mix ideally, the primary driving force for lipid phase separation stems from lipid-lipid interaction, although the presence of proteins can modulate this process, as commented below.

Lipid lateral phase separation was first detected and characterized in binary phospholipid lipid mixtures, in which the two components differed significantly in their main transition temperatures (T_m). This difference could be the result of a discrepancy in acyl chain length (e.g., 12:0,12:0 (*m*:*n* denotes a chain with *m* carbon atoms and *n* double bonds) phosphatidylcholine (PC)/18:0,18:0 PC; [85]), degree of unsaturation of the acyl chains (e.g., *trans*-Δ^9 18:1, *trans*-Δ^9 18:1 PC/18:0,18:0 PC; [146]) or headgroup structure (e.g., *trans*-Δ^9 18:1, *trans*-Δ^9 18:1 PC /16:0, 16:0 phosphatidylethanolamine (PE); [146]) of the two mixture components. The fact that phase separation occurs for mixtures of lipids coexisting in cell membranes under conditions close to physiological (e.g., [86]) made the detection and characterization of this kind of lateral heterogeneity of considerable interest in the biophysical and biochemical communities (e.g., [148]). The composition-temperature phase diagrams, derived under equilibrium conditions at constant pressure (and ionic strength, etc.), are a convenient way to represent this type of behavior for each pair of phospholipids. The classic book by Marsh [86] collects several of these examples, and a second, much expanded edition is expected around the time of publication of this work.

Another group of lipid mixtures that have been intensely used are phospholipid (mainly PC)/cholesterol (chol) binary systems. These mixtures' properties and interaction with peptides and proteins are of importance due to the high abundance of chol in mammalian plasma membranes. From the studies in model systems, important conclusions have been inferred about the role of sterols in biomembranes. The well-known effects of chol on the bilayer properties (see, e.g., [7, 96]) have been rationalized considering that in the presence of high amounts of chol in a PC bilayer, the membrane is in a liquid ordered (lo) phase (using the nomenclature introduced by Ipsen et al. [58]), with intermediate properties between those of the gel and the fluid. This designation highlights the facts that the translational diffusion is closer to the fluid phase (the diffusion coefficient in the lo phase is only two to three times lower than for the pure fluid phase of the PC), but the acyl chains are in a much more ordered configuration. In this nomenclature, the gel and fluid phases are designated by solid ordered (so) and liquid disordered (ld), respectively. The phase diagram is monotectic, and for intermediate chol concentrations, phase coexistence occurs: so and lo, below the monotectic temperature (which is close to T_m), and ld with lo, above the monotectic temperature (a recent collection of these diagrams has been published; [87]). The latter corresponds to fluid-fluid phase separation, which is thought to be of biological relevance, namely, to the raft phenomenon (e.g., Brown and London [16]). These were at first operationally defined as insoluble membrane fractions upon detergent extraction (the so-called detergent resistant membranes, DRM), and their composition revealed that they were rich in chol, spingomyelin, and saturated lipids. Later, these membrane patches, which were intensively studied both in membrane biophysics and cell biology, were described as liquid-ordered on the framework of the above-mentioned type of phases postulated in the presence of chol. Although phase separation was not a novelty in itself, the very pictorial raft concept bridged the fields of membrane biophysics and cell biology, and the communication between these two communities was instrumental to develop a very active research on both natural membranes and their model systems.

It should be mentioned that for PC/chol systems, the most studied being 16:0,16:0 PC/chol, there are several phase diagrams reported that differ considerably (e.g., [72, 92, 140]). The discrepancies are probably related to the similarity between the two phases, which makes differentiation between them difficult [149]. Nevertheless, some of those discrepancies have been rationalized in a study of one of the archetypal raft model systems, the 16:0,18:1 PC/*N*-palmitoylsphingomyelin (PSM)/chol mixture [33]. Following this work, ternary lipid mixtures of high melting/low melting/chol (proposed as minimal model systems of raft-forming lipid mixtures) have been studied and described in terms of fixed-temperature, fixed pressure, ternary phase diagrams [52].

Several interconnected concepts can be misdealing, among them phases and membrane domains, and also the problem of their sizes. Phases and phase diagrams only apply to systems under thermodynamic equilibrium, and it should be stressed that this situation can be a very slow process. In a situation of very strong mismatch of two lipid components, i.e., when the driving force is very high, the time that it takes to attain equilibrium upon a system perturbation is on the scale of hours for gel-fluid phase separation [32, 63]. However, even considering that a cell membrane is not under equilibrium, phase diagrams are instrumental to the rationalization of the processes that can occur in a natural membrane. Also according to thermodynamic considerations, phase separation would proceed until completion in order to decrease line tension, so this would imply that the observed domains, whatever the methodology used, should be very large. This is observed for gel-fluid phase separation where the phase boundary is very steep, and defects are present.

However for liquid-ordered/liquid-disordered phase coexistence, the domains are small (submicron, "nanodomains"), as will be discussed later in detail in this chapter, and chol or specified lipid configurations have been invoked to be able to reduce line tension, thus preventing their growth [121]. It should be stressed that increasing the number of domains also leads to an entropic compensation.

The existing discussion in the literature about domain sizes was prompted by the different information obtained according to the different type of experimental approach used to detect and study them. While in natural membranes there is clear evidence that domains are in general small (10–100 nm, nanodomains) (see e.g., [59, 68, 90]), in model systems different answers were obtained depending on the methodology used. Under a confocal microscope, large (micron size) domains can be observed in suitable systems, namely, ternary systems with ld/lo phase separation, but up to now no phase separation was detected for binary systems with cholesterol [136]. This is in general attributed to the very small size of these lo domains, below the lateral resolution of the microscope (~300 nm), and it prompts a global discussion about the type of phase diagrams (and thus the phases and domain sizes) that are described in the literature. There is a clear disagreement when comparing the ones coming from microscopy data, with the ones obtained from spectroscopic approaches such as fluorescence and ESR as described by, e.g., Veatch and Keller [137]. From fluorescence spectroscopy, and in particular using FRET with adequate modeling, as described later in this chapter, clear evidence for nanoscopic domains is obtained and the same happens, e.g., from residence times data in the submicrosecond regime in ESR spectroscopy [22]. Therefore, more detail is obtained from spectroscopic approaches which do not miss the existence of small domains (and so phase separation). Along this way the use of "spectroscopic diagrams" was suggested [4]. Microscopy is however invaluable in the way that it allows direct visualization of lipid domains, and we can foresee in the future fast development in this area, such as near-field scanning optical microscopy (see [38] for a review), stimulated emission depletion microscopy, and related "super-resolution microscopy" approaches (see [102] for a review), which allow a closer-to-molecular resolution (~30 nm) as compared to standard confocal. Also, recently, in a powerful conjugation with AFM (e.g., [25]), this technique has been instrumental in the study of membrane domains.

Another subject under discussion is whether the nanodomains (assuming they are not transient density fluctuations) should be considered as phases, otherwise the phase diagrams based on their detection, as well as the respective tie-lines, would have no meaning. Apart from the problem that phase separation will not go up to completion as discussed before (but the same happens for the larger domains micron size domains detected by microscopy), there is no critical restriction on size that would prevent the phase concept of being applied to them. In addition, they are formed at least by a few thousand molecules, i.e., they are large enough for their rationalization as a phase in the framework of statistical thermodynamics. Another relevant aspect is that spectroscopic methodologies allow the derivation of phase diagrams that are thermodynamically consistent.

It should be mentioned that the structure and dynamics of biological membranes and their model systems constitute a very lively research area, and also the concept of chol-induced lo and ld phases mentioned above has been questioned in recent literature using different approaches, e.g., X-ray diffraction [64] and NMR data [138]. However, in this chapter the discussion about phases and chol will be carried out in the framework of ld/lo phase separation, as this conceptualization proved to be instrumental in the rationalization of membrane biophysics.

Another source of lateral distribution heterogeneity is preferential interaction of proteins with selected lipids. These interactions are able to drive enrichment of the bilayer around the protein in these lipids and impoverishment in others, creating local heterogeneities that can potentially extend to several lipid shells around the protein. Some superficial membrane proteins demonstrate specific binding to some lipid classes, a phenomenon that can control protein recruitment to the membrane and activate signaling cascades [30]. In addition, transmembrane proteins display differential interactions with lipids of different acyl-chain lengths due to packing constraints in the lipid/protein hydrophobic interface, that have a significant effect on the activity of several proteins. Membrane proteins also have been shown to present binding sites for lipids in hydrophobic pockets away from the protein/lipid interface and binding of specific lipids to such sites is essential for activity in several cases [71]. In this chapter, we dedicate a subsection covering studies concerning protein-induced lipid heterogeneous distributions.

In all cases, we will focus on the application of Förster resonance energy transfer (FRET) to the systems at hand. For this purpose, it is useful to describe the basic formalisms of FRET and how they can be used in situations where molecule distribution is not uniform. This is done in the following section, which is organized in terms of increasing complexity (from the fixed donor-acceptor distance case to non-random fluorophore distribution formalisms, and from one-component, one-phase lipid bilayers to two- and three-component, phase-separated systems). Lipid distribution heterogeneity arising from protein-lipid interactions is described in a separate subsection. A brief finalizing section is dedicated to quantitative applications of FRET under the microscope to the study of heterogeneous lipid distribution in very large model systems (such as giant unilamellar vesicles (GUV)) and live cell membranes. As we focus on quantitative uses of FRET, phenomenological applications are not addressed in this chapter. On the other hand, unless explicitly mentioned otherwise, all formalisms and examples refer to the cases where donor and acceptor species are distinct species (hetero-FRET).

2 FRET Formalisms and Their Application in Spectroscopic Studies

2.1 One Donor – One Acceptor

FRET is a photophysical process by which an initially electronically excited fluorophore, termed “donor,” transfers its excitation energy (and thus becomes quenched) to another chromophore, termed “acceptor,” whose electronic absorption spectrum overlaps that of the emission of the donor. The latter, initially in the electronic ground state, becomes excited upon transfer, and may (or may not) fluoresce. FRET involves neither photon emission nor molecular contact between the two species, but is highly dependent on the distance between them. For an isolated donor-acceptor pair, the (first-order) rate coefficient for the FRET interaction is proportional to the inverse sixth power of this distance [44]:

$$k_T = \frac{1}{\tau_0}\left(\frac{R_0}{R}\right)^6 \tag{1}$$

In this equation, τ_0 is the donor excited state lifetime in the absence of acceptor, R is the donor-acceptor distance, and R_0 is the so-called Förster radius. The latter is a characteristic length for FRET, defined as the donor/acceptor distance for which FRET within a given donor/acceptor pair is 50 % efficient (i.e., as probable as the other processes of donor excitation decay). In practice, the distance range for which FRET is sensitive is between 0.5 R_0 and 2 R_0, as FRET efficiency varies from 98.5 % to 1.5 % in this interval. The value of R_0 is characteristic of each donor/acceptor pair in a given environment, but usually lies in the 1.5–6 nm range. It can be calculated from spectral data using the following equation:

$$R_0 = 0.2108\left[\kappa^2 \cdot \Phi_\mathrm{D} \cdot n^{-4} \cdot \int_0^\infty I(\lambda)\cdot \varepsilon(\lambda)\cdot \lambda^4 d\lambda\right]^{1/6} \tag{2}$$

where κ^2 is the FRET orientation factor (for a definition and discussion of κ^2, see [134]), Φ_D is the donor fluorescence quantum yield, n is the medium refractive index, $I(\lambda)$ is the donor normalized emission spectrum, and $\varepsilon(\lambda)$ is the acceptor absorption spectrum.

A convenient metric of the extent of FRET is the FRET efficiency, E, defined as the ratio between the rate of FRET and that of donor decay considering all pathways:

$$E = k_T \Big/ \sum_i k_i = k_T\tau \tag{3}$$

where τ is the donor lifetime in the presence of acceptor. For an experimental situation where all donor/acceptor pairs are identical and fluorophore diffusion can be neglected during the donor excited state lifetime (so that R has a fixed determined value),

$$E = 1 - \tau/\tau_0 = R_0^6/(R^6 + R_0^6) \tag{4}$$

An identical expression can be written for the fluorescence quantum yield. By rearranging this equation, one obtains R:

$$R = R_0 \left(\frac{1 - E}{E} \right)^{1/6} \tag{5}$$

These equations explain the use of FRET as a "spectroscopic ruler" [128], as it allows the measurement of distances of the order of R_0 from knowledge of the latter and measurement of E. As mentioned above, they apply strictly to a situation where all donor/acceptor pairs have identical separating distance. In case of conformational flexibility, modifications can be introduced that allow recovery of distance distributions or characterization of diffusion in a linked donor/acceptor pair (e.g., [70], and references therein). Additionally, expressions that relate the efficiency of FRET to geometrical parameters in a cluster of a finite number of donors and/or acceptors can be derived by taking into account the multiple FRET rate terms (Eq. 1) in the calculation of the overall FRET efficiency (which may be relevant to characterize oligomerization, e.g., [73]). However, it must be stressed that the most common situation in membranes is that of donor and acceptor fluorophores scattered in a quasi-two-dimensional geometrical arrangement, leading to a multitude of undetermined donor/acceptor separation distances. Even though for each particular pair Eq. 1 is still valid, the overall complexity rules out the use of Eq. 5 for distance determination (a distance retrieved from Eq. 5 would be a very complex average from which no useful information can be recovered). These scenarios require modeling using formalisms for infinite planar or bilayer geometry, as outlined in the following section.

2.2 *One Donor – Multiple Acceptors in Bidimensional Media*

2.2.1 Uniform Probe Distribution

We now move to the situation where all donors and acceptors are scattered in a plane and each donor fluorophore is surrounded by a uniform distribution of acceptors. We will assume for now that all donors are equivalent (so that one needs only to consider one donor molecule), no lateral diffusion is operative during the donor lifetime, and all donor/acceptor pairs have the same R_0 value. From Eq. 1, the total rate of FRET from the donor molecule under consideration to the N acceptor molecules located inside a disk of radius R_d (and centered on this particular donor) is

$$k_T = \tau_0^{-1} \left[1 + \sum_{i=1}^{N} (R_0/R_i)^6 \right] \tag{6}$$

where R_i is the distance between the donor and the i-th acceptor inside the disk. As shown previously [46, 145], this approach leads to closed-form analytical solutions. If the minimum donor-acceptor distance R_e (also termed "exclusion distance") is much smaller than R_0 (in practice, if $R_e < R_0/4$), the time evolution of donor fluorescence in presence of acceptor (i_{DA}) is simply given by

$$i_{DA}(t) = \exp(-t/\tau_0)\exp\left[-\pi\Gamma(2/3)R_0^2 c(t/\tau_0)^{1/3}\right] \tag{7}$$

In this equation, Γ is the complete gamma function and $c = N/(\pi R_d^2)$ is the average number of acceptors per unit area. In case that R_e and R_0 are of the same order, the donor decay becomes more complex [145]:

$$i_{DA}(t) = \exp\left(-\frac{t}{\tau_0}\right)\exp\left\{-\pi R_0^2 c\gamma\left[\frac{2}{3},\left(\frac{R_0}{R_e}\right)^6\left(\frac{t}{\tau_0}\right)\right]\left(\frac{t}{\tau_0}\right)^{1/3}\right\}\cdot$$
$$\cdot\exp\left\{\pi R_e^2 c\left(1-\exp\left[-\left(\frac{R_0}{R_e}\right)^6\left(\frac{t}{\tau_0}\right)\right]\right)\right\} \tag{8}$$

Here γ is now the incomplete gamma function. Equation 8 also applies to the situation where the planes of donors and acceptors are distinct but parallel, separated by a transverse distance w [145]. This condition is often met in membrane studies, as donor and acceptor depths of location in the bilayer are frequently distinct. In this case, $R_e = (R_l^2 + w^2)^{1/2}$, where R_l is the minimum donor-acceptor lateral distance (i.e., parallel to the membrane plane).

Depending on the experimental setup, acceptors may locate in two parallel planes rather than in a single plane. This is expected if, for example, they are evenly distributed between the two bilayer leaflets. In this case, there will be two R_e distances, one for each plane of acceptors. Computation of i_{DA} from Eq. 8 requires multiplying by two additional exponential terms, identical to the last two of the right-hand side, differing only in the value of R_e.

If the FRET measurement is carried out under photostationary conditions, the decay is not measured, and it is convenient to compare the experimental value of FRET efficiency (Eq. 4) with that expected from the above formalism. The latter can be computed from numerical integration:

$$E=1-\int_0^\infty i_{DA}(t)dt \Big/ \int_0^\infty i_D(t)dt \tag{9}$$

Here i_D (equal to $\exp(-t/\tau_0)$ for an exponentially decaying donor) is the time-resolved donor fluorescence in the absence of acceptor.

The formalism of two-dimensional FRET was first applied to bilayer systems (unilamellar vesicles of egg yolk PC) in the pioneering study of Fung and Stryer [46]. Four donor/acceptor pairs (using dansyl- and eosin-head-labeled phosphatidylethanolamine derivatives), with spectroscopic R_0 values ranging from 2.3 to 5.1 nm, were considered. The authors analyzed steady-state FRET efficiencies with a formalism

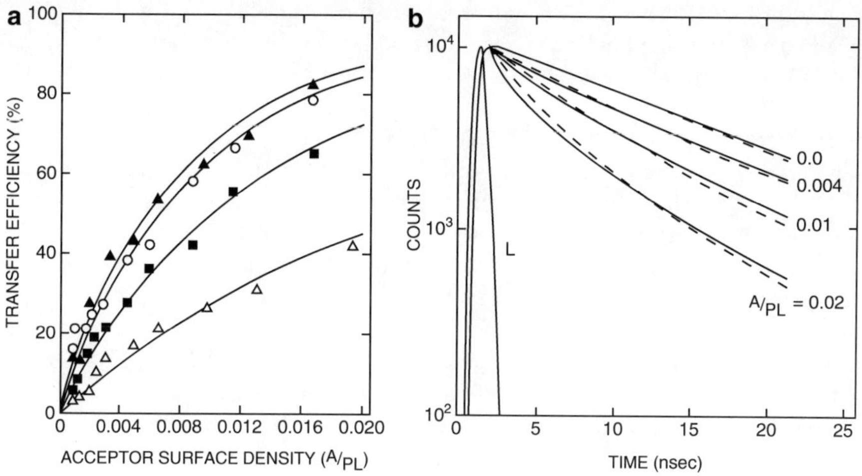

Fig. 1 (**a**) FRET efficiency as a function of the surface density of energy acceptor (acceptors per phospholipid) for four donor-acceptor pairs in egg yolk PC vesicles: *N*-(2-dimethyl-aminonaphthalene-6-sulfonyl)phosphatidylethanolamine (2,6-DPE) to *N*-(2-dimethylamino-naphthalene-5-sulfonyl)phosphatidylethanolamine (2,5-DPE) (*open triangles*); 2,6-DPE to *N*-eosin-*N*'-phosphatidylethanolaminothioureia (EPE) (*filled squares*); *N*-(l-dimethylamino-naphthalene-5-sulfonyl)phosphatidylethanolamine (1,5-DPE) to EPE (*open circles*); and 2,5-DPE to EPE (*filled triangles*). The lines are best-fit curves using the model of uniform probe distribution with R_0 values of 25.5, 37.5, 49, and 46 Å, respectively (compared with the values calculated from spectral data, 22.8, 39.1, 51.2, and 48.7 Å, respectively). (**b**) Fluorescence emission kinetics of 2,6-DPE in PC vesicles containing 0,0.004,0.01, and 0.02 molecules of EPE per molecule of PC (*dashed lines*). Synchrotron radiation was used as a pulsed light source (λexc = 350 nm, emission was viewed through a 440-nm interference filter). The decay curves were calculated using $R_0 = 39.5$ Å. $R_e = 8.4$ Å and $\tau_0 = 13.2$ ns, and then convoluted with the light pulse (*solid lines*) (Reprinted from Fung and Stryer [46] with permission. Copyright 1978 American Chemical Society)

equivalent to the combination of Eqs. 8 and 9 above, using R_0 as the sole optimized parameter. The recovered and calculated R_0 values differed less than 0.3 nm in all cases (Fig. 1a). The authors also measured donor fluorescence decays in the presence of varying concentration of acceptor for one of the studied FRET pairs. They verified that even though there was excellent agreement between the experimental decays and the theoretical expectations for low acceptor concentrations, deviations become apparent for higher acceptor loads (Fig. 1b). The observation of slower decays than predicted at short times was interpreted by the authors as possibly reflecting fewer short-distance donor-acceptor pairs than statistically expected, owing to both donor and acceptor probes being negatively charged. This shows that, right from the very first FRET application to bilayer systems, several important features were revealed: first, FRET in fluid bilayers was overall well described by the analytical two-dimensional formalism; second, deviations can be interpreted in terms of non-homogeneous fluorophore distribution; third, these deviations are much more evident in time-resolved experiments compared to steady-state conditions. The authors rightly predicted that FRET would become a most valuable tool for the study of phenomena including lipid phase separation and protein oligomerization.

Following this study, it took some time for quantitative applications of FRET (rather than phenomenological studies, which are not considered here) to membrane heterogeneity to fully blossom. This can be understood by noting that laboratories equipped with laser excitation sources, capable of adequate time resolution, were relatively scarce during much of the 1980s and even the 1990s, and analysis of time-resolved FRET data with formalisms such as described above was overwhelmingly slow before the development of affordable modern computers. At first, following Fung and Stryer's study, applications of the uniform distribution formalism to one-component bilayers aimed at verifying the applicability of Förster FRET theory to model membranes using different FRET pairs. From time-resolved data of FRET from *N*-(7-nitrobenz-2-oxa-1,3-diazol-4-yl) (NBD)-16:0,16:0 PE to *N*-(lissamine–rhodamine B) (Rh)-16:0,16:0 PE in 18:1,18:1 PC [69], a linear variation of the recovered *c* parameter as a function of the acceptor concentration was verified, as expected, allowing the calculation of the area per lipid molecule. This dependence was also verified in FRET from octadecylrhodamine B (ORB) to 1,1′,3,3,3′,3′-hexamethylindotricarbocyanine [$DiIC_1(7)$] in fluid 16:0,16:0 PC large unilamellar vesicles (LUV) [76]. In this study, a modified Eq. 7 was derived for biexponentially decaying donors, and the decays were globally analyzed, with linkage of donor lifetimes and preexponential ratio. However, analysis of the decays for the same system but below the main transition temperature was not successful, pointing to probe aggregation in the gel phase, possibly in line defects in the gel-phase structure. In this situation, the traditional framework, derived assuming a random distribution of probes, is no longer valid. This situation was also verified for the *N*-NBD-16:0,16:0 PE/*N*-Rh-16:0,16:0 PE pair in the same system [77]. In the latter study, a generalized FRET model, assuming distribution of acceptor concentrations [74], was also applied.

Presently, the uniform distribution formalism is still used as a test of whether addition of a new component to a given one-phase lipid bilayer system induces compartmentalization and/or phase separation. This would be detected in the failure to analyze FRET kinetics with uniform probe distribution formalisms. Recent examples of this kind include a study that demonstrated the absence of clustering of phosphatidylinositol-(4,5)-bisphosphate ($PI(4,5)P_2$) in a fluid PC matrix at slightly above physiological pH, following the satisfactory description of FRET between 1,6-diphenylhexatriene (DPH) and NBD-labeled $PI(4,5)P_2$, in 16:0, 18:1 PC vesicles with 5 mol% of total $PI(4,5)P_2$ [43] at pH 8.4. On the other hand, time-resolved FRET between the tryptophan residues of acetyl-GWW$(LA)_8$LWWA-amide peptide (WALP23) to the fluorescent chol analog dehydroergosterol (DHE), both with and without added equimolar amounts of chol, could be satisfactorily globally analyzed assuming uniform DHE distribution in the bilayer [51]. This FRET pair (tryptophan/DHE) was also used in a study of the hypothetical affinity of the γM4 peptide from the muscle acetylcholine receptor (donor: Trp453) for chol (acceptor: DHE) in the lo phase of 16:0, 18:1 PC/Chol. The measured FRET efficiencies were significantly lower than expected, which was interpreted on the basis of formation of peptide-rich, sterol-depleted patches [34]. Higher FRET efficiency than expected was observed between M13 major coat protein labeled with *N*-(iodoacetyl)aminoethyl-1-sulfonaphthylamine (IAEDANS, donor) and *N*-(4,4-difluoro-5,7-dimethyl-4-bora-3a,4a-diaza-s-indacene-3-yl)methyl iodoacetamide (BODIPY, acceptor) in supposedly monophasic (fluid) bilayers of *cis*-Δ^{13} 22:1, *cis*-Δ^{13} 22:1 PC/18:1, 18:1 PC and *cis*-Δ^{9} 14:1, *cis*-Δ^{9} 14:1 PC/18:1, 18:1 PC, due to formation of domains enriched in the protein and the matching lipid (18:1, 18:1 PC; [41]).

Sometimes addition of a component leads to changes in FRET efficiency that are not related to phase separation, but to other morphological changes in the lipid organization. This was the case in recent studies of mixed PC/anionic lipid (phosphatidylserine, PS) vesicles incubated with a basic peptide (K_6W; [80]) or protein (lysozyme; [27]), where formation of stacked lipid multilayers, bridged by peptide or protein, was concluded. Whereas the observed FRET efficiency variations could be due to either lateral demixing or multilayer formation, global analysis of time-resolved data can clearly distinguish between the two situations. In the mentioned studies, no significant lateral phase separation takes place. The variations in the extent of FRET result from multilayer formation, and it was even possible to measure the spacing repeat distance in these structures.

2.2.2 Phase Separation into Large Domains

Although the above equations refer to uniform fluorophore distribution, the generalization to a biphasic environment is straightforward, provided that the two phases involved are organized in large domains on the FRET length scale (i.e., much larger than R_0). The key idea is that in such a situation there are two donor and two acceptor populations, each characteristic of one of the coexisting phases. Probe concentrations are different in the two phases, but it is assumed that distribution remains uniform within each type of domains. Additionally, the hypothesis of large domains implies that boundary effects are negligible, that is, donors located in one phase only sense acceptors located in the same phase, and interphasic FRET may be ignored. With these assumptions, the donor decay in presence of acceptor is a linear combination of the donor decays $i_{\mathrm{DA,i}}$ inside the coexisting phases (labeled $i = 1, 2$):

$$i_{\mathrm{DA}}(t) = A_1 i_{\mathrm{DA,1}}(t) + A_2 i_{\mathrm{DA,2}}(t) \tag{10}$$

In this equation, the coefficients A_{i} are proportional to the amount of donor in the corresponding phase (provided that molar absorptivity is the same in the two phases), and $i_{\mathrm{DA,i}}$ is given by Eq. 7 or 8 above. For each phase, characteristic values of τ_0, c, R_0, and possibly R_{e} apply. Considering this, it follows that the total number of parameters in Eq. 10 is very large, and donor decays in presence of acceptor are best analyzed simultaneously with linked values of common parameters (global analysis) with decays in absence of acceptor for a more accurate retrieval of the best fitting parameter values. The latter contain information relative to the amount of donor and acceptor in each phase, from which the partition coefficients of these probes may be calculated. These are defined as (e.g., [31]):

$$K_p = (P_2/X_2)/(P_1/X_1) \tag{11}$$

where P_1 is the probe mole fraction in lipid phase 1, and X_1 is the lipid phase 1 mole fraction (therefore $P_2 = 1 - P_1$ and $X_2 = 1 - X_1$). The partition coefficients of donor (K_{pD}) and acceptor (K_{pA}) probes can be calculated straightforwardly from the FRET decay parameters [78],

$$K_{\mathrm{pD}} = (A_2/X_2)/(A_1/X_1) \tag{12}$$

$$K_{\mathrm{pA}} = \left(c_2 a_2/\tau_2^{1/3}\right)\Big/\left(c_1 a_1/\tau_1^{1/3}\right) \tag{13}$$

where a_{i} is the area per lipid molecule in phase i. These equations were originally applied to a situation of gel/fluid phase separation in the aforementioned 12:0,12:0 PC/18:0,18:0 PC system. Two different temperatures and compositions inside the phase coexistence range were studied. The short-tailed FRET donor, *N*-NBD-12:0,12:0 PE, and a short-tailed FRET acceptor, 1,1′-didodecil-3,3,3′,3′′-tetramethylindocarbocyanine (DiIC$_{12}$(3)), were shown to prefer the fluid phase (rich in short-tailed phospholipid) by both intrinsic anisotropy, lifetime and FRET measurements, in agreement with published reports. The other studied FRET acceptor, long-tailed probe 1,1′-dioctadecil-3,3,3′,3′-tetramethylindocarbocyanine (DiIC$_{18}$(3)), was expected to prefer the gel (rich in long-tailed phospholipid), on account of hydrophobic matching considerations [95]. While intrinsic lifetime studies indeed indicated preferential partition of DiIC$_{18}$(3) into a rigidified environment, FRET analysis pointed to an increased donor-acceptor proximity as a consequence of phase separation. These apparently conflicting results were rationalized on the basis of segregation of DiIC$_{18}$(3) to the gel/fluid interphase. In order for fluid-located donors to sense these interphase-located acceptors, fluid domains should be small (not exceeding ~10–15 nm). This work shows that membrane probes which apparently prefer the gel phase may show a nonrandom distribution in this medium (in agreement with the study described above for pure DPPC gel phase LUV) and tend to locate in an environment which simultaneously leads to less strict packing constraints and to favorable hydrophobic matching interactions.

The dynamics of domain growth was also studied in the 12:0,12:0 PC/18:0,18:0 PC (equimolar composition) [32]. LUV were first equilibrated at 65 °C, a temperature above the T_{m} of both lipids, where the system is in the one fluid (supposedly homogeneous) phase situation. The lipid vesicles contain also a probe that partitions preferentially to the gel (*trans*-parinaric acid, *t*-PnA) and another that prefers the fluid phase (*N*-NBD-12:0,12:0 PE). Then, a sudden thermal quench to 20 °C is carried out, and the lipid mixture is rapidly taken to the gel/fluid phase coexistence region of the phase diagram (Fig. 2a). The FRET efficiency as a function of time was measured and, as expected, it decreased, because as domains

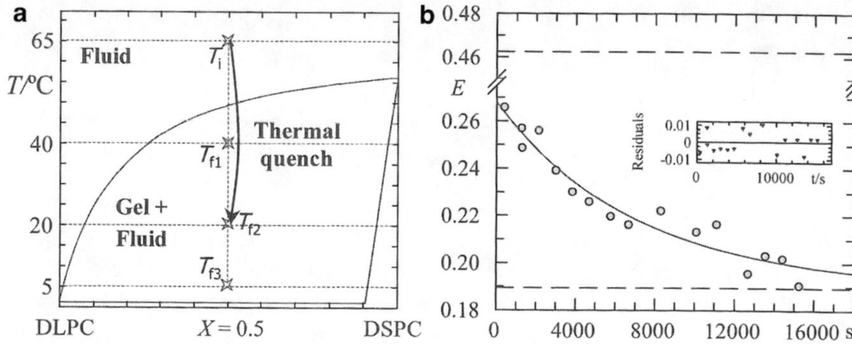

Fig. 2 (**a**) Phase diagram 12:0,12:0 PC/18:0,18:0 PC (adapted from [3]). The initial temperature (T_i) and the three final temperatures (T_f) after the thermal quenches are shown. (**b**) FRET efficiency E from *t*-PnA to *N*-NBD-12:0,12:0 PE vs. time equimolar 12:0,12:0 PC/18:0,18:0 PC LUV after a sudden thermal quench from $T_i = 65$ °C (fluid phase region) to $T_f = 20$ °C in the gel/fluid coexistence region. The line is the best fit of an exponential function with a nonzero infinite value, and the residuals are shown in the insert. The two limiting E values were calculated assuming uniform distribution of both probes at zero-time (*dashed line* above in the plot) and complete phase separation at infinite time (*dashed line* below in the plot), respectively (Reprinted from de Almeida et al. [32] with permission. Copyright 2002 Biophysical Society)

form and grow due to the differential portioning of the probes, the donor is sensing a decreasing local concentration of acceptors. The process has a dynamics on the time-scale of hours. The trend of FRET efficiency with time could be well described by an exponential function with a nonzero value at infinite time ($E = 0.19$), equal to the calculated value considering infinite phase separation (Fig. 2b). This shows that domains are at least 5–10 times R_0 [79], in apparent contradiction with the study described above with $\text{DiIC}_{18}(3)$ as acceptor. These two sets of results can be reconciled by assuming that $\text{DiIC}_{18}(3)$, by probably accumulating in the gel/fluid interface, lowers the line tension of the interface separating the two coexisting phases, promoting bilayer reorganization into smaller domains. This effect would not occur with *t*-PnA, which is able to accommodate its single chain in the bulk gel.

Besides probe partition, the composition/temperature (x, T) phase diagram boundaries (x_1 and x_2, corresponding to pure 1 and 2 phases, respectively) may also be obtained from the time-resolved FRET parameters [79]. To show this, we let F be the overall acceptor mole fraction and F_i be the acceptor mole fraction within phase i. The latter is related to c_i according to

$$F_i = c_i a_i \tag{14}$$

By inserting the F_i values into the acceptor mass balance equation,

$$F = F_2(1 - X_1) + F_1 X_1 \tag{15}$$

X_1 and X_2 can be straightforwardly calculated, even for an unknown phase diagram. If this is carried out for two points, A(x_A, T) and B(x_B, T), and combined with the lever rule, one obtains the following simple expressions for the phase boundaries:

$$x_1 = (x_A X_{2B} - x_B X_{2A})/(X_{1A} - X_{1B}) \quad (16)$$

$$x_2 = (x_B X_{1A} - x_A X_{1B})/(X_{1A} - X_{1B}) \quad (17)$$

If this procedure is repeated for several temperatures, the phase diagram is obtained.

More recently, Buboltz [17] developed an experimental approach (Steady-State Probe-Partitioning FRET or SP-FRET) for characterization of phase separation in lipid membranes, based on acceptor steady-state sensitized emission, and assuming that the coexisting phases are much larger than R_0. The procedure requires measurements with different FRET pairs exhibiting complementary partitioning. Sensitized emissions in each phase are expressed as a function of local donor and acceptor mole fractions, and two constants including all photophysical effects in the respective phase. On the low acceptor concentration regime, this relationship was significantly simplified [18]. Information on the phase boundaries can be obtained from the detection of the lipid compositions for which the gradient of sensitized acceptor emission relative to composition was maximal. In turn, model fitting to the sensitized fluorescence data, together with knowledge of the phase boundaries of the lipid system under study, allows for determination of the probes' interphasic partition coefficients as the sole fitting parameters. The author successfully applied this approach to the study of gel/fluid phase separation in 12:0,12:0 PC/18:0,18:0 PC mixtures by globally fitting data obtained from two different combinations of donor/acceptor pairs. The need for multiple donor/acceptor pairs is a consequence of the degeneracy of the model with respect to the partition coefficients of the probes in each set of measurements.

In another study, the same methodology for detection of phase boundaries was used for the characterization of the 18:1,18:1 PC/16:0,16:0 PC/chol lipid mixture [19]. Fluorescence data from 1294 independently prepared samples were analyzed and phase boundaries were obtained from the gradients of acceptor sensitized emission from 3,3′-dioctadecyloxacarbocyanine (18:0-DiO), using DHE as the FRET donor (Fig. 3). Three different regions of phase coexistence were clearly identified (gel/fluid, lo/ld, and lo/Chol crystals), presenting some discrepancies relative to confocal fluorescence microscopy (CFM) and solid-state NMR studies [136, 139]. Notably, the lo/ld coexistence range is narrower at room temperature (not extending beyond 33 mol% Chol, compared to 50 mol% for the CFM boundary). Also, as the temperature increases, coexistence in that region extends in its entirety to higher 16:0, 16:0 PC content, whereas for CFM measurements only the ld boundary is affected. The authors attribute this discrepancy to the use of different methodologies in sample preparation. CFM and solid-state NMR require film deposition of lipid to produce GUV or oriented membranes, and this might result in increased susceptibility to demixing of lipid components, while in this study, polydisperse multilamellar vesicle suspensions were prepared through the rapid

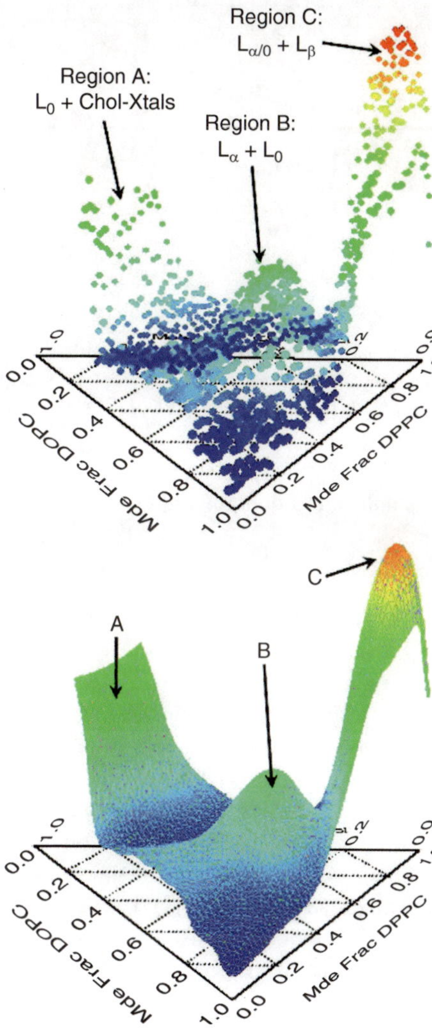

Fig. 3 Derivation of the phase boundaries of the 18:1,18:1 PC/16:0,16:0 PC/chol phase diagram using SP-FRET. Changes in sensitized acceptor fluorescence are plotted versus lipid composition in triangular coordinates. In the scatter plot on top, each data point corresponds to an independently prepared sample (1,294 total). The bottom plot shows a smooth surface fit to the same data (Reprinted from Buboltz et al. [19]. Copyright 2007 American Chemical Society)

solvent exchange method, which does not require formation of this intermediate lipid film. Interestingly, two other studies on this ternary system were recently published, neither of which restricted to GUV. De Almeida et al. [36], using a combined time-resolved fluorescence microspectroscopic approach (i.e., fluorescence lifetime imaging microscopy and microscopic fluorescence decays measured in GUV, and macroscopic fluorescence decays measured in large unilamellar vesicles), established the existence of the three-phase triangle near the 16:0/16:0 PC corner, thus narrowing the lo/ld range previously reported [139]. In the same year, an NMR study of multilamellar vesicles by Veatch and coworkers [138] indicated that lo/ld coexistence region does not extend beyond 35 mol% chol for 10–37 °C. SP-FRET was also the basic tool used recently in the study of 18:0,18:0

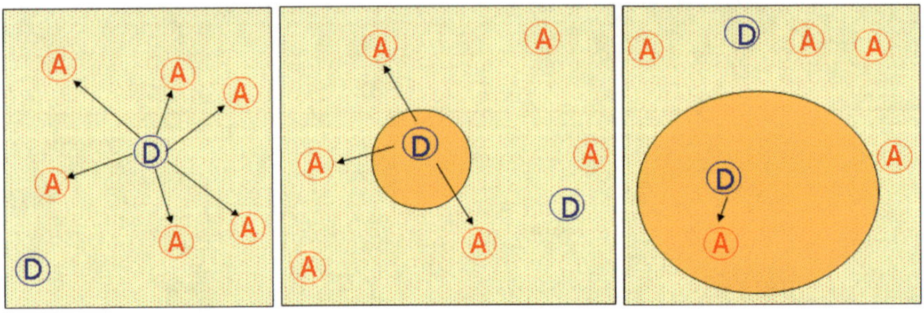

Fig. 4 Pictorial view of FRET between donor and acceptor preferring different phases in situation of: *left panel* – no phase separation (random distribution); *middle panel* – phase separation with small domains; *right panel* – phase separation with large domains (infinite phase separation limit) (Reprinted from de Almeida et al. [37] with permission. Copyright 2009 Elsevier)

PC/16:0,18:1 PC/chol and 18:0,18:0 PC/18:0,18:1 PC/chol, revealing the existence of small liquid domains, not observable by optical microscopy [48].

2.2.3 The Intermediate Case: Nanometer-Sized Domains

Many spectroscopic and other (calorimetry, X-ray diffraction) techniques have been informative in the identification of the phases present in a lipid mixture and in the derivation of phase diagrams (e.g., [52, 87]), and certainly are better suited than FRET to this purpose. On the other hand, it is certainly more convenient to use the variation of fluorescence parameters such as steady-state intensity, anisotropy, or lifetime to characterize the partition of membrane probes and fluorophore-bearing biomolecules, rather than to estimate them from the FRET decay fitting parameters. The major advantage of FRET is that its distance dependence, as expressed in Eq. 1, warrants a unique sensitivity to compartmentalization in its characteristic length scale, that is, of the order of R_0. Concerning membrane organization in particular, this feature allows FRET to report on formation of nanometer-sized domains, smaller than the limit of conventional optical microscopy. Consider the case of probes with complementary phase preference (Fig. 4). Phase separation will increase donor-acceptor separation and hence render FRET less efficient. However, if the domains formed are of the order of R_0, FRET between donors and acceptors located in distinct phases is significant, and FRET efficiency will fall between the values expected for the single-phase and two-infinite-phases scenarios. However, the obvious loss of symmetry and topological complexity of this intermediate regime has precluded the derivation of an exact solution of the FRET equations that allows convenient retrieval of the domain size for a given system. Several more or less approximate methodologies have been developed since the last decade to estimate domain sizes, which are succinctly described below.

Loura et al. [79] applied the infinite phase separation formalism described in the previous section to FRET between *N*-NBD-14:0, 14:0 PE (donor) and *N*-Rh-14:0,

14:0 PE (acceptor) dispersed in mixed 14:0, 14:0 PC/chol vesicles. The phase diagram for this system was known from discontinuities in electron spin resonance spectroscopy experiments and breaks in the behavior of the diffusion coefficient as determined from fluorescence recovery after photobleaching [3]. Six temperature/composition points were explored inside the lo/ld phase coexistence region: chol mole fraction $x_{chol} = 0.15$, 0.20, and 0.25, at temperatures $T = 30$ °C and 40 °C. The partition coefficients of the probes were determined from variation of steady-state intensity (*N*-Rh-14:0, 14:0 PE) or anisotropy (*N*-NBD-14:0,14:0 PE), revealing complementary phase preference (*N*-NBD-14:0,14:0 PE prefers the lo phase, whereas *N*-Rh-14:0, 14:0 PE prefers the ld phase). Donor decays (both in presence and absence of acceptor) were successfully analyzed, from a statistical point of view, using the formalism outlined above. For $x_{chol} = 0.25$, corresponding to high lo phase fraction X_{lo}, acceptor coefficient values calculated from FRET decay parameters using Eq. 13 compared well with those retrieved from variation of fluorescence intensity, indicating validity of the infinite-phase-separation assumption. However, for $x_{chol} = 0.15$ and 0.20, corresponding to lower X_{lo}, the FRET-derived K_{pA} values were closer to unity than the non-FRET size-independent values from fluorescence intensity variation. This behavior was also observed in the analysis of simulated FRET decay data obtained from numerical computation using a distribution of donor and acceptor probes in a lattice made up of small (~3.5–10 R_0) domains. The conclusion was that at variance with the high lo fraction limit, the lo domains distributed in majority ld phase have small size, of the order of a few nm. Support for this also comes from the application of Eqs. 16 and 17 to estimate the phase boundaries from FRET decay data. Whereas the pure lo phase boundary was estimated with excellent agreement with the published phase diagram ($x_{lo} = 0.28$ at $T = 30$ °C for both studies), this was not the case at all for the pure ld phase boundary ($x_{ld} = 0.18$ at $T = 30$ °C from FRET, compared with $x_{ld} = 0.075$ from the aforementioned phase diagram). The significance of this is that FRET is unable to detect phase separation in the $0.075 < x_{chol} < 0.18$ range because lo domains are very small (of the order of R_0 or smaller) for these compositions.

This study served as a blueprint from subsequent estimates of domain size in more complex systems by our group. In these studies, FRET efficiency is typically measured for several points along the phase coexistence tie-line. If the variation in E follows that predicted taking into account probe partition and assuming infinite phase separation, then formation of large domains (in the FRET scale) is inferred. Otherwise (e.g., if FRET between probes that have complementary phase preference fails to decrease upon entering the phase coexistence region), formation of small domains is deduced. Using the *N*-NBD-16:0,16:0 PE/*N*-Rh-18:1, 18:1 PE FRET donor/acceptor pair, de Almeida et al. [35] arrived at similar conclusions (small domains in the low X_{lo} range, large domains in the high X_{lo} range) for the 16:0,18:1 PC/*N*-palmitoylsphingomyelin (PSM)/chol raft-mimicking system, allowing rationalization of apparent discrepancies observed in the literature (e.g., 16:0,18:1 PC/chol heterogeneity was detected from size-independent fluorescence spectroscopy data [33, 89], but not from resolution-limited CFM [136]). On the other hand, Silva et al. [119] detected formation of large ceramide platforms in the 16:0,18:1 PC/palmitoylceramide system using FRET from ceramide gel-located *t*-PnA donor to fluid phase probe *N*-NBD-18:1,18:1 PE.

The approach of measuring the variation of FRET efficiency along a tie-line may also be used to assess the eventual perturbation induced by the addition of a foreign molecule to a given lipid mixture. The effect of physiologically relevant ceramide (Cer) concentrations ($\leq$4 mol %) on the lo/ld coexistence range of the POPC/PSM/chol system was also investigated [120]. For these systems, three donor/acceptor pairs were selected to obtain information regarding (1) ld-lo phase separation, i.e., alteration in lipid raft organization (*N*-NBD-16:0,16:0 PE/*N*-Rh-18:1,18:1 PE); (2) gel-lo phase separation, thus, the organization between the so-called Cer-platforms and lipid rafts (*t*-PnA/ *N*-NBD-16:0,16:0 PE); and (3) gel-fluid (ld + lo) phase separation (*t*-PnA/ *N*-NBD-18:1,18:1 PE). Data obtained for the first pair gave support to the hypothesis of ability of Cer to form gel domains only in the low chol range. Because in the presence of Cer, and in the low chol range, FRET efficiency increased, the inability of Cer to induce the coalescence of raft domains was concluded. Additional topological information of this complex system is obtained with the two other pairs, namely, the size of the gel domains. Because *t*-PnA (donor) has a strong preference toward Cer-enriched gel phases, while both acceptors (*N*-NBD-16:0,16:0 and *N*-NBD-18:1,18:1 PE) are excluded, when gel domains are formed, FRET efficiency decreases. Once again, this was observed for Cer-containing raft mixtures in the low chol range. Because the acceptors are completely excluded from Cer-gel domains, it is possible to estimate the size of the latter assuming that (1) FRET within gel phase does not occur because acceptors are excluded, (2) there is FRET from the gel to the fluid phase, (3) FRET occurs within the fluid phase. Furthermore, it is necessary to take into account the concentration of the donor in each of the phases (determined according to its partition coefficient), the amount of gel phase formed, and the existence of a gel-located donor/fluid-located acceptor exclusion distance, R_e, which was interpreted as an estimate of the gel domain size. It was found that Cer associates with PSM to form small, ~ 4 nm, gel domains. By noting that higher efficiencies were obtained for the *t*-PnA/*N*-NBD-16:0,16:0 pair compared to the *t*-PnA/ *N*-NBD-18:1,18:1 PE pair, and taking into account that *N*-NBD-16:0,16:0 PE prefers lo phase to ld, whereas *N*-NBD-18:1,18:1 PE does not discriminate between these two phases, it was concluded that Cer/PSM-enriched gel domains were surrounded by lo (rather than ld) phase. The foreign molecule may be a peptide or protein, as recently exemplified in a study of the effect of FAS death receptor's transmembrane domain on the domain size of the 16:0,18:1/PSM/chol phase diagram [21].

More recently, other authors have proposed methodologies aimed at the estimation of domain size using FRET. Towles and collaborators published two different analytical approaches recently [15, 132]. These formalisms rely on subtle (but restricting) approximations, which are discussed in detail elsewhere [81].

A convenient way to study the relationship between the FRET observables (donor decay, efficiency) and the domain organization of phase-separated bilayers is the use of numerical simulations. This has been done as a test of the abovementioned analytical formalisms [74, 79, 131, 132]. In these cases, the simulations were used to test the authors' analytical models and not with the intention of providing fitting equations. This is easily understood noting that there are too many input parameters (domain size and size distribution, domain shape, fraction of each phase, donor and acceptor partition coefficients) to be accommodated by a useful fitting scheme.

Recent simulation works have addressed the limits of FRET in the determination of domain sizes. Using stochastic simulations, Kiskowski and Kenworthy [66] calculated the dependence of FRET efficiency on acceptor concentration for a planar geometry with disklike domains. Two scenarios were considered: co-localization of donors and acceptors inside the domains and segregation of acceptors to the other phase with donors inside the domains. The authors showed that the local acceptor concentration inside the domains (and hence the domain fractional area) could be recovered from the first type of probe distribution, unlike the domain radius. However, the latter could be estimated from the acceptor segregation scenario, namely, for small and intermediate-sized domains. The authors chose to study the extreme situations of co-localization inside the domains (corresponding to domain/continuous phase partition coefficients $K_{pD} = K_{pA} = \infty$) and total acceptor segregation ($K_{pD} = \infty$, $K_{pA} = 0$). However, to model the effect of physical finite K_p values, simulations in which the probes are distributed taking them into account must be performed.

This was fully taken into account in a recent report by Šachl et al. [115]. These authors investigated, using numerical simulations, the feasibility of resolving domain sizes (judged by the resulting ratio between steady-state donor fluorescence intensity in the phase-separated system and that expected for uniform probe distribution) for a variety of K_{pD} and K_{pA} possible values, grouped into three categories: (1) donor/acceptor pairs reside inside lo nanodomains, (2) donor/acceptor pairs are excluded from lo nanodomains, and (3) donors and acceptors exhibit an increased affinity to the different phases. This latter situation (e.g., with lo/ld partition coefficients $K_{pD} = 5$ and $K_{pA} = 0.01$) was revealed as the most favorable, being able to resolve a broad spectrum of nanodomain sizes. However, even in this case, domains that occupy <2.5 % of the overall area and domains with radii $< R_0$ and occupying <10 % of the area were shown to lie beyond experimental resolution (unless K_{pD} increases to values not possible with current probes). On the other hand, currently available donor/acceptor pairs in which both probes prefer the same phase are not feasible for determination of domain sizes < 20 nm.

An altogether different approach was recently applied to the study of the Brain sphingomyelin (BSM) /16:0,18:1 PC/chol study based on a combination of FRET (between 1-[[(6,8-difluoro-7-hydroxy-4-methyl-2-oxo-2 H-1-benzopyran-3-yl)acetyl] oxy]- (Marina Blue-) and NBD-head labeled 16:0,18:1 PE) and statistical mechanical lattice Monte Carlo simulations [45]. For the purpose of FRET efficiency calculation from the simulations, the actual distance dependence of the FRET interaction was replaced by a step function, meaning that FRET was considered to occur if the donor-acceptor distance in a given pair is less than R_0 (4.6 nm). With this simplification and using unlike nearest-neighbor interaction parameters (the only potential fitting parameters in their Monte Carlo methodology) based on experimental data (and fine-tuned by comparison with the experimental FRET), the authors were able to observe extensive phase separation for BSM/chol/16:0,18:1 PC mole ratio 35:35:30. This agrees with the results of de Almeida et al. [35] described above, which indicate the existence of large ld domains in this lo-rich area of the phase coexistence range. No extensive phase separation is observed in the Monte Carlo simulation of either of the binary mixtures BSM/16:0,18:1 PC 70:30, chol/16:0,18:1 PC 70:30, and BSM/chol 50:50, which, as argued by the authors, agrees with the lack of observation by fluorescence microscopy in GUV of micron-scale phase separation in the binary sphingomyelin/chol, sphingomyelin/16:0,18:1 PC, and chol/16:0,18:1 PC systems, unlike some ternary mixtures of these components.

2.3 *Protein-Induced Lipid Distribution Heterogeneity*

The studies described in the previous subsections refer mainly to heterogeneity stemming from lipid-lipid interactions. However, proteins are ubiquitous in biological membranes and, as mentioned in Sect. 1, differential protein-lipid interactions may cause nonuniform distribution of lipid components in the bilayer. Several approaches described in the literature that use FRET to characterize this effect were critically reviewed recently [82]. Here we focus on the models proposed by our group and their applications to three different protein systems.

2.3.1 M13 Major Coat Protein

M13 major coat protein (M13 MCP) is the main component of the M13 bacteriophage coat and in its mature form is a polypeptide chain 50 amino acids long, presenting three domains which are expected to be required for the multiple interactions that this protein establishes during the bacteriophage reproductive cycle: (1) a single hydrophobic transmembrane segment of 20 amino acid residues, (2) an amphipathic N-terminal arm, (3) and a heavily basic C-terminus with a high density of lysine residues [49, 127]. ESR and fluorescence studies making use of site-directed labeling of MCP [124, 126] allowed to conclude that Thr36 is located in the center of the bilayer in 1,2-dioleoyl-*sn*-glycerol-3-phosphocholine (DOPC) and 1,2-dioleoyl-*sn*-glycerol-3-[phospho-*rac*-(1-glycerol)] (DOPG) bilayers, while aminoacids 25 and 46 delimit the transmembrane domain boundaries of MCP.

Insertion of M13 MCP in the bilayer milieu is expected to induce a packing stress at the protein-lipid interface and this stress is lessened by some adaptations of the protein [67]. However, the ability of M13 MCP to adapt to situations of hydrophobic mismatch is limited, and in extreme cases aggregation can occur [41]. In the case of complex lipid mixtures, it is energetically favorable to have a distinct lipid composition in the immediate vicinity of the protein that minimizes mismatch stress. Additionally, electrostatic effects can also drive enrichment of particular lipids around the protein. The lipid composition of the inner membrane of noninfected *Escherichia coli* is about 70 % of phosphatidylethanolamine (PE), 25 % of phosphatidylglycerol (PG), and 5 % cardiolipin (CL). During the infection of *Escherichia coli* by the M13 bacteriophage, the levels of anionic lipids in the cell membrane are slightly increased [109], suggesting that anionic phospholipids assist in the maintenance of a functional state for M13 MCP [49].

From ESR studies, it was known that some regularity exists in the number of immobilized lipids per protein transmembrane segment. The value recovered for this stoichiometry was 12, meaning that only the first layer of lipids around a transmembrane segment is expected to be significantly immobilized by interaction with the protein (assuming a hexagonal arrangement) [88]. Lipids in this layer are entitled annular lipids. Spin-labeled lipids found further away from this shell, although possibly affected by the presence of the protein segment, are not submitted to sufficient dynamic restriction to be detected by the ESR technique. ESR studies with M13 MCP were unable to detect significant immobilization of spin-labeled phospholipids by monomeric M13 MCP [150]. On the other hand, oligomeric MCP immobilized a population of phospholipids, especially at very high protein concentrations [104]. This suggests that a single transmembrane segment inserted in the membrane is unable to sequester a long living lipid shell around it, at least in the ESR timescale.

The hydrophobic surface of a membrane protein is not smooth and the interface between the protein and the lipids surrounding it is likely to be heterogeneous [71]. On the other hand, the largely fixed stoichiometry for annular lipids denotes some ordering in the protein-lipid interface. In this way, it is possible to describe this annular shell in terms of a uniform surface for which 12 identical binding sites are available [88], and the process has been described as competitive binding of lipids to the protein surface binding sites [71, 98, 111, 144].

FRET experiments assuming this model were applied to the study of the affinity of M13 MCP to different lipid classes and chain lengths [42]. The M13 MCP behavior in the membrane is extremely well resolved, especially in regard to its positioning in the lipid environment, and for that reason M13 MCP was a particularly attractive subject for the application of FRET as recovering quantitative information for protein-lipid selectivity requires some degree of parameterization. For the experiments, a M13 MCP mutant for which the Thr36 (located in the center of the bilayer) was changed to a cysteine was specifically labeled at this position with a coumarin fluorophore. Protein was then incorporated in liposomes composed of unsaturated phosphocholines of different thickness, loaded with different concentrations of *N*-NBD-18:1,18:1 PE, which acts as a FRET acceptor for coumarin, the FRET donor. FRET efficiencies were calculated from the integration of the donor fluorescence decays and the data were analyzed on the basis of a model which assumed two populations of energy transfer acceptors, one located in the annular shell around the protein, whose composition is determined by protein-lipid interactions, and the other outside it, with a random distribution unaffected by the protein. Due to the large R_0 of the coumarin and NBD pair (~39.3 Å), the donor fluorescence decay curve had energy transfer contributions from both of these acceptor populations:

$$i_{DA}(t) = i_D(t)\rho_{\text{annular}}(t)\rho_{\text{random}}(t) \tag{18}$$

here, ρ_{annular} and ρ_{random} are the FRET contributions arising from energy transfer to annular labeled lipids and to randomly distributed labeled lipids outside the annular shell, respectively. All annular acceptors were assumed to be at the same distance (*d*) to the coumarin fluorophore in the center of the transmembrane domain, and FRET to each of these acceptors was associated with the rate constant given by Eq. 1, with $R = d$. The probability of each of the 12 annular sites to be occupied by an acceptor depends on the acceptor molar fraction and on a relative selectivity constant (K_S) which quantifies the relative affinity of the labeled to unlabeled phospholipids:

$$\mu = K_S \frac{N_{\text{Acceptor}}}{N_{\text{Acceptor}} + N_{\text{Unlabeled lipid}}} \tag{19}$$

where N_{Acceptor} is the concentration of labeled lipid and $N_{\text{Unlabeled Lipid}}$ is the concentration of unlabeled lipid. A binomial distribution describing the probability of each occupation number assuming a given μ (0–12 sites occupied simultaneously by labeled lipid) is considered for the calculation of ρ_{annular}:

$$\rho_{\text{annular}} = \sum_{n=0}^{12} e^{-nk_T t} \binom{12}{n} \mu^n (1-\mu)^{12-n} \tag{20}$$

On the other hand, the FRET contribution arising from energy transfer to non-annular lipids, ρ_{random}, was dictated by Eq. 8, where R_e was calculated from the expected exclusion distance between the protein and lipids outside the annular shell (sum of protein and 1.5 lipid radii to account for exclusion effect of annular lipids) and the separation between donor and acceptor planes in the bilayer. NBD fluorophore position in the membrane is known to be close to the surface for phospholipid labeling in both headgroup and acyl-chains and has been determined through several techniques [1, 23, 75, 91].

The value for NBD-labeled lipid concentration outside the annular region (*c*) was also corrected for the presence of acceptors inside the annular region. Eqs. 1, 8, 18, 19, and 20 were then used to simulate donor decay curves under FRET, and FRET efficiencies were calculated from numerical integration of these curves (Eq. 9). During fitting of this model to experimental data, the only variable was K_S. This model was applied to the analysis of protein preference for acyl chain thickness and headgroup selectivity. In the first set of experiments, the same headgroup-labeled lipid *N*-NBD-18:1,18:1 PE, with perfect hydrophobic matching to the M13 MCP, was added to proteoliposomes presenting different lipid membrane thickness and coumarin-MCP. Coumarin-MCP quenching due to FRET to NBD-labeled lipids was measured and the selectivity model was fitted to the data. Different K_S values were recovered for *N*-NBD-18:1,18:1 PE depending on the thickness of the bulk lipid used. M13 MCP presented higher affinity for *N*-NBD-18:1,18:1 PE when there was considerable mismatch between the protein and the bulk lipid, reflecting an enrichment of the hydrophobically equivalent lipid in the annular shell of the protein, minimizing in this manner the hydrophobic mismatch stress in the protein-lipid interface.

The same method was used to determine the selectivity of the protein for different phospholipid headgroups in a second set of experiments. This time, different acceptors were used and all studies were carried out in 18:1,18:1 PC. The probes used as acceptors were phospholipids of identical acyl-chains (18:1 and 12:0) belonging to different phospholipids classes (PE, PC, phosphatidylglycerol (PG), phosphatidylserine (PS), and phosphatidic acid (PA)) labeled with NBD at the shorter acyl-chain (18:1,NBD-12:0 PE, PC, PG, PS, and PA). Larger K_S values were recovered for anionic labeled phospholipids, particularly for PA ($K_S = 3.0$) and PS probes ($K_S = 2.7$). The PG lipid presented an intermediate affinity for the annular shell of lipids ($K_S = 2.3$) whereas PC and PE probes induced less energy transfer reflecting lower K_S values ($K_S = 2.0$ for both). Selectivity for anionic phospholipids is a consequence of electrostatic interaction of these with the lysine-rich C-terminal domain of the protein. Even though the protein is shown to present higher affinity for the labeled lipid than for the bulk lipid ($K_S > 1$), possibly as a result of electrostatic interactions with the NBD group, this method was further validated by the fact that the relative association constants $[K_S/K_{S(\text{PC labeled lipid})}]$ obtained were almost identical to the values obtained with ESR and the aggregated form of the protein [104].

In this experiment, as a result of the low selectivity character of the M13 MCP protein–annular lipid interaction and the large coumarin-NBD R_0, FRET contribution from noninteracting acceptors clearly dominates over annular acceptor contribution. This significantly reduced the sensitivity of the method, and nearly error-free measurements of FRET efficiency and acceptor concentration are required for the recovery of accurate selectivity constants. This limitation can be bypassed through the use of donor-acceptor pairs presenting lower R_0 for which FRET due to acceptors in the annular shell will become more predominant.

2.3.2 Lactose Permease

Lactose permease (LacY) of *Escherichia coli*, one of the most intensively studied membrane proteins, is often taken as a paradigm for secondary transport [54]. It consists of 12 transmembrane α-helices, crossing the membrane in a zigzag fashion. LacY translocates the substrate (specifically disaccharides containing a D-β-galactopyranosyl ring) with H^+, in a symport (cotransport) reaction. It was pointed out early [24] that the amino groups of phospholipids such as PE are of crucial importance for LacY function. More recently, it has been conclusively demonstrated that LacY requires the presence of PE for its correct folding in the membrane during biogenesis [9], its function in vivo, [8] and to maintain its correct topology [10, 11]. In particular, LacY is fully functional when reconstituted in mixtures of 16:0,18:1 PE/16:0,18;1 PG. FRET experiments were used to elucidate whether one of these phospholipids is enriched in the annular region of LacY [108]. W151 of LacY was used as donor and two different pyrene-labeled phospholipids (a PG analog, 16:0, pyrene-10:0 PG (Pyr-PG), and a PE analog, 16:0, pyrene-10:0 PE (Pyr-PE)), were chosen as acceptors. Additionally, the effect of cardiolipin (CL) on the annular lipid composition was also investigated.

FRET efficiencies were analyzed essentially as described for M13 MCP, with the difference that, due to the size of the protein, the number N of annular sites available is now much larger (46 (23 in each leaflet) instead of 12; [108]). The donor LacY W151 was assumed to be located in the axis of the cylindrically symmetrical protein, near the membrane interface. All acceptor fluorophores were assumed to be located near the center of the bilayer. In this way, the distance between donor and annular acceptor molecules was taken as identical for all annular acceptors, independently of their membrane leaflet. Using l =1.2 nm as the transverse distance between W151 and the acceptor plane and $R_e = 3.0$ nm as the exclusion distance along the bilayer plane between the protein axis and the annular lipid molecules, the donor–annular acceptor distance is given by $R = (l^2 + R_e^{\,2})^{1/2} = 3.2$ nm. For the Förster radius, the value $R_0 = 3$ nm, reported for the Trp/pyrene pair [130], was used, whereas for the calculation of n_2, area/lipid values of 0.56, 0.56, and 1.26 nm^2 were assumed for 16:0,18;1 PE; 16:0,18;1 PG; and CL, respectively [55, 112].

Experimental FRET efficiencies between the single Trp151 of LacY and either Pyr-PG or Pyr-PE as acceptors were measured in the different lipid systems (16:0,18:1 PE/16:0,18:1 PG; 18:1,18:1 PE/16:0,18:1 PG; and 16:0,16:0 PE/16:0,18:1 PG, all 3:1) at 37 °C, in order to recover the best fit values of the model parameter μ. This parameter represents the probability of finding a given phospholipid in the annular region of LacY. The fact that higher FRET efficiencies are obtained for transfer to Pyr-PE (0.232 ± 0.028) compared to Pyr-PG (0.165 ± 0.018) in the 16:0,18:1 PE/16:0,18:1 PG system, and, in a similar way for Pyr-PE (0.231 ± 0.023) compared to Pyr-PG (0.211 ± 0.038) in the 18:1,18:1 PE/16:0,18:1 PG system, is a first indicator of the selectivity of LacY for PE relative to PG in these systems. This is confirmed by the quantitative model calculations. Best agreement with experimental values requires an annular region composed of approximately ~90 mol% PE in these systems, whereas 75 mol% would be expected for random distribution of both phospholipids. In the 16:0,18:1 PE/16:0,18:1 PG mixture, the FRET data are compatible with complete PG exclusion from the annular layer, which is therefore composed solely of PE (μ(PE) = 1.00, μ(PG) = 0.00). Notably, when LacY is reconstituted in 18:1,18:1 PE/16:0,18:1 PG, the experimental FRET efficiencies indicate an enrichment of PG in the annular region (μ(PE) = 0.86, μ(PG) = 0.14), but still in lower concentration than expected for uniform lipid distribution. Regarding the 16:0,16:0 PE/16:0,18:1 PG mixture, gel/fluid phase coexistence is expected, with 16:0,18:1 PG-enriched fluid domains coexisting with 16:0,16:0 PE-enriched gel-phase bilayer regions. In this system, an increase in the efficiency of FRET to Pyr-PG and a decrease in that to Pyr-PE are verified, to an extent that the efficiency of FRET to Pyr-PG now clearly surpasses that to Pyr-PE. This is a clear indication that LacY is preferably located in the fluid domains, where the PG acceptor probe is more abundant.

In addition to PE and PG, the lipid composition of *E. coli*'s membrane contains 5 % to 7 % of CL. To test the effects of this phospholipid on the annular region, 14:0 CL and 18:1 CL were incorporated in the 16:0,18:1 PE/16:0,18:1 PG matrix. Incorporation of CL decreases the efficiencies of FRET when comparing to the same phospholipid mixtures containing no CL, especially when the acceptor is Pyr-PG. This suggests that CL displaces 16:0,18:1 PE and, more extensively, 16:0,18:1 PG from the annular region of LacY. The fact that the effect is more pronounced for PG than for PE is probably related to the preference of the protein for PE species as described above for the binary systems. Upon applying the FRET quantitative model, when the acceptor is Pyr-PG, even by imposing segregation of this probe from the first annular layer ($\mu(\mathrm{PG}) = 0$), it is still not possible to conciliate the theoretical (0.162) and the experimental values (0.143 for 18:1 CL, 0.142 for 14:0 CL). This indicates that besides being totally excluded from the first layer, PG is also somewhat rarefied beyond it. On the other hand, when the acceptor is Pyr-PE, a model matching to the experimental efficiencies (0.183 for 18:1 CL, 0.196 for 14:0 CL) requires only partial replacement of PE by CL. When the CL lipid is 18:1 CL, the retrieved composition of the annular layer is 40 mol% PE and 60 mol% CL. On the other hand, when the CL lipid is 14:0 CL, the composition of the annular layer is 68 mol% PE and 32 mol% CL, indicating that in this case PE is kept in close proximity of the protein, in the same proportion as in the bulk. In the latter case, CL enrichment in the annular layer is solely produced by replacing PG. The fact that 14:0 CL is not able to displace PE in the same way that 18:1 CL does is probably due to the hydrophobic mismatch between the short 14:0 acyl chains and the protein. On the whole, this study confirms that PE is the most relevant component of the annular region and that, because it is not displaced by PG or (completely) by CL, it appears to be tightly bounded to LacY. Selectivity of LacY for PE and predominance of this phospholipid at the annular region, verified and characterized by FRET measurements and modeling, provide support for a hypothetical coupling between this lipid and LacY during the transport cycle.

The work described above was carried out mostly in PE/PG 3:1 mixtures, a ratio identical to that found in the inner membrane of *E. coli* [40]. This creates a dilution problem. For example, in an experiment where the acceptor is labeled PE, even if the annular region would be solely comprised of PE lipid, the enrichment of labeled PE in this layer would be only of a factor 4/3. Adding to the fact that unspecific FRET to acceptors outside the annular layer is always present, this would imply a rather modest increase in the expected FRET efficiency. Finally, the simple fact that a pyrene acyl chain–labeled lipid behaves identically to unlabeled lipid of the same class is questionable, and this cannot be resolved in an experiment where the host lipid matrix is a two-component mixture. For these reasons (to gain increased sensitivity and to assess the extent of correct reporting by the acceptor probes of each class), this protein system was readdressed in a recent work [129], using different one-component host lipid matrices. Besides, the influence of headgroup and acyl chain composition was also investigated.

In Table 1, the experimental FRET values are listed along with the calculated μ and K_s values. As can be seen, the probability of finding labeled phospholipids at

Table 1 Probabilities of each site in the annular ring being occupied by a pyrene-labeled phopholipid and relative association constant toward LacY, as determined by FRET. Reprinted from Suárez-Germà et al. [129] with permission. Copyright 2012 American Chemical Society

		16:0,18:1 PE matrix			16:0,18:1 PG matrix		
	Acceptor	μ	K_s	K_s/K_s(PE)	μ	K_s	K_s/K_s(PE)
25 °C	Pyr-PE	0.10	6.53	1.00	0.03	2.00	1.00
	Pyr-PG	0.00	0.00	0.00	0.00	0.00	0.00
	Pyr-PC	0.00	0.00	0.00	0.00	0.00	0.00
37 °C	Pyr-PE	0.08	5.53	1.00	0.04	2.47	1.00
	Pyr-PG	0.00	0.00	0.15	0.00	0.00	0.00
	Pyr-PC	0.02	1.40	0.25	0.03	2.00	0.81

the annular regions (μ) is always the highest, irrespectively of the matrix, for Pyr-PE. This behavior reflects the values of the FRET efficiency mentioned above. By inspecting the outcome for K_s values, we notice that the largest values are obtained for Pyr-PE in all matrices and both temperatures. It is worth mentioning that ideally $K_s = 1$ for any probe that mimics the non-labeled phospholipids and values between 1 and 3 have been reported (as measured for FRET between labeled M13 MCP and NBD lipids as described above; [42]). Therefore, these high values of K_s obtained for Pyr-PE may indicate either an annular region extremely enriched in the label or that Pyr-PE does not mimic well the unlabeled phospholipid. Although this may be a handicap, if one compares across probes in the same host lipid, it becomes clear that there is an effect of preference of Lac Y for PE over PG and PC. Since the probes are all equal except for the headgroup, and for comparing the different probes in the same host lipid, K_s/K_s(PE) ratios are provided in Table 1.

In the 16:0,18:1 PE matrix, μ values indicate that Pyr-PG is excluded at both temperatures and that 16:0, pyrene-10:0 PC (Pyr-PC) is excluded at 25 °C and shows a small enrichment at 37 °C. Since K_s =0 means no acceptor in the annular region, it becomes clear that at 25 °C, Pyr-PG and Pyr-PC are completely excluded. While Pyr-PG behaves in the same way at 37 °C, LacY shows an increased preference for Pyr-PC at this temperature. The overall results in the 16:0,18:1 PE matrix, where LacY is folded closely to the in vivo conditions, point to the fact that Pyr-PE should be in closer proximity than the other labels. On the other hand, μ and K_s for Pyr-PE in the POPG matrix are compatible with a moderate enrichment of the label in the annular region. Pyr-PG is depleted from the annular region at both temperatures when the host phospholipid is 16:0,18:1 PG. Similarly, we can observe that Pyr-PC is also depleted when hosted by 16:0,18:1 PG at 25 °C and that a very slight enrichment is observed at 37 °C. All these observations may be likely related with the inverted topology of domains C6 and P7 of LacY when reconstituted in POPG proteoliposomes [12]. Our FRET measurements in 16:0,18:1 PE and 16:0,18:1 PG matrices confirm the preference of LacY for PE and its probable predominance in the annular ring [108]. This may support, indirectly, a hypothetical interaction between the PE headgroup and some specific residue of the protein [24, 123]. Importantly, recent observations have shown that uphill transport occurs in *E. coli* in which PE has been completely exchanged by PC [13]. Since in PE and PC matrices LacY exhibits its natural topology, this intriguing observation pointed out to a more complex molecular interaction between the protein and the annular phospholipids. Hence FRET measurements in PC matrices become of interest given the fact that despite its natural topology in these matrices, LacY only shows downhill transport in 18:1,18:1 PC proteoliposomes [142]. Pyr-PG is slightly enriched in the annular region when the matrix is 18:1,18:1 PC ($K_s > 1$) but excluded from it in a 16:0,18:1 PC matrix ($K_s \sim 0$). However, the most interesting result is possibly that, according to the K_s values, Pyr-PC is enriched in the annular region in a 16:0,18:1 PC matrix ($K_s > 1$) and excluded from it in a 18:1,18:1 PC matrix ($K_s \sim 0$).

Given that 18:1,18:1 PC and 16:0,18:1 PC share the same headgroup and have very similar hydrophobic lengths in the bilayer (2.48 vs. 2.54 nm, respectively; [123]), this difference is probably related to the different specific curvature of the two lipid species. It has been reported that whereas proper topology of LacY depends on a dilution of high negative surface charge density (and hence probably the decreased affinity of the protein for PG), rather than on spontaneous curvature (C_0; [11]), the latter appears to be crucial regarding uphill transport of lactose by LacY in vivo [143], with negative curvature lipids like PE being required. C_0 (16:0,18:1) is essentially zero, while 18:1,18:1 PC, due to its additional unsaturated acyl chain, has a negative specific curvature (C_0(18:1,18:1 PC) $= -0.11$ nm^{-1}; [123]). Although its value is still far from the non-bilayer lipid 18:1,18:1 PE (C_0(18:1,18:1 PE) $= -0.35$ nm^{-1}; [123]), it may justify the preference of properly reconstituted LacY for 18:1,18:1 PC rather than 16:0,18:1 PC, and hence the differential behavior in these two PC matrices regarding selectivity for labeled probes.

2.3.3 Pulmonary Surfactant Protein SP-B

Pulmonary surfactant is composed of roughly 90 % lipids and 8–10 % of specific surfactant-associated proteins, termed in chronological order of discovery SP-A, SP-B, SP-C, and SP-D [61]. The most abundant phospholipid species in surfactant is 16:0,16:0 PC, which is also the main surface-active component [26, 97]. However, it is clear that surfactant lipids are not able by themselves to reach rapidly the air/liquid interface as they are secreted by type II pneumocytes, to form operative surface-active films. Presence of hydrophobic surfactant proteins SP-B and SP-C is strictly required to facilitate an efficient transfer of phospholipids from surfactant stores (in the form of bilayers) at the aqueous hypophase into the interfacial film, along the breathing cycles [105, 106]. However, the molecular mechanisms by which surfactant proteins participate in the assembly, transport, and reorganization of surfactant lipids at the respiratory surface are still not well understood.

Mature SP-B is a 79-residue polypeptide, which comes from the proteolytic maturation of a longer precursor of 381 amino acids (42 kDa) produced in type II cells. SP-B processing is coupled with the assembly of surfactant membranes into lamellar bodies, the specific surfactant-storing organelles in pneumocytes. In the native form, SP-B contains intramolecular and intermolecular disulphide bonds stabilizing a homodimeric structure [62]. The three-dimensional structure of SP-B has not been yet determined.

The essential role of SP-B in surfactant has been related with the ability of the protein to promote a rapid transfer of phospholipids into air-water interfaces. SP-B, therefore, could be required to establish an operative surface active film from the earliest air-liquid respiratory interface. SP-B has been described to promote more or less deep perturbations in membranes leading to lipid exchange and eventual fusion between liposome membranes and rapid leakage of their content [100, 110, 114]. The elucidation of the mode and extent of interaction of SP-B with surfactant membranes and films is therefore important to fully understand its structure-function relationships. The in-depth location and orientation of SP-B in phospholipid membranes as reconstituted in vitro has been a matter of controversy. On the one hand, evidence suggests that SP-B is located in a shallow region of membranes, with the polar positively charged sides of the helical segments interacting with anionic phosphatidylglycerol (PG). This superficial lipid-protein interaction would produce little perturbations on the acyl chain packing of surfactant phospholipids [94, 99, 133]. Other experiments, in contrast, have reported significant effects of SP-B on acyl chain order, consistent with a deeper penetration of SP-B in membranes and a direct perturbation by the protein of their hydrophobic core [39, 107, 118]. On the other hand, the extent of perturbation by SP-B of the structure and thermotropic properties of phospholipid membranes resulted to be critically dependent on the method used to reconstitute the lipid/protein complexes [28]. A matter of discussion has also been the occurrence of selective interactions between SP-B and anionic phospholipids in surfactant, such as phosphatidylglycerol (PG). Inclusion of different phospholipid spin probes in membranes allowed to determine that SP-B shows a preferential interaction with PG in membranes, compared with other zwitterionic or anionic species, as analyzed by electron spin resonance (ESR) spectroscopy [29, 107]. Other studies, however, have suggested that SP-B may prefer to partition into DPPC-enriched rather than DPPG-enriched regions in interfacial films, as detected by TOF-SIMS analysis of lipid/protein films transferred onto solid supports [14, 116].

Fluorescence from the single tryptophan (W9) in the sequence of SP-B was used to get further insight on the location, orientation, and structural dynamics of SP-B in different membrane environments, as well as on the existence of possible selective interactions between SP-B and particular phospholipid species, using FRET to NBD-labeled phospholipid probes [20]. Two different theoretical models could describe the two main possible protein arrangements in the membrane.

In model I (Fig. 5a), a deep embedment of the protein in the membrane would create an area of exclusion of phospholipid molecules equivalent to the surface taken by the protein. A topologically equivalent situation would be that originated if the protein dimer could span the whole bilayer thickness, as proposed in certain models [147]. Donor quenching by FRET in this first model could arise from two distinct acceptor populations: one located in a single circular layer of annular lipid surrounding the protein, and another uniformly distributed beyond the annular region.

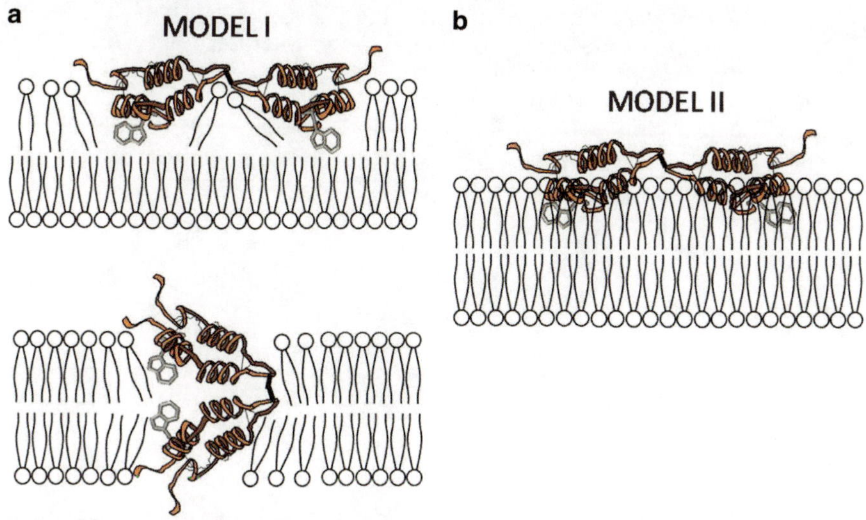

Fig. 5 Theoretical molecular models for the topology of SP-B in phospholipid membranes. *Panel a* represents model I, where the protein insertion leads to an area volume exclusion in the membrane occupied by the protein instead of lipids (and therefore represents an exclusion area for FRET acceptors). *Panel b* represents model II, where the protein exhibits a shallow interaction with the membrane surface, with no lateral lipid excluded area. Single FRET Donors (Trp residues) per SP-B monomer are indicated (Reprinted from Cabré et al. [20] with permission. Copyright 2012 Elsevier)

The formalism for this molecular arrangement is readily adaptable from the one used in the previous examples. However, it was verified that the theoretical efficiency obtained with this formalism is very low when compared with the experimental measurements for reasonable values of the model parameters. For this reason, this model was discarded in the quantitative analysis of lipid selectivity.

Model II (Fig. 5b) assumes that the protein adsorbs to the membrane surface. FRET can again occur to two distinct acceptor populations, one located directly below the protein (where the possibility of acceptor enrichment due to protein-lipid selectivity is considered) and another, located beyond this region (bulk bilayer). Derivation details of this model are presented elsewhere [20]. Despite the difference in topology, the main fitting parameter in this formalism is again a selectivity constant K_S, which is a measure of the preference of the protein for a given acceptor to be located underneath, over the host lipid.

The experimental FRET efficiencies were obtained in these experiments from the measured quenching of donor fluorescence, in 50 mM Hepes buffer, pH 7.0, in the absence of salt or in the presence of 150 mM NaCl. Fig. 6a illustrates how at low salt content, FRET from SP-B tryptophan is less efficient toward zwitterionic NBD-PC than to the negatively charged species NBD-PG or NBD-PS. Similar

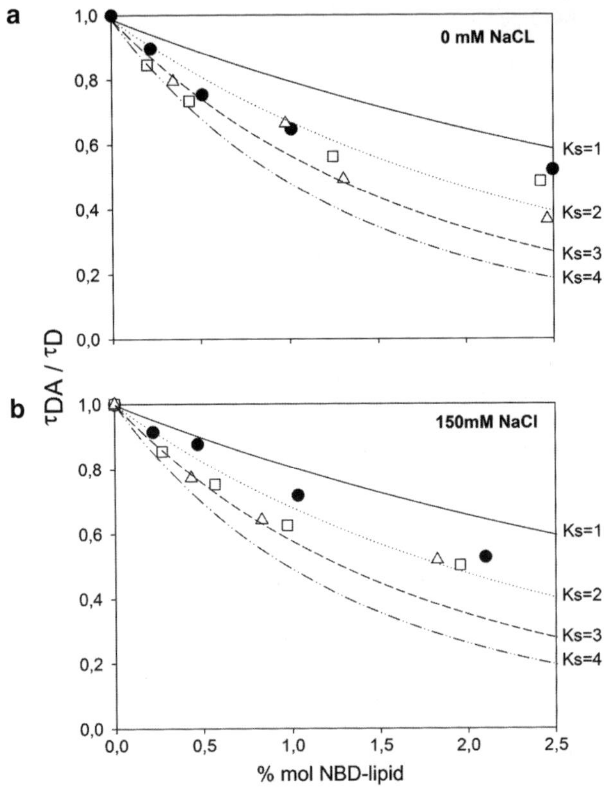

Fig. 6 Donor (SP-B tryptophan) fluorescence quenching by FRET acceptor (NBD-lipids) in a 16:0,18:1 PC membrane matrix. Experimental time-resolved FRET data have been obtained in 50 mM Hepes buffer, pH 7.0, at low (0 mM NaCl, *panel a*) or physiological (150 mM NaCl, *panel b*) ionic strength. Acceptors were NBD-PC (*closed circles*), NBD-PG (*squares*), or NBD-PS (*triangles*). Lines are the theoretical curves for the different indicated selectivity constant K_s values, which consider the topology of the protein in the membrane (model II; see Fig. 3) (Reprinted from Cabré et al. [20] with permission. Copyright 2012 Elsevier)

trends are observed at physiological ionic strength (Fig. 6b). In contrast to the data previously reported using spin-labeled lipids [107], significant differences in selectivity between PG and PS could not be detected. The experimental data obtained for FRET from SP-B to NBD-labeled lipids have been compared in Fig. 6 with the theoretical behavior expected considering a random distribution of donor and acceptor probes in the membranes. As shown in Figs. 6a and b, the selectivity parameter $K_s = 1$ plots still predict less FRET efficiency than measured experimentally, even for the zwitterionic probe NBD-PC, and some degree of probe preference must be invoked (the experimental data are in fact closer to the $K_s = 2$ curves), similarly to the previous examples.

3 Microscopy Studies Using FRET to Probe Lateral Heterogeneity

Recent developments in multiwavelength and polarization-resolved imaging have led to a widespread use of FRET imaging in studies of functional assemblies in cell membranes. The experimental methods for visualizing membrane microdomains and quantifying FRET efficiencies in FRET microscopy with emphasis on novel strategies have been reviewed elsewhere [60, 101, 103, 113]. Several approaches were developed in order to explore, on the nanoscale range, specific protein-protein, lipid-lipid, or lipid-protein interactions in live cells, both using homo- and hetero-FRET.

Cell membranes are characterized by a large number of lipid and protein components in a nonequilibrium state. One common simplification is to assume two types of domains, e.g., raft/non-raft or ordered/disordered. The results can then be compared to, e.g., the ld/lo coexistence on a lipid phase diagram in a ternary model system. Due to intrinsic limitations such as cell stability, and because usually in cells microscopy studies are carried out (in order to control cell state, to know the fluorophore localization, and use the signal coming only from the membrane of interest) fluorescence intensity decays with a high number of photons and low background signal (necessary to the applications of most of the formalisms described above) are generally unfeasible. Usually, steady-state data is obtained and compared to an integrated FRET formalism. Even when Fluorescence Lifetime Imaging Microscopy (FLIM; see Stöckl and Herrmann [125] for a review of its applications to membrane heterogeneity) lifetime data is obtained (FRET-FLIM), a relatively low number of counts is often obtained, which implies that the decay is traditionally used to calculate FRET efficiency using Eq. 9, rather than being directly analyzed with the underlying FRET kinetic model. However, with instrumental improvements as well as development of novel analysis approaches [53], this trend is being reversed. Selected works combining FRET and microscopy are listed in Table 2, which succinctly describes illustrative literature reports in which FRET was used in characterization of membrane domains, protein/lipid selectivity, or protein oligomerization.

In addition to the usual advantages of time-resolved methodologies such as independence on local probe (donor) concentration, under the microscope FRET-FLIM is a method of choice, since it is much less affected by artifacts, and direct information on the FRET efficiency is obtained, without the more complex approaches used in steady state, such as the so-called filter cube FRET microscopy. Additionally, this one also relies on FLIM for the determination of correction factors related to intensity determinations. One example of a relevant FRET-FLIM application, even in the framework of a low time-resolution approach (time-gated detection),

Table 2 Selected examples of FRET microscopy membrane studies

[65]	Clustering of the GPI-anchored protein 5′ nucleotidase was not detected using FRET between labeled antibodies
[135]	Clustering in domains with less than 70 nm of a glycosylphosphatidylinositol (GPI) anchored protein at the cell surface detected by homo-FRET
[56]	FRET-FLIM study of the raft-dependent interaction of tetanus neurotoxin with Thy-1
[57]	Suggestion of preferential interaction of phospholipase D with PC, rather than PE – a qualitative FLIM study
[117]	Characterization of size of lipid-dependent organization of GPI-anchored proteins in live cells, using homo- and hetero-FRET
[141]	FRET-FLIM revealed interaction between BACE (β site of amyloid precursor protein-cleaving enzyme) and the LDL receptor–related protein occurring on lipid rafts at the cell surface
[2]	Improved model for analysis of FRET adapted to the case where D and A label two probing proteins. Application to the data of [65] gave quantitative support to the presence of lipid rafts
[93]	Quantitative study of the distribution of functional neurokinin-1 receptors in the plasma membrane. The receptors are found to be monomeric and reside in membrane microdomains of size below optical resolution
[5]	Derivation of a model considering intramolecular and/or intermolecular FRET and oligomerization, and its experimental verification. Discussion of the effect of cell fixation.
[47]	Cortical actin activity regulates spatial organization of nanoclusters of GPI-anchored proteins at the cell surface, as shown by homo-FRET
[50]	FRET-FLIM revealed that ganglioside GM1 co-localizes with EGF receptor, but not with the non-raft transferrin receptor

is the study related to the localization of the epidermal growth factor receptor (EGFR) in membrane rafts [50]. These authors observed a decrease on fluorescence lifetime of the receptor derivatized with fluorescent antibodies (Alexa 488), in the presence of the typical raft marker GM1(labeled with cholera toxin-A594; Fig. 7). The study was carried out with the two fluorescent antibodies (Fig. 7a, b) in order to discard any specific effect that could be invoked regarding the first antibody, and similar FRET efficiencies were observed in both cases. The result depicted in Fig. 7c is the negative control of the previous experiments. In this case, the labeled donor is the transferrin receptor (Tf-A488), which is known not to partition into rafts, and this rules out the hypothesis that in the previous experiments, the known artifact of GM1 clustering was biasing the data. Interestingly, no significant interaction of EGFR with another well-known raft marker, GPI (glycosylphosphatidylinositol), in this case with GFP (green fluorescent protein), was observed (Fig. 7d). It should be stressed that FRET is essential to clearly demonstrate interaction at the molecular scale, since pixel co-localization cannot be used due to the restricted low lateral resolution of conventional microscopy, as compared to the molecular range of distances.

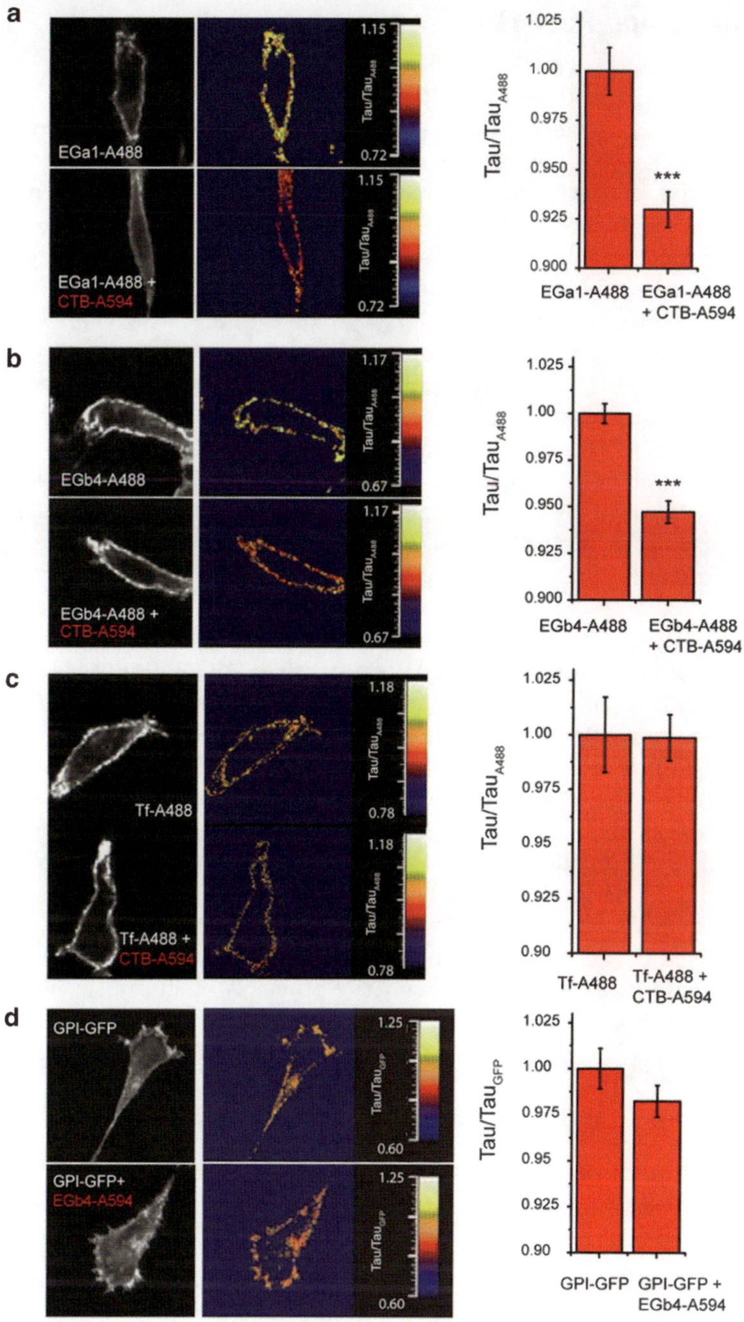

Fig. 7 Nanoscale co-localization of EGFR with GM1 gangliosides from FRET-FLIM data. (**a–c**) HER14 cells grown on coverslips were incubated on ice with 100 nM (**a**) of the donor probes anti-EGFR nanobody EGa1 or (**b**) EGb4 directly conjugated to Alexa-Fluor-488 or with (**c**) 20 μg/ml

4 Concluding Remarks

In this chapter, applications of FRET in membrane biophysics are described, comprising studies of lateral heterogeneity (membrane domains) and determination of protein/lipid selectivity (preference of a specific lipid for the protein vicinity). The complexity of FRET in membranes was addressed, and detailed topological information can be obtained from this methodology, once adequate modeling is taken into account. Examples of relevant works in this area are described. On the whole, the power of FRET as a tool in the characterization of membrane component lateral distribution is emphasized.

FRET combines the unique sensitivity of fluorescence with a steep dependence on distance and local concentration, which in turn is reflected on an ability to detect heterogeneity on a length-scale unavailable to other techniques. However, to make the most of FRET's potential, some points should be taken into account. Simulation studies described in this chapter have revealed that for optimal characterization of domain structure, probe partition between the coexisting domains should be complementary. In this way, when designing a FRET experiment for investigation of domain structure, it is important to consider/anticipate the nature of the coexisting phases and select them accordingly. Most probes prefer fluid, disordered phases (e.g., Rh-PE, NBD-PC, *N*-NBD-PE with short or unsaturated acyl chains). A good gel-phase probe is *t*-PnA, whereas head-labeled NBD-PE probes with long saturated acyl chains are convenient lo reporters. Therefore, *t*-PnA/NBD-PC (gel/ld), *t*-PnA/*N*-NBD-PE (gel/ld, gel/lo), and *N*-NBD-PE/Rh-PE (lo/ld) constitute useful FRET pairs.

Regarding domain detection/characterization, different applications can be envisaged, ranging from the identification of the onset of phase separation/construction of previously unknown phase diagrams to the estimation of domain organization in systems where the underlying phase diagram is known. Naturally, pertinent to this is the question addressed in Section 1 of what constitutes a phase. As illustrated in Sect. 2.2, regarding both binary and ternary mixtures exhibiting ld/lo phase coexistence, FRET cannot detect extremely small domains which may be sensed by distance-independent spectroscopic techniques (such as variation of fluorescence lifetime, quantum yield or anisotropy along a tie-line). However, even in this case, FRET is useful, as it allows concluding that the domains perceived by other techniques have necessarily very small size (~R_0 or smaller). On the other

Fig. 7 (continued) transferrin-A488 (Tf-A488) in the absence or presence of 1 μg/ml of the acceptor probe CTB-A594. (**d**) HER14 cells expressing GPI-GFP were incubated (GPI-GFP + EGb4-A594) or not (GPI-GFP) for 1 h on ice with 100 nM EGb4-A594 (acceptor). After fixation with 4 % formaldehyde, average fluorescence lifetime values of GFP were determined. *Left panels* represent the distributions of the donor probes with or without acceptor probe. The lifetimes are shown in the *middle panels* in false colors. The histograms on the right are mean fluorescence lifetimes of the probes. See text for details (Reprinted from Hofman et al. [50] with permission. Copyright 2008 The Company of Biologists Limited)

hand, whereas the formalisms presented allow, in favorable cases, estimation of average domain size, polydispersivity and shape cannot be normally resolved.

Generally, measurement of donor decay in absence and presence of acceptor is preferred to measurement of FRET efficiency from steady-state. Despite being a rapid way of obtaining a first glimpse into the possible membrane organization, the steady-state FRET detection is always limited. The FRET efficiency parameter results from time integration of the actual decay law, and in this process considerable detail regarding the probes distribution is lost. Thus, whenever possible, a complete parameter analysis of the time-resolved fluorescence donor decays should be carried out (ideally with global analysis of samples with and without acceptor. Moreover, time-resolved measurements are less prone to artifacts such as inner filter effects or variations in donor concentration across samples.

The dependence of FRET upon donor-acceptor distance is reflected in a sensitivity to geometry and topology of probe distribution, and, by extension, of the underlying system. Modeling of FRET in most problems in membrane biophysics therefore requires knowledge of structural parameters such as area/lipid, transverse location of fluorophores, or protein dimensions. In favorable cases (i.e., situations in which uniform distribution of donor and acceptor can be safely assumed), the FRET experiment can itself be used to determine some of these parameters. However, in the context of this chapter, which concerns deviations to uniform distribution of membrane components, it is especially important that as much FRET-independent information is used (otherwise precise recovery of the heterogeneity features is impossible). Of course, for many membrane systems this is unfeasible, because of lacking lipid and/or protein structural information. In these situations, estimates are used, which may limit the usefulness of the FRET experiment.

In addition to structure, the dynamic properties of the system under study may also influence the FRET process. In most cases, translational diffusion during the donor excited state is negligible. However, relative fluorophore orientation and rotational diffusion, as expressed in the κ^2 parameter, affects R_0 (Eq. 2) and hence FRET kinetics. Because there is no experimental technique suited to a definite measurement of κ^2, the theoretical value for the so-called dynamic isotropic limit ($<\kappa^2> = 2/3$) is often used. However, the $< \kappa^2 >$ uncertainty is still widely regarded as an inconvenience that may be especially important in membranes, because of their intrinsic anisotropic nature and the restricted rotational mobility experienced by fluorophores incorporated inside the bilayer. Recently, we used atomistic molecular dynamics simulations to calculate $< \kappa^2> \approx 0.87 \pm 0.06$ for homo-FRET between NBD-PC probes in fluid 16:0, 16:0 bilayers [83]. In this case, and taking into account the dependence of R_0 on κ^2 (Eq. 2), an error of ~4–5 % or ~1 Å (taking R_0(NBD-NBD) = 22–24 Å; [77]) would ensue by calculating this parameter using $< \kappa^2 > = 2/3$. This relatively modest effect in ld bilayers is probably much increased in more ordered systems, such as lo- or gel-containing membranes. To this effect, and also for a more precise estimation of fluorophore transverse location, MD simulations may be an invaluable aid in design and analysis of FRET experiments. Additionally, they allow the determination of the potential perturbation by the probes of the bilayer structure and dynamics (for a review, see Loura and Prates Ramalho [84]). After all, because the vast majority of natural lipids do not fluoresce, membrane FRET studies most often employ fluorescent probes which might behave differently in membranes when compared to the lipids they are supposed to emulate. Examples of the latter situation are the protein-lipid selectivity constants $K_s \neq 1$ recovered for probes incorporated in pure bilayers of supposedly identical (bar the fluorescent label) phospholipids, as described in Sect. 2.3. However, even in this case, adequate controls (e.g., calculating relative selectivity constants by dividing K_s for a given probe by the value recovered for the closest analog) may allow circumvention of this inconvenience.

FRET became a standard methodology under the microscope, and the most relevant application of FLIM is in energy transfer studies (FRET-FLIM). This is going to become more relevant with the foreseen instrumental developments, namely, super-resolution techniques, such as STED, now carried out up to the level of a living higher animal [6].

Acknowledgments The authors acknowledge funding by FEDER (COMPETE program) and by FCT (Fundação para a Ciência e Tecnologia); projects references: PTDC/QUI-BIQ/119494/2010, PTDC/QUI-BIQ/112067/2009, PTDC/QUI-BIQ/099947/2008, and FCOMP-01-0124-FEDER-010787 (FCT PTDC/QUI-QUI/098198/2008).

References

1. Abrams FS, London E (1993) Extension of the parallax analysis of membrane penetration depth to the polar region of model membranes: use of fluorescence quenching by a spin-label attached to the phospholipid polar headgroup. Biochemistry 32:10826–10831
2. Acasandrei MA, Dale RE, VandeVen M, Ameloot M (2006) Two-dimensional Förster resonance energy transfer (2-D FRET) and the membrane raft hypothesis. Chem Phys Lett 419:469–473
3. Almeida PFF, Vaz WLC, Thompson TE (1992) Lateral diffusion in the liquid-phases of dimyristoylphosphatidylcholine cholesterol lipid bilayers - a free-volume analysis. Biochemistry 31:6739–6747
4. Almeida PF, Pokorny A, Hinderliter A (2005) Thermodynamics of membrane domains. Biochim Biophys Acta 1720:1–13
5. Anikovsky M, Dale L, Ferguson S, Petersen N (2008) Resonance energy transfer in cells: a new look at fixation effect and receptor aggregation on cell membrane. Biophys J 95:1349–1359
6. Berning S, Willig KI, Steffens H, Dibaj P, Hell SW (2012) Nanoscopy in a living mouse brain. Science 335:551
7. Bloom M, Mouritsen OG (1995) The evolution of membranes. In: Lipowsky R, Sackmann E (eds) Handbook of biological physics, vol 1A, Structure and dynamics of membranes - from cells to vesicles. Elsevier, Amsterdam, pp 65–95
8. Bogdanov M, Dowhan W (1995) Phosphatidylethanolamine is required for in vivo function of the membrane-associated lactose permease of *Escherichia coli*. J Biol Chem 270:732–739
9. Bogdanov M, Dowhan W (1998) Phospholipid-assisted protein folding: phosphatidylethanolamine is required at a late step of the conformational maturation of the polytopic membrane protein lactose permease. EMBO J 17:5255–5264
10. Bogdanov M, Heacock PN, Dowhan W (2002) A polytopic membrane protein displays a reversible topology dependent on membrane lipid composition. EMBO J 21:2107–2116

11. Bogdanov M, Xie J, Heacock P, Dowhan W (2008) To flip or not to flip: lipid-protein charge interactions are a determinant of final membrane protein topology. J Cell Biol 182:925–935
12. Bogdanov M, Xie J, Dowhan W (2009) Lipid-protein interactions drive membrane protein topogenesis in accordance with the positive inside rule. J Biol Chem 284:9637–9641
13. Bogdanov M, Heacock P, Guan Z, Dowhan W (2010) Plasticity of lipid-protein interactions in the function and topogenesis of the membrane protein lactose permease from *Escherichia coli*. Proc Natl Acad Sci USA 107:15057–15062
14. Breitenstein D, Batenburg JJ, Hagenhoff B, Galla HJ (2006) Lipid specificity of surfactant protein B studied by time-of-flight secondary ion mass spectrometry. Biophys J 91:1347–1356
15. Brown AC, Towles KB, Wrenn SP (2007) Measuring raft size as a function of membrane composition in PC-based systems: Part I - binary systems. Langmuir 23:11180–11187
16. Brown DA, London E (2000) Structure and function of sphingolipid- and cholesterol-rich membrane rafts. J Biol Chem 275:17221–17224
17. Buboltz JT (2007) Steady-state probe-partitioning FRET: A simple and robust tool for the study of membrane phase behavior. Phys Rev E 76:021903
18. Buboltz JT, Bwalya C, Reyes S, Kamburov D (2007) Stern-Volmer modeling of steady-state Forster energy transfer between dilute, freely diffusing membrane-bound fluorophores. J Chem Phys 127:215101
19. Buboltz JT, Bwalya C, Williams K, Schutzer M (2007) High resolution mapping of phase behavior in a ternary lipid mixture: do lipid-raft phase boundaries depend on sample-prep procedure? Langmuir 23:11968–11971
20. Cabré EJ, Loura LMS, Fedorov A, Pérez-Gil J, Prieto M (2012) Topology and lipid selectivity of pulmonary surfactant protein SP-B in membranes: answers from fluorescence. Biochim Biophys Acta. doi:10.1016/j.bbamem.2012.03.008
21. Castro BM, de Almeida RF, Goormaghtigh E, Fedorov A, Prieto M (2011) Organization and dynamics of Fas transmembrane domain in raft membranes and modulation by ceramide. Biophys J 101:1632–1641
22. Chachaty C, Rainteau D, Tessier C, Quinn PJ, Wolf C (2005) Building up of the liquid-ordered phase formed by sphingomyelin and cholesterol. Biophys J 88:4032–4044
23. Chattopadhyay A (1990) Chemistry and biology of N-(7-nitrobenz-2-oxa-1,3-diazol-4-yl)-labeled lipids: fluorescent probes of biological and model membranes. Chem Phys Lipids 53:1–15
24. Chen CC, Wilson TH (1984) The phospholipid requirement for activity of the lactose carrier of *Escherichia coli*. J Biol Chem 259:10150–10158
25. Chiantia S, Kahya N, Schwille P (2007) Raft domain reorganization driven by short- and long-chain ceramide: a combined AFM and FCS study. Langmuir 23:7659–7665
26. Clements JA (1977) Functions of the alveolar lining. Am Rev Respir Dis 115:67–71
27. Coutinho A, Loura LMS, Fedorov A, Prieto M (2008) Pinched multilamellar structure of aggregates of lysozyme and phosphatidylserine-containing membranes revealed by FRET. Biophys J 95:4726–4736
28. Cruz A, Casals C, Plasencia I, Marsh D, Pérez-Gil J (1998) Depth profiles of pulmonary surfactant protein B in phosphatidylcholine bilayers, studied by fluorescence and electron spin resonance spectroscopy. Biochemistry 37:9488–9496
29. Cruz A, Marsh D, Pérez-Gil J (1998) Rotational dynamics of spin-labelled surfactant-associated proteins SP-B and SP-C in dipalmitoylphosphatidylcholine and dipalmitoylphosphatidylglycerol bilayers. Biochim Biophys Acta 1415:125–134
30. Czech MP (2000) PIP2 and PIP3: complex roles at the cell surface. Cell 100:603–606
31. Davenport L (1997) Fluorescence probes for studying membrane heterogeneity. Meth Enzymol 278:487–512
32. de Almeida RFM, Loura LMS, Fedorov A, Prieto M (2002) Nonequilibrium phenomena in the phase separation of a two-component lipid bilayer. Biophys J 82:823–834

33. de Almeida RFM, Fedorov A, Prieto M (2003) Sphingomyelin/Phosphatidylcholine/Cholesterol phase diagram: boundaries and composition of lipid rafts. Biophys J 85:2406–2416
34. de Almeida RFM, Loura LMS, Prieto M, Watts A, Fedorov A, Barrantes FJ (2004) Cholesterol modulates the organization of the γM4 transmembrane domain of the muscle nicotinic acetylcholine receptor. Biophys J 86:2261–2272
35. de Almeida RFM, Loura LMS, Fedorov A, Prieto M (2005) Lipid rafts have different sizes depending on membrane composition: a time-resolved fluorescence resonance energy transfer study. J Mol Biol 346:1109–1120
36. de Almeida RFM, Borst J, Fedorov A, Prieto M, Visser AJWG (2007) Complexity of lipid domains and rafts in giant unilamellar vesicles revealed by combining imaging and microscopic and macroscopic time-resolved fluorescence. Biophys J 93:539–553
37. de Almeida RF, Loura LMS, Prieto M (2009) Membrane lipid domains and rafts: current applications of fluorescence lifetime spectroscopy and imaging. Chem Phys Lipids 157:61–77
38. Dickenson NE, Armendariz KP, Huckabay HA, Livanec PW, Dunn RC (2010) Near-field scanning optical microscopy: a tool for nanometric exploration of biological membranes. Anal Bioanal Chem 396:31–43
39. Dico AS, Hancock J, Morrow MR, Stewart J, Harris S, Keough KM (1997) Pulmonary surfactant protein SP-B interacts similarly with dipalmitoylphosphatidylglycerol and dipalmitoylphosphatidylcholine in phosphatidylcholine/phosphatidylglycerol mixtures. Biochemistry 36:4172–4177
40. Dowhan W (1997) Molecular basis for membrane phospholipid diversity: why are there so many lipids? Annu Rev Biochem 66:199–232
41. Fernandes F, Loura LMS, Prieto M, Koehorst R, Spruijt R, Hemminga MA (2003) Dependence of M13 major coat protein oligomerization and lateral segregation on bilayer composition. Biophys J 85:2430–2441
42. Fernandes F, Loura LM, Koehorst R, Spruijt RB, Hemminga MA, Fedorov A, Prieto M (2004) Quantification of protein-lipid selectivity using FRET: application to the M13 major coat protein. Biophys J 87:344–352
43. Fernandes F, Loura LMS, Fedorov A, Prieto M (2006) Absence of clustering of phosphatidylinositol-(4,5)-bisphosphate in fluid phosphatidylcholine. J Lipid Res 47:1521–1525
44. Förster T (1949)) Experimentelle und theoretische Untersuchung des Zwischenmolekularen übergangs von Elektrinenanregungsenergie. Z Naturforsch 4a:321–327
45. Frazier ML, Wright JR, Pokorny A, Almeida PF (2007) Investigation of domain formation in sphingomyelin/cholesterol/POPC mixtures by fluorescence resonance energy transfer and Monte Carlo simulations. Biophys J 92:2422–2433
46. Fung BK, Stryer L (1978) Surface density determination in membranes by fluorescence energy transfer. Biochemistry 17:5241–5248
47. Goswami D, Gowrishankar K, Bilgrami S, Ghosh S, Raghupathy R, Chadda R, Vishwakarma R, Rao M, Mayor S (2008) Nanoclusters of GPI-anchored proteins are formed by cortical actin-driven activity. Cell 135:1085–1097
48. Heberle FA, Wu J, Goh SL, Petruzielo RS, Feigenson GW (2010) Comparison of three ternary lipid bilayer mixtures: FRET and ESR reveal nanodomains. Biophys J 99:3309–3318
49. Hemminga MA, Sanders JC, Spruijt RB (1992) Spectroscopy of lipid-protein interactions: structural aspects of two different forms of the coat protein of bacteriophage M13 incorporated in model membranes. Prog Lipid Res 31:301–333
50. Hofman EG, Ruonala MO, Bader AN, van den Heuvel D, Voortman J, Roovers RC, Verkleij AJ, Gerritsen HC, van Bergen en Henegouwen PMP (2008) EGF induces coalescence of different lipid rafts. J Cell Sci 121:2519–2528
51. Holt A, de Almeida RFM, Nyholm TK, Loura LMS, Daily AE, Staffhorst RW, Rijkers DT, Koeppe RE 2nd, Prieto M, Killian JA (2008) Is there a preferential interaction between cholesterol and tryptophan residues in membrane proteins? Biochemistry 47:2638–2649

52. Goñi FM, Alonso A, Bagatolli LA, Brown RE, Marsh D, Prieto M, Thewalt JL (2008) Phase diagrams of lipid mixtures relevant to the study of membrane rafts. Biochim Biophys Acta 1781:665–684
53. Grecco HE, Roda-Navarro P, Verveer PJ (2009) Global analysis of time correlated single photon counting FRET-FLIM data. Opt Express 17:6493–6508
54. Guan L, Smirnova IN, Verner G, Nagamori S, Kaback HR (2006) Manipulating phospholipids for crystallization of a membrane transport protein. Proc Natl Acad Sci USA 103:1723–1726
55. Gutberlet T, Dietrich U, Bradaczek H, Pohlentz G, Leopold K, Fischer W (2000) Cardiolipin, alpha-D-glucopyranosyl, and L-lysylcardiolipin from gram-positive bacteria: FAB MS, monofilm and X-ray powder diffraction studies. Biochim Biophys Acta 1463:307–322
56. Herreros J, Ng T, Schiavo G (2001) Lipid rafts act as specialized domains for tetanus toxin binding and internalization into neurons. Mol Biol Cell 12:2947–2960
57. Hughes WE, Larijani B, Parker PJ (2002) Detecting protein-phospholipid interactions. J Biol Chem 277:22974–22979
58. Ipsen JH, Karlström G, Mouritsen OG, Wennerström H, Zuckermann MJ (1987) Phase equilibria in the phosphatidylcholine-cholesterol system. Biochim Biophys Acta 905:162–172
59. Jacobson K, Mouritsen OG, Anderson RG (2007) Lipid rafts: at a crossroad between cell biology and physics. Nat Cell Biol 9:7–14
60. Jares-Erijman EA, Jovin TM (2006) Imaging molecular interactions in living cells by FRET microscopy. Curr Opin Chem Biol 10:409–416
61. Johansson J, Curstedt T (1997) Molecular structures and interactions of pulmonary surfactant components. Eur J Biochem 244:675–693
62. Johansson J, Curstedt T, Jornvall H (1991) Surfactant protein B: disulfide bridges, structural properties, and Kringle similarities. Biochemistry 30:6917–6921
63. Jørgensen K, Klinger A, Biltonen RL (2000) Nonequilibrium lipid domain growth in the gel-fluid two phase region of a $DC_{16}PC$-$DC_{22}PC$ lipid mixture investigated by Monte-Carlo computer simulation, FT-IR and fluorescence spectroscopy. J Phys Chem 104:11763–11773
64. Karmakar S, Raghunathan VA, Mayor S (2005) Phase behaviour of dipalmitoylphospatidylcholine (DPPC)-cholesterol membranes. J Phys Condens Matter 17:S1177–S1182
65. Kenworthy AK, Edidin M (1998) Distribution of a glycosylphosphatidylinositol-anchored protein at the apical surface of MDCK cells examined at a resolution of <100 Å using imaging fluorescence resonance energy transfer. J Cell Biol 142:69–84
66. Kiskowski MA, Kenworthy AK (2007) In silico characterization of resonance energy transfer for disk-shaped membrane domains. Biophys J 92:3040–3051
67. Koehorst RB, Spruijt RB, Vergeldt FJ, Hemminga MA (2004) Lipid bilayer topology of the transmembrane alpha-helix of M13 Major coat protein and bilayer polarity profile by site-directed fluorescence spectroscopy. Biophys J 87:1445–1455
68. Kusumi A, Nakada C, Ritchie K, Murase K, Suzuki K, Murakoshi H, Kasai RS, Kondo J, Fujiwara T (2005) Paradigm shift of the plasma membrane concept from the two-dimensional continuum fluid to the partitioned fluid: high-speed single-molecule tracking of membrane molecules. Annu Rev Biophys Biomol Struct 34:351–378
69. Lantzsch G, Binder H, Heerklotz H (1994) Surface area per molecule in lipid/$C12E_n$ membranes as seen by fluorescence resonance energy transfer. J Fluoresc 4:339–343
70. Lakowicz JR (2006) Principles of fluorescence spectroscopy. Kluwer Academic/Plenum, New York
71. Lee AG (2003) Lipid-protein interactions in biological membranes: a structural perspective. Biochim Biophys Acta 1612:1–40
72. Lentz BR, Barrow DA, Hoechli M (1980) Cholesterol-phosphatidylcholine interactions in multilamellar vesicles. Biochemistry 19:1943–1954

73. Li M, Reddy LG, Bennett R, Silva ND Jr, Jones LR, Thomas DD (1999) A fluorescence energy transfer method for analyzing protein oligomeric structure: application to phospholamban. Biophys J 76:2587–2599
74. Loura LMS, Prieto M (2000) Resonance energy transfer in heterogeneous planar and bilayer systems: theory and simulation. J Phys Chem B 104:6911–6919
75. Loura LM, Ramalho JP (2007) Location and dynamics of acyl chain NBD-labeled phosphatidylcholine (NBD-PC) in DPPC bilayers. A molecular dynamics and time-resolved fluorescence anisotropy study. Biochim Biophys Acta 1768:467–478
76. Loura LMS, Fedorov A, Prieto M (1996) Resonance energy transfer in a model system of membranes: application to gel and liquid crystalline phases. Biophys J 71:1823–1836
77. Loura LMS, Fedorov A, Prieto M (2000) Membrane probe distribution heterogeneity: a resonance energy transfer study. J Phys Chem B 104:6920–6931
78. Loura LMS, Fedorov A, Prieto M (2000) Partition of membrane probes in a gel/fluid two-component lipid system: a fluorescence resonance energy transfer study. Biochim Biophys Acta 1467:101–112
79. Loura LMS, Fedorov A, Prieto M (2001) Fluid-fluid membrane microheterogeneity: a fluorescence resonance energy transfer study. Biophys J 80:776–788
80. Loura LMS, Coutinho A, Silva A, Fedorov A, Prieto M (2006) Structural effects of a basic peptide on the organization of dipalmitoylphosphatidylcholine/dipalmitoylphosphatidylserine membranes: a fluorescent resonance energy transfer study. J Phys Chem B 110:8130–8141
81. Loura LMS, Fernandes F, Prieto M (2010) Membrane microheterogeneity: Förster resonance energy transfer characterization of lateral membrane domains. Eur Biophys J 39:589–607
82. Loura LMS, Prieto M, Fernandes F (2010) Quantification of protein-lipid selectivity using FRET. Eur Biophys J 39:565–578
83. Loura LMS, Palace Carvalho AJ, Prates Ramalho JP (2010) Direct calculation of Förster orientation factor of membrane probes by molecular simulation. J Mol Struct THEOCHEM 946:107–112
84. Loura LMS, Prates Ramalho JP (2011) Recent developments in molecular dynamics simulations of fluorescent membrane probes. Molecules 16:5437–5452
85. Mabrey S, Sturtevant JM (1976) Investigation of phase transitions of lipids and lipid mixtures by sensitivity differential scanning calorimetry. Proc Natl Acad Sci U S A 73:3862–3866
86. Marsh D (1990) Handbook of Lipid Bilayers. CRC Press, Boca Raton
87. Marsh D (2010) Liquid-ordered phases induced by cholesterol: a compendium of binary phase diagrams. Biochim Biophys Acta 1798:688–699
88. Marsh D, Horváth LI (1998) Structure, dynamics and composition of the lipid-protein interface. Perspectives from spin-labelling. Biochim Biophys Acta 1376:267–296
89. Mateo CR, Acuna AU, Brochon J-C (1995) Liquid-crystalline phases of cholesterol lipid bilayers as revealed by the fluorescence of *trans*-parinaric acid. Biophys J 68:978–987
90. Mayor S, Rao M (2004) Rafts: scale-dependent, active lipid organization at the cell surface. Traffic 5:231–240
91. Mazères S, Schram V, Tocanne JF, Lopez A (1996) 7-nitrobenz-2-oxa-1,3-diazole-4-yl-labeled phospholipids in lipid membranes: differences in fluorescence behavior. Biophys J 71:327–335
92. McMullen TP, McElhaney RN (1995) New aspects of the interaction of cholesterol with dipalmitoylphosphatidylcholine bilayers as revealed by high-sensitivity differential scanning calorimetry. Biochim Biophys Acta 1234:90–98
93. Meyer BH, Segura J-M, Martinez KL, Hovius R, George N, Johnsson K, Vogel H (2006) FRET imaging reveals that functional neurokinin-1 receptors are monomeric and reside in membrane microdomains of live cells. Proc Natl Acad Sci USA 103:2138–2143
94. Morrow MR, Pérez-Gil J, Simatos G, Boland C, Stewart J, Absolom D, Sarin V, Keough KM (1993 Apr 27) Pulmonary surfactant-associated protein SP-B has little effect on acyl chains in dipalmitoylphosphatidylcholine dispersions. Biochemistry 32(16):4397–402

95. Mouritsen OG, Bloom M (1984) Mattress model of lipid-protein interactions in membranes. Biophys J 46:141–153
96. Needham D, Nunn RS (1990) Elastic deformation and failure of lipid bilayer membranes containing cholesterol. Biophys J 58:997–1009
97. Notter RH, Finkelstein JN (1984) Pulmonary surfactant: an interdisciplinary approach. J Appl Physiol 57:1613–1624
98. O'Keefe AH, East JM, Lee AG (2000) Selectivity in lipid binding to the bacterial outer membrane protein OmpF. Biophys J 79:2066–2074
99. Oosterlaken-Dijksterhuis MA, Haagsman HP, van Golde LM, Demel RA (1991) Characterization of lipid insertion into monomolecular layers mediated by lung surfactant proteins SP-B and SP-C. Biochemistry 30:10965–10971
100. Oosterlaken-Dijksterhuis MA, van Eijk M, van Golde LM, Haagsman HP (1992) Lipid mixing is mediated by the hydrophobic surfactant protein SP-B but not by SP-C. Biochim Biophys Acta 1110:45–50
101. Owen DM, Neil MAA, French PMW, Magee AI (2007) Optical techniques for imaging membrane lipid microdomains in living cells. Sem Cell Develop Biol 18:591–598
102. Owen DM, Gaus K, Magee AI, Cebecauer M (2010) Dynamic organization of lymphocyte plasma membrane: lessons from advanced imaging methods. Immunology 131:1–8
103. Padilla-Parra S, Auduge N, Coppey-Moisan M, Tramier M (2008) Quantitative FRET analysis by fast acquisition time domain FLIM at high spatial resolution in living cells. Biophys J 95:2976–2988
104. Peelen SJ, Sanders JC, Hemminga MA, Marsh D (1992) Stoichiometry, selectivity, and exchange dynamics of lipid-protein interaction with bacteriophage M13 coat protein studied by spin label electron spin resonance. Effects of protein secondary structure. Biochemistry 31:2670–2677
105. Pérez-Gil J (2001) Lipid-protein interactions of hydrophobic proteins SP-B and SP-C in lung surfactant assembly and dynamics. Pediatr Pathol Mol Med 20:445–469
106. Pérez-Gil J, Keough KM (1998) Interfacial properties of surfactant proteins. Biochim Biophys Acta 1408:203–217
107. Pérez-Gil J, Casals C, Marsh D (1995) Interactions of hydrophobic lung surfactant proteins SP-B and SP-C with dipalmitoylphosphatidylcholine and dipalmitoylphosphatidylglycerol bilayers studied by electron spin resonance spectroscopy. Biochemistry 34:3964–3971
108. Picas L, Suárez-Germà C, Montero MT, Vázquez-Ibar JL, Hernández-Borrell J, Prieto M, Loura LM (2010) Lactose permease lipid selectivity using Förster resonance energy transfer. Bio-chim Biophys Acta 1798:1707–1713
109. Pluschke G, Hirota Y, Overath P (1978) Function of phospholipids in *Escherichia coli*. Charac-terization of a mutant deficient in cardiolipin synthesis. J Biol Chem 253:5048–5055
110. Poulain FR, Allen L, Williams MC, Hamilton RL, Hawgood S (1992) Effects of surfactant apolipoproteins on liposome structure: implications for tubular myelin formation. Am J Physiol 262:L730–L739
111. Powl AM, East JM, Lee AG (2003) Lipid-protein interactions studied by introduction of a tryp-tophan residue: the mechanosensitive channel MscL. Biochemistry 42:14306–14317
112. Rand RP, Parsegian VA (1989) Hydration forces between phospholipid-bilayers. Biochim Biophys Acta 988:351–376
113. Rao M, Mayor S (2005) Use of Förster's resonance energy transfer microscopy to study lipid rafts. Biochim Biophys Acta 1746:221–233
114. Ryan MA, Qi X, Serrano AG, Ikegami M, Perez-Gil J, Johansson J, Weaver TE (2005) Mapping and analysis of the lytic and fusogenic domains of surfactant protein B. Biochemistry 44:861–872
115. Šachl R, Humpolíčková J, Stefl M, Johansson LB, Hof M (2011) Limitations of electronic energy transfer in the determination of lipid nanodomain sizes. Biophys J 101:L60–L62
116. Seifert M, Breitenstein D, Klenz U, Meyer MC, Galla HJ (2007) Solubility versus electrostatics: what determines lipid/protein interaction in lung surfactant. Biophys J 93:1192–1203

117. Sharma P, Varma R, Sarasij RC, Gousset K, Ira RC, Krishnamoorthy G, Rao M, Mayor S (2004) Nanoscale organization of multiple GPI-anchored proteins in living cell membranes. Cell 116:577–589
118. Shiffer K, Hawgood S, Haagsman HP, Benson B, Clements JA, Goerke J (1993) Lung surfactant proteins, SP-B and SP-C, alter the thermodynamic properties of phospholipid membranes: a differential calorimetry study. Biochemistry 32:590–597
119. Silva L, de Almeida RF, Fedorov A, Matos AP, Prieto M (2006) Ceramide-platform formation and -induced biophysical changes in a fluid phospholipid membrane. Mol Membr Biol 23:137–148
120. Silva LC, de Almeida RF, Castro BM, Fedorov A, Prieto M (2007) Ceramide-domain formation and collapse in lipid rafts: membrane reorganization by an apoptotic lipid. Biophys J 92:502–516
121. Simons K, Vaz WL (2004) Model systems, lipid rafts, and cell membranes. Annu Rev Biophys Biomol Struct 33:269–295
122. Singer SJ, Nicolson GL (1972) The fluid mosaic model of the structure of cell membranes. Science 175:720–731
123. Soubias O, Teague WE Jr, Hines KG, Mitchell DC, Gawrisch K (2010) Contribution of membrane elastic energy to rhodopsin function. Biophys J 99:817–824
124. Spruijt RB, Wolfs CJ, Verver JW, Hemminga MA (1996) Accessibility and environment probing using cysteine residues introduced along the putative transmembrane domain of the major coat protein of bacteriophage M13. Biochemistry 35:10383–10391
125. Stöckl MT, Herrmann A (2010) Detection of lipid domains in model and cell membranes by fluorescence lifetime imaging microscopy. Biochim Biophys Acta 1798:1444–1456
126. Stopar D, Jansen KA, Páli T, Marsh D, Hemminga MA (1997) Membrane location of spin-labeled M13 major coat protein mutants determined by paramagnetic relaxation agents. Biochemistry 36:8261–8268
127. Stopar D, Spruijt RB, Wolfs CJ, Hemminga MA (2003) Protein-lipid interactions of bacteriophage M13 major coat protein. Biochim Biophys Acta 1611:5–15
128. Stryer L (1978) Fluorescence energy transfer as a spectroscopic ruler. Annu Rev Biochem 47:819–846
129. Suárez-Germà C, Loura LMS, Prieto M, Domènech O, Montero MT, Rodríguez-Banqueri A, Vázquez-Ibar JL, Hernández-Borrell J (2012) Membrane protein-lipid selectivity: enhancing sensitivity for modeling FRET data. J Phys Chem B 116:2438–2445
130. Tahara Y, Murata M, Ohnishi S, Fujiyoshi Y, Kikuchi M, Yamamoto Y (1992) Functional signal peptide reduces bilayer thickness of phosphatidylcholine liposomes. Biochemistry 31:8747–8754
131. Towles KB, Dan N (2007) Determination of membrane domain size by fluorescence resonance energy transfer: effects of domain polydispersity and packing. Langmuir 23:4737–4739
132. Towles KB, Brown AC, Wrenn SP, Dan N (2007) Effect of membrane microheterogeneity and domain size on fluorescence resonance energy transfer. Biophys J 93:655–667
133. Vandenbussche G, Clercx A, Clercx M, Curstedt T, Johansson J, Jörnvall H, Ruysschaert JM (1992) Secondary structure and orientation of the surfactant protein SP-B in a lipid environment. A Fourier transform infrared spectroscopy study. Biochemistry 31:9169–9176
134. Van Der Meer B, Coker G 3rd, Chen S-YS (1994) Resonance energy transfer: theory and data. VCH Publishers, New York
135. Varma R, Mayor S (1998) GPI-anchored proteins are organized in submicron domains at the cell surface. Nature 394:798–801
136. Veatch SL, Keller SL (2003) Separation of liquid phases in giant vesicles of ternary mixtures of phospholipids and cholesterol. Biophys J 85:3074–3083
137. Veatch SL, Keller SL (2005) Miscibility phase diagrams of giant vesicles containing sphingomyelin. Phys Rev Lett 94:148101–148104
138. Veatch SL, Keller SL, Gawrisch K (2007) Critical fluctuations in domain-forming lipid mixtures. Proc Natl Acad Sci USA 104:17650–17655

139. Veatch SL, Polozov IV, Gawrisch K, Keller SL (2004) Liquid domains in vesicles investigated by NMR and fluorescence microscopy. Biophys J 86:2910–2922
140. Vist MR, Davis JH (1990) Phase equilibria of cholesterol/dipalmitoylphosphatidylcholine mixtures: 2 H nuclear magnetic resonance and differential scanning calorimetry. Biochemistry 29:451–464
141. Von Arnim CAF, Kinoshita A, Peltan ID, Tangredi MM, Herl L, Lee BM, Spoelgen R, Hshieh TT, Ranganathan S, Battey FD, Liu CX, Bacskai BJ, Sever S, Irizarry MC, Strickland DK, Hyman BT (2005) The low density lipoprotein receptor-related protein (LRP) is a novel beta-secretase (BACE1) substrate. J Biol Chem 280:17777–17785
142. Wang X, Bogdanov M, Dowhan W (2002) Topology of polytopic membrane protein subdomains is dictated by membrane phospholipid composition. EMBO J 21:5673–5681
143. Wikström M, Kelly AA, Georgiev A, Eriksson HM, Klement MR, Bogdanov M, Dowhan W, Wieslander A (2009) Lipid-engineered *Escherichia coli* membranes reveal critical lipid headgroup size for protein function. J Biol Chem 284:954–965
144. Williamson IM, Alvis SJ, East JM, Lee AG (2002) Interactions of phospholipids with the potassium channel KcsA. Biophys J 83:2026–2038
145. Wolber PK, Hudson BS (1979) An analytical solution to the Förster energy transfer problem in two dimensions. Biophys J 28:197–210
146. Wu SH, McConnell HM (1975) Phase separations in phospholipid membranes. Biochemistry 14:847–854
147. Zaltash S, Palmblad M, Curstedt T, Johansson J, Persson B (2000) Pulmonary surfactant protein B: a structural model and a functional analogue. Biochim Biophys Acta 1466:179–186
148. Edidin M (2003) Lipids on the frontier: a century of cell-membrane bilayers. Nat Rev Mol Cell Biol 4:414–418
149. London E, Brown DA (2000) Insolubility of lipids in triton X-100: physical origin and relationship to sphingolipid/cholesterol membrane domains (rafts). Biochim Biophys Acta 1508:182–195
150. Sanders JC, Ottaviani MF, van Hoek A, Visser AJ, Hemminga MA (1992) A small protein in model membranes: a time-resolved fluorescence and ESR study on the interaction of M13 coat protein with lipid bilayers. Eur Biophys J 21:305–311

FRET Analysis of Protein-Lipid Interactions

Galyna Gorbenko and Paavo K.J. Kinnunen

Abstract Förster resonance energy transfer (FRET) is an old but constantly developing spectroscopic tool possessing enormous potential for studies on structure and dynamics of biological macromolecules and their assemblies. One of the main advantages of FRET technique is the possibility of measuring the nanometer-scale distances between donor and acceptor fluorophores. This chapter highlights some aspects of FRET-based monitoring of intermolecular interactions in membrane systems. Analytical model of energy transfer between membrane-associated donors and acceptors randomly distributed over parallel planes separated by a fixed distance is presented. The factors determining the efficiency of energy transfer are considered with special attention to orientational behavior of the donor emission and acceptor absorption transition dipoles. It is demonstrated that FRET can provide proof for specific orientation of the protein molecule relative to lipid-water interface. The applications of FRET to quantification of protein-lipid binding parameters and membrane position of protein fluorophores are exemplified. It is illustrated how FRET may help in obtaining evidence for protein aggregation in a membrane environment and domain formation.

Keywords Domain formation · Förster resonance energy transfer · Protein aggregation · Protein-lipid interactions · Structural and binding parameters

G. Gorbenko
Department of Biological and Medical Physics, V.N. Karazin Kharkov National University, Svobody Sq. 4, 61022 Kharkov, Ukraine

P.K.J. Kinnunen (✉)
Department of Biomedical Engineering and Computational Science, School of Science, Aalto University, Otakaari 3, 02150 Espoo, Finland
e-mail: paavo.kinnunen@aalto.fi

Y. Mély and G. Duportail (eds.), *Fluorescent Methods to Study Biological Membranes*, Springer Ser Fluoresc (2013) 13: 115–140, DOI 10.1007/4243_2012_45,

Contents

1 Introduction

Förster resonance energy transfer is one of the most powerful fluorescence techniques and has found innumerable applications in biomedical research [1–3]. Due to explicit dependence on the distance between the fluorophores acting as energy donor and acceptor, FRET represents a unique tool for establishing nanometer-scale proximity relationships both in vivo and in vitro, thereby giving structural information complementary to that provided by other mighty physical methods, such as NMR, X-ray and neutron scattering, and electron and optical microscopy [4, 5]. FRET has proven to be especially powerful in structural characterization of a wide variety of macromolecular assemblies, biological membranes, in particular [3, 6–8]. The interactions between two major membrane constituents, proteins, and lipids have long been a focus of FRET studies, greatly contributing to understanding the mechanisms of the membrane binding of proteins [9, 10], conformational transitions of polypeptide chains in a membrane environment [11, 12], lipid-mediated protein aggregation [13, 14], domain formation [15, 16], etc. Among the principal advantages of FRET technique are (1) the possibility of experimentation within protein and lipid concentration range where most other methods appear powerless; (2) revealing the subtle peculiarities of protein-membrane interactions; (3) detection of small amounts of molecular clusters, unaccessible to other techniques; (4) the opportunity for obtaining information about the structural state of a protein and lipid bilayer under the same experimental conditions; and (5) high sensitivity and relative simplicity of the experiment.

This chapter is intended to give a concise overview of the possibilities provided by FRET in the studies of protein-lipid interactions. Along with considering the fundamental principles of FRET, basic formalism for description of energy transfer in membranes, and some problems encountered in quantification of FRET data, we present some examples illustrating how steady-state FRET can be used to determine protein-membrane binding characteristics, orientation of the protein molecule relative to lipid-water interface, its transverse bilayer location, and aggregation state.

2 Basics of FRET

Förster resonance energy transfer (FRET) is a process by which energy is passed between molecules over long distances (from about 1 to 10 nm) [17]. The donor molecule, which must be a fluorophore, is excited by incident light and transfers the absorbed energy nonradiatively to the acceptor molecule, which, in the majority of cases, is capable of fluorescing. Since energy transfer occurs without emitting a photon and results from long-range dipole-dipole interactions between the donor and acceptor, it seems inaccurate to decipher the frequently used acronym FRET as fluorescence resonance energy transfer. Instead, according to the recommendations of IUPAC, it is more correct to use the term "Förster resonance energy transfer," paying tribute to the German scientist Theodor Förster who first proposed the mechanism of this phenomenon.

Förster developed a theoretical basis of FRET using both classical physics and quantum mechanics approaches [18]. Classical physics considers the transfer of excitation energy from one molecule to another as resulting from the resonance interactions between closely spaced oscillating dipoles, similar to two oscillating pendulums that are mechanically coupled. Quantum mechanical description of FRET is based on Fermi's golden rule which is used to calculate the probability of transition from initial state ***i*** ($\Psi_i = \Psi_{D_1}\Psi_{A_0}$, donor is in an excited state, while acceptor is in a ground state,) to final state ***f*** ($\Psi_f = \Psi_{D_0}\Psi_{A_1}$, donor is in a ground state, acceptor is in an excited state) due to a perturbation caused by light absorption:

$$\frac{dP_f}{dt} = \frac{1}{2\hbar^2}\left|\langle\Psi_f|\tilde{\mu}|\Psi_i\rangle E_0\right|^2 \tag{1}$$

The rate constant of energy transfer between donor and acceptor separated by a distance R is given by the following equations:

$$k_T(\nu) \propto \left|\langle\Psi_{D_1}\Psi_{A_0}|\tilde{V}|\Psi_{D_0}\Psi_{A_1}\rangle\right|^2 \tag{2}$$

$$\tilde{V} = \frac{1}{\varepsilon}\left[\frac{(\tilde{\mu}_D\cdot\tilde{\mu}_A)}{R^3} - 3\frac{(\tilde{\mu}_D\cdot R)(R\cdot\tilde{\mu}_A)}{R^5}\right] \tag{3}$$

$$k_T(\nu) \propto \frac{\kappa^2}{\varepsilon^2 R^6}|\langle\Psi_{D_1}|\tilde{\mu}_D|\Psi_{D_0}\rangle|^2|\langle\Psi_{A_0}|\tilde{\mu}_A|\Psi_{A_1}\rangle|^2 \tag{4}$$

$$|\langle\Psi_{A_0}|\tilde{\mu}_A|\Psi_{A_1}\rangle|^2 \propto \frac{\varepsilon_A}{\nu}, \quad |\langle\Psi_{D_1}|\tilde{\mu}_D|\Psi_{D_0}\rangle|^2 \propto \frac{1}{\nu^3\tau_R} = \frac{\varphi_D}{\nu^3\tau_D} \tag{5}$$

Here $\tilde{V}$ is the operator for dipole-dipole interaction between donor and acceptor. Remarkably, both classical and quantum mechanical theories of energy transfer yield similar final expression for the rate constant of FRET:

$$k_T \propto \frac{\kappa^2 \varphi_D}{n^4 R^6 \tau_D} \int \frac{\varepsilon_A(\nu) f_D(\nu)}{\nu^4} d\nu = \frac{\kappa^2 \varphi_D}{R^6 \tau_D} J \tag{6}$$

Förster theory predicts that the rate of energy transfer depends on the donor-acceptor distance (R); quantum yield of the donor in the absence of acceptor (φ_D); refractive index of the medium (n), which is typically taken as 1.4 for biomolecules in aqueous solution; donor lifetime in the absence of acceptor (τ_D); orientation factor (κ^2), determined by the relative orientation of the donor and acceptor dipoles; and overlap between donor emission ($f_D(\nu)$) and acceptor absorption spectra ($\varepsilon_A(\nu)$). By combining distance-independent terms, the expression for rate constant can be rewritten as

$$k_T = \frac{1}{\tau_D}\left(\frac{R_0}{R}\right)^6; \quad R_0^6 = \frac{\kappa^2 \varphi_D}{n^4}\left(\frac{9{,}000(\ln 10)}{128\pi^5 N_A}\right) \int \frac{\varepsilon_A(\nu) f_D(\nu)}{\nu^4} d\nu \tag{7}$$

here, R_0 is an important parameter called Förster distance. The efficiency of energy transfer (E) is the fraction of photons absorbed by the donor that are transferred to the acceptor, which is given by the ratio of the transfer rate to the total decay rate of the donor. On the other hand, transfer efficiency can be represented as a function of Förster radius and donor-acceptor distance:

$$E = \frac{k_T}{k_T + \tau_D^{-1}} = \frac{R_0^6}{R_0^6 + R^6} \tag{8}$$

As follows from this equation, Förster radius defines the distance at which energy transfer efficiency is 50%, and the transfer rate is equal to the donor decay rate. This is a characteristic parameter of each donor-acceptor pair, which typically falls in the 2–6-nm range [19]. A very steep distance dependence of FRET efficiency, being inversely proportional to the sixth power of donor-acceptor separation, along with the fact that Förster radii of most donor-acceptor pairs are comparable to the size of proteins and membranes, makes FRET the technique of choice, a spectroscopic ruler, to quantify spatial relationships between intrinsic and extrinsic fluorophores in biological macromolecules and their assemblies [20].

3 FRET in Membranes

To determine FRET transfer efficiency experimentally, three different approaches are generally used, based on measuring the decrease of donor fluorescence intensity (I_D) or average lifetime ($\langle\tau_D\rangle$) in the presence of acceptor (I_{DA},$\langle\tau_{DA}\rangle$):

$$E = 1 - \frac{\varphi_{DA}}{\varphi_D} = 1 - \frac{I_{DA}}{I_D}; \qquad E = 1 - \frac{\langle \tau_{DA} \rangle}{\langle \tau_D \rangle} \tag{9}$$

or enhancement of acceptor fluorescence (I_A) after donor excitation:

$$E = \frac{\varepsilon_A(\lambda_D^{ex})C_A}{\varepsilon_D(\lambda_D^{ex})C_D}\left(\frac{I_{AD}(\lambda_A^{em})}{I_A(\lambda_A^{em})} - 1\right) \tag{10}$$

Here $I_A(\lambda_A^{em})$, $I_{AD}(\lambda_A^{em})$ are acceptor fluorescence intensities in the absence and presence of donor at emission wavelength λ_A^{em}; $\varepsilon_D(\lambda_D^{ex})$, $\varepsilon_A(\lambda_D^{ex})$ are extinction coefficients of the donor and acceptor at the donor excitation wavelength λ_D^{ex}; and C_D,C_A are the concentrations of donor and acceptor, respectively.

It is important to emphasize that the simple relationship 8 describing distance dependence of energy transfer efficiency is valid only for one donor and one acceptor separated by a fixed distance, as may be the case, for instance, in labeled proteins. However, this relationship is inapplicable for donors and acceptors distributed in solution or in a membrane phase. In these cases, more complex expressions are required, which are commonly derived from averaging the transfer rate over multiple donor-acceptor pairs. This problem is addressed in a number of models developed for two-dimensional systems [21–27]. To illustrate principal features of FRET in membranes, we consider here one of these models proposed by Fung and Stryer for the case of donors and acceptors randomly distributed in a plane [21]. It should be noted in this regard that if the diameter of membrane vesicles is more than twofold greater than Förster radius, the curvature effect is negligible so that energy transfer can be considered as occurring in a plane. This is valid for the majority of membrane systems since the Förster distances do not exceed 10 nm, while the radius of vesicles formed by the model or isolated native membranes is usually greater than 20 nm. In terms of the model of Fung and Stryer, the efficiency of energy transfer is given by

$$E = 1 - \int_0^\infty \exp(-\lambda)\exp\left(-C_a^s S(\lambda)\right)d\lambda \tag{11}$$

$$S(\lambda) = \int_{r_c}^\infty \left[1 - \exp\left(-\lambda\left(\frac{R_o}{R}\right)^6\right)\right]2\pi R dR, \quad \lambda = \frac{t}{\tau_D} \tag{12}$$

where τ_D is the lifetime of excited donor in the absence of acceptor, r_c is the distance of closest approach between the donor and acceptor, and C_a^s is the concentration of acceptors per unit area related to the molar concentrations of lipids accessible to acceptor (L_a) and bound acceptor (B_a):

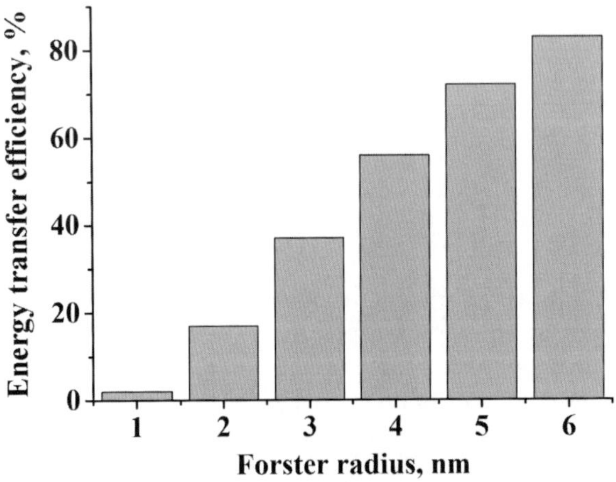

Fig. 1 Efficiencies of energy transfer between donors and acceptors randomly distributed in a plane. Acceptor surface density is one acceptor molecule per 100 lipid molecules. The distance of closest approach between donor and acceptor is taken as 0.8 nm

$$C_a^s = \frac{B_a}{L_a \sum f_i S_i} \tag{13}$$

Here f_i, S_i are the fraction and mean surface area of the *i-th* membrane constituent.

Shown in Fig. 1 are energy transfer efficiencies calculated by numerical integration of the Eq. 11. It can be seen that a relatively low acceptor surface density (one acceptor molecule per 100 lipid molecules) results in rather high energy transfer at Förster radii exceeding 2 nm. Remarkably, the above formalism is applicable to various geometric conditions of energy transfer. Particularly, for donors and acceptors uniformly distributed over different planes separated by a distance d_a (Fig. 2a), Eq. 12 takes the form

$$S(\lambda) = \int_{d_a}^{\infty} \left[1 - \exp\left(-\lambda\left(\frac{R_o}{R}\right)^6\right)\right] 2\pi R dR \tag{14}$$

Another possible configuration involves one donor plane located at a distance d_c from the membrane center and two acceptor planes with identical C_a^s, separated by a distance d_t (Fig. 2b). Given that for the outer acceptor plane, $d_a = |d_c - 0.5d_t|$, while for the inner plane, $d_a = d_c + 0.5d_t$, the following relationships hold:

$$S_1(\lambda) = \int_{|d_c - 0.5d_t|}^{\infty} \left[1 - \exp\left(-\lambda\left(\frac{R_o}{R}\right)^6\right)\right] 2\pi R dR \tag{15}$$

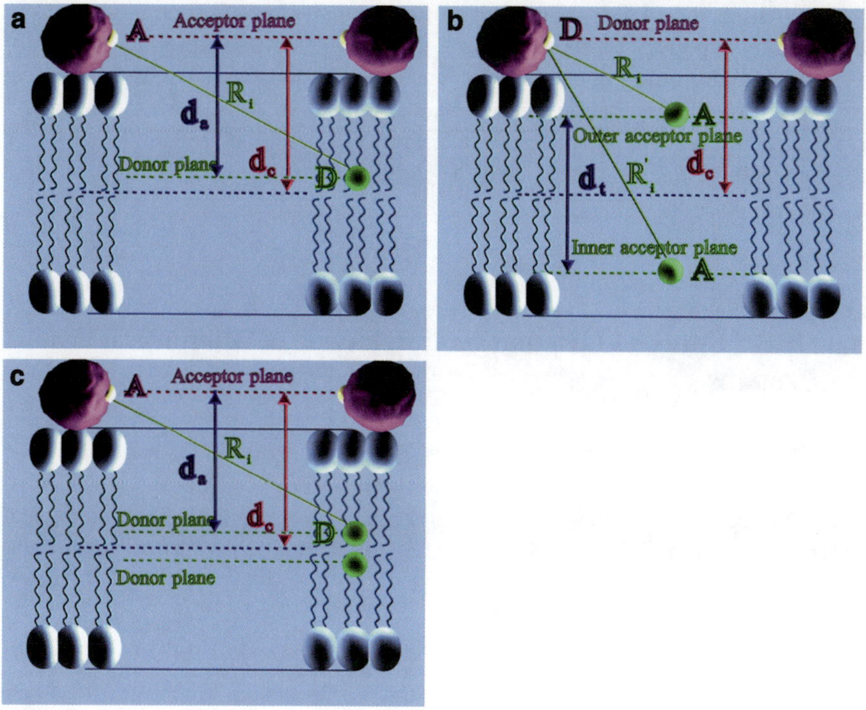

Fig. 2 Schematic representation of planar arrangement of donors and acceptors in a lipid bilayer

$$S_2(\lambda) = \int_{d_c+0.5d_t}^{\infty} \left[1 - \exp\left(-\lambda\left(\frac{R_o}{R}\right)^6\right)\right] 2\pi R dR \tag{16}$$

$$E = 1 - \int_0^{\infty} \exp(-\lambda)\exp\left[-C_a^s(S_1(\lambda) + S_2(\lambda))\right] d\lambda \tag{17}$$

where S_1 and S_2 are the quenching contributions describing energy transfer to the outer and inner acceptor planes, respectively. Alternatively, if there exist one acceptor plane located at a distance d_c from the membrane center and two donor planes separated by a distance d_t (Fig. 2c), energy transfer efficiency can be written as

$$E = 1 - 0.5 \times \left(\int_0^{\infty} \exp(-\lambda)\exp\left[-C_a^s S_1(\lambda)\right] d\lambda + \int_0^{\infty} \exp(-\lambda)\exp\left[-C_a^s S_2(\lambda)\right] d\lambda\right) \tag{18}$$

Next, it seems of importance to illustrate what kind of information can be obtained by applying the above uniform distribution formalism to the analysis of FRET in protein-lipid systems. To this end, we will consider some representative examples of experimental design and data treatment strategy, concerning protein (1) orientation relative to lipid-water interface, (2) membrane binding parameters, (3) the depth of bilayer penetration, (4) aggregation state, and (5) effect on lipid lateral distribution.

4 FRET Study of Orientational Behavior of Membrane-Bound Proteins

As follows from the above considerations, the efficiency of FRET is a function of the donor-acceptor distance, the donor quantum yield, overlap of the donor emission and acceptor absorption spectra, and relative orientation of the donor and acceptor transition dipoles. Of these, the most problematic parameter is the orientation factor κ^2 defined as [28]:

$$\kappa^2 = (\sin\theta_D \sin\theta_A \cos\varphi - 2\cos\theta_D \cos\theta_A)^2 \tag{19}$$

where θ_D and θ_A are the angles between the donor emission (***D***) or acceptor absorption (***A***) transition moments and the vector ***R*** joining the donor and acceptor and φ is the dihedral angle between the planes (***D***, ***R***) and (***A***, ***R***). Orientation factor can vary from 0 to 4, the minimum value corresponds to perpendicularly oriented donor and acceptor dipoles, while the maximum one characterizes the case when these dipoles are parallel and identically directed [19]. It is a common practice in FRET studies to use the isotropic value of orientation factor, 0.67, which is valid in the case when donors and acceptors can adopt all orientations (isotropic condition) in a time short compared to the transfer time (dynamic averaging condition) [28]. However, in a highly anisotropic membrane environment, rotational mobility of donors and acceptors is usually restricted, and the isotropic condition is hardly satisfied. Besides, if the donor and acceptor dipoles exhibit certain preferable orientations, for instance, when protein-lipid interactions involve specific binding sites, κ^2 value may substantially differ from 0.67. Such an uncertainty in the choice of orientation factor is regarded as the main source of uncertainty in the quantitative interpretation of FRET data. In the following, we will consider some possible ways for circumventing this problem.

First example is intended to illustrate how FRET can help to answer the question of whether there exists certain specific orientation of the protein molecule relative to lipid-water interface. This question has been addressed in our early study of lipid bilayer interactions of cytochrome *c*, a peripheral mitochondrial protein which plays important role in respiration process and apoptosis. Cytochrome *c* is a globular protein with the size about 3 nm, containing heme group located 1.5 nm from its

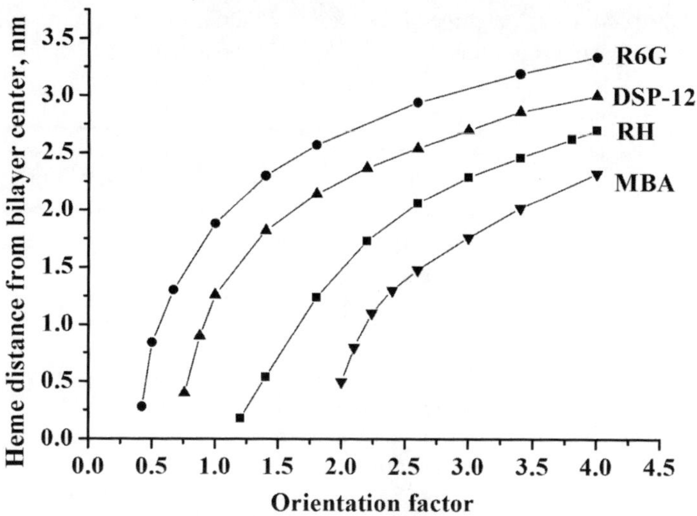

Fig. 3 Dependence of the distance between heme group of cytochrome *c* and membrane midplane on orientation factor

surface. This group is an effective energy acceptor for a number of fluorophores. We employed FRET technique to examine cytochrome *c* association with the model membranes, composed of zwitterionic lipid phosphatidylcholine (PC) and anionic lipid cardiolipin (CL) [29]. Several fluorescent probes, including (3-methoxybenzanthrone (MBA), *N*,*N*′-bishexamethylenrhodamine (RH), rhodamine 6G (R6G), and 4-(dimethylaminostyryl)-1-dodecylpiridine (DSP-12)), were used as donors, while heme group of the protein was recruited as acceptor. The donors are distributed between the outer and inner bilayer leaflets, thus forming two donor planes, separated by a certain distance d_t, while protein molecules are supposed to reside at the outer membrane side, forming one acceptor plane, located at a distance d_c from the membrane center (Fig. 2c). In this case, relative quantum yield of the donor can be written as the sum of two terms, where lower integration limits correspond to the separation of acceptor plane from the outer and inner donor planes. The data treatment strategy involved fitting of the relative quantum yields calculated by numerical integration of the Eqs. 15, 16 and 18 to those measured experimentally with the distance between the acceptor plane and membrane center (d_c) as the optimizing parameter. This procedure was repeated for all FRET profiles with κ^2 being varied in the widest possible range, from 0 to 4. As a result, we obtained numerous sets of two parameters—acceptor distance from the bilayer midplane and orientation factor – providing the best fit of the experimental data. Several features of $d_c(\kappa^2)$ dependencies are worthy of mention. As can be seen in Fig. 3, for each of the employed donors, there exists certain minimum κ^2 value characterizing the situation when heme groups reside at the bilayer midplane ($d_c = 0$), and maximum heme distance from the bilayer center ($d_c^{\max}$), estimated at $\kappa^2 = 4$. In the case of random reorientation of donors and acceptors, one should expect the intersection or closing of the plots $d_c(\kappa^2)$, obtained for different donors, at a point, corresponding to the isotropic value of orientation factor ($\kappa^2 = 0.67$) and actual heme separation from the bilayer midplane, since this parameter appears to be invariant over a series of donors. The absence of such an intersection implies that orientation of heme dipole cannot be considered as random. Another support for this assumption comes from the observation that for some donors, specifically, MBA and RH, the minimum possible κ^2 value is greater than 0.67. Taken together, these findings suggest that there exists certain specific orientation of the heme group relative to lipid-water interface. It is known that acidic and basic groups on the surface of cytochrome *c* molecule are segregated into two positively charged patches with negative patch between them [30]. The involvement of one of the positively charged patches in the interactions with lipids may account for specific protein disposition with respect to the membrane surface. This example demonstrates that uncertainty in the orientation factor value can be turned to our advantage by the proper experimental design, employing the donors distributed symmetrically between the outer and inner membrane leaflets and protein-associated acceptors. The observations, such as noncrossing dependencies of the acceptor distance from the membrane center on the orientation factor obtained for different donors and/or successful data fitting only for κ^2 exceeding the isotropic value, may serve as the indications for the existence of specific lipid-binding site on the surface of protein molecule.

Importantly, FRET can provide not only qualitative but also quantitative topological information about membrane-associated proteins. For instance, using a series of site-directed cysteine mutants fluorescently labeled with AEDANS, Nazarov et al. succeeded in determining the topology and bilayer embedment of M13 major coat protein in dioleoylphosphatidylcholine/dioleoyl phosphatidylglycerol (4:1 mol/mol) vesicles. The authors developed novel simulation-based fitting approach that was applied to analyzing the steady-state FRET between tryptophan as a donor and AEDANS as an acceptor. The tilt of the transmembrane helix of M13 major coat protein was found to be around 18° [31].

5 Membrane Location of Proteins Determined from FRET

One of the most widespread applications of FRET in membrane studies involves characterization of protein disposition in a lipid bilayer. Both analytical and numerical approaches have been used to obtain quantitative estimates for membrane embedment of a variety of structurally different proteins, as summarized in Table 1. It should be noted in this regard that the choice of a certain theoretical strategy for analyzing FRET data is dictated by specific structural features of the system under study, such as randomness of fluorophore distribution, size of the donor- or acceptor-bearing protein domain, and spatial relationships between donor and acceptor arrays. However, in all cases, the uncertainty in orientation factor value may substantially reduce or even eliminate the reliability of FRET estimates of the protein bilayer location. In view of this, we found it reasonable to briefly

Table 1 Illustrative examples of FRET-based evaluation of protein position in a membrane

Protein	Membrane	Structural information	Reference
GPI-anchored alkaline phosphatase (PLAP)	PC	The distance between PLAP and lipid-water interface is smaller than 1–1.4 nm	[32]
Acetylcholine receptor (AChR)	Native AChR-rich membrane	Minimum distance between Trp and Laurdan is ca. 1.4 nm	[33]
C-terminal sterile α motif (SAM) domain of human p73	PC, phosphatidic acid	Trp is located at 1.1–2 nm from the bilayer center	[34]
Perfringolysin	PC, cholesterol	Domain 1 is located ca. 11 nm above the membrane surface	[35]
Cytochrome b_5	PC	Heme moiety is located about 1.5 nm from the membrane surface, Trp is buried at the depth ca. 2 nm	[36]
α-toxin	PC, PG, cholesterol	Cys130 and Cys69 distances from polar/nonpolar transition region are ca. 0.8 and 8 nm, respectively	[37]
P-glycoprotein	PC, phosphatidylethanolamine	The distance of Cys residues from the lipid-water interface is ca. 3.1–3.5 nm	[38]
Hemoglobin	Red blood cell membrane	Distance between heme groups and 12-(9-anthroyl)stearic acid is ca. 4–6 nm	[39]
(Ca^{2+} + Mg^{2+})-ATPase	PC	Ca^{2+} binding sites of ATPase reside at 2 nm distance from the lipid-water interface	[40]
Acetylcholine receptor (AChR)	AChR membrane	Ethidium binding site is positioned ca. 5 nm from the membrane surface	[41]
G-protein	PC	Distances between protein α, β, and γ subunits and lipid headgroups are ca. 4.6, 3.8, and 3.7 nm, respectively	[42]
Epidermal growth factor (EGF)	Plasma membranes of epidermoid carcinoma cells	Amino terminus of EGF is about 6.7 nm away from the cell surface	[43]
Lysozyme	PC, PG	Average distance between Trp residues and bilayer center is ca. 2 nm	[44]
Cytochrome *c*	PC, CL	Heme distance from the bilayer midplane lies between 3.6 and 4 nm	[45]
M13 major coat protein	PC, PG	Trp separation from the membrane center is ca. 8.5 nm	[31]

describe here one approach to reducing this problem, based on the early work of Davenport et al. [46].

The applicability of Eq. 19 is limited to the case when the vectors $\boldsymbol{D}$ and $\boldsymbol{A}$ do not undergo any reorientation during the transfer time. Alternatively, Förster radius should be calculated using the dynamic average value of orientation factor $(\langle\kappa^2\rangle)$. If the donor emission and acceptor absorption transition moments are symmetrically distributed within the cones about certain axes $\boldsymbol{D_x}$ and $\boldsymbol{A_x}$, $\langle\kappa^2\rangle$ is given by [28]:

$$\begin{aligned}\langle\kappa^2\rangle = &(\sin\Theta_D \sin\Theta_A \cos\Phi - 2\cos\Theta_D \cos\Theta_A)^2\langle d_D^x\rangle\langle d_A^x\rangle \\ &+ 1/3(1-\langle d_D^x\rangle)+1/3(1-\langle d_A^x\rangle) \\ &+ \cos^2\Theta_D\langle d_D^x\rangle(1-\langle d_A^x\rangle) + \cos^2\Theta_A\langle d_A^x\rangle(1-\langle d_D^x\rangle)\end{aligned} \quad (20)$$

where Θ_D and Θ_A are the angles made by the axes $\boldsymbol{D_x}$ and $\boldsymbol{A_x}$ with the vector $\boldsymbol{R}$, Φ is the angle between the planes containing the cone axes and the vector $\boldsymbol{R}$, and $\langle d_D^x\rangle$ and $\langle d_A^x\rangle$ are so-called axial depolarization factors:

$$\left\langle d_{D,A}^x\right\rangle = 3/2\langle\cos^2\psi_{D,A}\rangle - 1/2 \quad (21)$$

where $\psi_{D,A}$ are the cone half-angles. These factors are related to the steady-state (r) and fundamental (r_0) anisotropies of the donor and acceptor [28]:

$$d_{D,A}^x = \pm(r_{D,A}/r_{0D,A})^{1/2} \quad (22)$$

When the donor and acceptor planar arrays are located at different levels across the membrane, separated by a distance d_a, multiple donor-acceptor pairs are involved in energy transfer, so that orientation factor appears to be a function of the donor-acceptor separation (R). Particularly, for the most probable membrane orientation of $\boldsymbol{D_x}$ and $\boldsymbol{A_x}$, parallel to the bilayer normal, the angles Θ_D and Θ_A made by $\boldsymbol{D_x}$ and $\boldsymbol{A_x}$ with $\boldsymbol{R}$ are equal and depend on the distance between donor and acceptor ($\Theta_A = \Theta_D = \theta,\ \theta = f(R)$). Under these circumstances, Eq. 20 can be rewritten in the form

$$\begin{aligned}\langle\kappa^2(\theta)\rangle = &\langle d_D^x\rangle\langle d_A^x\rangle(3\cos^2\theta - 1)^2 + 1/3(1-\langle d_D^x\rangle) + 1/3(1-\langle d_A^x\rangle) \\ &+ \cos^2\theta(\langle d_D^x\rangle - 2\langle d_D^x\rangle\langle d_A^x\rangle + \langle d_A^x\rangle)\end{aligned} \quad (23)$$

where $\cos^2\theta = (d_a/R)^2$. Next, by representing Förster radius as the product of distance-dependent and distance-independent terms $R_o = [\kappa^2(R)]^{1/6} \cdot R_o^r$, one obtains

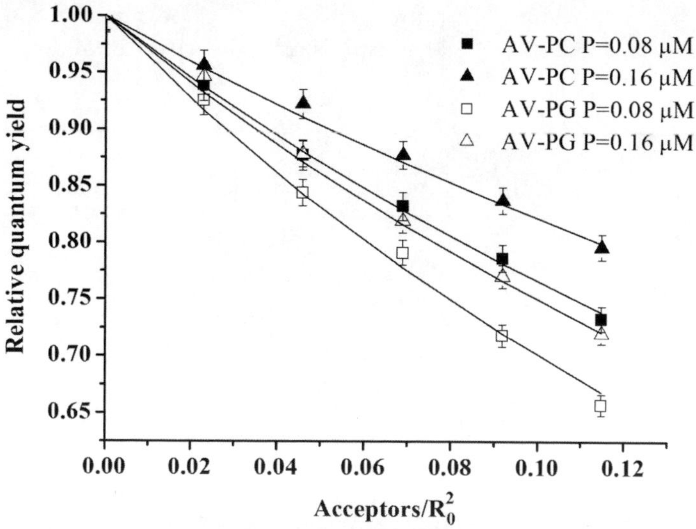

Fig. 4 Förster energy transfer profiles for lysozyme Trp-AV donor-acceptor pair in PC/PG liposomes (40 mol% PG). Lipid concentration was 5 μM

$$S(t) = \int_{d_a}^{\infty} \left[1 - \exp\left(-\lambda\kappa^2(R) \left(\frac{R_o^r}{R} \right)^6 \right) \right] 2\pi R dR \tag{24}$$

$$R_o^r = 979\left(n_r^{-4} Q_D J\right)^{1/6}, \qquad J = \frac{\int_0^\infty F_D(\lambda)\varepsilon_A(\lambda)\lambda^4 d\lambda}{\int_0^\infty F_D(\lambda) d\lambda} \tag{25}$$

This formalism was employed to describe FRET between tryptophan residues of lysozyme as donors and anthrylvinyl-labeled phosphatidylcholine (AV-PC) or phosphatidylglycerol (AV-PG) as acceptors in lipid vesicles composed of phosphatidylcholine and varying proportions of phosphatidylglycerol (10, 20, or 40 mol%) [44]. The efficiency of energy transfer was calculated from the sensitized emission of the acceptor by measuring tryptophan fluorescence in the absence and presence of donors. Shown in Fig. 4 are typical dependencies of energy transfer efficiency on the acceptor concentration. AV fluorophore is distributed between the outer and inner membrane leaflets, residing at the level of terminal methyl groups, thus forming two acceptor planes, separated by a certain distance d_t. Taking into account distance dependence of the orientation factor, energy transfer to the outer and inner acceptor planes can be described by the following equations:

$$S_1(\lambda) = \int\limits_{|d_c - 0.5d_t|}^{\infty} \left[1 - \exp\left(-\lambda\kappa_1^2(R)\left(\frac{R_o^r}{R}\right)^6\right)\right] 2\pi R dR \tag{26}$$

$$S_2(\lambda) = \int\limits_{d_c + 0.5d_t}^{\infty} \left[1 - \exp\left(-\lambda\kappa_2^2(R)\left(\frac{R_o^r}{R}\right)^6\right)\right] 2\pi R dR \tag{27}$$

$$\kappa_{1,2}^2(R) = \langle d_D^x \rangle \langle d_A^x \rangle \left(3\left(\frac{d_c \mp 0.5d_t}{R}\right)^2 - 1\right) + \frac{1 - \langle d_D^x \rangle}{3} + \frac{1 - \langle d_A^x \rangle}{3} + \left(\frac{d_c \mp 0.5d_t}{R}\right)^2 \left(\langle d_D^x \rangle - 2\langle d_D^x \rangle \langle d_A^x \rangle + \langle d_A^x \rangle\right) \tag{28}$$

$$E = 1 - \int\limits_0^{\infty} \exp(-\lambda) \exp\left[-C_a^s(S_1(\lambda) + S_2(\lambda))\right] d\lambda \tag{29}$$

The relationships 26–29 are valid when the donor and acceptor transition moments are distributed about the axes $\boldsymbol{D_x}$ and $\boldsymbol{A_x}$ parallel to the bilayer normal $\boldsymbol{N}$. If this is not the case, additional depolarization factors accounting for the deviations of $\boldsymbol{D_x}$ and $\boldsymbol{A_x}$ from $\boldsymbol{N}$ should be introduced: $d_{D,A}^a = \frac{3}{2}\cos^2\alpha_{D,A} - \frac{1}{2}$, where $\alpha_{D,A}$ are the angles made by $\boldsymbol{D_x}$ and $\boldsymbol{A_x}$ with $\boldsymbol{N}$. By applying the Soleillet's theorem stating the multiplicativity of depolarization factors, Eq. 28 may be rewritten in a more general form

$$\kappa_{1,2}^2(R) = d_D d_A \left(3\left(\frac{d_c \mp 0.5d_t}{R}\right)^2 - 1\right) + \frac{1 - d_D}{3} + \frac{1 - d_A}{3} + \left(\frac{d_c \mp 0.5d_t}{R}\right)^2 (d_D - 2d_D d_A + d_A) \tag{30}$$

where $d_{D,A} = \left\langle d_{D,A}^x \right\rangle d_{D,A}^a$. Allowing for the size of AV fluorophore (ca. 0.7 × 0.3 nm) and high mobility of the terminal groups of hydrocarbon chains, d_t value was slightly varied in the range from 0.3 to 0.7 nm, while parameter d_c, characterizing the distance between the donor plane and membrane center, was optimized in the data-fitting procedure. The axial depolarization factors were calculated using the results of steady-state fluorescence anisotropy measurements. The recovered values of the donor separation from bilayer center are consistent with the location of lysozyme Trp residues in the interfacial bilayer region. Our data agree with the viewpoint that helix-loop-helix domain (87–114 residues) located at the upper lip of the active site cleft accounts for lysozyme insertion into lipid bilayer. The nonpolar portion of this domain (residues 87–95) penetrating into hydrophobic bilayer region serves as a membrane anchor, while terminal basic residues form electrostatic contacts with anionic phospholipid headgroups. This mode of membrane binding of lysozyme causes Trp62 and Trp108 to be accommodated at the interface.

This example demonstrates that for donors and acceptors residing at different depths within the membrane, orientation factor can be incorporated as a distance-dependent parameter at the integration over R step in the theoretical calculation of FRET efficiency. Knowing the steady-state and fundamental anisotropies of donor and acceptor, allows to raise the accuracy of FRET-based estimates of the transverse location of membrane-associated proteins.

6 FRET in Characterization of Protein-Lipid Binding

A specific feature of energy transfer in membranes is the dependence of FRET efficiency on acceptor surface concentration (Eq. 17). This makes FRET technique suitable for determination of not only structural parameter (position relative to lipid-water interface) but also the binding parameters (association constant and the number of lipids per bound protein) [47–49]. In the case when acceptor is intrinsic or extrinsic protein chromophore, for example, tryptophan residue or heme group, and protein is distributed between aqueous and lipid phases, the amount of membrane-bound acceptor (B_a) can be related to the total acceptor concentration (A), using the appropriate binding model. In terms of conventional Langmuir model, this relationship is given by

$$K_a = \frac{B_a}{(A - B_a)(L_a/n - B_a)},$$
$$B_a = \frac{1}{2}\left[A + \frac{L_a}{n} + \frac{1}{K_a} - \sqrt{\left(A + \frac{L_a}{n} + \frac{1}{K_a}\right)^2 - 4\frac{AL_a}{n}}\right] \tag{31}$$

where K_a is the equilibrium association constant and n is the binding stoichiometry.

By combining the FRET and binding models, one obtains the relative quantum yield as a function of five variables $Q_r = f(A, L_a, K_a, n, d_c)$. Of these, total acceptor concentration, A, and amount of accessible lipid L_a(usually taken as a half of total lipid concentration for large unilamellar vesicles and nonpermeating acceptor) are independent parameters that can be varied in the experiment, while K_a, n, and d_c are unknown parameters which can be derived from least-squares fitting of the experimental data. However, unambiguous evaluation of the structural and binding characteristics is complicated by a strong cross-correlation between optimizing parameters, which makes it impossible to judge whether the changes in donor quantum yield result from the variation of the acceptor transverse location or from alterations in the extent of its association with the membrane. To overcome this problem, we have tried to ascertain what format of FRET measurements

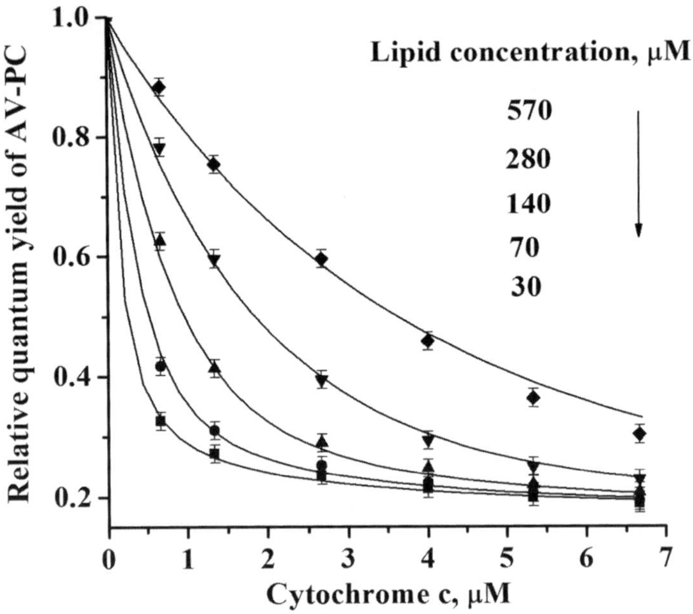

Fig. 5 Förster energy transfer profiles for cytochrome *c* heme-AV donor-acceptor pair in PC/CL liposomes (10 mol% CL)

permits unequivocal resolution of the structural and binding parameters [50]. Using computer simulations relative quantum yield of a donor was calculated in terms of the above analytical model for various combinations of acceptor (A) and lipid concentrations (L) with preset values of optimizing parameters K_a, n, and d_c. Gaussian noise was added to the calculated data to mimic the experimental errors. Three characteristic FRET datasets were analyzed: (1) univariate data (1D data) obtained at a fixed lipid concentration by varying total acceptor concentration only, (2) two-dimensional data array (2D data) obtained by varying both lipid and total acceptor concentrations for the case of saturable binding, and (3) 2D data for the case of unsaturable binding. To assess how the quality and uniqueness of the least-squares fitting depend on the format of FRET data, we analyzed the projections of the corresponding error surfaces (χ^2 statistic for a given dataset) with respect to a specific parameter. Simulation results clearly indicated that only analysis of two-dimensional data arrays with saturable binding allows the model parameters to be recovered with high accuracy and statistical significance.

The proposed global analysis approach has been employed to analyze FRET between anthrylvinyl-labeled phosphatidylcholine (AV-PC) as an energy donor and heme group of cytochrome *c* as an acceptor [50]. AV-PC was incorporated into lipid vesicles composed of phosphatidylcholine and 10 mol% of cardiolipin (CL). AV fluorophore is localized in a lipid bilayer close to the terminal methyl groups, preferentially orienting parallel to the lipid acyl chains. Shown in Fig. 5 is relative quantum yield of AV-PC as a function of the total cytochrome c concentration, recorded at five different lipid concentrations. As the lipid concentration increases, acceptor molecules tend to distribute over a greater bilayer area resulting in the decrease of acceptor surface concentration and lowered energy transfer. It is this peculiarity of energy transfer dependence on total acceptor and lipid concentrations that makes it possible to resolve structural and binding parameters by the least-squares fitting of the expanded data array. Solid lines in Fig. 5 represent best fit of the FRET data by the global analysis procedure yielding the following estimates of the fitting parameters: $d_c = 3.6 \pm 0.2$, nm; $K_a = 3.6 \pm 1.1$, μM^{-1}; and $n = 27 \pm 2$. Thus, in the case where energy transfer occurs from donors localized in the membrane at a known depth to acceptors distributed between aqueous and lipid phases, analysis of two-dimensional data arrays can provide unambiguous information on both binding parameters and transverse membrane location of protein fluorophore.

The above example reflects only one of the possibilities offered by FRET technique in elucidating the quantitative and qualitative aspects of protein-lipid binding. Most of these aspects are highlighted in the brilliant recent works from Loura and Prieto group [51–54]. The authors discuss how FRET may help in establishing protein preference to a certain type of lipids and make a critical comparison of the existing theoretical approaches to quantification of protein-lipid selectivity.

7 Protein Aggregation in a Membrane Environment Detected by FRET

FRET has been conventionally used in membrane studies for monitoring the changes in protein aggregation state [55–58]. The enhancement of energy transfer between protein derivatives fluorescently labeled with donor and acceptor is regarded as a qualitative proof of protein self-association. Further quantitative analysis can, in principle, yield the estimates of aggregate size using appropriate data treatment formalism. To perform such kind of analysis most correctly, it is important to differentiate FRET between donors and acceptors confined to protein aggregates from energy transfer involving nonaggregated species [54]. The contribution of this interfering FRET can be neglected only in the case of very low concentration of the tagged protein. Furthermore, fluorescent labeling can modify self-associating propensity of polypeptide chain [59]. Along with the complexity of aggregation process per se, these factors may limit the amount of reliable quantitative information on oligomeric structure. A versatility of the problems addressed by FRET in the field of protein aggregation is illustrated by a number of works reviewed, particularly, in [3]. To exemplify, in the pioneering work of Vanderkooi et al., Ca^{2+} ATPase of sarcoplasmic reticulum was covalently labeled with N-iodoacetyl-N'-(5-sulfo-naphthyl)ethylenediamine (IAEDANS) as a donor and iodoacetamidofluorescein (IAF) as an acceptor [60]. While analyzing the results of FRET measurements in reconstituted vesicles, the authors reasonably considered two possibilities: (1) energy transfer occurring when the donor- and acceptor-tagged molecules collide or come within the Förster distance due to Brownian motion and (2) energy transfer in ATPase oligomers in which the donor and acceptor come into close proximity. The observations that FRET was not altered with increasing lipid-to-protein molar ratio but was abolished by the addition of unlabeled protein have been regarded as the arguments in favor of ATPase self-association. By measuring energy transfer between Trp and its nonfluorescent 2-hydroxy-5-nitrobenzylbromide (HNB) derivative, John and Jahnig obtained evidence for melittin aggregation in dimyristoylphosphatidylcholine (DMPC) bilayer [61]. Investigation of FRET between fluorescent tags 2,6-dansyl chloride and dabsyl chloride led Adair and Engelman to conclusion that peptide mimicking transmembrane region of glycophorin A forms helical dimers in DMPC model membranes [62]. Using the formalism of site-directed cross-linking approach to the determination of oligomeric structure [63] and assuming equal energy transfer to all subunits in oligomer, the authors proposed the method allowing to distinguish dimers from higher oligomers. The linear dependence of FRET efficiency on acceptor mole fraction was shown to be a characteristic sign of dimer formation. Further steps to solving this problem have been made by Li et al. [64]. A simulation-based approach was developed to resolve the number of subunits in oligomer, the distance between labeled sites on different subunits, and the fraction of monomers. This approach was used to characterize the oligomeric structure of phospholamban (PLB) in dioleoylphosphatidylcholine (DOPC) bilayer.

Sparr et al. employed FRET technique to ascertain how lipid bilayer thickness affects the aggregation propensity of a series of Trp-flanked peptides specifically tagged with pyrene [65]. Analysis of energy transfer from Trp to pyrene indicated that hydrophobic mismatch promotes helix-helix association. Using fluorescent labels 5-((2-aminoethyl)amino) naphthalene-1-sulfonic acid (EDANS) as a donor and fluorescein isothiocyanate (FITC) as an acceptor, Fernandes et al. obtained the arguments in favor of membrane-mediated dimerization of N-terminal amphipathic α-helix peptide (H0-NBAR) of the BAR (Bin, amphiphysin, Rvs) domain. It was demonstrated that simultaneous fitting of two datasets corresponding to different lipid/protein ratios permits quantification of the FRET contributions from aggregated and nonaggregated peptide [66]. FRET measurements performed by Fung and coworkers with Cy3-Cy5 donor-acceptor pair offered insights into the mechanisms of membrane-mediated oligomerization of β_2-adrenoceptor [67]. Tetrameric species of this protein were hypothesized to dominate in the lipid bilayer from DOPC and cholesterol hemisuccinate.

One of the main tendencies in the past decade is the increasing use of FRET in monitoring the process of protein aggregation in vivo. For instance, Rajan et al. found evidence for high specificity of protein-protein interactions in living cell [68]. For this purpose, aggregation-prone proteins were biosynthetically tagged with mutant forms of green fluorescent protein (GFP), cyan (CFP), and yellow fluorescent proteins (YFP), with spectral overlap suitable for FRET. Similar donor-acceptor pair was employed by Mihai et al. who investigated oligomerization of discoidin domain receptor on the cell surface [69] and Woehler et al. who analyzed the oligomeric state of the 5-HT1A receptor within living cells [70]. Liu and coauthors recruited GFP as a donor and red fluorescence protein (RFP) as an acceptor in confocal FRET microscopy study of T5P gC-crystallin aggregation in cell cytoplasm [71].

Remarkably, combined use of different FRET formats allows determining the structure of complex protein-lipid aggregates, as was demonstrated in recent studies of Coutinho et al. [72, 73]. By measuring energy transfer in two types of donor-acceptor pairs, involving BODIPY-labeled lipid or Alexa-488-labeled protein as donors and rhodamine-labeled lipids as acceptors, the authors succeeded in structural characterization of supramolecular aggregates formed by lysozyme and negatively charged membranes. The recovered pinched multilamellar motif was hypothesized to govern the structure of amyloid-like fibrils found in protein-lipid systems.

8 Protein-Induced Lipid Demixing Monitored by FRET

Another essential aspect in which FRET technique may prove useful concerns membrane heterogeneity, viz., lipid lateral redistribution in response to protein binding followed by domain formation [74, 75]. Two underlying processes are considered: (1) local lipid demixing implicating molecular-scale deviation from the average lipid composition within and around the protein-membrane interaction zone and (2) the formation of macroscopic protein-lipid domains enriched in specific lipid [76, 77]. The characteristic distance scale of energy transfer makes this technique particularly sensitive to nanometer-size domains. Pros and cons of different approaches to FRET analysis of lateral membrane domains are given in exhaustive recent reviews [15, 78]. Therefore, here, we restrict ourselves to simplest illustrations of how steady-state FRET can be used to detect domain formation in the model protein-lipid systems.

We employed FRET to define demixing-favoring conditions and the extent of lipid redistribution produced by the basic proteins lysozyme and cytochrome *c* in the binary negatively charged model membranes composed of PC and PG [44, 79]. To answer the question of whether these proteins are capable of inducing the formation of lateral domains enriched in anionic lipids, we compared energy transfer efficiencies in the systems containing either AV-PC or AV-PG as energy acceptors for lysozyme tryptophans or energy donors for heme group of cytochrome *c*. As shown in Fig. 4, the use of AV-PG instead of AV-PC as energy acceptor for tryptophan residues of lysozyme resulted in higher efficiencies of energy transfer. This finding can be interpreted in terms of lysozyme ability to promote lipid lateral redistribution followed by the accumulation of anionic lipid in the vicinity of bound protein. As positively charged lysozyme approaches the membrane surface, PG molecules migrate toward interaction zone, thereby replacing PC molecules. To describe this process, the FRET model considered above was complemented by two additional parameters, characterizing the radius of lateral domains (r_{dm}) and the ratio of PG concentrations in the interaction zone at nonrandom and random acceptor distribution (k). Thus, the expressions for FRET efficiency take the following form:

$$E = 1 - \int_0^{\infty} \exp(-\lambda)\exp\left[-C_a^s\left(\left[\frac{L_{out}S_L - Pf_b\pi r_{dm}^2 k}{L_{out}S_L - Pf_b\pi r_{dm}^2}\right]S_{11}(\lambda) + kS_{12}(\lambda) + S_2(\lambda)\right)\right]d\lambda \tag{32}$$

$$S_{11}(\lambda) = \int_{\left[r_{dm}^2 + (d_c - 0.5d_t)^2\right]^{0.5}}^{\infty} \left[1 - \exp\left(-\lambda\kappa_1^2(R)\left(\frac{R_o^r}{R}\right)^6\right)\right]2\pi R dR \tag{33}$$

$$S_{12}(\lambda) = \int_{|d_c - 0.5d_t|}^{\left[r_{dm}^2 + (d_c - 0.5d_t)^2\right]^{0.5}} \left[1 - \exp\left(-\lambda\kappa_1^2(R)\left(\frac{R_o^r}{R}\right)^6\right)\right]2\pi R dR \tag{34}$$

Here L_{out} is the lipid concentration in the outer monolayer, $f_b = B/P$ is the fraction of bound protein, B is the molar concentration of bound protein, and P is the total protein concentration. The FRET profiles obtained with AV-PG as energy acceptor were approximated by Eqs. 32–34, using the following data treatment strategy: (1) parameter k was allowed to vary from 1 to its maximum value (2.5) possible for the model membranes under study, in which PG proportion was 40 mol%; (2) parameter r_{dm} was taken from the limits dictated by the requirement that the area occupied by protein-induced lateral domains must be less than the total membrane area $Pf_b\pi r_{dm}^2 k \le L_{out}S_L$; and (3) parameter d_c, denoting the distance between the plane of donors, that is, lysozyme tryptophan residues, and membrane midplane, was optimized for every pair of r_{dm} and k. In this way, we recovered the sets $\{r_{dm}, k, d_c\}$ providing the best fit of the experimental data. Good agreement between theory and experiment was found for conditions where size of the region with increased PG concentration (r_{dm}) was less than 3.4 nm. The fact that this estimate is comparable with lysozyme radius (ca. 3 nm) is indicative of a local deviation from the average lipid composition.

Analogous conclusion was reached for another basic protein – cytochrome c using simulation-based data analysis methodology. We examined FRET between AV-PC or AV-PG as donors and the heme group of cytochrome c as an acceptor in PC/PG model membranes containing 10-, 20-, or 40-mol% PG [79]. The differences between AV-PC and AV-PG FRET profiles were observed only at PG content 10 mol%, suggesting a segregation of anionic lipids into lateral domains.

To interpret this effect quantitatively, the Monte-Carlo simulation was designed, in which AV-PG donors were considered as being randomly distributed within disk-shaped domains centered at each acceptor location. Our main goal was to determine characteristic domain size, that is, dimension of the protein-affected region where PG concentration is k times higher than that for a random lipid distribution. Positions of donors and acceptors were generated randomly in a square cell with the side length taken as $10R_o$, assuming periodic boundary conditions to avoid edge effects. The relative quantum yield averaged over all donors was calculated from the fluorophore coordinates using the following equation:

$$Q_r = \frac{1}{N_D}\sum_{j=1}^{N_D}\left[1+\sum_{i=1}^{N_{AC}}\left(\frac{R_o^r\kappa^2(r_{ij})}{r_{ij}}\right)^6\right]^{-1} \tag{35}$$

where N_D, N_{AC} stand for the number of donors and acceptors in a square cell. The number of acceptors was determined by multiplying protein surface density by the cell square. By operating with the quantities, such as domain radius, the ratio of PG concentrations in the domain region at nonrandom and random distribution of charged lipids and total number of disk-shaped domains, which was taken equal to the number of membrane-bound protein molecules, we calculated the number of donors (i.e., AV-PG or AV-PC) in domain and nondomain regions of a cell. The simulation procedure was repeated for multiple fluorophore configurations until the standard deviation in relative quantum yield was less than 2 %. Simulation-based fitting of the FRET data with k being varied from 1 to 10 (the value corresponding to complete replacement of PC with PG) revealed domain radius not to exceed 4 nm, implying that cytochrome-induced lipid demixing takes place locally, in the immediate vicinity of the adsorbed protein.

9 Concluding Remarks

The past few decades have seen tremendous upsurge in the use of FRET technique in biomedical studies, especially in structural characterization of membrane systems. This chapter illuminates only a minor fraction of FRET applications in this research area. Our goal was to illustrate that even the simplest steady-state formats of FRET measurements feasible for most laboratories can provide valuable information about the structure of protein-lipid assemblies. FRET between multiple donors and acceptors distributed in lipid phase or confined to membrane-associated protein molecules can be described analytically only for the case of random fluorophore distribution. Therefore, the accent is now shifting toward the development of simulation-based data analysis approaches, particularly, Monte-Carlo simulations, which offer much more versatility and can be applied for any complex geometric conditions [80, 81]. Another recent trend is the extension of time-resolved FRET measurements having serious advantages in determining the size of lateral membrane domains [15]. Nevertheless, potential of steady-state FRET is far from being fully exhausted. The above examples demonstrate that this kind of FRET may prove useful in answering the questions of (1) whether there exists certain specific orientation of the protein molecule relative to lipid-water interface; (2) what is the transverse membrane location of protein fluorophores; (3) what are the protein-membrane binding characteristics; (4) how membrane environment affects protein aggregation state; and (5) whether lipid molecules undergo lateral redistribution in response to protein adsorption.

In general, the most sensible experimental strategies for FRET-based analysis of protein-lipid interactions seem to lie in correlating the data from a multitude of donor-acceptor pairs (involving both intrinsic and extrinsic protein fluorophores and lipid tags with defined bilayer location) and different formats of FRET measurements. The perspectives of FRET are associated with synergistic progress in the development of novel (1) instrumentation and more sophisticated data treatment methods, (2) genetically encoded fluorescent proteins and site-specific fluorescent labels, and (3) fluorophores with improved characteristics (higher signal/background noise ratio and photostability, smaller size, wider range of excited state lifetimes). There is no doubt that continuous refinement of FRET technique will substantially expand the area of its application in membrane studies.

Acknowledgments GG gratefully acknowledges a visiting scientist award by the Sigrid Juselius Foundation. This work was supported by the grants from European Social Fund (project number 2009/0205/1DP/1.1.1.2.0/09/APIA/VIAA/152) and Fundamental Research State Fund of Ukraine (project number F.41.4/014).

References

1. Selvin PR (2000) The renaissance of fluorescence resonance energy transfer. Nat Struct Biol 7:730–734
2. Giepmans BNG, Adams SR, Ellisman MH, Tsien RY (2006) The fluorescent toolbox for assessing protein location and function. Science 312:217–224
3. Loura LMS, Prieto M (2011) FRET in membrane biophysics: an overview. Front Physiol. doi:10.3389/fphys.2011.00082
4. Wu P, Brand L (1994) Resonance energy transfer: methods and applications. Anal Biochem 218:1–13
5. Selvin PR (1995) Fluorescence resonance energy transfer. Method Enzymol 246:300–334
6. Matko J, Edidin M (1997) Energy transfer methods in detecting molecular clusters on cell surfaces. Method Enzymol 278:444–462
7. Wong AP, Groves JT (2002) Molecular topography imaging by intermembrane fluorescence resonance energy transfer. Proc Natl Acad Sci U S A 99:14147–14152
8. Hoppe A, Christensen K, Swanson JA (2002) Fluorescence resonance energy transfer-based stoichiometry in living cells. Biophys J 83:3652–3664
9. Subramanian M, Jutila A, Kinnunen PKJ (1998) Binding and dissociation of cytochrome c to and from membranes containing acidic phospholipids. Biochemistry 37:1394–1402

10. Corbalan-Garcia S, Sanchez-Carrillo S, Garcia-Garcia J, Gomez-Fernandez JC (2003) Characterization of the membrane binding mode of the C2 domain of PKCε. Biochemistry 42:11661–11668
11. Calleja V, Ameer-Beg SM, Vojnovic B, Woscholski R, Downward J, Larijani B (2003) Monitoring conformational changes of proteins in cells by fluorescence lifetime imaging microscopy. Biochem J 372:33–40
12. Chigaev A, Buranda T, Dwyer DC, Prossnitz ER, Sklar LA (2003) FRET detection of cellular α_4-integrin conformational activation. Biophys J 85:3951–3962
13. Yano Y, Takemoto T, Kobayashi S, Yasui H, Sakurai H, Ohashi W, Niwa M, Futaki S, Sugiura Y, Matsuzaki K (2002) Topological stability and self-association of a completely hydrophobic model transmembrane helix in lipid bilayers. Biochemistry 41:3073–3080
14. You M, Li E, Wimley WC, Hristova K (2005) Förster resonance energy transfer in liposomes: measurements of transmembrane helix dimerization in the native bilayer environment. Anal Biochem 340:154–164
15. Loura LMS, Fernandes F, Prieto M (2010) Membrane microheterogeneity: Förster resonance energy transfer characterization of lateral membrane domains. Eur Biophys J 39:589–607
16. Brown AC, Towles KB, Wrenn SP (2007) Measuring raft size as a function of membrane composition in PC-based systems: part I- binary systems. Langmuir 23:11180–11187
17. Scholes GD (2003) Long-range resonance energy transfer in molecular systems. Annu Rev Phys Chem 54:57–87
18. Förster T (1948) Intermolecular energy migration and fluorescence. Ann Phys 2:55–75
19. Lakowicz JR (1999) Principles of fluorescence spectroscopy. Kluwer/Plenum, New York
20. Valeur B (2001) Molecular fluorescence: principles and applications. Wiley-VCH Verlag GmbH, Weinheim
21. Fung B, Stryer L (1978) Surface density determination in membranes by fluorescence energy transfer. Biochemistry 17:5241–5248
22. Estep T, Thompson T (1979) Energy transfer in lipid bilayers. Biophys J 26:195–208
23. Wolber P, Hudson B (1979) An analytic solution to the Förster energy transfer problem in two dimensions. Biophys J 28:197–210
24. Dewey T, Hammes G (1980) Calculation of fluorescence resonance energy transfer on surfaces. Biophys J 32:1023–1036
25. Snyder B, Freire E (1982) Fluorescence energy transfer in two dimensions. A numeric solution for random and nonrandom distributions. Biophys J 40:137–148
26. Doody M, Sklar L, Pownall H, Sparrow J, Gotto A, Smith L (1983) A simplified approach to resonance energy transfer in membranes, lipoproteins and spatially restricted systems. Biophys Chem 17:139–152
27. Gutierrez-Merino G, Munkonge F, Mata A, East J, Levinson B, Napier R, Lee A (1987) The position of ATP binding site on the (Ca^{2+} + Mg^{2+})-ATPase. Biochim Biophys Acta 897:207–216
28. Dale R, Eisinger J, Blumberg W (1979) The orientational freedom of molecular probes. The orientation factor in intramolecular energy transfer. Biophys J 26:161–194
29. Gorbenko GP (1999) Structure of cytochrome c complexes with phospholipids as revealed by resonance energy transfer. Biochim Biophys Acta 1420:1–13
30. Dickerson RE, Takano T, Eisenberg D, Kallai OB, Samson L, Cooper A, Margoliash E (1971) Ferricytochrome c. General features of the horse and bonito proteins at 2.8 Å resolution. J Biol Chem 246:1511–1535
31. Nazarov PV, Koehorst RB, Vos WL, Apanasovich VV, Hemminga MA (2006) FRET study of membrane proteins: simulation-based fitting for analysis of membrane protein embedment and association. Biophys J 91:454–466
32. Lehto MT, Sharom FJ (2002) Proximity of the protein moiety of a GPI-anchored protein to the membrane surface: a FRET study. Biochemistry 41:8368–8376
33. Antollini SS, Soto MA, de Romanelli IB, Gutierrez-Merino C, Sotomayor P (1996) Physical state of bulk and protein-associated lipid in nicotinic acetylcholine receptor-rich membrane

studied by laurdan generalized polarization and fluorescence energy transfer. Biophys J 70:1275–1284
34. Barrera FN, Poveda JA, Gonzalez-Ros JM, Neira JL (2003) Binding of the C-terminal sterile α motif (SAM) domain of human p73 to lipid membranes. J Biol Chem 278:46878–46885
35. Ramachandran R, Tweten RK, Johnson AE (2005) The domains of a cholesterol-dependent cytolysin undergo a major FRET-detected rearrangement during pore formation. Proc Natl Acad Sci U S A 102:7139–7144
36. Kleinfeld AM, Lukacovic MF (1985) Energy transfer study of cytochrome b5 using the anthroyloxy fatty acid membrane probes. Biochemistry 24:1883–1890
37. Ward RJ, Palmer M, Leonard K, Bhakdi S (1994) Identification of a putative membrane-inserted segment in the α -toxin of *Staphylococcus aureus*. Biochemistry 33:7411–7484
38. Liu R, Sharom FJ (1998) Proximity of the nucleotide binding domains of the P-glycoprotein multidrug transporter to the membrane surface: a resonance energy transfer study. Biochemistry 37:6503–6512
39. Shaklai N, Yguerabide J, Ranney HM (1977) Interaction of hemoglobin with red blood cell membranes as shown by a fluorescent chromophore. Biochemistry 16:5585–5592
40. Munkonge F, East JM, Lee AG (1989) Positions of the sites labeled by *N*-cyclohexyl-*N'*-(4-dimethylamino-1-naphthyl)carbodiimide on the (Ca^{2+} + Mg^{2+})-ATPase. Biochim Biophys Acta 979:113–120
41. Johnson DA, Nuss JM (1994) The histrionicotoxin-sensitive ethidium binding site is located outside of the transmembrane domain of the nicotinic acetylcholine receptor: a fluorescence study. Biochemistry 33:9070–9077
42. Remmers AE, Neubig RR (1993) Resonance energy transfer between guanine nucleotide binding protein subunits and membrane lipids. Biochemistry 32:2409–2414
43. Carraway KL, Koland JG, Cerione RA (1990) Location of the epidermal growth factor binding site on the EGF receptor. A resonance energy transfer study. Biochemistry 29:8741–8747
44. Gorbenko GP, Ioffe VM, Molotkovsky JG, Kinnunen PKJ (2008) Resonance energy transfer study of lysozyme-lipid interactions. Biochim Biophys Acta 1778:1213–1221
45. Gorbenko GP, Molotkovsky JG, Kinnunen PKJ (2006) Cytochrome c interaction with cardiolipin/phosphatidylcholine model membranes: effect of cardiolipin protonation. Biophys J 90:4093–4103
46. Davenport L, Dale R, Bisby R, Cundall R (1985) Transverse location of the fluorescent probe 1,6-diphenyl-1,3,5-hexatriene in model lipid bilayer membrane systems by resonance excitation energy transfer. Biochemistry 24:4097–4108
47. Dorn IT, Neumaier KR, Tampe R (1998) Molecular recognition of histidine-tagged molecules by metal-chelating lipids monitored by fluorescence energy transfer and correlation spectroscopy. J Am Chem Soc 120:2753–2763
48. Wang T, Pentyala S, Rebecchi MJ, Scarlata S (1999) Differential association of the pleckstrin homology domains of phospholipases C-β_1, C-β_2, and C-δ_1 with lipid bilayers and the $\beta\gamma$ subunits of heterotrimeric G proteins. Biochemistry 38:1517–1524
49. Komander D, Fairservice A, Deak M, Kular GS, Prescott AR, Downes CP, Safrany ST, Alessi DR, van Aalten DMF (2004) Structural insights into the regulation of PDK1 by phosphoinositides and inositol phosphates. EMBO J 23:3918–3928
50. Domanov YA, Gorbenko GP, Molotkovsky JG (2004) Global analysis of steady-state energy transfer measurements in membranes: resolution of structural and binding parameters. J Fluoresc 14:49–55
51. Fernandes F, Loura LMS, Koehorst R, Spruijt RB, Hemminga M, Fedorov A, Prieto M (2004) Quantification of protein-lipid selectivity using FRET: application to the M13 major coat protein. Biophys J 87:344–352
52. Capeta RC, Poveda JA, Loura LMS (2006) Non-uniform membrane probe distribution in resonance energy transfer: application to protein-lipid selectivity. J Fluoresc 16:161–172

53. Picas L, Suarez-Germa C, Montero MT, Vazquez-Ibar JL, Hernandez-Borrell JH, Prieto M, Loura LMS (2010) Lactose permease lipid selectivity using Förster resonance energy transfer. Biochim Biophys Acta 1798:1707–1713
54. Loura LMS, Prieto M, Fernandes F (2010) Quantification of protein-lipid selectivity using FRET. Eur Biophys J 39:565–578
55. Hillger F, Nettels D, Dorsch S, Schuler B (2007) Detection and analysis of protein aggregation with confocal single molecule fluorescence spectroscopy. J Fluoresc 17:759–765
56. Li E, You M, Hristova K (2005) Sodium dodecyl sulfate – polyacrylamide gel electrophoresis and Förster resonance energy transfer suggest weak interactions between fibroblast growth factor receptor 3 (FGFR3) transmembrane domains in the absence of extracellular domains and ligands. Biochemistry 44:352–360
57. Floyd DH, Geva A, Bruinsma SP, Overton MC, Blumer KJ, Baranski TJ (2003) C5a receptor oligomerization. II. Fluorescence resonance energy studies of a human G protein-coupled receptor expressed in yeast. J Biol Chem 278:35354–35361
58. Agirre A, Barco A, Carrasco L, Nieva JL (2002) Viroporin-mediated membrane permeabilization: pore formation by nonstructural poliovirus 2B protein. J Biol Chem 277:40434–40441
59. Moens PDJ, Yee DJ, Remedios CG (1994) Determination of the radial coordinate of Cys-374 in F-actin using fluorescence resonance energy transfer spectroscopy: effect of phalloidin on polymer assembly. Biochemistry 33:13102–13108
60. Vanderkooi JM, Ierokomas A, Nakamura H, Martonosi A (1977) Fluorescence energy transfer between Ca^{2+} transport ATPase molecules in artificial membranes. Biochemistry 16:1262–1267
61. John E, Jahnig F (1991) Aggregation state of melittin in lipid vesicle membranes. Biophys J 60:319–328
62. Adair BD, Engelman DM (1994) Glycophorin A helical transmembrane domains dimerize in phospholipid bilayers: a resonance energy transfer study. Biochemistry 33:5539–5544
63. Milligan DL, Koshland DE (1988) Site-directed cross-linking: establishing the dimeric structure of the aspartate receptor of bacterial chemotaxis. J Biol Chem 263:6268–6275
64. Li M, Reddy LG, Bennett R, Silva ND, Jones LR, Thomas DD (1999) A fluorescence energy transfer method for analyzing protein oligomeric structure: application to phospholamban. Biophys J 76:2587–2599
65. Sparr E, Ash WL, Nazarov PV, Rijkers DT, Hemminga MA, Tieleman DP, Killian JA (2005) Self-association of transmembrane-helices in model membranes: importance of helix orientation and role of hydrophobic mismatch. J Biol Chem 280:39324–39331
66. Fernandes F, Loura LMS, Chichon FJ, Carrascosa JL, Fedorov A, Prieto M (2008) Role of helix 0 of the N-BAR domain in membrane curvature generation. Biophys J 94:3065–3073
67. Fung JJ, Deup X, Pardo L, Yao XJ, Velez-Ruiz GL, DeVree BT, Sunahara RK, Kobilka BK (2009) Ligand-regulated oligomerization of β2-adrenoceptors in a model lipid bilayer. EMBO J 28:3315–3328
68. Rajan SR, Illing ME, Bence NF, Kopito RR (2001) Specificity in intracellular protein aggregation and inclusion body formation. Proc Natl Acad Sci U S A 98:13060–13065
69. Mihai C, Chotani M, Elton TS, Agarwal G (2009) Mapping of DDR1 distribution and oligomerization on the cell surface by FRET microscopy. J Mol Biol 385:432–445
70. Woehler A, Wlodarczyk J, Ponimaskin EG (2009) Specific oligomerization of the 5-HT1A receptor in the plasma membrane. Glycoconj J 26:749–756
71. Liu BF, Song S, Hanson M, Liang JJN (2008) Protein-protein interactions involving congenital cataract T5P gC-crystallin mutant: a confocal fluorescence microscopy study. Exp Eye Res 87:515–520
72. Coutinho A, Loura LMS, Fedorov A, Prieto M (2008) Pinched multilamellar structure of aggregates of lysozyme and phosphatidylserine-containing membranes revealed by FRET. Biophys J 95:4726–4736

73. Coutinho A, Loura LMS, Prieto M (2011) FRET studies of lipid-protein aggregates related to amyloid-like fibers. J Neurochem 116:696–701
74. Kenworthy AK, Petranova N, Edidin M (2000) High-resolution FRET microscopy of cholera toxin B-subunit and GPI-anchored proteins in cell plasma membranes. Mol Biol Cell 11:1645–1655
75. Loura LMS, de Almeida RFM, Prieto M (2001) Detection and characterization of membrane microheterogeneity by resonance energy transfer. J Fluoresc 11:197–209
76. Sperotto MM, Mouritsen OG (1993) Lipid enrichment and selectivity of integral membrane proteins in two-component lipid bilayers. Eur Biophys J 22:323–328
77. Mbamala EC, Ben-Shaul A, May S (2005) Domain formation induced by the adsorption of charged proteins on mixed lipid membranes. Biophys J 88:1702–1714
78. Loura LMS, de Almeida RFM, Silva LC, Prieto M (2009) FRET analysis of domain formation and properties in complex membrane systems. Biochim Biophys Acta 1788:209–224
79. Gorbenko GP, Trusova VM, Molotkovsky JG, Kinnunen PKJ (2009) Cytochrome c induces lipid demixing in weakly charged phosphatidylcholine/phosphatidyl-glycerol model membranes as evidenced by resonance energy transfer. Biochim Biophys Acta 1788: 1358–1365
80. Berney C, Danuser G (2003) FRET or no FRET: a quantitative comparison. Biophys J 84:3992–4010
81. Corry B, Jayatilaka D, Rigby P (2005) A flexible approach to the calculation of resonance energy transfer efficiency between multiple donors and acceptors in complex geometries. Biophys J 89:3822–3836

Hydration and Mobility in Lipid Bilayers Probed by Time-Dependent Fluorescence Shift

Sarka Pokorna, Agnieszka Olżyńska, Piotr Jurkiewicz, and Martin Hof

Abstract Biological membranes as an indispensable part of living organisms are permanently surrounded by the molecules of water. The presence of water is essential for maintaining their structure and functionality. Therefore, lipid bilayer hydration, mobility of the hydrated lipids, and their changes upon perturbations are appealing characteristics in the lipid membrane research. Time-dependent fluorescent shift (TDFS) measurements enable probing these properties in biologically relevant fully hydrated liquid crystalline lipid bilayers with a simple instrumentation and easy data treatment. Since the native lipid molecules do not fluoresce naturally, the extrinsic probing with a suitable fluorescent dye is necessary. There are a number of fluorescent membrane polarity probes designed for this purpose with different spectral properties and locations within the lipid bilayer. The basics of the technique are explained together with some useful additional considerations. The convenience of the TDFS method is demonstrated with examples from recent research on the study of the interactions of ions with lipid bilayers, and the monitoring of mobility and hydration changes along the bilayer normal upon addition of the oxidized phospholipids.

Keywords Hofmeister series · Model lipid membranes · Oxidized lipids · Solvent relaxation · Specific ionic effects · TDFS

Contents

S. Pokorna, A. Olżyńska, P. Jurkiewicz (✉) and M. Hof
J. Heyrovský Institute of Physical Chemistry of the ASCR, v. v. i., Dolejškova 3, 182 23 Prague 8, Czech Republic
e-mail: piotr.jurkiewicz@jh-inst.cas.cz

Y. Mély and G. Duportail (eds.), *Fluorescent Methods to Study Biological Membranes*, Springer Ser Fluoresc (2013) 13: 141–160, DOI 10.1007/4243_2012_46,

1 Introduction

Biological membranes are an integral part of all organisms and as such are fully immersed in the aqueous media. Water is essential for preserving their structure and function. Water molecules vividly interact with the molecules of lipids, mostly via hydrogen bonding. The very existence of lipid membranes rests on the hydrophobic effect, which organizes the amphiphilic lipid molecules with their hydrophilic headgroups facing aqueous solution and the hydrophobic tails hidden inside the bilayer. Also, the fluidity of the membranes depends on their hydration [1, 2]. It was shown that 70% of dimyristoylphosphatidylcholine molecules are clustered via water molecules, which form an intermolecularly hydrogen-bonded network, thus hindering the lipid headgroup mobility [3, 4]. Both hydration and mobility are the appealing characteristics of the biological membranes.

Under physiological conditions, lipid bilayer is a dynamic, fluidlike, self-sealing film comprising thousands of lipids, proteins, and sugars. Due to this complex chemical composition, phase heterogeneity (liquid ordered, liquid disordered, and gel phases), and the dominant role of weak interactions, the structure-function paradigm, successfully used to study many aspects of molecular biology, fails when applied to the lipid membranes. Instead, physical methods based on experimental and computational techniques should be used to reveal the subtle structural organization of membranes. There are a number of techniques that provide valuable information about the structure, dynamics, and hydration of the lipid bilayers, e.g., NMR [5–8], neutron and x-ray scattering [9–11], ESR [12], infrared spectroscopy [13, 14], fluorescence [15–17], or molecular dynamic simulations [1, 4, 18, 19]. The time-dependent fluorescent shift (TDFS), which utilizes fluorescent polarity probes, can shed light on the properties directly related to the hydration: the extent of lipid bilayer hydration, penetration and permeation of water molecules, interaction of lipids with water molecules and ions, the strength and dynamics of such interactions and their influence on the lipid bilayer structure, and local mobility of the fully hydrated lipid headgroups. The big advantage of this method is the possibility of studying the fully hydrated liquid crystalline, therefore, biologically relevant, bilayers with simple instrumentation and convenient data treatment.

The lipids, which arc commonly present in the cellular membranes, do not fluoresce naturally, and thus, an incorporation of an extrinsic, synthetic fluorescent probe is necessary. The choice of the right probe is a crucial step of the experiment, as the probe molecule could specifically interact with the lipids and influence the properties of a membrane. Once a suitable probe is chosen, TDFS was proven to be a reliable and elegant method for measuring the membrane hydration and local mobility.

2 What Is Time-Dependent Fluorescence Shift (TDFS)?

Fluorescence emission can be influenced by the processes occurring in the vicinity of the fluorophore. The dipoles of the fluorescent probes used in the TDFS measurements interact with the dipoles of the surrounding molecules, and hence, the probes are sensitive to the polarity of their environment. By applying the time-resolved fluorescence measurements, one can obtain not only the polarity but also the knowledge about the kinetics in the vicinity of the fluorescent probe.

The TDFS method is based on the observation of the solvent dipole reorientation (solvent relaxation) around the fluorescent dye in its excited state. In the ground state, solvent molecules surrounding the probe are spatially ordered, so as to compensate for the dipole moment of the probe, and thus occupy a state with the lowest possible energy. Rapid electronic excitation of the dye leads to redistribution of its charge, while the solvent nuclei remain in the spatial arrangement commensurate with the ground-state dipole moment of the dye following the Franck-Condon principle. To reach the energetically favorable state again, solvent molecules have to reorient; this process is called solvent relaxation (SR). If the solvent relaxation occurs within the timescale of the fluorescence lifetime of a probe, emission from the states between the Franck-Condon and the relaxed S_1 state can be observed (Fig. 1) [20–22].

2.1 Reconstruction of the Time-Resolved Emission Spectra (TRES)

Emitted photons carry less and less energy in time following excitation, because of the ongoing relaxation process, so the continuous red shift of the emission spectra in time is observed. The time-dependent red shift of the fluorescence spectra contains complex information about solvent relaxation dynamics. Since direct recording of the fluorescence spectra with time has serious limitations [22], the commonly used procedure involves an indirect measurement followed by the reconstruction of the time-resolved emission spectra (TRES). From a further analysis of TRES, two main parameters as well as other characteristics can be obtained.

TRES are reconstructed from the experimentally obtained fluorescence decays. Due to the nonideality of the instrumentation, the measured decays $D_{\mathrm{m}}(t)$ are in fact convolutions of the real fluorescence decays $D(t)$ with the so-called instrument response function, $IRF(t)$. Since deconvolution is an ill-posed problem, an iterative

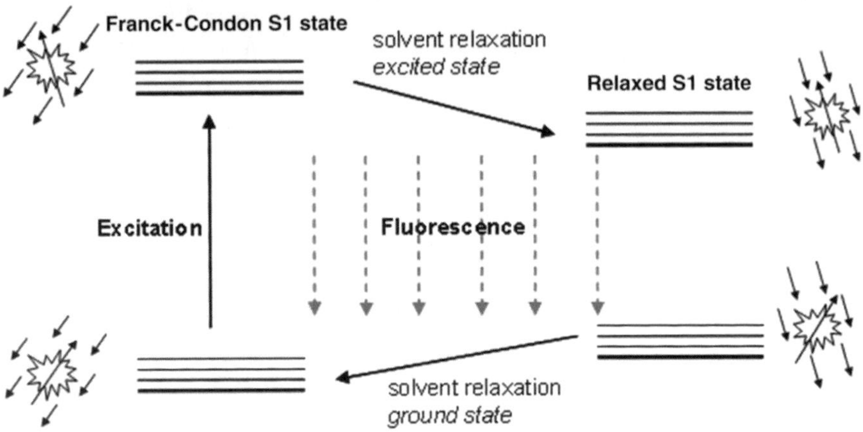

Fig. 1 *Solvent relaxation process*. In the ground state of the probe, the arrangement of the dipoles of the fluorophore and the solvent molecules minimizes the free energy of the system. Upon an electronic excitation, the dipole moment of the probe changes rapidly, while the solvent dipoles remain in the ground-state order (so-called Franck-Condon state). The reorientation of the solvent molecules, which are adapting to this new situation, is known as solvent relaxation. This relaxation leads to the solvent-relaxed S_1 state, where the dipoles are oriented so as to lower the free energy of the system. When the solvent relaxation occurs on a similar timescale to fluorescence lifetime, emission from the states between the Franck-Condon and the relaxed one is observed. Relaxation of the solvent molecules happens also in the ground state, but in the TDFS experiments, the excited state process is the one that is observed

reconvolution fitting procedure is used, where each of the measured fluorescence decays is fitted with multi-exponential function convoluted with the instrument response function:

$$D_m(t) = IRF(t) * \sum_i A_i(\lambda).\exp(-\frac{t}{\tau_i}). \quad (1)$$

IRF is measured as Rayleigh scattering at the excitation wavelength that was used for recording the fluorescence decays. The fitting is performed as a way of "deconvolution" and to interpolate the decay originally recorded in a limited number of channels. Thus, the obtained multi-exponential function is a parametrized form of $D(t)$.

A set of fluorescence decays, $D_m(\lambda, t)$, is measured over a series of emission wavelengths, λ, spanning the steady-state emission spectrum of the probe, and is recorded for each sample. In Fig. 2a, decays measured for $\lambda = 400$ nm and 540 nm are depicted. It can be seen that for the shorter wavelength, the fluorescence emission decays are significantly faster than for the longer one. At shorter wavelengths, the photons with a higher energy (close to the Franck-Condon state) are emitted predominantly. On the contrary, the photons emitted from the solvent-relaxed state carry less energy and, thus, are detected at longer wavelengths. The system needs time

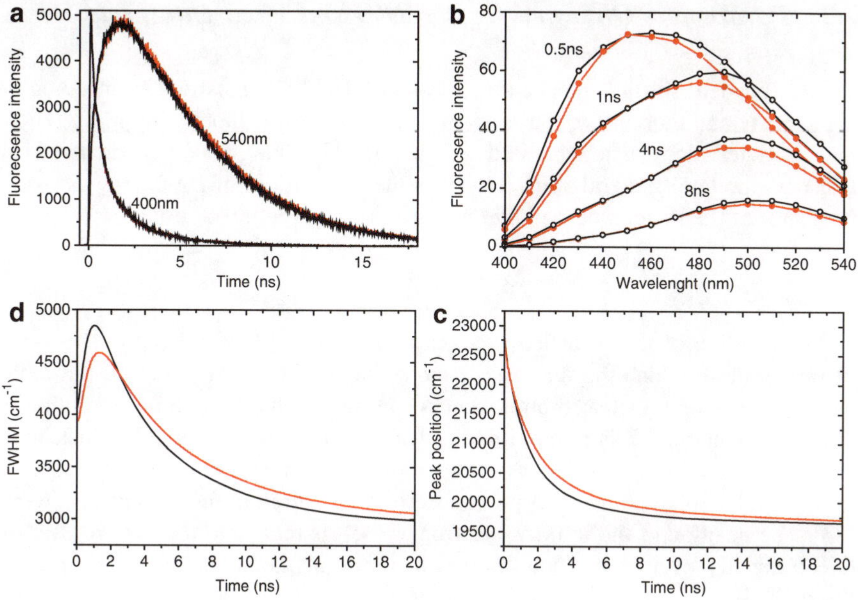

Fig. 2 *Process of data analysis*. Illustration of data treatment for two samples: dioleoylphosphatidylcholine (DOPC) large unilamellar vesicles in 150-mM NaCl (*black solid lines*, *open circles*) and TBACl (*red dotted lines, filled circles*); the final values were $\tau_{NaCl} = 1.59$ ns, $\Delta\nu_{NaCl} = 4{,}136\ cm^{-1}$ and $\tau_{TBACl} = 1.89$ ns, $\Delta\nu_{TBACl} = 4{,}004\ cm^{-1}$. (**a**) Fluorescence decays for both NaCl (*black line*) and TBACl (*red line*) were measured for wavelengths from 400 to 540 nm with 10-nm step. The first (400 nm) and last (540 nm) decays are shown. (**b**) The TRES calculated from the series of decays together with steady-state emission spectra. Fitting the TRES provides the information about (**c**) the position of the spectra maxima $\nu(t)$ and (**d**) the full width at half maximum (FWHM)

to relax to that lower energy state; thus, the decays monitored at 540 nm have pronounced rising beginnings.

The decays measured at different emission wavelengths are put together and normalized to the steady-state emission spectrum, $S_0(\lambda)$, in order to reconstruct the $TRES(\lambda, t)$:

$$TRES(\lambda, t) = \frac{D(\lambda, t) \cdot S_0(\lambda)}{\int_0^{\infty} D(\lambda, t)dt}. \tag{2}$$

An example of the reconstructed TRES is given in Fig. 2b. Spectra are transformed to the wave number domain and fitted with an appropriate function (log-normal function is often used) in order to determine their position, $\nu(t)$, and width (full width at half maximum, $FWHM(t)$).

2.2 Parameters Obtained from TRES and Their Interpretation

Time-dependent shift of the peak maximum of TRES, $\nu(t)$, carries information on the polarity and mobility of the microenvironment of the fluorescent probe. These two characteristics can be separated and quantified by defining two parameters. The first one, the total spectral shift, describes the overall loss of the energy (in wave numbers) of the system, caused by the process of solvent relaxation:

$$\Delta\nu = \nu(0) - \nu(\infty). \tag{3}$$

The position of the so-called time-zero spectrum, $\nu(0)$, is related to the hypothetical state in which the fluorescent molecule is vibrationally relaxed but emits before any solvent nuclear motion occurs. Because of the finite time resolution of the instrumentation, it is impossible to measure this spectrum directly. Instead, it can be estimated using the steady-state absorption and emission spectra measured in a nonpolar solvent and absorption spectrum measured in the system of interest [23]. The position of the infinity spectrum, $\nu(\infty)$, is related to the relaxed S_1 state (electronically excited but equilibrated in terms of the solvation of the excited fluorophore) and usually can be easily extrapolated from the measurement. Provided the relaxation is the only action connected with the dissipation of energy, $\Delta\nu$ should represent the microenvironment polarity. The linear proportionality between $\Delta\nu$ and polarity of the neat solvent, in which the probe was dissolved, was verified experimentally [20].

Moreover, $\nu(t)$ contains information about the kinetics of the relaxation. After a simple normalization of $\nu(t)$,

$$C(t) = \frac{\nu(t) - \nu(\infty)}{\nu(0) - \nu(\infty)}, \tag{4}$$

the resulting correlation function, $C(t)$ (also referred to as the spectral response function), can be easily analyzed giving a measure of the mobility of the solvent molecules in the vicinity of the probe. Examples of two different correlation functions are presented in Fig. 2c. It is worth noting that, although the differences between the two samples are hardly visible in the decays (Fig. 2a), the corresponding correlation functions can be easily distinguished (Fig. 2c). Correlation function can be fitted or integrated to obtain a numerical measure of the solvent relaxation kinetics.

The mean integrated relaxation time can be calculated as follows:

$$\tau \equiv \int_0^\infty C(t)dt \tag{5}$$

Relaxation time describes the resistance in the vicinity of the fluorophore against the rearrangement of molecules or group of atoms. Macroscopically, it was shown to be proportional to the viscosity of neat solvents [20].

The interpretation of the above-defined TDFS parameters, when measured in lipid membranes, may be the following. The spectral shift is usually attributed to the extent of membrane hydration at the site probed, since the dipoles of water molecules are the main contributors to the relaxation process. A detailed knowledge about the location of the fluorescent probe in the membrane allows for the determination of the hydration at a precise depth in the bilayer; in particular, usage of probes with different locations in the bilayer aids in determining the whole hydration profile of the bilayer (see Sect. 3.2). However, it is important to remember that $\Delta\nu$ reflects the polarity of the probe environment, and, when the other polar moieties can participate in the relaxation process, the relation between $\Delta\nu$ and membrane hydration may no longer be valid.

Relaxation time measured in lipid membranes is usually related to the local mobility of the hydrated lipids. It is remarkable that while in bulk water at room temperature the solvent relaxation is a subpicosecond process [20], in lipid membranes, the relaxation has a pronounced nanosecond component [24, 25]. This significant retardation of solvent relaxation by ~ 1,000-fold is believed to be a result of the bonding of water molecules to the lipids [25]. When probing the outer part of the lipid headgroup region, a part of the relaxation process is very fast and originates from the water molecules that can easily exchange with the bulk water molecules. But already at the glycerol level, the relaxation is purely a nanosecond process, since the water molecules are fully bound to the carbonyls and phosphate groups, and, thus, the mobility of these hydrated lipid moieties is the one observed in the TDFS experiment. Water molecules can also penetrate down to the hydrophobic backbone region of the lipid bilayer, although these are rather rare events and the TDFS response in this region becomes more complex [26]. The recent studies, combining TDFS experiment with computer simulations, can greatly help with the interpretation of the relaxation processes in membranes on a molecular level [27].

2.3 *Additional Information Contained in TRES*

The two above-mentioned parameters ($\Delta\nu$ and τ) are used regularly to characterize the properties of phospholipid membranes in TDFS measurements. However, as the solvent relaxation is a complex phenomenon even in pure solvents, not to mention such an anisotropic chemically complex aggregates as lipid bilayers, one should be careful not to misinterpret the data. Thus, the TRES should be analyzed in a more complete way.

The time course of the full width at the half maximum of the TRES, $FWHM(t)$, is a very useful tool when evaluating the results. It was shown that in the lipid bilayers [28, 29] as well as in the supercooled liquids [30], the $FWHM(t)$ passes through a maximum. In those heterogeneous systems, the solvent microenvironments are spatially distributed and respond to the change in the electric dipole moment of the dye independently, thus allowing some solvent molecules to relax faster than others. This means that, while at the beginning and at the end of the relaxation the microenvironments of all fluorophore molecules are similar, during the relaxation, a variety of different microenvironments is present. This causes a broadening of the spectra at intermediate times of the solvent relaxation process, which is reflected in the shape of *FWHM* [24, 30, 31]. A typical example of the *FWHM* is depicted in Fig. 2d. In this example, both the initial increase of the *FWHM* and the decrease are completely portrayed, which means that the relaxation process was occurring on the timescale of the experiment and was fully captured. The position of the well-pronounced *FWHM* maximum is related to the mean relaxation time. When only a decrease of the *FWHM* is present, it indicates that the relaxation was too fast for the time resolution of the instrument and that only a part of the process was captured. On the other hand, if only the increase is observed, the relaxation process was too slow, and due to limited lifetime of the S_1 state, only the beginning of the relaxation process was detected.

FWHM is proportional to the heterogeneity of the probe environments, so it gives important additional information about the possible nonunimodal distribution of the probe, e.g., in the case of two locations of the probe within the lipid bilayer. *FWHM* also helps in determining the existence of different modes of relaxation. When the relaxation process consists of different modes, the kinetics of which substantially differs, it is often manifested in the presence of two or more maxima in the *FWHM* curve.

A comparison of the position of the peak maximum in the reconstructed TRES at $t = 0$ to that of the estimated time-zero spectrum provides quantitative information on the extent of the relaxation process that was captured in the experiment. Under the assumption that the $\nu(\infty)$ was recorded correctly, the percentage of the relaxation observed in the experiment is given by

$$\%\ of\ relaxation\ observed = \Delta\nu^{\text{estimated}} - \Delta\nu^{\text{reconstructed}}, \tag{6}$$

where $\Delta\nu^{\text{estimated}}$ and $\Delta\nu^{\text{reconstructed}}$ are calculated based on the ν_0 values obtained from the estimation of the time-zero spectrum and reconstructed TRES, respectively.

2.4 TDFS in Lipid Membranes: Gradient of Polarity and Relaxation Timescale Along the Bilayer Normal

It was demonstrated that the ultrafast subpicosecond response dominates the process in aqueous solvation [20, 32]. In contrast, a substantial part of the relaxation in supramolecular assemblies, such as phospholipid bilayers, occurs on the nanosecond timescale [33–38]. This creates an enormous gradient in the relaxation times.

With respect to their structure, two main regions could be distinguished within the bilayers: the headgroup and the backbone regions. The first consists of lipid headgroups, i.e., phosphate, choline, and serine moieties, as well as glycerol and carbonyl groups. This part is very heterogeneous in its composition and consequently in its hydration. For instance, around one choline group, 25–30 water molecules form a clathrate shell; on the other hand, ethanolamine moiety is surrounded by 10–12 water molecules, which are able to interact directly with its ammonium group [39–42]. Also the carbonyl and phosphate groups interact with water molecules directly, via the hydrogen bonds [43, 44]. Fluorescent probes embedded in this region probe partially also the bulk water, the relaxation of which occurs on a much faster timescale (sub- to picosecond). Therefore, the solvent dynamics in the headgroup region may occur on a rather broad timescale, with reported picosecond as well as (sub)-nanosecond components [25, 33, 45]. Probes located deeper, e.g., close to the carbonyl level, show quite a homogenous response on the nanosecond timescale [25, 28].

The backbone region consists of hydrocarbon chains of fatty acids. Surprisingly, it was reported that water is able to penetrate even this hydrophobic part of the bilayer [16, 46]. Monitoring the backbone region using anthroyloxy (AS) dyes revealed hindered mobility and a decrease of water content when approaching the bilayer center. The relaxation process of AS dyes is known to be rather complex mostly because of an intramolecular relaxation of the molecule, which is responsible for the ultrafast (picosecond) component in the fluorescence decay. Nevertheless, it was shown that these dyes can serve as a valuable source of information on the diffusion-controlled part of the solvent relaxation process, which occurs within nanoseconds [25, 26, 36].

2.5 Lipid Membrane Probes Suitable for TDFS Measurements

As for all extrinsic molecular probes, the first requirement for a suitable probe is not to interfere with the event studied. In TDFS measurement, the red shift originating from an interaction between the dipoles of the fluorescent molecule and solvent molecules is detected, so that the dye molecule should be characterized by large charge redistribution upon the electronic excitation. Such molecules often contain electron donor and acceptor groups (e.g., amino and carbonyl group of Prodan, Laurdan, and Patman; Fig. 3). Fluorophores are commonly attached to a hydrocarbon chain, in order to anchor the dye molecule in the phospholipid bilayer.

Lipid bilayers are very complex systems. Their hydration and mobility varies significantly along the bilayer normal, from hydrophilic exterior often comprising charged groups (choline, phosphate, etc.) to hydrophobic interior consisting of fatty acid chains. Thus, the precise location of fluorescent probe is crucial to imparting knowledge about the hydration and mobility of lipid bilayers through TDFS measurements.

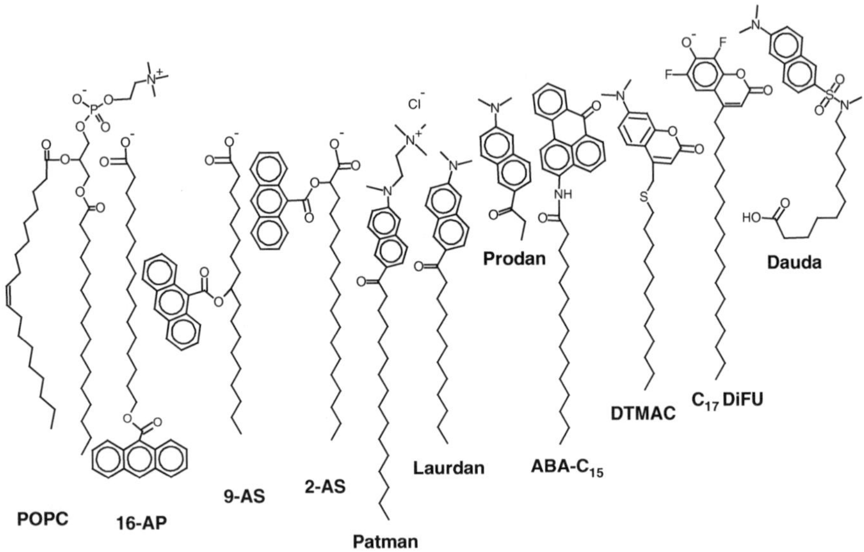

Fig. 3 *Location of fluorescent dyes*. Chemical structures of the common fluorescent membrane polarity probes are shown in their approximate position within the phospholipid bilayer in respect to palmitoyloleoylphosphocholine (POPC) molecule

A set of fluorescent probes, which enable monitoring the mobility and hydration at different depths in lipid bilayer through TDFS measurements, is shown in Fig. 3. The outer headgroup region can be probed by Dauda (11-((5-dimethylamino-naphthalene-1- sulfonyl)amino)undecanoic acid) or C17DiFU (6,8-difluoro-4-heptadecyl-7-hydroxycoumarin). DTMAC (4-[(*n*-dodecylthio)methyl]-7-(*N,N*-dimethylamino)-coumarin), ABA-C_{15} (*N*-palmitoyl-3-aminobenzanthrone), and Prodan (6-propionyl-2-dimethylaminonaphthalene) are located close to the phosphate group of the phospholipid. Laurdan (2- dimethylamino-6-lauroylnaphthalene), Patman (6-hexadecanoyl-2-(((2-(trimethylammonium)ethyl)methyl) amino)naphthalene chloride), and 2-AS (9-(9-anthroyloxy)stearic acid) are located close to the carbonyl region. Prodan, Laurdan, and Patman are widely used in TDFS measurement. All three molecules have the same fluorophore: electron donating (amino) and electron accepting (carbonyl) groups are separated by the naphthalene ring, ensuring large dipole moment. They differ in the length of their hydrocarbon chains and, therefore, in their location within the bilayer. Prodan with the shortest propionyl chain is located closest to the lipid water interface, whereas Laurdan reflects the sn-1 glycerol level and Patman is embedded even slightly deeper (Fig. 3) [25, 33, 34, 47–49]. Such probes, with the same fluorophore and different location, can be advantageously used to determine which region is more affected by, e.g., protein binding [34], or by changing the lipid composition [28] with good resolution.

The backbone region is often examined using n-anthroyloxy fatty acids. These molecules are composed from anthroyloxy ring attached to the fatty acid chain at different positions ($n = 2, 3, 6, 9, 12, 16$) [26, 36]. Since the fluorophore could be attached to the hydrocarbon chain at different positions, these dyes are used to monitor membrane hydration at different depths. Unfortunately, the nontrivial solvation response of anthroyloxy dyes caused by the pronounced intramolecular relaxation makes their application for TDFS measurements demanding [26, 36].

3 Applications

3.1 Interactions of Ions with Model Lipid Membranes

Under physiological conditions, biological membrane is always in contact with aqueous solution of various salts. While the distinct biological functions of the ions are often related to membrane proteins (e.g., ionic channels), their selectivity is also displayed in their interactions with lipids. These specific interactions are repeatedly linked to the Hofmeister series [50], an ordering of the effectiveness of ions to precipitate proteins. While the same series (sometimes reversed or slightly altered [51]) are displayed in a wide range of phenomena, their origin remains vague [52].

We have shown that the TDFS method can be successfully applied to study specific effects of monovalent cations and anions in model membranes composed of electrically neutral lipid dioleoylphosphatidylcholine (DOPC) [53]. While the Hofmeister ordering has been clearly demonstrated for the series of monovalent sodium salts (NaCl, NaBr, NaI, NaClO4, NaSCN), for which both the TDFS parameters measured for Laurdan were significantly affected [54], little changes were observed between different chlorides (NH_4Cl, CsCl, KCl, NaCl) (unpublished data). This is a common observation that Hofmeister effects are more sensitive to anions than to cations.

Our study showed that big soft anions, like SCN^-, easily incorporate into the bilayer and locate at the hydrophilic/hydrophobic interface destabilizing the organization of the lipid molecules, which is manifested in faster relaxation kinetics and increased hydration at the level of lipid carbonyls, where Laurdan is located. Cations have an opposite effect – they all seem to rigidify the membrane, and the relaxation becomes slightly slowed down, although no significant differences between Na^+, K^+, and Cs^+ were registered. The molecular level interpretation of the TDFS results can be greatly facilitated by the use of molecular dynamics simulations [27], and a number of mechanisms of ion-lipid bilayer interactions have been proposed (i.e., ion pairing, presence of the water/hydrophobic interface, steric hindrance, and the effect of counterion); see [54] for details.

In a recent study, we have investigated the interactions of 1M NaCl, KCl, and CsCl with a negatively charged lipid bilayer, where the specific effects of cations were expected to be enhanced. The lipid bilayer consisted of neutral palmitoyloleoylphosphatidylcholine (POPC) (i.e., DOPC and POPC headgroups are zwitterionic) and 20 mol% of negatively charged palmitoyloleoylphosphatidylserine (POPS), the mixture that is often used as a model of the inner lipid leaflet of a

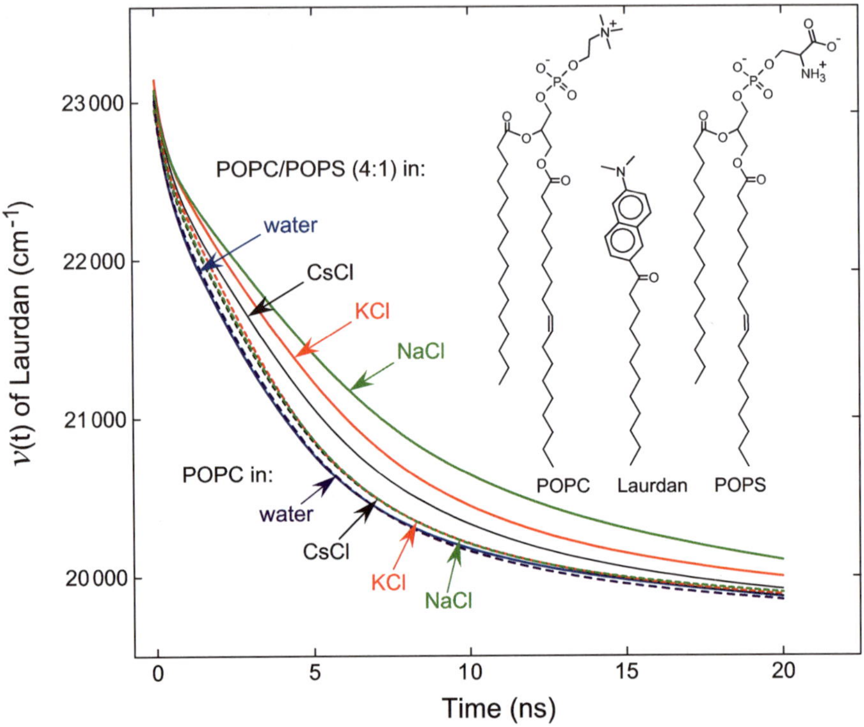

Fig. 4 Position of the maxima of TRES of Laurdan embedded in POPC or POPC/POPS (4:1) large unilamellar vesicles suspended in water or 1 M salt solutions (Adapted from [55])

plasma membrane. For a comparison, TDFS in pure POPC membrane was also measured. Both membranes were labeled with Laurdan. The TDFS results presented in Fig. 4 show that indeed the addition of POPS strengthens adsorption of monovalent cations and, moreover, reveal their specificity.

Relaxation curves (position of the maxima of TRES, $\nu(t)$) for all POPC samples and for the POPC/POPS bilayer in pure water are very similar. The mobility of the hydrated carbonyls of POPC was only slightly slowed down by the presence of salt, but no specificity was observed. On the contrary, strong specific cationic effects were observed in the POPC/POPS bilayer. The mobility at the carbonyls level was restricted by the salts in the following order: water > CsCl > KCl > NaCl; with $\tau = 3.17$, 3.90, 4.16, and 4.50 ns, respectively (measured at 10 °C). For all of the measured samples, the FWHM of TRES exhibited single maxima at times corresponding to the integrated relaxation time, and no signs of any heterogeneity in Laurdan relaxation were observed.

Also the second TDFS parameter, $\Delta\nu$, which for Laurdan is usually attributed to the hydration of the lipid carbonyls, was sensitive to the cations when measured for POPC/POPS vesicles. Hydration of POPC/POPS carbonyls decreased in the same order as the mobility: water > CsCl > KCl > NaCl, with $\Delta\nu = 3{,}925$, 3,850, 3,800, and 3,700 cm^{-1}, respectively. Dehydration was not observed for pure POPC. It is important to mention that $\Delta\nu$ is usually less sensitive than the second SR parameter, τ. In this view, the observed level of dehydration, namely, a 225-cm^{-1} difference between water and NaCl solution, is remarkably large.

The above-discussed experimental findings were supported by molecular dynamics simulations. Cations were shown to penetrate down to the glycerol level of the lipid bilayer, where they pair with oxygen atoms of lipid carbonyls clustering the neighboring lipids. This, in turn, compresses the membrane laterally, hinders lipid dynamics, and dehydrates the membrane. All of these effects followed the same Hofmeister ordering as the TDFS results. Please refer to the original publication for a more detailed molecular interpretation provided by MD [55].

3.2 *Changes of Lipid Bilayer Mobility and Polarity Upon Lipid Oxidation*

Upon oxidative stress, polyunsaturated phospholipids that are present in the living organisms relatively easily degenerate producing different forms of oxidized phospholipids (oxPLs) [56]. The most dramatic change in the lipid molecule morphology can be seen for the so-called truncated oxPLs, hydrocarbon chain of which is shorter and ended with an aldehyde or a carboxylic group (e.g., see the oxPLs in Fig. 6). Changed geometry of the truncated oxPLs together with the introduction of those polar groups in their backbone has consequences for the biophysical properties of oxidized lipid membranes [57].

In our early work, TDFS method was applied to study hydration and mobility in POPC bilayer that was modified by replacing 10 mol% of POPC lipids with synthetic well-defined oxPLs (POVPC and PGPC, both with truncated *sn*-2 chains at their 5th carbons, terminated with aldehyde and carboxyl, respectively) [58]. Both oxPLs significantly increase membrane hydration and mobility (see $\Delta\nu$ and τ values in Fig. 5). While lateral diffusion of lipids measured using z-scan FCS method was also slightly faster in oxPL-containing bilayers, probing the mobility at the carbonyl level locally using TDFS method proved to be much more sensitive to the presence of truncated oxPLs. Moreover, the changes depended on the chemical nature of the truncated chains. The terminal carboxylic groups (PGPC) influenced more the lateral diffusion than the local mobility. In contrast, the chains terminated with aldehydic group (POVPC) affected the local mobility more than the lateral diffusion. The explanation of the measured differences was provided by the structural changes postulated previously in the monolayer study of Kinnunen and coworkers [59], namely, by the looping back of the truncated chains toward the aqueous solution. Figure 5 depicts the orientation of *sn*-2 chains of the studied lipids as obtained from MD simulations: carboxyl-terminated chains tend to loopback fully into the aqueous phase, while aldehyde-terminated ones acquire all orientations including that parallel to the bilayer [58]. The later produce voids

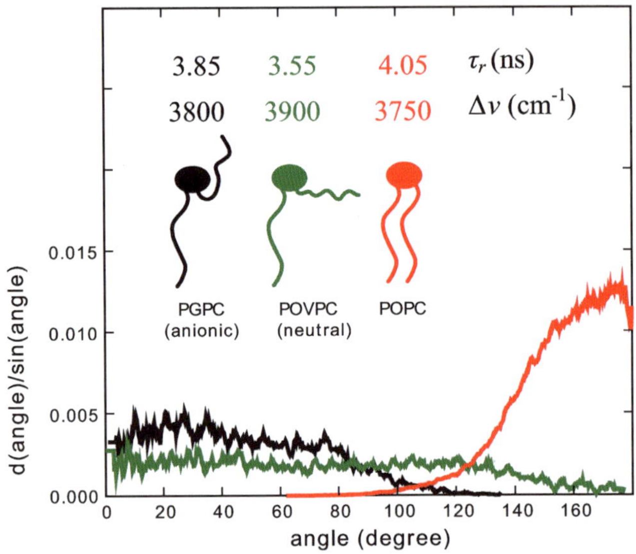

Fig. 5 Tiltangle distributions of *sn*-2 chains of PGPC, POVPC, and POPC molecules obtained from MD simulations together with TDFS results measured at 10 °C using Laurdan. Tiltangle of 180° corresponds to chains oriented in parallel to the membrane normal and pointing toward the membrane interior, 0° represents chains pointing toward water phase, and 90° represent chains oriented perpendicular to the membrane normal. Typical configurations observed for the three considered molecules are schematically depicted (Adapted from [58])

below the glycerol level, which facilitates relaxation probed by Laurdan, while the lateral diffusion remains limited by the larger area of the lipid in such conformation and the interaction of the truncated chains with the headgroups of neighboring lipids. In conclusion, both experiments and simulations showed that even a modest addition (10 mol%) of oxPLs changes the structural properties of the bilayers, leading to the increased lipid mobility and hydration. In addition, not only the conical geometry of the studied oxPL molecules but also the chemical character of their truncated chains can alter the mechanistic properties of the oxidized lipid bilayers.

The above-discussed TDFS results were obtained for Laurdan, the fluorophore of which is precisely located at the carbonyl level of the lipid bilayer. In order to study alterations of the hydration/polarity and mobility along the z axis of the oxidized bilayer (i.e., perpendicular to the surface of the membrane), fluorescent polarity probes located at different depths of the bilayer were used. Lipid bilayers consisted of POPC and 10 or 14 mol% of PAzPC or 10 or 20 mol% of PONPC labeled with 1 mol% of one the following probes: Dtmac, Laurdan, Patman, 9-AS, and 16-AP (see Fig. 6 for the structures of these molecules and the depth of the location of the probes within the bilayer). It is important to mention that due to the gradient of hydration and pressure within the lipid bilayer the spectral shift, $\Delta\nu$ (for a given fluorophore) becomes smaller and the mobility slower, when

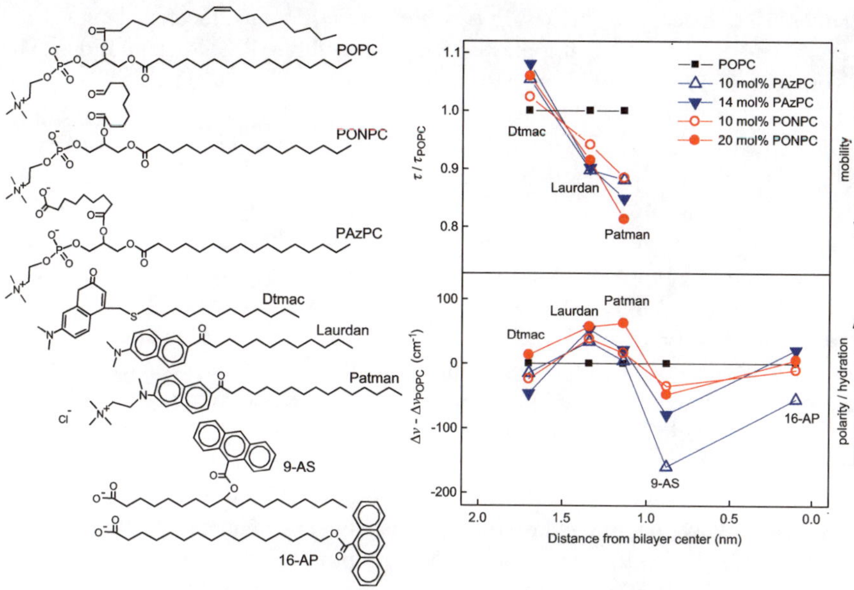

Fig. 6 Relative mobility and polarity profiles in oxidized lipid membranes composed of POPC and one of the synthetic oxPCs (PAzPC or PONPC). Figure adapted from [61]. The values of the mean integrated relaxation time and the total response shift are normalized relative to the values obtained for pure POPC. The fluorescent probes are depicted according to their location in the bilayer known from the literature (see [61])

moving from the aqueous phase toward bilayer interior. The values of TDFS parameters measured in POPC for Dtmac, Laurdan, Patman, 9-AS, and 16-AP were $\Delta\nu = 2{,}100, 3{,}950, 3{,}350, 2{,}500$, and $2{,}350\ \mathrm{cm}^{-1}$, $\tau = 1.17, 3.12, 4.15$, -, -, respectively. Relaxation times for 9-AS and 16-AP were not calculated because of the difficulties connected with the complex TDFS response of those dyes, i.e., interplay between solvent relaxation and intramolecular relaxation [26]. Due to the above-mentioned differences in the TDFS parameters measured for different probes, the results are present in values relative to those obtained for pure POPC. The changes upon addition of both oxPLs, although similar for all concentrations, are nonmonotonic with respect to the distance from the bilayer center, i.e., the relative profiles are sinusoidal in shape, Fig. 6.

Below a short analysis of the obtained profiles is given starting from Dtmac – the most distant from the bilayer center, moving toward the bilayer interior. Dtmac senses a slight dehydration and a hindrance of the mobility in oxPL-containing bilayers (the most pronounced the effects were for 14 mol% PAzPC). This result complies with the looped-back conformation postulated for PAzPC [60]. The truncated chains of PAzPC, similarly to the ones of PGPC discussed above, when looped back, are present in the vicinity of Dtmac, which reduces both hydration and mobility probed in this region. Both Laurdan and Patman report an increased hydration and a considerably augmented mobility at the glycerol level of oxidized membranes. The changes in mobility are stronger for Patman located about 0.1 nm deeper than Laurdan [28] and are proportional to the oxPLs concentrations. The kinetics of TDFS measured for anthroyloxy chromophores located in the hydrophobic region (9-AS, 16-AP, respectively, at 9th and 16th carbons of the lipid hydrocarbon chains) displayed a complex behavior, making evaluation of mean relaxation time ambiguous; thus, only $\Delta\nu$ was calculated for these probes. 9-AS reported lowering of $\Delta\nu$ in oxPL-containing bilayers, suggesting that oxidatively truncated chains reduce the penetration of water into the vicinity of this chromophore. For 16-AP, embedded in the center of the bilayer, this effect was detected only for 10 mol% PAzPC. The effects for both probes were more pronounced for carboxylic oxPC (PAzPC). The observed phenomenon might be the effect of tighter lipid backbone packing. In fact, molecular dynamic simulations predicted that oxPL can induce a thinning of the bilayer with the increased density in its center [61]. Moreover, the reduced polarity observed for 10 mol% of PAzPC, which was to a large extent absent at 14 mol% of PAzPC, could be caused by a concentration dependent segregation of PAzPC and formation of oxPL-enriched domains, from which negatively charged anthroyloxy probes would be electrostatically repelled.

The above-discussed changes in the polarity profiles can explain the facilitated flip-flop in oxPL membranes. Flip-flop of phosphatidylserine (PS) was experimentally shown and numerically calculated to be much faster in the presence of truncated oxPLs [61], which might lead to a cell death (apoptosis) as a result of PS exposure in the outer membrane leaflet. Membrane thinning together with deeper water penetration into the bilayer, shown in TDFS experiments, lowers the energetic barrier for the passing of the polar PS headgroup through the hydrophobic core of the bilayer, thus speeding up the process.

Acknowledgments The authors acknowledge support from the Czech Science Foundation via grants 208/10/1090 (SP and AO) and EUROMEMBRANES project MEM/09/E006 (MH, PJ). Moreover, MH acknowledges the Praemium Academie Award (Academy of Sciences of the Czech Republic).

References

1. Hogberg CJ, Lyubartsev AP (2006) A molecular dynamics investigation of the influence of hydration and temperature on structural and dynamical properties of a dimyristoylphosphatidylcholine bilayer. J Phys Chem B 110(29):14326–14336
2. Milhaud J (2004) New insights into water-phospholipid model membrane interactions. Biochim Biophys Acta 1663(1–2):19–51
3. Berntsen P, Svanberg C, Swenson J (2011) Interplay between hydration water and headgroup dynamics in lipid bilayers. J Phys Chem B 115(8):1825–1832
4. Pasenkiewicz-Gierula M, Takaoka Y, Miyagawa H, Kitamura K, Kusumi A (1997) Hydrogen bonding of water to phosphatidylcholine in the membrane as studied by a molecular dynamics simulation: location, geometry, and lipid-lipid bridging via hydrogen-bonded water. J Phys Chem A 101(20):3677–3691

5. Volke F, Eisenblatter S, Galle J, Klose G (1994) Dynamic properties of water at phosphatidylcholine lipid-bilayer surfaces as seen by deuterium and pulsed field gradient proton NMR. Chem Phys Lipids 70(2):121–131
6. Borle F, Seelig J (1983) Hydration of *Escherichia coli* lipids: deuterium T1 relaxation time studies of phosphatidylglycerol, phosphatidylethanolamine and phosphatidylcholine. Biochim Biophys Acta 735(1):131–136
7. Sparrman T, Westlund PO (2003) An NMR line shape and relaxation analysis of heavy water powder spectra of the L-alpha, L-beta ′ and P-beta ′ phases in the DPPC/water system. Phys Chem Chem Phys 5(10):2114–2121
8. Finer EG, Darke A (1974) Phospholipid hydration studied by deuteron magnetic resonance spectroscopy. Chem Phys Lipids 12(1):1–16
9. Franks NP, Lieb WR (1979) Structure of lipid bilayers and the effects of general anesthetics: an X-ray and neutron-diffraction study. J Mol Biol 133(4):469–500
10. Simon SA, Mcintosh TJ, Latorre R (1982) Influence of cholesterol on water penetration into bilayers. Science 216(4541):65–67
11. Nagle JF, Tristram-Nagle S (2000) Structure of lipid bilayers. Biochim Biophys Acta 1469(3):159–195
12. Ge MT, Freed JH (2003) Hydration, structure, and molecular interactions in the headgroup region of dioleoylphosphatidylcholine bilayers: an electron spin resonance study. Biophys J 85(6):4023–4040
13. Volkov VV, Nuti F, Takaoka Y, Chelli R, Papini AM, Righini R (2006) Hydration and hydrogen bonding of carbonyls in dimyristoyl-phosphatidylcholine bilayer. J Am Chem Soc 128(29):9466–9471
14. Volkov VV, Takaoka Y, Righini R (2009) What are the sites water occupies at the interface of a phospholipid membrane? J Phys Chem B 113(13):4119–4124
15. Mazeres S, Schram V, Tocanne JF, Lopez A (1996) 7-nitrobenz-2-oxa-1,3-diazole-4-yl-labeled phospholipids in lipid membranes: differences in fluorescence behavior. Biophys J 71(1):327–335
16. Chattopadhyay A (2003) Exploring membrane organization and dynamics by the wavelength-selective fluorescence approach. Chem Phys Lipids 122(1–2):3–17
17. Bernik DL, Zubiri D, Tymczyszyn E, Disalvo EA (2001) Polarity and packing at the carbonyl and phosphate regions of lipid bilayers. Langmuir 17(21):6438–6442
18. Zhou F, Schulten K (1995) Molecular dynamics study of a membrane-water interface. J Phys Chem 99(7):2194–2207
19. Zubrzycki IZ, Xu Y, Madrid M, Tang P (2000) Molecular dynamics simulations of a fully hydrated dimyristoylphosphatidylcholine membrane in liquid-crystalline phase. J Chem Phys 112(7):3437–3441
20. Horng ML, Gardecki JA, Papazyan A, Maroncelli M (1995) Subpicosecond measurements of polar solvation dynamics – Coumarin-153 revisited. J Phys Chem 99(48):17311–17337
21. Maroncelli M, Fleming GR (1987) Picosecond solvation dynamics of Coumarin-153 – the importance of molecular aspects of solvation. J Chem Phys 86(11):6221–6239
22. Lakowicz JR (2006) Principles of fluorescent spectroscopy, 3rd edn. Springer, Singapore
23. Chapman CF, Fee RS, Maroncelli M (1995) Measurements of the solute dependence of solvation dynamics in 1-propanol: the role of specific hydrogen-bonding interactions. J Phys Chem 99(13):4811–4819
24. Jurkiewicz P, Sýkora J, Olżyńska A, Humplickova J, Hof M (2005) Solvent relaxation in phospholipid bilayers: principles and recent applications. J Fluoresc 15(6):883–894
25. Sýkora J, Kapusta P, Fidler V, Hof M (2002) On what time scale does solvent relaxation in phospholipid bilayers happen? Langmuir 18(3):571–574
26. Sýkora J, Slavíček P, Jungwirth P, Barucha J, Hof M (2007) Time-dependent stokes shifts of fluorescent dyes in the hydrophobic backbone region of a phospholipid bilayer: combination of fluorescence spectroscopy and ab initio calculations. J Phys Chem B 111(21):5869–5877

27. Jurkiewicz P, Cwiklik L, Jungwirth P, Hof M (2012) Lipid hydration and mobility: an interplay between fluorescence solvent relaxation experiments and molecular dynamics simulations. Biochimie 94(1):26–32
28. Jurkiewicz P, Olżyńska A, Langner M, Hof M (2006) Headgroup hydration and mobility of DOTAP/DOPC bilayers: a fluorescence solvent relaxation study. Langmuir 22(21):8741–8749
29. Sýkora J, Mudogo V, Hutterer R, Nepras M, Vanerka J, Kapusta P, Fidler V, Hof M (2002) ABA-C-15: a new dye for probing solvent relaxation in phospholipid bilayers. Langmuir 18 (24):9276–9282
30. Richert R (2002) Heterogeneous solvent dynamics and time-resolved optical linewidths. J Non-Cryst Solids 307:50–56
31. Yang M, Richert R (2001) Observation of heterogeneity in the nanosecond dynamics of a liquid. J Chem Phys 115(6):2676–2680
32. Jimenez R, Fleming GR, Kumar PV, Maroncelli M (1994) Femtosecond solvation dynamics of water. Nature 369(6480):471–473
33. Hutterer R, Schneider FW, Sprinz H, Hof M (1996) Binding and relaxation behaviour of Prodan and Patman in phospholipid vesicles: a fluorescence and H-1 NMR study. Biophys Chem 61(2–3):151–160
34. Hutterer R, Schneider FW, Hermens WT, Wagenvoord R, Hof M (1998) Binding of prothrombin and its fragment 1 to phospholipid membranes studied by the solvent relaxation technique. Biochim Biophys Acta 1414(1–2):155–164
35. Hof M, Hutterer R, Perez N, Ruf H, Schneider FW (1994) Influence of vesicle curvature on fluorescence relaxation kinetics of fluorophores. Biophys Chem 52(2):165–172
36. Hutterer R, Schneider FW, Lanig H, Hof M (1997) Solvent relaxation behaviour of n-anthroyloxy fatty acids in PC-vesicles and paraffin oil: a time-resolved emission spectra study. Biochim Biophys Acta 1323(2):195–207
37. Krishna MMG (1999) Excited-state kinetics of the hydrophobic probe Nile red in membranes and micelles. J Phys Chem A 103(19):3589–3595
38. Pal SK, Sukul D, Mandal D, Sen S, Bhattacharyya K (2000) Solvation dynamics of DCM in dipalmitoyl phosphatidylcholine lipid. Tetrahedron 56(36):6999–7002
39. Langner M, Kubica K (1999) The electrostatics of lipid surfaces. Chem Phys Lipids 101(1):3–35
40. Lis LJ, Mcalister M, Fuller N, Rand RP, Parsegian VA (1982) Interactions between neutral phospholipid bilayer membranes. Biophys J 37(3):657–665
41. Damodaran KV, Merz KM (1993) Head group-water interactions in lipid bilayers: a comparison between DMPC- and DLPE-based lipid bilayers. Langmuir 9(5):1179–1183
42. McIntosh TJ, Simon SA (1986) Area per molecule and distribution of water in fully hydrated dilauroylphosphatidylethanolamine bilayers. Biochemistry 25(17):4948–4952
43. Grdadolnik J, Kidric J, Hadzi D (1991) Hydration of phosphatidylcholine reverse micelles and multilayers – an infrared spectroscopic study. Chem Phys Lipids 59(1):57–68
44. Blume A, Hubner W, Messner G (1988) Fourier transform infrared spectroscopy of C13:O labeled phospholipids hydrogen bonding to carbonyl groups. Biochemistry 27(21):8239–8249
45. Bursing H, Ouw D, Kundu S, Vohringer P (2001) Probing solvation dynamics in liquid water and at phospholipid/water interfaces with femtosecond photon-echo spectroscopies. Phys Chem Chem Phys 3(12):2378–2387
46. Perochon E, Lopez A, Tocanne JF (1992) Polarity of lipid bilayers. A fluorescence investigation. Biochemistry 31(33):7672–7682
47. Hof M (1999) Solvent relaxation in biomembranes. In: Rettig W, Strehmel B, Schrader S (eds) Applied fluorescence in chemistry, biology, and medicine. Springer, Berlin, pp 439–456
48. Krasnowska EK, Gratton E, Parasassi T (1998) Prodan as a membrane surface fluorescence probe: partitioning between water and phospholipid phases. Biophys J 74(4):1984–1993
49. Parasassi T, De Stasio G, Ravagnan G, Rusch RM, Gratton E (1991) Quantitation of lipid phases in phospholipid vesicles by the generalized polarization of Laurdan fluorescence. Biophys J 60(1):179–189

50. Hofmeister F (1888) Arch Exp Pathol Pharmacol 24(4–5):247–260
51. Leontidis E (2002) Hofmeister anion effects on surfactant self-assembly and the formation of mesoporous solids. Curr Opin Colloid Interface Sci 7(1–2):81–91
52. Parsons DF, Bostrom M, Lo Nostro P, Ninham BW (2011) Hofmeister effects: interplay of hydration, nonelectrostatic potentials, and ion size. Phys Chem Chem Phys 13(27):12352–12367
53. Vacha R, Siu SWI, Petrov M, Bockmann RA, Barucha-Kraszewska J, Jurkiewicz P, Hof M, Berkowitz ML, Jungwirth P (2009) Effects of alkali cations and halide anions on the DOPC lipid membrane. J Phys Chem A 113(26):7235–7243
54. Vacha R, Jurkiewicz P, Petrov M, Berkowitz ML, Bockmann RA, Barucha-Kraszewska J, Hof M, Jungwirth P (2010) Mechanism of interaction of monovalent ions with phosphatidylcholine lipid membranes. J Phys Chem B 114(29):9504–9509
55. Jurkiewicz P, Cwiklik L, Vojtiskova A, Jungwirth P, Hof M (2012) Structure, dynamics, and hydration of POPC/POPS bilayers suspended in NaCl, KCl, and CsCl solutions. Biochim Biophys Acta 1818(3):609–616
56. Fruhwirth GO, Loidl A, Hermetter A (2007) Oxidized phospholipids: from molecular properties to disease. Biochim Biophys Acta 1772(7):718–736
57. Jurkiewicz P, Olżyńska A, Cwiklik L, Conte E, Jungwirth P, Megli FM, Hof M (2012) Biophysics of lipid bilayers containing oxidatively modified phospholipids: insights from fluorescence and EPR experiments and from MD simulations. Biochim Biophys Acta-Biomembranes 1818(10):2388–2402, doi: 10.1016/j.bbamem.2012.05.020
58. Beranova L, Cwiklik L, Jurkiewicz P, Jungwirth P, Hof M (2010) Oxidation changes physical properties of phospholipid bilayers: fluorescence spectroscopy and molecular simulations. Langmuir 26(9):6140–6144
59. Sabatini K, Mattila JP, Megli FM, Kinnunen PKJ (2006) Characterization of two oxidatively modified phospholipids in mixed monolayers with DPPC. Biophys J 90(12):4488–4499
60. Khandelia H, Mouritsen OG (2009) Lipid gymnastics: evidence of complete acyl chain reversal in oxidized phospholipids from molecular simulations. Biophys J 96(7):2734–2743
61. Volinsky R, Cwiklik L, Jurkiewicz P, Hof M, Jungwirth P, Kinnunen PKJ (2011) Oxidized phosphatidylcholines facilitate phospholipid flip-flop in liposomes. Biophys J 101 (6):1376–1384

Part II
Exploring Membrane Organization, Dynamics and Interactions by Advanced Fluorescence-Based Imaging Techniques

Visual Discrimination of Membrane Domains in Live Cells by Widefield Microscopy

Claire E. Butler, Guy Wheeler, Jeremy Graham, and Kevin M. Tyler

Abstract Membrane dynamics is a fast-evolving field with the many new methods and probes being developed each year affording ever increased insights into how membranes behave in the laboratory. Typically, these developments are first tested in model membranes using high-cost, bespoke microscopes which often employ confocal and two-photon systems and which give little consideration to preservation of cellular integrity and homeostasis during experiments. This chapter addresses the clear need to rapidly apply and deploy this work into mainstream biological laboratories by development of economical, four-dimensional imaging on user-friendly low-cost systems using widefield optics and simultaneous capture of multiple fluorescent markers. Such systems are enabling biologists to consider the coordinated processes triggered from signalling platforms during cellular interaction with the environment. In this chapter, we describe the progress made to date and in particular we focus on the Laurdan family of fluorescent probes, which are being used to image whole cells and tissues using widefield epifluorescence microscopy and which can be usefully combined with simultaneous capture at longer wavelengths (yellow through far red) for imaging of cell morphology or for following expressed markers such as fluorescent adaptor proteins.

Keywords Laurdan · Membrane · Microdomains · Widefield microscopy

---

C.E. Butler, G. Wheeler, and K.M. Tyler (✉)
Biomedical Research Centre, Norwich Medical School, University of East Anglia, Norwich NR4 7TJ, England
e-mail: K.Tyler@uea.ac.uk

J. Graham
Cairn Research Ltd, Graveney Road, Faversham Kent ME13 8UP, UK

Y. Mély and G. Duportail (eds.), *Fluorescent Methods to Study Biological Membranes*,
Springer Ser Fluoresc (2013) 13: 163–184, DOI 10.1007/4243_2012_47,

Contents

1 Introduction

1.1 Plasma Membrane Composition

The plasma membrane is a dynamic interface between the internal cytoplasm and the extracellular milieu. As a lipid bilayer, it comprises a heterogeneous mix of lipid species including sterols, glycolipids and phospholipids. Phospholipids are defined by their hydrophilic "head" groups and by their hydrophobic "tails" which can vary in length and are generally either saturated or unsaturated fatty acids. Saturated fatty acid chains form long straight chains which pack closely excluding water (and some detergents) thereby reducing membrane hydration and fluidity. Conversely, unsaturated fatty acids contain double bonds which introduce kinks disrupting packing, increasing hydration and reducing membrane viscosity. There is believed to be a thermodynamically driven propensity for saturated fatty acids to coalesce to the exclusion of unsaturated fatty acids, giving rise to lateral organisation and regions or subcompartments of different biochemical composition with different biophysical characteristics [1].

Subcompartmentalisation of the plasma membrane has profound implications for cellular interaction with the environment, particularly with regard to cell signalling and cellular trafficking. Discrete regions within the plasma membrane, rich in saturated fatty acids and sterols which resist detergent treatment, are commonly referred to as lipid rafts, a term coined to graphically describe them in the context of, and as exceptions to, the fluid mosaic model [2]. GPI-linked (outer leaflet) and acylated (inner leaflet) proteins, as well as sphingolipids, glycosphingolipids (including gangliosides) and sterols (typically cholesterol and/or ergosterol), are found to accumulate in these regions, which are usually <100 nm in size giving rise to the alternative name, microdomain. Surprisingly, transmembrane proteins including G-protein-coupled receptors (GPCRs) and ion channels are also associated in lipid rafts, typically this association is generally dependent on multiple acylations (typically palmitoylation)[3].

1.2 The Lipid Raft Debate

Two types of lipid rafts are widely described: detergent-resistant microdomains (DRMs) and caveolae. DRMs have been conceptualised as planar regions of the membrane, whereas caveolae are invaginations which require caveolin-1 to form [4]. Lipid rafts are also described as membrane attachment sites for cholera toxin which binds to ganglioside G_{M1} located in these regions allowing them to easily be localised [5].

There is however some debate over the very existence of lipid rafts, partly due to the large range in sizes reported for these regions, 10–200 nm, and their potentially transient nature [6]. The reported size variation may reflect the methods of analysis employed, particularly where cross-linkage of these domains as preparation for their observation may result in their translocation and fusion to form larger more stable structures [7]. Fixation of the plasma membrane generally alters its structure and integrity, thus creating a dilemma for the effective visualisation of microdomains. The detergent method of extraction is also problematic, as it leads to the production of large sheets of membrane, contamination and loss of components; however, it is still considered a useful tool, and a large number of lipid raft components were first isolated using this method [8].

The timescale of existence for lipid rafts is also still under discussion. Visualisation by FRET between acylated proteins and caveolin has suggested that they can be relatively stable structures [9], but others have described them as transient and only detectable in response to a stimulus [8]. Despite this controversy, a considerable number of studies have been published on the subject of lipid rafts. To date, it is widely accepted that eukaryotic plasma membranes are not simple liquids but two-dimensional nonideal liquids containing dynamic structures [7].

1.3 Membrane Domains and Membrane Polarisation: Cilia and Synapses, Infection and Immunity

Sterols and sphingolipids are almost exclusively associated with eukaryotic cells, and the peculiar association between sterols and sphingolipids in eukaryotes potentiates lateral segregation, subcompartmentalisation and sorting within contiguous membranes, upon which membrane trafficking and plasma membrane polarisation are contingent [10]. More than 200 proteins have been associated with lipid rafts [11]; in some instances, cell surface receptors are permanently localised in these regions, with other receptors believed to relocate into lipid rafts after ligand binding [12, 13]. Although small, lipid rafts are thought to be able to fuse together upon binding of certain membrane receptors to create large signalling platforms and are therefore considered to be dynamic as well as transient [2]. Lipid rafts can also be stabilised by galectin lattices, structures formed by endogenous lectins which bind branches of the N-glycan domain on glycoproteins. These lattices likely increase the capacity for concerted actions such as signalling by concentrating receptors and signalling intermediates and also contribute to the regulation of the lifetime of lipid rafts [14]. Lipid rafts as signalling platforms have been described in GPCR signalling where such receptors are thought to cluster together within caveolae (described below) or to translocate upon activation to lipid rafts as summarised by [12].

Subcompartmentalisation of the plasma membrane in polarised cells into contiguous but discrete membrane domains with distinct biochemical compositions and which are large enough to be readily visualised is well documented. Even at a steady state, the apical and basolateral membranes of epithelial cells are known to have different constitutions reflecting membrane trafficking, which is believed to be dependent on lipid rafts. Similarly in ciliated cells, the ciliary membranes have a distinct biochemical composition from the somatic regions of the plasmalemma, and this may reflect not only trafficking but sorting at the ciliary junction and the propensity of lipid rafts to accumulate in membrane protrusions [15, 16]. For budding yeast, lipid rafts are fundamentally associated with apical sorting for polarised growth and cell-cell interaction during mating [17]. In other cell-cell and cell-pathogen interactions, the plasma membranes also become effectively polarised at the contact site or synapse, and a discrete functional domain is formed in response to contact with other cells or during contact with a pathogen be it viral (e.g. dengue and HIV) [18, 19], bacterial (e.g. Shigella and Campylobacter) [20] or parasitic (e.g. malaria and *Trypanosoma cruzi*) [21, 22]. Common to all of these interactions is the ligation of raft-associated surface receptors which are commonly glycoproteins or glycolipids and the requirement for this interaction to generate signalling in order to effect cellular function – supporting the prevailing view of a key role for lipid raft formation in environmental sensing by formation of signalling platforms [2]. With this in mind, increasing the accessibility of membrane imaging will help drive forward the field of pathogen-cell interaction.

2 Lipid Domain-Binding Probes

2.1 Cholera Toxin

A variety of molecular markers have been produced for the visualisation and functional analysis of lipid rafts. Cholera toxin B (CtxB) from *Vibrio cholerae* is known to bind the ganglioside G_{M1} which generally localises to microdomains.

Fluorescent fusions of CtxB can therefore be used to image the interaction between CtxB and G_{M1} specifically [23], for example, G_{M1} localisation and patching have been used to identify pathways for MHC endocytosis [24]. Although this association is a widely used method for identifying lipid rafts, there are some limitations; G_{M1} is also sometimes found outside of the lipid raft fraction, and the binding of CtxB and G_{M1} may also result in the clustering of sphingolipids [25, 26]. CtxB is capable of simultaneously binding five separate G_{M1} molecules which may be located in five different lipid rafts, consequently dynamic domains will be stabilised, and there is some evidence to suggest that CtxB can cross-link other molecules such as glycoproteins resulting in a loss of specificity of the detection. CtxB can also be conjugated to quantum dots for live cell imaging as well as observing the progression between generations [27]. Quantum dots are semiconductor nanocrystals which can be used for a wide range of applications but usually in conjunction with confocal microscopy [28].

2.2 *Fluorescent Sphingolipids and Sphingolipid-Binding Proteins*

Sphingolipids are another component of lipid rafts, and the fluorescent sphingolipid-binding domain (SPD) appears to interact with cholesterol, sphingomyelin and glycosphingolipids enabling live cell imaging of the endocytic pathways of these lipid raft components [29, 30]. SPD is derived from amyloid β-peptide and interacts with the glycophospholipid head groups. Sphingomyelins (a species of sphingolipid) are around 50 % more prevalent in lipid rafts than in the rest of the membrane [31] making them an ideal target for fluorescent probes. Lysenin, a toxin which has an affinity for sphingomyelin-rich domains [32], has shown a plasma membrane distribution separate from that of G_{M1} microdomains [33] suggesting that there may be a separate genre of sphingomyelin-associated lipid rafts. Sphingomyelin synthase, which catalyses its synthesis from ceramide and phosphotidylcholine, is thought to regulate clustering of the product [34] and so may play a role in one or other genre formation. Lysenin is more specific than other probes in that it only recognises sphingomyelin and no other sphingolipids; thus, potentially lysenin may be a candidate to specifically study lipid rafts [35]. The main problem with lysenin as an imaging tool is that it is cytotoxic in most cells, and thus, it was necessary to develop a truncated form without the cytotoxic N terminus [36] before it could be used reliably.

2.3 *Fluorescently Labelled Cholesterol*

Cholesterol appears 50 % more concentrated in lipid rafts than in the plasma membrane [31] and therefore may be used as a microdomain marker. Polyene antibiotics such as filipin bind free cholesterol at a 1:1 stoichiometry at the cell surface, resulting in emission of light in the ultraviolet range of the spectrum [37], and thus could be used as a fluorescent cholesterol marker. Filipin can also diffuse across the membrane and incorporate into the inner leaflet providing additional structural information as well as details about the Golgi membrane [38]. Despite its widespread usage as a lipid raft marker, filipin does not necessarily discriminate between lipid rafts, and the cholesterol at the rest of the cell surface and accumulation of filipin over time has proven cytotoxic; thus, alternative cholesterol probes have been developed.

Perfingolysin O is another cholesterol-binding cytolysin which can bind large cholesterol structures and therefore detect cholesterol-enriched domains. A perfringolysin O domain D4 fusion protein has also been shown to bind cholesterol specifically localised in lipid rafts [39] and has revealed the presence of lipid rafts on the inner leaflet of the plasma membrane [40].

Dehydroergosterol (DHE) is a naturally occurring cholesterol analogue which differs from cholesterol only in possessing two double bonds and an extra methyl group and is able to incorporate into the cell membrane at a concentration of up to 85% before any cytotoxic effects are observed [41]. Absorption and emission spectra are in the UV region but can be imaged with epifluorescence after some minor modifications [42]. DHE can also be used to visualise the sterol distribution in living cells although this method does not exhibit a particular affinity for G_{M1} discernable microdomains [43] and may not behave functionally as cholesterol.

Polyethylene glycol-derived cholesterol ether (PEG-Chol) has a high affinity for cholesterol-rich domains, and addition of a fluorescent ester (e.g. fluorescein) has led to the production of a lipid raft-specific probe (fPEG-chol). Its lack of cytotoxity meant that it was initially used to help disperse otherwise insoluble antibiotics [44]. Its size prevents movement to the inner leaflet of the membrane, but it is still a useful tool to monitor reorganisation of lipid rafts at the cell surface [45], exhibiting a similar distribution to that of filipin and being internalised along with lipid raft components [46].

The enrichment of sphingolipids in lipid raft regions also provides the opportunity for labelling sphingolipids and their analogues with BODIPY (boron dipyrromethane [(4,4-difluoro-4,7-dimethyl-4-bora-3a,4a,diaza-s-indacene)]) to investigate structure and function. BODIPY-labelled sphingolipids have interesting spectral properties and show a shift in emission spectra from green to red at increased molar densities [47]. Both BODIPY-Cer in vitro and BODIPY-LacCer in live cells can therefore be used to differentiate specific regions of the cell membrane by wavelength. Furthermore, BODIPY can be conjugated to cholesterol in a manner that does not inhibit normal cholesterol function and allows its partitioning into regions of high lipid order [48] and also to sphingomyelin, where it has been observed incorporating into lipid rafts in an endogenous sphingomyelin-dependent manner [49]. A derivative, BODIPY-D-*erythro*-LacCer, has also shown that sphingolipids can be induced to redistribute into specific microdomains on the plasma membrane in response to CtxB.

Sphingolipid positioning within the plasma membrane is controlled by palmitoylation, a process affecting membrane microlocalisation that is regulated during sphingolipid biosynthesis by a family of aspartate-histidine-histidine-cysteine palymitoyl transferases (DHHCs). Labelling DHHC's allows the monitoring of their positions in the cell membrane in response to different stimuli, and the live imaging of DHHC recycling to and from the plasma membrane has been recently described [50].

3 Adaptor Proteins

3.1 *Caveolin*

A number of adaptor proteins are found to cluster in lipid-rich microdomains, and fluorescent fusion proteins have become widespread in live cell imaging for caveolin and flotillin. Caveolae, which have been mentioned previously, are distinct cell surface membrane invaginations associated with endocytosis (reviewed by [51]). These structures are rich in cholesterol, and indeed, the presence of cholesterol is required for efficient formation of these structures, leading to them being proposed as a form of lipid raft. Each caveola contains clusters of GPI-linked proteins and acylated proteins packed to a density of some 30,000 molecules per square micrometre [52].

As well as cholesterol, the formation of caveolae also requires the presence of a scaffolding protein called caveolin which coats the cytoplasmic surface of the invagination [53]. There are three subtypes of caveolin with caveolin-1 and caveolin-2 the most abundant, being present in a wide range of cell types especially epithelial lines [54], whereas caveolin-3 is confined to muscle cells [55]. All three subtypes appear to exhibit similar functions in cell signalling, lipid raft formation, trafficking and possibly heterotrimeric G-protein signalling [56] although of these, caveolin-2 has proven non-essential for caveolae formation.

Fusion proteins for all three isomers of caveolin have been used to elucidate the structure of caveolae and their function as well as the mechanism of caveolin transport to and from the plasma membrane [57]. In fact, the use of fusion proteins has been integral to all the major steps forward in caveolin protein-protein understanding, suggesting such roles as a negative-signalling molecule, the base of a signalling platform and antigen presentation.

3.2 *Flotillin*

Another protein which is thought to be lipid raft associated is flotillin-1 (flot1) which is sometimes referred to as reggie-2. This protein is important in an endocytic pathway separate to that of the caveolae [58]. Flotillin-based structures are uncoated invaginations in the cell surface membrane that are dependent on the coassembly of flot1 and 2 to promote membrane curvature [59]. Internalisation in this pathway is regulated by the src family kinase, fyn [60], which is considered to be involved in endocytosis of GPI-linked signalling molecules [61]. Again, its fluorescent fusion proteins have been integral in the above microscopy studies, also showing both caveolin and flotillin to be found only in specific subsets of cells and in their own types of membrane microdomains.

3.3 GPI-Linked Proteins

An alternative approach may be to use fluorescent protein-coupled GPI-adaptor proteins (GPI-APs) which would insert directly into the lipid raft, which, as previously stated, are rich in GPI-anchored proteins. However, the size of these molecules may cause perturbation within such a small system, and expression of these fluorescent proteins may also lead to their fluorescence in their biosynthetic pathway.

4 Environment Sensitive Probes

4.1 Laurdan

The fluorescent probe Laurdan (6-dodecanyl-2-dimethylaminonaphthalene) has been used to study the lateral organisation of membrane domains since it was first synthesised in 1979 [62], due to its ability to incorporate into the membrane with an even distribution, without detectably altering the physiology of the cell and without being altered by cell surface interactions such as the binding of lipoproteins. The Laurdan dipole aligns with the phospholipid bilayer between the lipid head groups and the first carbon atom of the acyl chains, and its emission spectra changes depending upon the fluid state of the surrounding lipid chains [63]. In a more fluid membrane domain, Laurdan fluoresces with greater intensity at green wavelengths, with a maximum centred around 490 nm. This shifts to 440 nm and a blue fluorescence, in more ordered membrane regions. A method for the ratiometric analysis of Laurdan fluorescence, the generalised polarisation (GP), was developed to help identify the fluid order of Laurdan-stained membranes [64, 65], giving a range for GP values between -1.0 and $+1.0$; the more positive the GP value, the more ordered the membrane.

Laurdan has been extensively used for studying membrane dynamics, with over 300 reports in the literature. However, the vast majority of these use either fluorescence spectrophotometry [66, 67], evaluating the GP of entire preparations, or two-photon [68] and confocal microscopy [69]. Recently, confocal microscopy has also been used to explore the liquid order of a zebra fish embryo initiating the development of novel whole organism techniques [70]. Alternatively, many studies have utilised artificially synthesised unilammelar vesicles or membrane extracts from biological systems. It is only in the last few years that reports have been published using Laurdan fluorescence with epifluorescence microscopy on live cells [71, 72].

The reasons for using microscopy rather than spectrophotometry have been described previously [73]. Two-photon microscopy is able to minimise the rapid photobleaching that Laurdan is susceptible to, particularly under widefield fluorescence microscopy; however, the cost of the system may render it inaccessible. Ultimately, the bleaching effect of Laurdan under widefield microscopy can be overcome [71, 74]. Taken together, such studies demonstrate clearly that the GP is higher for surface membranes compared to intracellular membranes and demonstrate the utility of their systems through seeing the expected shift of emission spectra and GP values in increasingly fluid membranes.

The key to using Laurdan-stained live cells under widefield conditions is to be able to utilise other markers and fluorescent proteins. This affords the ability to identify interactions and monitor responses and to co-localise membrane markers with membrane phases and gauge the cellular response whilst monitoring the extent of membrane disruption in cases such as cholesterol depletion. Widefield Laurdan microscopy has been used to demonstrate that temperature can directly affect membrane fluidity but that living cells (in this case the ciliated protozoan *Paramecium*) can respond to changing environmental conditions by rapid adaption – restoring fluidity to a cooled membrane [75]. Further, Sitrin and colleagues have reported the utility of Laurdan using widefield microscopy, with a time-lapse assay of neutrophil migration [76]. Their widefield setup used an emission splitter to isolate the emission spectra of Laurdan for GP analysis. They showed over a 10-min time course with 10-s intervals that the migration tip of the neutrophils, the uropod, had consistently higher GP values compared to the rest of the cell but that this was a dynamic system, with high-GP regions within the uropod-shifting position even between the 10-s time points. They also showed that application of the cholesterol depletory MβCD, which therefore lowers GP, marginally slowed the speed of the neutrophil migration but also reduced its ability to migrate towards chemotactic signals. These studies demonstrate utility of widefield Laurdan microscopy in assessing membrane dynamics in live cells distinct and from spectrophotometry or the two-photon system; photobleaching can be minimised because epifluorescent sources can be of low intensity compared with confocal and exposure times can be low when combined with a high-sensitivity CCD. Whilst the latter study reported that receptors known to be involved in cell movement pathways localised to higher GP regions, they were unable to co-localise structures using fluorescent markers or to observe the cell directly using condenser-based techniques such as phase-contrast or DIC microscopy. Weber and colleagues combined widefield Laurdan microscopy with fluorescence anisotropic microscopy and total internal reflection (TIR) microscopy using a flexible setup that allowed them to look at membrane stiffness of the plasma membrane in living cells and compare it to internal membranes whilst varying temperature and cholesterol content [74]. This flexible setup afforded the ability to image cells at long wavelength but not to capture and compare simultaneously and in real time.

Laurdan microscopy for live imaging of cultured cells can be set up straightforwardly at modest cost on an inverted fluorescent microscope using the setup described in Fig. 1. Key to the system is the triple emission beam splitter which

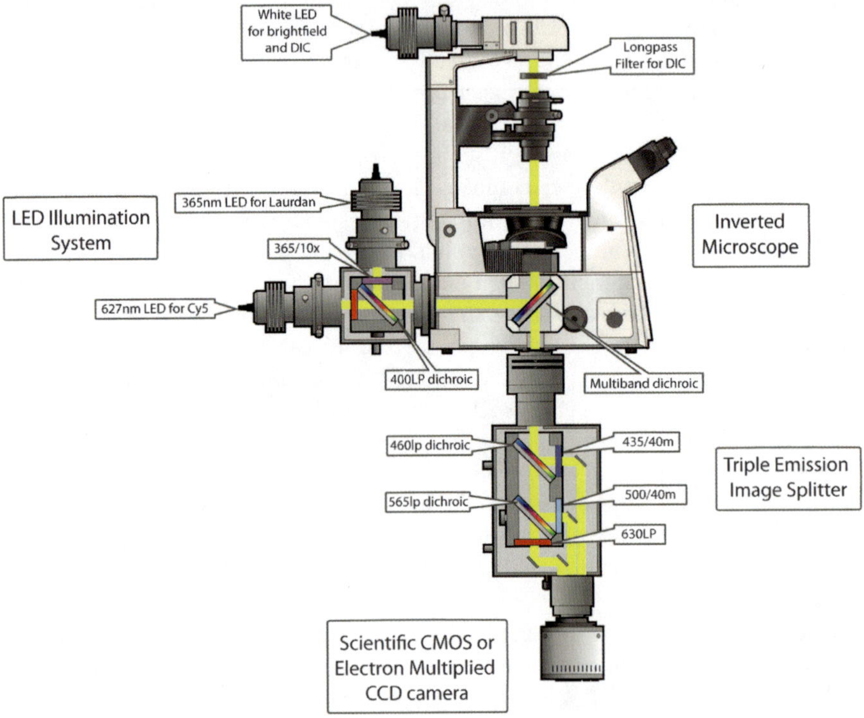

Fig. 1 Widefield Laurdan microscopy setup. Figure shows an experimental setup for imaging Laurdan simultaneously with (*1*) Cy5 or other red/far-red fluorophore, (*2*) infrared DIC or other transmitted light technique using red or infrared illumination, and (*3*) both (*1*) and (*2*), alternating in time. The configuration consists of a research-grade inverted fluorescence microscope with a high numerical aperture objective lens and an image splitter capable of separating the emission pathway into three spectrally distinct channels using either a splitter device (e.g. Cairn Research TripleSplit, Photometrics QuadView) or three independent cameras and appropriate beam-splitting optics (e.g. Cairn Research TriCam). In order to ensure high stability and minimise photobleaching, a rapidly modulated LED-based light source is preferred for both fluorescence and transmitted light illumination (Available from Cairn Research, CoolLED, 89 North, Lumencor, Prior Scientific). A back-illuminated electron-multiplied camera will give maximum sensitivity allowing illumination levels to be minimised; alternatively an SCMOS camera might be preferred to give higher resolution and a larger field of view (suitable cameras available from Andor, Photomterics, Hamamatsu, PCO). For truly simultaneous imaging and to avoid vibration or registration problems, the microscope dichroic must be multiband so that it can remain in the lightpath at all times (typically a Quad Sedat 4-band dichroic). The excitation and emission filters and dichroic mirrors need to be carefully selected to fit with this multiband device (Available from Chroma Corporation, Semrock, Omega); suggested wavelengths and bandwidths are indicated in the diagram. In protocol (*1*), the 365-nm and 627-nm LEDs on the epifluorescence port should be triggered to be on whenever the camera is exposing. In protocol (*2*), the 365-nm fluorescence and white-transmitted LED should be triggered with the camera. In protocol (*3*), the 365-nm LED is triggered every frame, and the 627-nm and white LED are triggered on alternating frames. The "red" emission channel should then be demultiplexed after acquisition

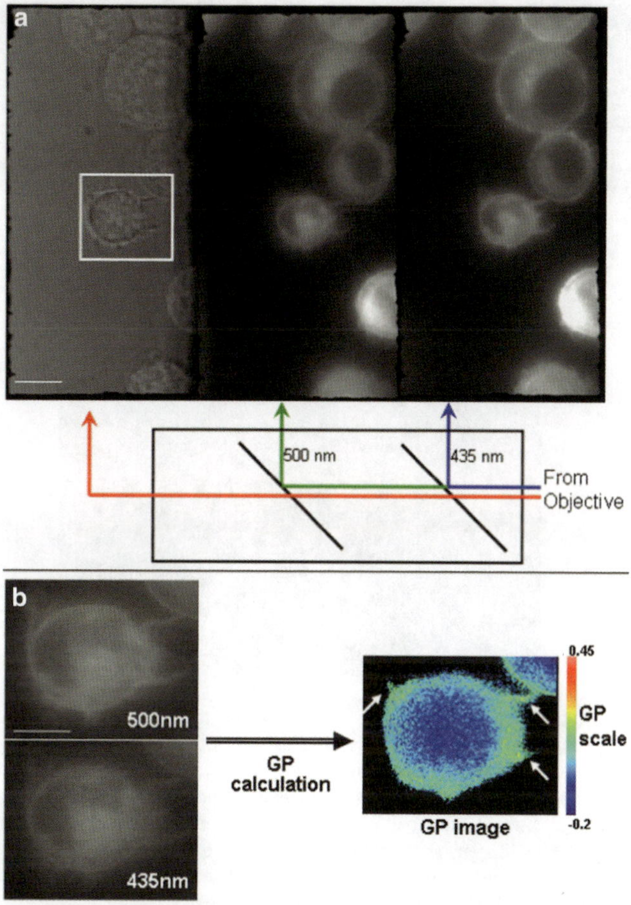

Fig. 2 Widefield Laurdan microscopy. Representation of the Optosplit III block with dichroic mirror set up for Laurdan staining. The two dichroic mirrors separate the fluorescence emission of the Laurdan-stained cells into the blue (435 nm) and green (500 nm) wavelengths, with a third channel showing light from the red end of the spectrum (N500 nm) (**a**). Calculation of the generalised polarisation of the cell outlined in (**a**) showing fluorescence at the blue and green wavelengths (**b**) and the GP image calculated from these. The range of values represented on the pseudo-coloured GP image is shown as a colourimetric scale to the right of the GP image. Filopodia are highlighted with *arrows* in the GP image. Scale bars = 10 μm (**a**) and 5 μm (**b**)

allows for acquisition of distinct images from three different wavelengths in parallel, in real time (Fig. 2). This allows for biological phenomena (such as filipodia formation, shown here) and processes to be followed. Laurdan technology can also be combined with fluorescent protein imaging to study the interaction of protein-coated microparticles with the cell surface prior to internalisation (Fig. 3). Laurdan emission is restricted to visible wavelengths in the blue/green area of the spectrum, and thus, by imaging with fluorescent markers or a

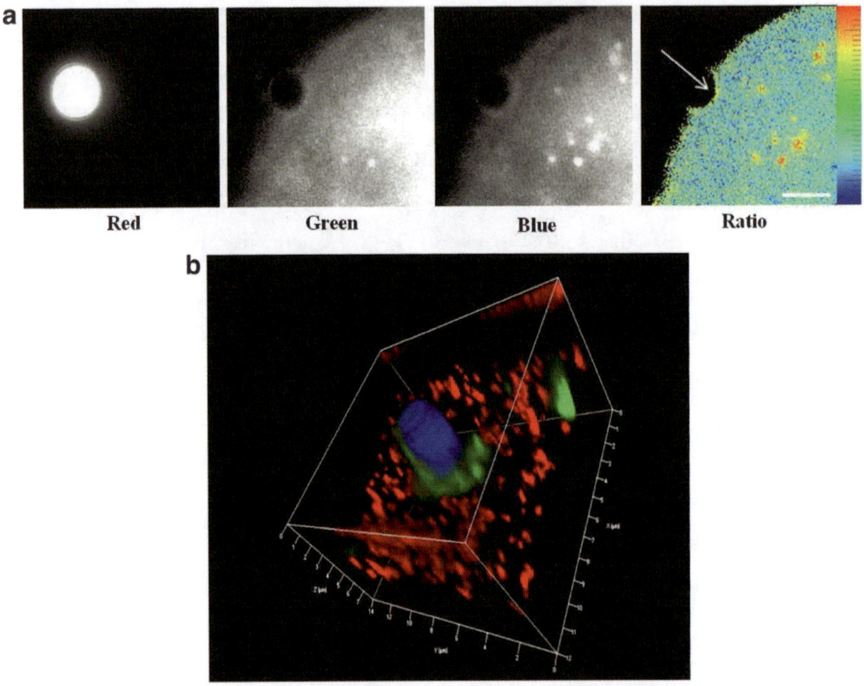

Fig. 3 Coated beads which bind to glycosylated moeities on the cell surface induce high lipid order and recruit raft markers such as caveolin to the bound area of the plasma membrane. Triplesplit images of a trans-sialidase-coated bead (*red*) and *green* and *blue* images of a Laurdan-labelled cell (**a**), GP analysis (*ratio*) shows a region more highly ordered at the area of bead attachment (*arrow*). Scale bar = 4 μm. Early endosome (*red*) and cav1 (*green*) immunolabelling of MDCK cells show an accumulation of cav1 at the bead (*blue*)-cell interface. Images were taken with a Zeiss Axioplan2 microscope and deconvolved before reconstructing into a 3D representation using Axiovision software (**b**)

transmitted light in the red/far red of the spectrum, high-resolution imaging of the cell or fluorescent cellular markers may be utilised to identify regions of interest. Having identified the region of interest, the GP of that area can be quantified directly (Fig. 4). In addition, it is possible to facilitate discrimination of areas of distinct GP by setting thresholds and masking areas outside the threshold as in the case of a living protozoan flagellum shown here as distinct from the rest of the plasmalemma (Fig. 5).

Another benefit to using a widefield setup may be the reduction of the photoselection effect. This arises from the way Laurdan molecules align parallel to the lipid chains in the membrane and that the light used for its excitation in confocal and two-photon microscopy is polarised giving rise to the strongest excitation in those molecules parallel to the plane of the light [73]. The irregular shape of cells, with regions inevitably in many different orientations relative to the plane of excitation, results in an uneven excitation of Laurdan molecules, skewing

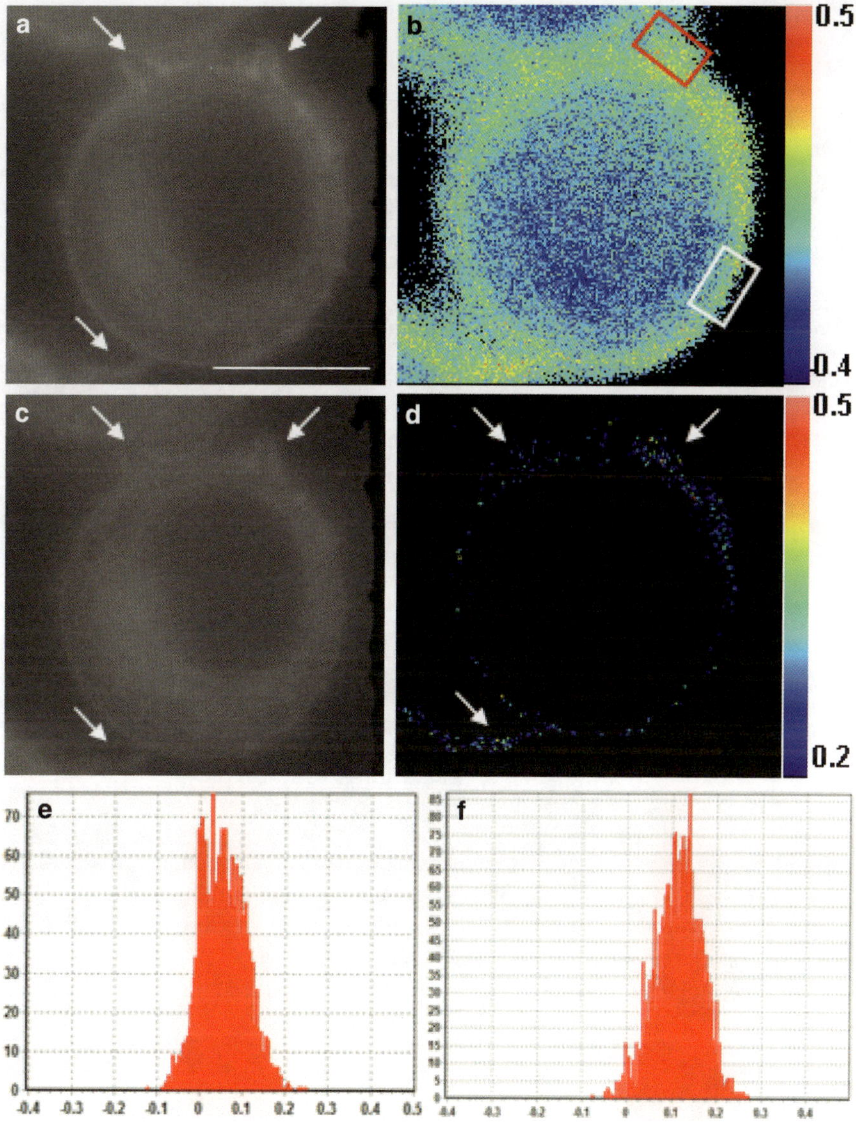

Fig. 4 Isolation of discrete high and low generalised polarisation regions within a RAW264.7 cell membrane. Filopodia are highlighted (*arrows*). Fluorescence of Laurdan stained. RAW264.7 cell in the blue (**a**) and green (**c**) wavelengths. (**b**) GP of the cell showing the range of the cell's calculated GP values. Regions of high (*red box*) and low (*white box*) membrane order are shown. (**d**) GP image showing only those pixels with a GP of 0.2 or more. The regions around the filopodia appear more ordered. GP histograms (x-axis = GP value, y-axis = number of pixels) of the high- (**e**) and low- (**f**) order regions from (**b**). Scale bar = 5 μm

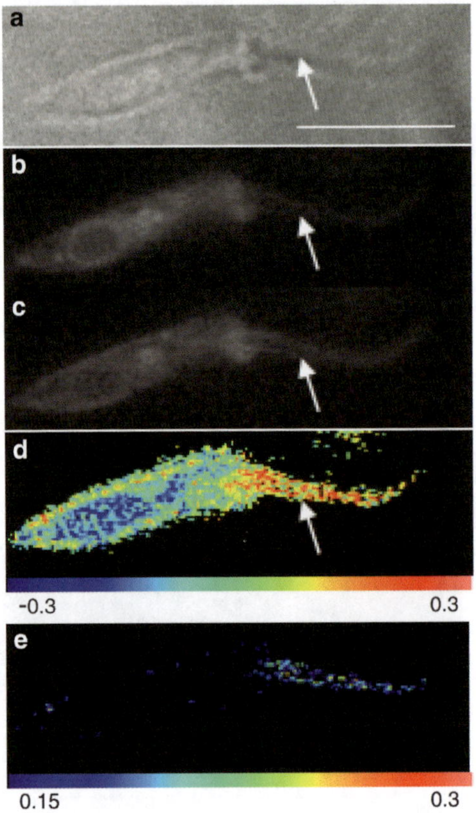

Fig. 5 Laurdan staining of *Trypanosoma rangeli*. Far-red illuminated brightfield (**a**) and fluorescent images (**b** and **c**) of a Laurdan-stained *T. rangeli* using the triplesplit widefield microscope. GP analysis (**d**) shows the more highly ordered flagellum (*arrow*) when compared to the rest of the cellular membrane. Exclusion of low-GP pixels (**e**) demonstrates the greater number of high-GP pixels in the flagellum. Scale bar = 4 μm

the GP measurements. This effect is particularly pronounced in more ordered regions, where the Laurdan molecule is even more rigidly aligned with the membrane. Although direct analysis of the magnitude of the photoselection effect and methods for its mitigation are not yet published, it is believed that non-polarised light sources such as those used in widefield setups reduce this artefact markedly [71, 77].

4.2 C-Laurdan

Despite its use in these studies, Laurdan is still limited as a tool for studying membrane dynamics in living cells due to its rapid photobleaching and hydrophobic nature. It is not readily soluble in water, but instead a solute such as DMSO or methanol must be used, which may itself have an effect on the target cell [15]. An improvement on Laurdan has recently been synthesised [78]: 6-dodecanoyl-2-[*N*-methyl-*N*-(carboxymethyl)amino]naphthalene, or C-Laurdan. This has proven to be more photostable and to have higher water stability arising from the substitution of one of Laurdan's methyl groups with a carboxylic group. The improvement of C-Laurdan over Laurdan in this regard has already been shown in two-photon microscopy (e.g. [79] with potentially greater benefits in widefield microscopy, allowing prolonged excitation and its application in a more neutral vehicle buffer).

In spite of the improvement offered by C-Laurdan, GP imaging of membrane domains remains far from optimal. The need for the highest possible resolution GP imaging in combination with the simultaneous imaging of the whole cell and additional markers of cellular behaviours (such as cell signalling) in order to obtain a maximum amount of biological information remains to be addressed. It is not clear to what extent such images can be subjected to standard deconvolution software to increase image quality and clarity, and this is an area which should be considered in the future. Further, the monopolisation of a broad (blue/green) area of the spectrum required by Laurdan significantly limits the scope for the number and types of processes and cellular behaviours that can currently be simultaneously imaged, typically restricting investigators to simply mapping the GP image onto a simultaneously captured image of the cell, or to follow just one or two additional cellular markers at the (red) end of the spectrum.

4.3 Di-4-ANEPPDHQ

Di-4-ANEPPDHQ is an environmentally sensitive styryl dye that can also be used to differentiate between liquid-ordered phases and liquid-disordered phases both in GUV's and polarised neutrophil membranes [80, 81]. Similar to Laurdan, di-4-ANEPPDHQ can indicate lipid packing without being affected by inserted peptides, [82] and both can be used in lower resolution techniques [83]. It incorporates into the membrane in a different manner to Laurdan, aligning with the acyl heads deeper into the membrane. Its two positive charges also confer a lesser ability to flip between the inner and outer leaflets, but its solubility in water does make it a more useful tool for studying live cells and thus in the future could be adapted for use with a widefield microscope.

5 FRET/FLIM

5.1 FRET

The development of Förster resonance energy transfer (FRET) has enabled the imaging of physical processes occurring at the cell surface in real time and to this end is a commonly used technique. Excitation of a fluorescent molecule occurs upon absorption of a photon leading to one of its electrons being promoted to a higher energy level. When this electron returns to its lower energy state, energy is released in the form of light. FRET essentially is the nonradioactive transfer of this energy to a nearby acceptor molecule instead of its loss by fluorescence. The extent of FRET is dependent primarily on three factors: the spectral overlap of the donor and acceptor fluorophores, their closeness (i.e. 1–10 nm apart for the ratio between the donor and acceptor fluorescence levels to be measured) and their dipole-dipole interaction. Initial utilisation of FRET was limited to purification of specific molecules and their reintroduction, the methodology being destructive and requiring microinjection. The development of green fluorescent protein (GFP) as a tool led to a non-invasive protocol for the introduction of the tagged molecules by transfection. Since the pioneering introduction of GFP, many other fluorescent analogues have been extracted, and the current fluorescent proteins of choice for FRET are YFP (yellow) and CFP (cyan), first seen in conjugation with retinoic, steroid receptor coactivator-1 and peroxisome proliferator-activated receptor-binding protein [84]. Although widely used, both YFP and CFP are pH sensitive [85, 86], thus restricting their use to biological processes which do not alter cellular pH. Recent work has seen the development of pH-stable versions of YFP, Venus [87] and Citrine [88] which should therefore prove more exploitable in living cells.

Similar to FRET, homo-FRET relies on two fluorophores being less than 10 nm apart for the exchange of energy; however, both fluorophores are the same, and therefore, interaction cannot be monitored by changes in fluorescence emission. Nevertheless, processes leading to close approach, for example, oligomerisation, can be followed using anisotropy measurements that reflect changes in fluorophore polarisation, both due to energy transfer and changes in molecular rotation. The use of these tools for studying lipid rafts has been neatly reviewed by [89]. The ability to detect the rotation of specific lipids within the membrane gives a general idea of the level of fluidity in specific regions of the membrane and therefore the level of lipid order. Similarly, this microscopy technique has also been used to show that BODIPY-cholesterol arranges itself into the membrane and into specific regions of higher order [90]. In addition to determining orientation, anisotropy can also be used to shed light on protein clustering: by calculating the rate of homo-transfer, the distance between fluorophores can be determined, and the number of molecules per cluster elucidated. The size of the protein clusters in lipid rafts and the spatial distribution has also been determined using anisotropy techniques [91]. The clustering of GPI-AP's into regions of the plasma membrane of approximately 4–5 nm has also been described using anisotropy [92]. All of these studies, however, are used with either confocal or two-photon microscopy, so an important question for the field is as follows: can FRET be used with widefield microscopes? Indeed it can; however, the resultant images can be affected by a significant amount of noise, blurring the signal from excited fluorophores. This additional noise can now be removed using 3D-FRET reconstruction and additional photobleaching correction [93] thus cost-effectively improving accessibility of this technique.

5.2 *FLIM*

FRET microscopy can also be combined with fluorescence lifetime imaging microscopy (FLIM) to monitor signalling across the whole plasma membrane [94]. There are two types of FLIM (time domain and frequency domain), but for the purposes of this chapter, we will discuss only the frequency-domain method which is best suited for widefield microscopy. When a fluorophore becomes excited, the transition from the excited state to the ground state occurs over a given average decay time that can be measured using FLIM techniques. FLIM in recent years has been employed in the study of lipid rafts which has been reviewed in [95]. Although the majority of these studies have been done using GUV's and other controlled model membrane systems to reduce the complex lipid-protein interactions, FLIM for live cells was demonstrated by Owen and co-workers in 2006 using di-4-ANEPPDHQ and normalising their system with unilamellar vesicles. Here, the FLIM contrast gave good resolution of regions of high lipid order indicative of lipid rafts, especially when concentrated in membrane protrusions [96]. To date, all FLIM work on lipid rafts has been done using confocal or two-photon systems, but widefield FLIM systems are now commercially available, and so it may only be a matter of time before this type of imaging is applied to the study of lipid rafts.

6 Conclusions

Membrane dynamics is a fast-evolving field with the many new methods and probes being developed each year affording ever increased insights into how membranes behave in the laboratory. However, the major limitations of these techniques lie typically with the high cost of bespoke microscopes which often employ confocal and two-photon systems and which give little consideration to preservation of cellular integrity and homeostasis during experiments. There is a clear need to rapidly apply and deploy this work into mainstream biological laboratories by development of economical, 4-dimensional imaging on user-friendly systems using widefield optics and simultaneous capture of multiple fluorescent markers. Such systems should enable biologists to consider the coordinated processes triggered from signalling platforms during cellular interaction with the environment. A plethora of microscopic methods and labelling molecules have been described, but many have some disadvantages for use in live cells or are confined in their use to very small regions of the cell membrane. The Laurdan family of fluorescent probes, however, can be used to image whole cells and tissues and can be combined with longer wavelength (yellow through far red) fluorescent adaptor proteins. Although, developed for use in confocal and two-photon systems, application to widefield-based investigation is now established and should facilitate future investigations seeking to relate modulation of plasma membrane dynamics to resultant change in cellular behaviour.

Acknowledgements We thank Dr. Paul Thomas at The Wellcome Laboratory for cell imaging for critical reading of the manuscript and advice and insight along the way and Rosanna Davies for her tremendous illustrative skills in producing the drawing of our widefield Laurdan microscopy setup.

References

1. Lingwood D, Simons K (2010) Lipid rafts as a membrane-organizing principle. Science 327 (5961):46–50
2. Simons K, Ikonen E (1997) Functional rafts in cell membranes. Nature 387(6633):569–572
3. Levental I et al (2010) Palmitoylation regulates raft affinity for the majority of integral raft proteins. Proc Natl Acad Sci U S A 107(51):22050–22054
4. Lisanti MP, Tang ZL, Sargiacomo M (1993) Caveolin forms a hetero-oligomeric protein complex that interacts with an apical GPI-linked protein: implications for the biogenesis of caveolae. J Cell Biol 123(3):595–604
5. Badizadegan K et al (2004) Trafficking of cholera toxin-ganglioside GM1 complex into Golgi and induction of toxicity depend on actin cytoskeleton. Am J Physiol Cell Physiol 287(5): C1453–C1462
6. Pike LJ (2006) Rafts defined: a report on the Keystone symposium on lipid rafts and cell function. J Lipid Res 47(7):1597–1598
7. Kusumi A, Suzuki K (2005) Toward understanding the dynamics of membrane-raft-based molecular interactions. Biochim Biophys Acta 1746(3):234–251
8. Lai EC (2003) Lipid rafts make for slippery platforms. J Cell Biol 162(3):365–370
9. Zacharias DA (2002) Sticky caveats in an otherwise glowing report: oligomerizing fluorescent proteins and their use in cell biology. Sci STKE 2002(131):pe23
10. Simons K, Sampaio JL (2011) Membrane organization and lipid rafts. Cold Spring Harb Perspect Biol 3(10):a004697
11. Lang T (2007) SNARE proteins and 'membrane rafts'. J Physiol 585(Pt 3):693–698
12. Insel PA et al (2005) Compartmentation of G-protein-coupled receptors and their signalling components in lipid rafts and caveolae. Biochem Soc Trans 33(Pt 5):1131–1134
13. Zhang Y et al (2009) Ceramide-enriched membrane domains–structure and function. Biochim Biophys Acta 1788(1):178–183
14. Lajoie P et al (2009) Caveolin-1 regulation of dynamin-dependent, raft-mediated endocytosis of cholera toxin-B sub-unit occurs independently of caveolae. J Cell Mol Med 13 (9B):3218–3225
15. Tyler KM et al (2009) Flagellar membrane localization via association with lipid rafts. J Cell Sci 122(Pt 6):859–866
16. Vieira OV et al (2006) FAPP2, cilium formation, and compartmentalization of the apical membrane in polarized Madin-Darby canine kidney (MDCK) cells. Proc Natl Acad Sci U S A 103(49):18556–18561
17. Bagnat M, Simons K (2002) Cell surface polarization during yeast mating. Proc Natl Acad Sci U S A 99(22):14183–14188
18. Reyes-Del Valle J et al (2005) Heat shock protein 90 and heat shock protein 70 are components of dengue virus receptor complex in human cells. J Virol 79(8):4557–4567
19. Liu NQ et al (2002) Human immunodeficiency virus type 1 enters brain microvascular endothelia by macropinocytosis dependent on lipid rafts and the mitogen-activated protein kinase signaling pathway. J Virol 76(13):6689–6700
20. Watson RO, Galan JE (2008) Campylobacter jejuni survives within epithelial cells by avoiding delivery to lysosomes. PLoS Pathog 4(1):e14

21. Silvie O et al (2006) Cholesterol contributes to the organization of tetraspanin-enriched microdomains and to CD81-dependent infection by malaria sporozoites. J Cell Sci 119(Pt 10):1992–2002
22. Fernandes MC et al (2007) Novel strategy in *Trypanosoma cruzi* cell invasion: implication of cholesterol and host cell microdomains. Int J Parasitol 37(13):1431–1441
23. Kuziemko GM, Stroh M, Stevens RC (1996) Cholera toxin binding affinity and specificity for gangliosides determined by surface plasmon resonance. Biochemistry 35(20):6375–6384
24. Knorr R, Karacsonyi C, Lindner R (2009) Endocytosis of MHC molecules by distinct membrane rafts. J Cell Sci 122(Pt 10):1584–1594
25. Antes P, Schwarzmann G, Sandhoff K (1992) Detection of protein mediated glycosphingolipid clustering by the use of resonance energy transfer between fluorescent labelled lipids. A method established by applying the system ganglioside GM1 and cholera toxin B subunit. Chem Phys Lipids 62(3):269–280
26. Janes PW, Ley SC, Magee AI (1999) Aggregation of lipid rafts accompanies signaling via the T cell antigen receptor. J Cell Biol 147(2):447–461
27. Chakraborty SK et al (2007) Cholera toxin B conjugated quantum dots for live cell labeling. Nano Lett 7(9):2618–2626
28. Rivera EM et al (2011) Imaging heterostructured quantum dots in cultured cells with epifluorescence and transmission electron microscopy. Proc SPIE 7909:79090N
29. Steinert S et al (2008) A fluorescent glycolipid-binding peptide probe traces cholesterol dependent microdomain-derived trafficking pathways. PLoS One 3(8):e2933
30. Hebbar S et al (2008) A fluorescent sphingolipid binding domain peptide probe interacts with sphingolipids and cholesterol-dependent raft domains. J Lipid Res 49(5):1077–1089
31. Pike LJ et al (2002) Lipid rafts are enriched in arachidonic acid and plasmenylethanolamine and their composition is independent of caveolin-1 expression: a quantitative electrospray ionization/mass spectrometric analysis. Biochemistry 41(6):2075–2088
32. Kiyokawa E et al (2004) Recognition of sphingomyelin by lysenin and lysenin-related proteins. Biochemistry 43(30):9766–9773
33. Kiyokawa E et al (2005) Spatial and functional heterogeneity of sphingolipid-rich membrane domains. J Biol Chem 280(25):24072–24084
34. Kidani Y et al (2012) Differential localization of sphingomyelin synthase isoforms in neurons regulates sphingomyelin cluster formation. Biochem Biophys Res Commun 417(3):1014–1017
35. Yamaji A et al (1998) Lysenin, a novel sphingomyelin-specific binding protein. J Biol Chem 273(9):5300–5306
36. Kobayashi H, Suzuki H, Ohta N (2006) Exfoliation of the epidermal cells and defecation by amphibian larvae in response to coelomic fluid and lysenin from the earthworm *Eisenia foetida*. Biomed Res 27(4):169–181
37. Bittman R, Fischkoff SA (1972) Fluorescence studies of the binding of the polyene antibiotics filipin 3, amphotericin B, nystatin, and lagosin to cholesterol. Proc Natl Acad Sci U S A 69 (12):3795–3799
38. Orci L et al (1981) Heterogeneous distribution of filipin–cholesterol complexes across the cisternae of the Golgi apparatus. Proc Natl Acad Sci U S A 78(1):293–297
39. Ohno-Iwashita Y et al (2004) Perfringolysin O, a cholesterol-binding cytolysin, as a probe for lipid rafts. Anaerobe 10(2):125–134
40. Hayashi M et al (2006) Detection of cholesterol-rich microdomains in the inner leaflet of the plasma membrane. Biochem Biophys Res Commun 351(3):713–718
41. Schroeder F et al (1991) Transmembrane distribution of sterol in the human erythrocyte. Biochim Biophys Acta 1066(2):183–192
42. Mukherjee S et al (1998) Cholesterol distribution in living cells: fluorescence imaging using dehydroergosterol as a fluorescent cholesterol analog. Biophys J 75(4):1915–1925
43. Wustner D (2007) Plasma membrane sterol distribution resembles the surface topography of living cells. Mol Biol Cell 18(1):211–228

44. Tasset C et al (1992) Comparison of nephrotoxicities of different polyoxyethyleneglycol formulations of amphotericin B in rats. Antimicrob Agents Chemother 36(7):1525–1531
45. Nagy E et al (2007) Hyperfluidization-coupled membrane microdomain reorganization is linked to activation of the heat shock response in a murine melanoma cell line. Proc Natl Acad Sci U S A 104(19):7945–7950
46. Sato SB et al (2004) Distribution and transport of cholesterol-rich membrane domains monitored by a membrane-impermeant fluorescent polyethylene glycol-derivatized cholesterol. J Biol Chem 279(22):23790–23796
47. Pagano RE et al (1991) A novel fluorescent ceramide analogue for studying membrane traffic in animal cells: accumulation at the Golgi apparatus results in altered spectral properties of the sphingolipid precursor. J Cell Biol 113(6):1267–1279
48. Marks DL, Bittman R, Pagano RE (2008) Use of Bodipy-labeled sphingolipid and cholesterol analogs to examine membrane microdomains in cells. Histochem Cell Biol 130(5):819–832
49. D'Auria L et al (2011) Segregation of fluorescent membrane lipids into distinct micrometric domains: evidence for phase compartmentation of natural lipids? PLoS One 6(2):e17021
50. Greaves J, Chamberlain LH (2011) Differential palmitoylation regulates intracellular patterning of SNAP25. J Cell Sci 124(Pt 8):1351–1360
51. Lajoie P, Nabi IR (2007) Regulation of raft-dependent endocytosis. J Cell Mol Med 11 (4):644–653
52. Anderson RG (1993) Plasmalemmal caveolae and GPI-anchored membrane proteins. Curr Opin Cell Biol 5(4):647–652
53. Rothberg KG et al (1992) Caveolin, a protein component of caveolae membrane coats. Cell 68 (4):673–682
54. Scherer PE et al (1997) Cell-type and tissue-specific expression of caveolin-2. Caveolins 1 and 2 co-localize and form a stable hetero-oligomeric complex in vivo. J Biol Chem 272 (46):29337–29346
55. Tang Z et al (1996) Molecular cloning of caveolin-3, a novel member of the caveolin gene family expressed predominantly in muscle. J Biol Chem 271(4):2255–2261
56. Parton RG (1996) Caveolae and caveolins. Curr Opin Cell Biol 8(4):542–548
57. Mundy DI et al (2002) Dual control of caveolar membrane traffic by microtubules and the actin cytoskeleton. J Cell Sci 115(Pt 22):4327–4339
58. Glebov OO, Bright NA, Nichols BJ (2006) Flotillin-1 defines a clathrin-independent endocytic pathway in mammalian cells. Nat Cell Biol 8(1):46–54
59. Frick M et al (2007) Coassembly of flotillins induces formation of membrane microdomains, membrane curvature, and vesicle budding. Curr Biol 17(13):1151–1156
60. Riento K et al (2009) Endocytosis of flotillin-1 and flotillin-2 is regulated by Fyn kinase. J Cell Sci 122(Pt 7):912–918
61. Blanchet MH et al (2008) Cripto recruits Furin and PACE4 and controls Nodal trafficking during proteolytic maturation. EMBO J 27(19):2580–2591
62. Weber G, Farris FJ (1979) Synthesis and spectral properties of a hydrophobic fluorescent probe: 6-propionyl-2-(dimethylamino)naphthalene. Biochemistry 18(14):3075–3078
63. Bagatolli LA et al (1999) A model for the interaction of 6-lauroyl-2-(N, N-dimethylamino) naphthalene with lipid environments: implications for spectral properties. Photochem Photobiol 70(4):557–564
64. Parasassi T et al (1990) Phase fluctuation in phospholipid membranes revealed by Laurdan fluorescence. Biophys J 57(6):1179–1186
65. Parasassi T et al (1991) Quantitation of lipid phases in phospholipid vesicles by the generalized polarization of Laurdan fluorescence. Biophys J 60(1):179–189
66. Buffone MG et al (2009) High cholesterol content and decreased membrane fluidity in human spermatozoa are associated with protein tyrosine phosphorylation and functional deficiencies. J Androl 30(5):552–558
67. Kaiser HJ et al (2011) Molecular convergence of bacterial and eukaryotic surface order. J Biol Chem 286(47):40631–40637

68. Gaus K et al (2003) Visualizing lipid structure and raft domains in living cells with two-photon microscopy. Proc Natl Acad Sci U S A 100(26):15554–15559
69. Khan NA et al (2011) Stability of fatty acids during wilting of perennial ryegrass (*Lolium perenne* L.): effect of bruising and environmental conditions. J Sci Food Agric 91 (9):1659–1665
70. Owen DM et al (2010) Imaging membrane lipid order in whole, living vertebrate organisms. Biophys J 99(1):L7–L9
71. Wheeler G, Tyler KM (2011) Widefield microscopy for live imaging of lipid domains and membrane dynamics. Biochim Biophys Acta 1808(3):634–641
72. Hansen JS, Helix-Nielsen C (2011) An epifluorescence microscopy method for generalized polarization imaging. Biochem Biophys Res Commun 415(4):686–690
73. Sanchez SA, Tricerri MA, Gratton E (2007) Interaction of high density lipoprotein particles with membranes containing cholesterol. J Lipid Res 48(8):1689–1700
74. Weber P, Wagner M, Schneckenburger H (2010) Fluorescence imaging of membrane dynamics in living cells. J Biomed Opt 15(4):046017
75. Toyoda T et al (2009) Thermo-sensitive response based on the membrane fluidity adaptation in *Paramecium multimicronucleatum*. J Exp Biol 212(17):2767–2772
76. Sitrin RG et al (2010) Migrating human neutrophils exhibit dynamic spatiotemporal variation in membrane lipid organization. Am J Respir Cell Mol Biol 43(4):498–506
77. Parasassi T et al (1997) Two-photon fluorescence microscopy of laurdan generalized polarization domains in model and natural membranes. Biophys J 72(6):2413–2429
78. Kim HM et al (2007) A two-photon fluorescent probe for lipid raft imaging: C-laurdan. Chembiochem 8(5):553–559
79. Klemm RW et al (2009) Segregation of sphingolipids and sterols during formation of secretory vesicles at the trans-Golgi network. J Cell Biol 185(4):601–612
80. Jin L et al (2005) Cholesterol-enriched lipid domains can be visualized by di-4-ANEPPDHQ with linear and nonlinear optics. Biophys J 89(1):L04–L06
81. Jin L et al (2006) Characterization and application of a new optical probe for membrane lipid domains. Biophys J 90(7):2563–2575
82. Dinic J et al (2011) Laurdan and di-4-ANEPPDHQ do not respond to membrane-inserted peptides and are good probes for lipid packing. Biochim Biophys Acta 1808(1):298–306
83. Owen DM, Gaus K (2010) Optimized time-gated generalized polarization imaging of Laurdan and di-4-ANEPPDHQ for membrane order image contrast enhancement. Microsc Res Tech 73 (6):618–622
84. Llopis J et al (2000) Ligand-dependent interactions of coactivators steroid receptor coactivator-1 and peroxisome proliferator-activated receptor binding protein with nuclear hormone receptors can be imaged in live cells and are required for transcription. Proc Natl Acad Sci U S A 97(8):4363–4368
85. Miyawaki A (2011) Development of probes for cellular functions using fluorescent proteins and fluorescence resonance energy transfer. Annu Rev Biochem 80:357–373
86. Miyawaki A, Tsien RY (2000) Monitoring protein conformations and interactions by fluorescence resonance energy transfer between mutants of green fluorescent protein. Methods Enzymol 327:472–500
87. Nagai T et al (2002) A variant of yellow fluorescent protein with fast and efficient maturation for cell-biological applications. Nat Biotechnol 20(1):87–90
88. Griesbeck O et al (2001) Reducing the environmental sensitivity of yellow fluorescent protein. Mechanism and applications. J Biol Chem 276(31):29188–29194
89. Rao M, Mayor S (2005) Use of Forster's resonance energy transfer microscopy to study lipid rafts. Biochim Biophys Acta 1746(3):221–233
90. Ariola FS et al (2009) Membrane fluidity and lipid order in ternary giant unilamellar vesicles using a new bodipy-cholesterol derivative. Biophys J 96(7):2696–2708
91. Bader AN et al (2009) Homo-FRET imaging enables quantification of protein cluster sizes with subcellular resolution. Biophys J 97(9):2613–2622

92. Sharma SD et al (2004) Radiochromic film measurement of anisotropy function for high-dose-rate Ir-192 brachytherapy source. Phys Med Biol 49(17):4065–4072
93. Hoppe AD et al (2008) Three-dimensional FRET reconstruction microscopy for analysis of dynamic molecular interactions in live cells. Biophys J 95(1):400–418
94. Verveer PJ, Squire A, Bastiaens PI (2000) Global analysis of fluorescence lifetime imaging microscopy data. Biophys J 78(4):2127–2137
95. de Almeida RF, Loura LM, Prieto M (2009) Membrane lipid domains and rafts: current applications of fluorescence lifetime spectroscopy and imaging. Chem Phys Lipids 157 (2):61–77
96. Owen DM et al (2006) Fluorescence lifetime imaging provides enhanced contrast when imaging the phase-sensitive dye di-4-ANEPPDHQ in model membranes and live cells. Biophys J 90(11):L80–L82

Quantitative Fluorescence Studies of Intracellular Sterol Transport and Distribution

Daniel Wüstner, Frederik W. Lund, and Lukasz M. Solanko

Abstract Unraveling the pathways of intracellular cholesterol transport is of great importance for biomedicine, since disturbed cholesterol trafficking is involved in many metabolic diseases. Most fluorescent probes for cholesterol, however, have physico-chemical properties deviating from the natural sterol. Intrinsically fluorescent sterols like dehydroergosterol (DHE) and the related cholestatrienol (CTL) have great potential for analysis of sterol trafficking due to their close resemblance of ergosterol and cholesterol, respectively. Excitation and emission of both sterols are in the ultraviolet (UV), which, together with high bleaching propensity and low brightness, make fluorescence imaging of DHE and CTL challenging. Here, we present an overview of how UV-sensitive wide field (UV-WF) and multiphoton (MP) microscopy can be applied to image both sterols in living cells and tissues. In addition, we show, for the first time, how applying advanced image denoising can dramatically enhance the signal-to-noise ratio in MP image sequences of DHE. This allowed us to track DHE-containing vesicles and surface protrusions in cells over prolonged time. We also discuss the properties of BODIPY-tagged cholesterol (BChol) compared to DHE and cholesterol and present an overview of fluorescence imaging techniques for analyzing cellular sterol dynamics.

Keywords Cholesterol · Diffusion · Dynamics · Ergosterol · Fluorescence · Intracellular · Membrane · Multiphoton · Non-vesicular · Vesicular

---

D. Wüstner (✉)
Department of Biochemistry and Molecular Biology, University of Southern Denmark, Odense, Denmark

Department of Biochemistry and Molecular Biology, University of Southern Denmark, Campusvej 55, DK-5230 Odense M, Denmark
e-mail: wuestner@bmb.sdu.dk

F.W. Lund and L.M. Solanko
Department of Biochemistry and Molecular Biology, University of Southern Denmark, Odense, Denmark

Y. Mély and G. Duportail (eds.), *Fluorescent Methods to Study Biological Membranes*, Springer Ser Fluoresc (2013) 13: 185–214, DOI 10.1007/4243_2012_57,

Contents

1 Introduction

Cholesterol is an indispensable lipid of mammalian cell membranes. Its unique physico-chemical properties regulate membrane fluidity, permeability, and bending resistance, while proper function of some membrane proteins critically depends on cholesterol [19, 84, 135]. Cholesterol is also the metabolic precursor for vitamin D, oxysterols, bile salts, and steroid hormones. The cholesterol biosynthesis pathway produces many important intermediates like the isoprenoids farnesyl and geranyl being required for covalent protein modification. Mammalian cells strictly regulate the cholesterol content of their membranes, since small changes in the required membrane sterol mole fraction can have deleterious effects on the biological function. For example, cholesterol depletion of the plasma membrane using cyclodextrin inhibits clathrin-dependent endocytosis, while cholesterol loading of macrophages induces membrane ruffling and inhibits cell migration probably by a disturbing effect on the membrane attached cytoskeleton [89, 98, 99, 117]. Similarly, cholesterol loading diminishes glucose-stimulated insulin secretion from pancreatic β-cells likely by blocking a transient increase of cytoplasmic calcium upon glucose addition [41, 42]. Due to its effects on plasma membrane structure, cholesterol also modulates the rate and mechanism of internalization of acetylcholine receptor [14]. Cholesterol increases the lateral packing of unsaturated phospholipids, a process known as cholesterol's condensing effect on membranes [84, 135]. Consequently, the permeability and area of the bilayer are reduced, while the thickness gets increased such that total membrane volume remains constant [49, 63, 135]. Alterations in membrane thickness and lateral pressure due to the presence of cholesterol are a modulating factor on membrane protein function [54]. In addition, some proteins, like the oxytocin receptor, rhodopsin, or the β-adrenergic receptor, are known to directly bind membrane cholesterol, an inherent requirement for their proper function or dimerization [19, 66].

Our knowledge about the pathways and molecular players of whole body cholesterol circulation and intercellular cholesterol trafficking has increased tremendously over the last two decades. This is a consequence of progress in deciphering the genome of mouse and humans but also of the provision of new experimental approaches like generation of knock-out mouse models of various human sterol-related diseases [25, 48, 50, 59]. Intracellular cholesterol transport is very difficult to analyze, but technical progress during the last 12 years made it possible to follow reliably the trafficking of fluorescent cholesterol analogs. Such studies have shown that vesicular and nonvesicular pathways contribute to establishing and maintaining a heterogeneous distribution of cholesterol between various cellular organelles [74, 75, 80]. How intracellular cholesterol trafficking can be studied using various fluorescent imaging approaches and how quantitative fluorescence microscopy contributes to unravel the kinetics and molecular basis of the different sterol transport modes are the topics of this chapter.

2 Fluorescent Probes for Studying Intracellular Cholesterol Transport

In this section, we will give a very brief overview of the most common fluorescence methods for monitoring intracellular sterol distribution and transport. Focus is on filipin, as the classical cholesterol-binding fluorophore, and on fluorescent sterol analogs with close resemblance of cholesterol's properties. For a more comprehensive discussion of various fluorescent sterol analogs and sterol-binding fluorophores, the reader is referred to several reviews published in that area during the last 5 years [35, 36, 76, 132, 137].

2.1 *Filipin as the Prototype of Cholesterol-Binding Probes*

Filipin is the most used sterol-binding polyene being first described in 1955 as a potent antifungal agent [8, 127]. Filipin has been isolated from bacteria *Streptomyces filipinensis,* which was found in a sample of Philippine soil. Originally, it was thought that filipin is a single type polyene macrolide antibiotic with nonaromatic groups and the sum formula of $C_{35}H_{58}O_{11}$ giving an approximate molecular weight of 654 [21]. Bergy and Eble showed in 1968 that filipin is actually a mixture of four major components and should therefore be considered as a filipin complex [10]. It took almost 30 years until scientists established the ratio between the four components of the filipin complex: filipin I (4 %), filipin II (25 %), filipin III (53 %), and filipin IV (18 %). Filipin I is a mixture of several different subcomponents of filipin III, and filipin II is 1′-deoxy-filipin III. The last filipin IV is an isomer of filipin III [100, 125]. All of them are fluorescent and have three characteristic excitation maxima at 322, 338, and 355 nm, and a shoulder starting at 305 nm. Filipin should be kept in molecular sieve-treated dimethylsulfoxide to remove water, since aqueous environments degrade filipin fluorescence [76]. Binding of filipin to cholesterol or other sterols with a free 3′ hydroxyl group causes dramatic changes in the ratio between peaks 1 and 3. The ratio changes from 0.7 in the case of free filipin up to 2.5 in the case of filipin-sterol complexes [91]. In contrast to free cholesterol, all photophysical parameters of filipin remained unchanged in the presence of cholesteryl palmitate, which indicates that filipin cannot bind to the ester form of sterols [109]. Filipin has been proposed to form 1:1 complexes with cholesterol and other free sterols in membranes and also to self-aggregate and to extract the sterols from bilayers [8]. In fact, alterations in absorption and emission spectra of filipin are likely a consequence of sterol-promoted self-interaction resulting in formation of excitons with parallel orientation of the transition dipoles [20]. Membrane association of filipin has been measured by changes in absorption, fluorescence intensity and polarization or by circular dichroism (see [11] and refs. therein). The rate of filipin III binding to liposomes made of egg yolk lecithin was determined by Bittman and co-workers using stopped-flow measurements of absorption at 360 nm [12]. The authors found that the initial rate of absorption change depended linearly on cholesterol concentration below 10 μM filipin [12, 13]. They could relate binding kinetics of filipin to the transbilayer distribution of cholesterol in model and cell membranes, for example, in erythrocytes, where a 50:50 % cholesterol distribution in the plasma membrane was measured by that approach [13]. Due to the problems mentioned above concerning filipin-induced alterations in bilayer structure but also due to unpredictable effects of membrane composition and protein content on filipin binding kinetics, these results must be related to more recent approaches using intrinsically fluorescent sterols or spin-labeled sterols [78, 83, 88, 115]. Visualization of fixed filipin-stained cells by fluorescence microscopy is a routine assay in clinical laboratories for assessment of free sterol content in cells as well as a diagnostic tool for identification of cholesterol storage in various diseases, like Niemann-Pick type C disease [23, 90, 94]. It has been shown that cholesterol content of cells determined by biochemical methods correlates with measured filipin fluorescence in labeled cells over a wide sterol concentration range, for example, when cholesterol content was altered in macrophages using cholesterol/cyclodextrin [98]. Absolute quantification of subcellular cholesterol distribution based on filipin fluorescence, however, is complicated by the fact that filipin disturbs membrane structure, and equal access of filipin to all sterol-containing cellular membranes is not guaranteed [8, 9, 17, 112, 116]. Also, cross-contamination is possible, since filipin has even been shown to bind gangliosides in addition to cholesterol [2]. Since cells need to be fixed, filipin does not allow for following inter-organelle sterol transport, and the assay does not work with living cells [35]. The total free sterol pool becomes stained by filipin, such that this method does not allow for discriminating endogenously synthesized from exogenous cholesterol taken up from the plasma membrane or delivered to cells via lipoproteins. In other words, it remains difficult to track a particular sterol pool in the cell by filipin. A recent exception is a study where uptake of cholesterol from atherogenic LDL has been visualized in filipin-stained macrophages [39]. Locally increased filipin staining was found in these cells when acetylated LDL (acLDL) with hydrolyzable cholesterol ester was used as sterol donor compared to nonhydrolyzable cholesteryl ether in the acLDL particle [39]. Filipin staining could be used in that study, since the detected increase in free sterol was massive and local as a consequence of creating an acidic extracellular compartment where degradation of the lipoprotein was initiated [40]. Adapting that assay to other cell types and research objectives will be challenging. Filipin is often used to report about cellular cholesterol content after sterol depletion or loading using the chelator cyclodextrin (CD). In a recent example, acute cholesterol depletion using methyl-β-CD in rat hepatoma cells expressing the NPC1L1 protein was followed by cholesterol loading from cholesterol-CD complexes, and the cellular cholesterol content was monitored by filipin staining [31]. In cells expressing functional NPC1L1 constructs, cholesterol reloading was more efficient than in control cells, and it was suggested that this observation resembles the physiological process of intestinal cholesterol absorption mediated by NPC1L1 [31, 56]. Acute cholesterol depletion using CD blocked clathrin-dependent endocytosis in these cells and thereby shifted the distribution of not only transferrin (Tf, a marker for that endocytic pathway) but also of NPC1L1 to the cell surface [45, 149]. Whether cholesterol-depletion-dependent cholesterol uptake by NPC1L1 takes place in enterocytes and hepatocytes, where this protein is expressed, is not clear.

2.2 *Fluorescent Cholesterol Analogs*

The general advantage of fluorescent cholesterol analogs is that these probes allow for live-cell imaging of sterol redistribution in pulse-chase experiments. This is not possible with stains like filipin, since binding of polyene macrolides grossly alters membrane structure and function [35, 76]. Most fluorescent sterols containing an attached fluorophore, like a nitrobenzoxadiazole (NBD) group, however, behave very different from cholesterol in model membranes and cells [76, 132]. For example, NBD-cholesterol with the fluorophore at carbon 25 of cholesterol was found to have the opposite membrane orientation than cholesterol and to be targeted to cholesterol-poor mitochondria in cells [86, 106]. Cholesterol is known to mediate formation of a liquid-ordered phase (lo) in model membranes containing saturated phospholipid species, for example, in dipalmitoylphosphatidylcholine (DPPC) bilayers at room temperature [53, 124]. In ternary mixtures, for example, composed of DPPC, an unsaturated lipid like dioleoylphosphatidylcholine (DOPC) and cholesterol, liquid-liquid immiscibility is induced by cholesterol creating liquid-disordered (ld)/lo phase coexistence [121]. The lo phase is known to be cholesterol-rich, since cholesterol has a higher packing capacity in phases containing phospholipids with saturated fatty acyl chains [5, 47, 82, 122, 152]. A decisive criterion for similarity of a fluorescent sterol with cholesterol (or the related ergosterol) is the ability of the probe to partition into the lo phase. While NBD-cholesterol analogs partition exclusively into the opposite, ld, phase, only the intrinsically fluorescent sterols dehydroergosterol (DHE) and the related synthetic cholestatrienol (CTL) have high partition preference for the lo phase [4, 30, 68, 113]. DHE has even been shown to induce formation of the lo phase in a similar concentration range as cholesterol and ergosterol [5, 30, 122]. For other fluorescent sterols, like dansyl cholesterol or pyrene-tagged cholesterol, no studies about the lo/ld phase partition have been performed; only their properties in liposomes made exclusively of either lo- or ld-type lipids have been reported [60, 61, 114]. That, however, cannot replace partition experiments in supported membrane or free-standing bilayers or in giant unilamellar vesicles (GUVs). A reasonable exception is the recently introduced BODIPY-cholesterol (BChol), which fulfills this criterion of being a good cholesterol mimic [1, 52, 113, 145]. BChol with the BODIPY moiety at carbon 24 but no other BODIPY-tagged cholesterols showed slight preference for the lo phase and co-distributed with DHE in baby hamster kidney (BHK) and HeLa cells [1, 145]. BChol shows a similar steady state distribution as [^{3}H]-cholesterol in purified membrane fractions of cells, but its esterification is lower than that of radiolabeled cholesterol [52]. BChol has been also used to monitor cellular cholesterol efflux but released from the cell more readily than cholesterol due to the disordering effect of the BODIPY group on surrounding lipid hydrocarbon chains [52]. Due to the exquisite fluorescence properties with high quantum yield, low environmental sensitivity, and large excitation coefficient, only spurious amounts of BChol need to be incorporated into cellular membranes for investigation of cellular sterol trafficking (0.1–0.5 %). Above 3 mol%, we found evidence for self-quenching of BChol likely by formation of ground state dimers, but no evidence for excimers, that is, excited state dimers [145]. Due to its partition into the ld and lo phase, BChol can be used as a sensitive liquid order/disorder microdomain fluorescence marker, for studying lipid-lipid interactions and membrane fluidity in model membranes [1, 145]. The mean fluorescence lifetime of the BODIPY moiety of BChol has been measured in ternary model membranes to be 5.57 ns and 5.34 ns in the lo and ld phase, respectively [1]. Lateral and rotational diffusion of BChol was slightly faster in the ld than in the lo phase supporting that the packing density is lower in the disordered phase [1]. Similar values of mean fluorescence lifetime as in model membranes were found for BChol in living cells [145].

Many studies have shown that the intrinsically fluorescent sterols DHE and CTL are the most suitable mimics of cholesterol for monitoring sterol distribution and transport kinetics in living cells [28, 43, 45, 78, 79, 108, 110, 133, 138–140, 142, 143]. Minimal chemical alterations of DHE compared to cholesterol (Fig. 1a) ensure that DHE has the same transport itineraries and intracellular distribution as cholesterol [29, 75, 132]. A serious limitation of DHE and CTL is their low brightness, excitation in the rather far ultraviolet and high bleaching propensity. DHE is synthesized by some yeast species and the red sea sponge (Biemna fortis) [78]. It bears minimal chemical alterations compared to ergosterol (one additional double bond) and to cholesterol (two additional double bonds in the steroid ring

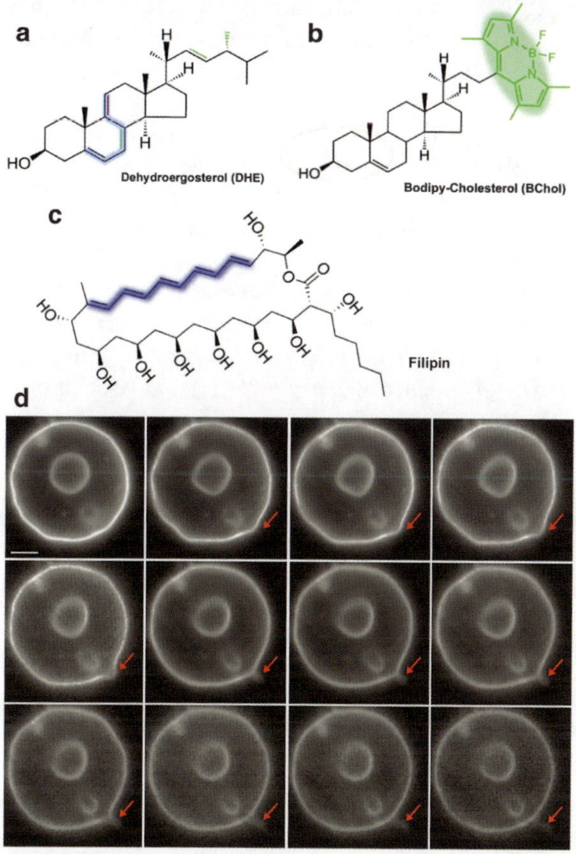

Fig. 1 Chemical structure of dehydroergosterol (DHE), BODIPY-cholesterol and filipin. (**a**) Structure of DHE. Differences to cholesterol are indicated in green and red, that is, an extra methyl group and double bond in the alkyl side chain and two additional double bonds in the steroid ring system. Only one additional double bond in the second ring distinguishes DHE from ergosterol (indicated in *red*). The three conjugated double bonds in the steroid ring (indicated in *light blue*) give DHE its slight fluorescence. The structure of CTL is identical to that of DHE in the steroid ring, while the isooctyl side chain of CTL is the same as in cholesterol. (**b**) Structure of BODIPY-cholesterol with the fluorophore (*green*, with *light green* underlay to highlight the fluorescent group) at carbon 24 of the sterol side chain. (**c**) Structure of filipin III with the conjugated double bonds of the polyene indicated in *dark blue*. (**d**) Giant unilamellar vesicles made of POPC and 40 mol% DHE were repeatedly imaged on an UV-WF microscope. Some vesicle undulations as well as formation and budding of a small daughter vesicle (*red arrows*) were observed in a time course of 2 s. Bar, 10 μm

system and one extra methyl group and double bond in the side chain, see Fig. 1). Many studies have shown that DHE's biophysical properties closely match those of ergosterol and cholesterol [76, 78, 132, 136]. This, however, comes along with very poor suitability as a fluorophore due to low extinction coefficient (11,000 M^{-1} cm^{-1}), low quantum yield ($\Phi = 0.004$ in ethanol) and high bleaching propensity [76, 78, 132, 136]. Visualization of DHE is possible in cells and model organisms using UV-sensitive wide field microscopy (UV-WF) or, alternatively, multiphoton (MP) excitation microscopy [73, 76, 78, 136, 141, 144]. UV-WF imaging requires an objective and collector lenses with high UV throughput as well as an array detector with good quantum yield between 360 and 400 nm. We use either a back-thinned piezo-cooled charge coupled device (CCD) camera from Hamamatsu Inc. Japan (model Orca 2.0) or, more recently, an electron multiplication CCD from Andor Technology, USA (EMCCD, model iXon Blue). The latter camera has not only high sensitivity in the UV with around 70 % quantum yield at 400 nm but also the great advantage that the charge generated by the released photoelectrons can be several hundred fold enhanced. This results in much higher signal-to-noise ratio (SNR) even for low numbers of incoming photons. Accordingly, one can strongly reduce the exposure time or excitation light intensity, thereby preventing high UV irradiation of cells and photobleaching of DHE or CTL [76]. An example is shown in Fig. 1d, where GUVs made of 1-palmitoyl PC, 2-oleoyl PC (POPC), and 40 mol% DHE were repeatedly imaged on our UV-WF microscope equipped with the iXon Blue EMCCD. Despite bleaching, which was corrected for here by intensity normalization, one can clearly discern a small vesicle budding from the large GUV as a consequence of temperature-induced membrane undulations (red arrows in Fig. 1d). In that example, the acquisition time was set to 100 ms with EM factor of 100 to visualize the budding event.

MP excitation microscopy is a nonlinear imaging technique, where almost simultaneous absorption of two or more photons causes an electronic transition in the fluorophore from the ground to the first excited state. The absorbed photons have approximately one-half (for two-photon excitation) or one-third of the energy (for three-photon excitation) compared to the one-photon excitation process, respectively. Thus, as a rule of thumb, excitation of a fluorescent probe by two- or three-photon microscopy can be efficiently stimulated using twice or three times the wavelength used for one-photon excitation. For example, DHE is excited around 320 nm by an one-photon process, while we and others were successful in exciting it with 920 nm by MP microscopy [28, 78, 79, 137, 141, 150]. The longer excitation wavelength has several advantages for UV microscopy including less cytotoxicity and photobleaching, deeper specimen penetration, and less light scattering. Apart from these advantages, the key benefit of using MP microscopy is its intrinsic sectioning capability, that is, the efficient rejection of out-of-focus light. There are several ways of getting rid of out-of-focus background in fluorescence microscopy; in WF microscopy, one can use deconvolution, a mathematical operation which reassigns out-of-focus fluorescence to the correct plane in a post-processing step [118]. Deconvolution combined with bleaching correction has been used several times in UV-WF imaging of DHE [133, 137, 138]. In confocal microscopy, a pinhole in front of the detector rejects out-of-focus light but also lowers the detected fluorescence. Confocal one-photon excitation laser scanning microscopy cannot be used for visualization of DHE or CTL, since the scanning laser beam will bleach the sterol probes long before enough photons could be collected at every pixel position. In addition, UV lasers emitting below 350 nm, as required for excitation of DHE or CTL, are hardly available. Further discussion of that point can be found in Wüstner and Færgeman [138]. Excitation (and thereby eventual photobleaching) occurs along the whole depth in confocal microscopy. Only by MP microscopy, one restricts excitation and thereby emission to the focal plane, and there is therefore no need for a pinhole or deconvolution, though both provisions can slightly improve the quality of MP images, given enough photon counts at the detector. In all forms of fluorescence microscopy, high-numerical-aperture objectives strongly focus the incident light resulting in photon densities at the focal plane several fold higher than slightly above and below that plane [130]. In one-photon excitation, the light intensity at these locations outside the center of the focus is still enough to excite the fluorophore, and this is causing the out-of-focus background in conventional fluorescence microscopy. In contrast, MP excitation is restricted to the focal plane due to the extremely low probability for the (almost) simultaneous absorption of several photons requiring very high photon densities [72, 146, 151]. This is experimentally realized using femtosecond pulsed lasers, for example, titanium-sapphire lasers, generating the required photon densities only in the focal plane. These light sources allow for extremely high peak intensities for triggering the MP excitation events, while keeping the mean intensity (averaged over the pulses and pauses) low and thereby preventing heating of the specimen [151]. We find a nonlinear (quadratic for two-photon excitation and cubic for three-photon excitation) dependency between incident light intensity and emitted fluorescence providing the sectioning effect in this imaging modality [107, 136]. Consequently, out-of-focus light is essentially absent in this method and photobleaching is restricted to the focal plane. The required infrared excitation is not only less harmful for cells than one-photon excitation with $\lambda_{ex} = 320$ nm in case of imaging DHE but allows also for deeper tissue penetration. This has been used to study DHE distribution in living animals – the nematode *Caenorhabditis elegans* (*C. elegans*) [141]. A major limitation for whole-animal imaging of DHE by MP microscopy, however, is the efficient excitation of autofluorescence emitting in the same spectral range as the sterol [33, 141]. UV-WF imaging together with computational detection of differences in photobleaching rates of DHE versus autofluorescence enabled us to selectively detect sterol-containing tissues in *C. elegans* [141, 144].

3 Analysis of Intracellular Sterol Dynamics by Quantitative Fluorescence Microscopy

With the increasing access of advanced imaging equipment and techniques, a number of methods for studying the dynamics of fluorescence-tagged biomolecules have emerged. Here, we describe some of these techniques and give examples of their application to studies of sterol dynamics. Progress in dynamic imaging of the weakly fluorescent DHE by MP excitation combined with image denoising will be presented.

3.1 Pulse-Chase Studies

By pulse-chase labeling, one can study how a fluophore is taken up by the cell and follow intracellular transport of the fluophore in real time. Typically, cells are plated on a microscope dish and (pulse) labeled with the fluophore for a few minutes. For a fluorescent ligand, that would result in binding to its receptor on the cell surface (e.g., fluorescent transferrin (Tf) to the Tf receptor). For fluorescent lipids, and in particular sterols, a short labeling pulse inserts the probe into the plasma membrane from which intracellular targeting can be studied over time [43, 44, 96]. Efficient labeling of cells with fluorescent sterols is possible using sterol complexes with methyl-β-cyclodextrin, as recently summarized with detailed protocols [76]. The cells may be prelabeled with one or more dyes which specifically stain a given organelle. The probe transport kinetics to the highlighted organelles can then be measured using suitable image analysis protocols [34]. Sterol transport has been often studied in pulse-chase experiments using DHE. For example, pulse-chase labeling experiments have shown that the distribution of DHE in CHO cells, nonpolarized HepG2 cells, and macrophages largely resembles that of cholesterol [43, 140, 143]. Colocalization with transferrin tagged with a green or red fluorescent dye, like the Alexa dye series, revealed that the ERC is a major sterol pool, while very little DHE was found in the trans-Golgi network [43, 140, 143]. Since the light path in the fluorescence microscope, and especially the objective, is not corrected for chromatic aberration between ultraviolet and visible green or red light, a correction for lateral and axial off-set has to be carried out. Multicolor fluorescent beads of sub-resolution size are well suited for that task, and they can serve in parallel as an internal standard for the shape of the microscope point spread function (PSF), as described previously [138]. By the pulse-chase approach, trafficking itineraries of two fluorescent sterols, DHE and BChol from the plasma membrane have been compared in a quantitative manner [145]. BHK and HeLa cells were loaded with oleic acid to induce lipid droplet (LD) formation, and the ERC was labeled with Alexa546-tagged Tf (a red probe), while the infrared droplet stain deep red LipidTox (from Invitrogen Inc.) was used to visualize LDs [145]. Four-color wide field fluorescence microscopy revealed the targeting of DHE and BChol to ERC and LDs in parallel. ERC and LD boundaries were determined from binary masks generated for each image set after intensity thresholding of the organelle-marker images. A whole-cell mask allowed for determining total cellular fluorescence of the sterols and for normalizing the measured organelle sterol to total intensity [145]. We observed that BChol is preferentially targeted to LDs on a time scale of a few minutes, while DHE became much more enriched in the ERC and only to a minor extent transported to LDs in those cells [145]. The affinity of BChol for oleic-acid-induced LDs is likely a consequence of the attached BODIPY group, since BODIPY dyes are used as prominent LD markers [65]. Our image analysis protocol can be easily implemented and automated in open-source imaging packages like ImageJ (http://rsbweb.nih.gov/ij/) using the ImageJ Macro – language being available with good documentation at http://rsbweb.nih.gov/ij/developer/macro/macros.

html. While treatment of cells with oleic acid induces formation of triacylglycerol (TAG)-rich LDs, incubation of macrophages with atherogenic LDL causes formation of cholesteryl-ester (CE)-rich LDs [119, 147]. DHE was found to be rapidly targeted to those LDs from the plasma membrane without significant esterification [143]. In adipocytes, large LDs containing mostly TAG are formed [62]. We found that DHE did not became enriched in these LDs but inserted to some extent into the LD rim, that is, the phospholipid monolayer surrounding LDs [138]. Macrophage foam cell LDs and the phospholipid monolayer of adipocyte LDs were found to contain nonesterified cholesterol, as shown by filipin staining [77, 95]. Thus, results obtained using UV-WF imaging of DHE are fully in line with these earlier reports. Droplet targeting of DHE and CTL paralleled by esterification of the sterols has been found after overnight incubation of CHO cells [81, 83].

Pulse-chase studies are not limited to fluorescence imaging of sterol transport in mammalian cells. In yeast, the oxysterol-binding proteins have been suggested to function as sterol transfer proteins [7]. A pulse-chase experiment with metabolically radiolabeled ergosterol from the ER to the plasma membrane showed that the Osh proteins were not required for sterol transport through the cytoplasm but were possibly involved in transport of sterol to methyl-β-CD extractable pools [32]. These experiments were complemented with imaging of DHE inserted into the cells under auxotrophic growth conditions. Redistribution of DHE from the PM to LDs was nonvesicular, since it was not affected in end4 deletion mutants, but it required ATP as well as the sterol acyltransferase Are2 [32]. LDs were co-stained with Nile red, a traditional droplet marker [38], and DHE transport was quantified by semi-automated image analysis [32]. A fully automated image analysis strategy for quantifying DHE transport to LDs based on prelabeling with Nile red has been recently published for macrophage foam cells and can be adapted to yeast as well [138]. Menon and colleagues found that Oh4 protein accelerates sterol redistribution without affecting the steady state distribution of DHE [32]. Recent in vitro experiments using purified and tagged Oh4 protein combined with DHE as transferable sterol suggest that Oh4 exchanges ergosterol against phosphatidylinositols in membranes [24]. This example illustrates that DHE imaging can be supplemented by spectroscopic investigation of carrier-mediated DHE transfer between membranes. Other examples are the cytosolic sterol transfer protein StARD4, the Niemann-Pick type C disease proteins 1 and 2 and the sterol chelator methyl-β-CD [22, 43, 57, 67, 81, 93, 148].

3.2 *Fluorescence Recovery After Photobleaching and Fluorescence Loss in Photobleaching*

Fluorescence recovery after photobleaching (FRAP) is a commonly used method to study intracellular dynamics of fluorescent macromolecules. Initially a region of interest (ROI) is bleached with high-intensity illumination, typically provided by an attenuable laser at a laser scanning microscope. Subsequently, the return of fluorescence intensity in the ROI is quantified over time as a measure of probe transport to the ROI. Incomplete recovery may indicate an immobile fraction of fluophores, that is, bleached molecules which do not exchange with fluorescent molecules from outside the ROI in the time course of the experiment [64]. FRAP data can be analyzed either purely qualitatively, semi-quantitatively by fitting with empirical functions for comparison or, finally, by physical modeling of the complex binding/diffusion process. Only by the latter approach, parameters such as the diffusion constant and/or binding constants for the fluophore may be determined. However, since cells have very irregular geometries and spatially varying diffusion properties, the modeling of the real reaction–diffusion problem is extremely challenging and not appropriately solved yet [87]. DHE cannot be imaged on a confocal laser scanning microscope (see above), but fluorescence recovery of DHE after photobleaching has been studied on a UV-WF microscope. Here, the ROI is defined by the size and position of the fully closed field aperture of the imaging system (approx. 13 μm in diameter). It takes less than 10 s with full illumination to bleach all DHE molecules of a cell within that ROI, which is comparable to bleaching times of conventional fluorophores for the same area on a scanning confocal system [64, 126]. Detailed modeling of DHE FRAP data acquired on an UV-WF microscope will be challenging, and so far, only semi-quantitative analysis has been performed for comparison of different experimental conditions. For example, DHE in LDs of J774 macrophage foam cells was bleached, and fluorescence recovery was measured in control and ATP-depleted cells. The transport process occurred with a half-time of 1.5 min and did not require ATP [143]. In another study, FRAP was used to show that the sterol carrier protein STARD4 is involved in sterol transport from the plasma membrane to the ERC [81]. Initially, cells were labeled with DHE and Tf to locate the ERC. Then the ERC was bleached in control cells and cells with overexpressed or silenced STARD4. This showed that upon STARD4 overexpression, the half-time of recovery decreased from 3.2 to 2.6 min, while the mobile fraction given by the amplitude of fluorescence recovery increased [81]. Thus, STARD4 enhances transport of sterols from the plasma membrane and ERC, and that was also found for cholesterol towards the sterol sensing machinery in the ER [81]. Interestingly, silencing of STARD4 had little effect on fluorescence recovery of DHE, suggesting the existence of STARD4 independent transport mechanisms [81]. Using a laser scanning confocal microscope, it is possible to bleach arbitrarily shaped ROIs, for example, an organelle, due to exact positioning of the laser with acoustic-optical tunable filters (AOTFs). However, in quantitative FRAP experiments performed on such as ROI, one must be careful in determining parameters. The FRAP method was originally performed by bleaching a spot with a size of the e^{-2} radius of the laser beam [3]. Hence, many analytical expressions for FRAP analysis were developed for the spot-FRAP approach, and those models are not strictly valid for FRAP experiments performed in scanning modus on a confocal microscope [15]. Analytical expressions exist for simple geometries such as a line or a circle, while recovery in more complicated geometries has been analyzed by computational simulations of diffusion in the geometry [3, 6, 104].

A method related to FRAP is fluorescence loss in photobleaching (FLIP). In FLIP, an ROI is bleached repeatedly, while the fluorescence intensity in the entire cell or organelle of interest is monitored with reduced laser intensity between the high-intensity laser bleaches. Thus, FLIP reveals which compartments in the cell continuously exchange the fluorescent molecule with the bleached region. For determining sterol dynamics, we use FRAP and FLIP of BChol on a confocal microscope (Lund et al., manuscript in preparation), but one can also use the FLIP method on an UV-WF system to determine the dynamic connectivity of DHE-containing compartments. For example, the exchange dynamics of DHE between the two plasma membrane domains of polarized HepG2 cells has been studied by FLIP combined with FRAP and time-lapse imaging [140]. DHE in the basolateral membrane was found to be in continuous exchange with that in the canalicular membrane of the same cell, while no DHE crossed the canalicular lumen from the second cell forming a couplet. Thus, UV-WF adapted FLIP revealed that no sterol is secreted into the canalicular lumen, since otherwise, the DHE would have been replenished in the bleached cell [140].

3.3 *Fluorescence Fluctuation Techniques*

According to the fluctuation-dissipation theorem, one can either deflect a system from the equilibrium (or steady state) and follow its relaxation to a new steady state over time, as in FRAP experiments, or one can observe a large number of particle fluctuations at steady state; both procedures will provide transport rate constants. Fluorescence correlation spectroscopy (FCS) was developed in the 1970s to follow the second approach, as nicely reviewed in [26]. In FCS, a femtoliter volume is illuminated by a stationary laser, and fluctuations in the photon counts are recorded as a function of time. The fluctuations arise from changes in fluorophore concentration but also from photophysical effects, such as blinking or triplet formation of the fluorophore within the focal volume. From the autocorrelation function of the fluctuation time series, information about triplet dynamics, bleaching, and fluorophore mobility can be inferred [26]. Similar as in FRAP analysis, a suitable physical model is required to extract reliable diffusion constants and binding parameters from the autocorrelation data. In contrast to FRAP experiments, classical FCS techniques rely on bright and photostable fluorophores, where each dye-tagged molecule crossing the focal volume causes significant fluorescence fluctuations for an overall low mean fluorescence. Thus, FCS techniques can be used to determine sterol diffusion using BChol, but not with intrinsically fluorescent sterols like DHE [71]. Image correlation spectroscopy (ICS) is an imaging analog to FCS, first described by Petersen et al. in 1993 [92]. From the ICS autocorrelation function measured for spatial lag variables, the number of fluophores per area and their degree of aggregation can be determined. In the classical ICS implementation, the autocorrelation function is calculated for spatial lag variables from a single image, and to obtain better statistics, one often performs the calculations on a stack of images of the same cell [92, 128]. Unlike FCS, ICS does not give any information about the fluophore dynamics. To study that, temporal image correlation spectroscopy (TICS) was introduced [129]. In TICS, which is a spatially extended version of FCS, the time autocorrelation function is determined for every pixel position, such that spatially heterogeneous probe dynamics can be resolved. The mobility of molecules detected by TICS must be lower than the time the scanning laser needs to return to the initial position between successive frames. Otherwise, any temporal correlation will be lost, which makes that TICS is typically used to analyze the dynamics of membrane-bound proteins or slowly moving cytoplasmic molecules [58]. If the fluophore is staying within the same location for a number of frames, the decay of the autocorrelation function depends on the type of dynamics. Therefore, TICS provides information about the type of dynamics, that is, whether a molecule is moving by pure diffusion, by flow, or a combination of both. However, TICS cannot reveal the direction of flow, since no correlation between individual pixel positions is entering the correlation function. Spatiotemporal ICS (STICS) calculates the full autocorrelation function, that is, spatial and temporal, and thereby provides information about diffusion, flow speed, and flow direction [46]. In STICS flow will be seen as movement of the center of the Gaussian correlation function, and from this, the direction of flow may be inferred. It is not in the scope of this article to describe the ICS techniques in more detail. Therefore, we guide the interested reader to the excellent review by Kolin and Wiseman [58].

Due to their weak fluorescence, individual DHE or CTL molecules cannot be detected by any of the fluorescence correlation techniques. However, assemblies of sterol, like moving DHE-containing vesicles, can be measured, since their movement through the laser focus will cause detectable fluorescence fluctuations. For example, the mobility of DHE-containing vesicles in HepG2 cells was analyzed by STICS from multiphoton microscopy image stacks. This showed that the vesicles moved by pure diffusion with a diffusion constant of $0.501 \pm 0.045 \times 10^{-5}\ \mu m^2/s$. Furthermore, since the autocorrelation function remained Gaussian over time, it was concluded that the vesicles were not subject to active transport [137]. Due to its much better fluorescence, BChol can be studied by fluorescence fluctuation techniques. We have recently applied two-photon TICS to measure cytoplasmic and membrane diffusion of BChol providing evidence for locally varying sterol dynamics in living CHO cells [71].

3.4 *Single Particle Tracking*

While FRAP, FLIP, and the ICS techniques allow for measuring diffusion constants of fluorescent biomolecules, these methods do not provide direct access to the particle trajectories. Single particle tracking (SPT) is a versatile method for monitoring the movement of individual particles like vesicles, protein aggregates, or supramolecular assemblies or even single molecules in the cell [101, 103, 105]. Since SPT determines trajectories for each particle separately, the method can reveal subpopulations with different dynamics which is not possible with FRAP or ICS, since these methods show the average dynamics of many molecules. SPT has been used to study the mobility of viruses or vesicles containing lipoproteins and of lipid granules [27, 51, 55, 105, 120]. By particle tracking, the trajectory (i.e., x- and y-coordinates as a function of time) of the particle is acquired. Subsequently, the mean square displacement (MSD) of the particle trajectory is calculated. For pure diffusion, the tracked particles or molecules will perform a random walk, whose probability density function (PDF) is a Gaussian function describing the distribution of step lengths for a given time increment [16, 103, 111]. The second moment of the PDF is the MSD, which is most often used to characterize diffusion processes and to determine a diffusion constant for the particles. From the shape of the MSD curve, various types of diffusion can be distinguished, all being observed in living cells. For normal Brownian diffusion, the MSD is linearly proportional with time [97, 102]. Anomalous subdiffusion caused by molecular crowding and/or binding to mobile or immobile intracellular objects is signified by downward curvature of the MSD plot, while anomalous superdiffusion caused by active transport or cellular flow gives an upward bent MSD curve. Using SPT of image sequences acquired on a UV-sensitive wide field microscope, the formation and subsequent dynamics of DHE-containing vesicles were observed in J774 cells [139]. This showed that DHE-containing vesicles were formed 2.5 min after labeling and that these generally stayed in the vicinity of the plasma membrane for an extended period of time. SPT of those vesicles over a period of 20 min revealed that their motion followed a model of confined diffusion with a diffusion constant of ~10^{-4} $\mu m^2/s$ in an area of 2.3 ± 0.13 μm^2. According to the Stokes-Einstein relation, a vesicle of radius 150 nm diffusing in a fluid ten times more viscous than water would have a diffusion constant of 0.14 $\mu m^2/s$. Thus, the diffusion constant of DHE-containing vesicles in the cell periphery is about 1,000 times slower than predicted for pure diffusion [139]. The rather large difference in diffusion constants found for DHE vesicles by SPT on a UV-WF set-up versus by TICS with the MP microscope (see above) is likely a consequence of two factors: (1) imaging modalities and analysis strategies differed for both experiments and (2) different cell types might have different transport properties and membrane dynamics. We found in recent simulations that TICS and SPT give similar values for particles performing pure diffusion, that is, a classical random walk (Lund and Wüstner, unpublished observations). The poor fluorescence properties of DHE somehow limit the analysis options for diffusion measurement of vesicles containing that sterol in living cells. Some improvement is possible by combining MP time-lapse imaging of DHE with advanced denoising routines before tracking of vesicles, as outlined in the next paragraph.

3.5 Multiphoton Excitation Fluorescence Microscopy Combined with PURE-LET Image Denoising

Due to its low absorption coefficient, quantum yield and MP cross section, single-frame MP images of DHE-stained cells are of poor quality (i.e., low-peak

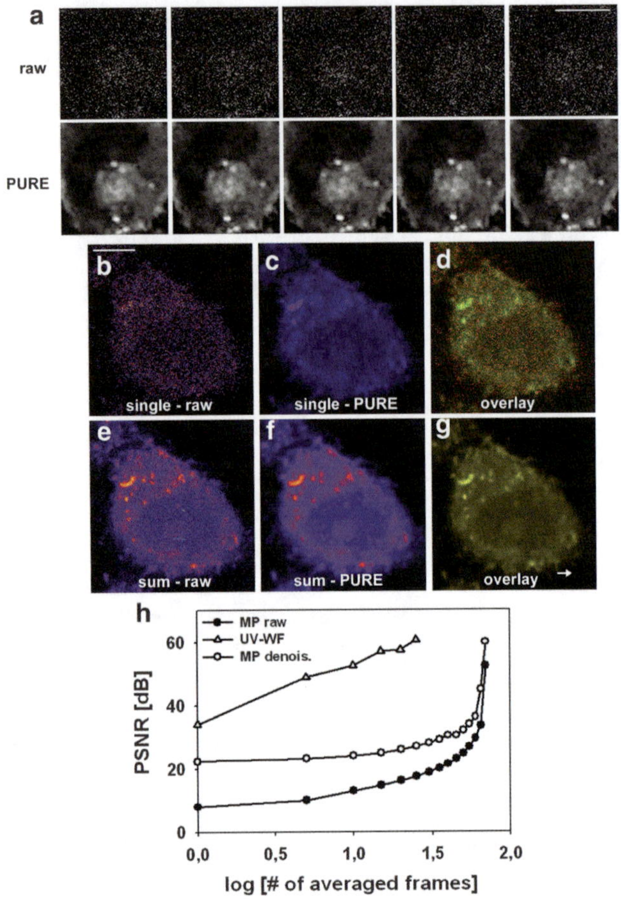

Fig. 2 PURE-LET denoising of multiphoton image series of DHE-stained cells. Chinese hamster ovarian (CHO) cells were labeled with DHE from a cyclodextrin complex for 1 min at 37 °C, washed with buffer medium and chased for 30 min at 37 °C. Cells were imaged on a home-built MP microscope with an excitation wavelength of 930 nm and a pixel dwell time of 50 μs. (**a**) *upper panel*, raw data of an image series; *lower panel*, same sequence denoised using the PURE-LET algorithm, as described in the text, below. (**b–g**), another cell imaged with the cell surface in focus. (**b**, **c**) Single frame unprocessed (**b**, "single – raw"), or denoised (**c**, "single – PURE"); (**d**) color overlay with unprocessed in red and denoised in green. (**e**, **f**) Sum projection of 30 unprocessed frames (**e**, "sum – raw") or of 30 denoised frames (**f**, "sum – PURE"); (**g**) color overlay with unprocessed sum projection in red and denoised sum projection in green. (**h**) Plot of PSNR as a function of the logarithm of successive frame averages compared to the total average of all frames from MP raw data (*closed circles*), a UV-WF sequence after bleaching correction (*open triangles*), both taken from ref. Wüstner et al. [137], and the same MP sequence after applying PURE-LET denoising (*open circles*). Bar, 5 μm. See text for further explanations

signal-to-noise ratio (PSNR)). This is exemplified in Fig. 2a, upper row for a Chinese hamster ovarian (CHO) cell labeled with DHE for 1 min and chased for 30 min at 37 °C. One can envision the cell outline and a perinuclear sterol enrichment, but this is dominated by a high noise level. The important aspect in these low-count images is that the noise is highly related to the signal. Since we used a photomultiplier tube (PMT) in photon counting mode as detector, the major source of noise is clearly shot noise, that is, the random number of photons hitting the detector in a given period of time. Due to this photon counting process, the statistical distribution of each pixel is well described by a Poisson law, whose underlying intensity is the actual *uncorrupted* measurement giving strongly *signal-dependent noise*. Consequently, the common independent additive white Gaussian noise model, which may be adequate for CCD detectors operating under standard illumination conditions, can clearly not be considered for the MP imaging of DHE. Denoising is an important image-processing step which can greatly improve the quality of the data for the subsequent analysis (feature segmentation, object tracking, etc.) of images. The purpose of denoising is to remove irrelevant information (noise), while preserving the structures of interest. To efficiently denoise those Poisson data, we relied on a recent wavelet-based denoising algorithm [69, 70] which consists of two main features: (1) a data-adaptive unbiased estimate of the mean-squared error (MSE) between the processed and the *unknown* noise-free data – this statistical quantity is called "Poisson's unbiased risk estimate" (PURE) – and (2) A linear parameterization of the denoising process expressed as a wavelet-domain "linear expansion of thresholds" (LET). While PURE allows to practically monitor the denoising quality, the LET ensures a low computational complexity. Applying 10 cycles of the multiframe (11 frames) wavelet-based PURE-LET denoising to the recorded MP time-lapse sequence resulted in a dramatic improvement of image quality. This can be inferred from the reduced level of noise, while sterol-containing vesicles are well preserved. The plasma membrane, the sterol-rich perinuclear recycling endosome and multiple sterol-containing vesicles can be clearly discerned in individual frames (Fig. 2a, lower panel). This improvement is mainly due to the adequacy of the considered Poisson noise model (confirmed by the automatic noise parameter estimation), which guarantees the mean square error (MSE) optimality of the PURE-LET denoising. As a result, the denoised image is ensured to have a significantly higher PSNR than the raw input and thus to be closer to the expected underlying noise-free image. In our recent work, we showed that calculation of an average or sum projection of multiple low-quality frames can result in a single high-quality MP image of DHE-stained specimen [137]. To directly assess the improvement in image fidelity brought by PURE-LET denoising, we calculated the PSNR (see Eq. 1 in [137]) for another DHE-stained CHO cell image series. The test image was either a single raw image (Fig. 2b) or the corresponding denoised image (Fig. 2c), and the reference image was the sum projection of 30 raw frames (Fig. 2e) or the sum projection of the corresponding 30 denoised frames (Fig. 2f), respectively. We found that the PSNR between a single raw image (Fig. 2b) and the sum projection of the raw image stack (Fig. 2e) was 10.99 dB, while that between the corresponding denoised single image (Fig. 2c) and the sum projection of the denoised image stack (Fig. 2f) was 24.59 dB. This underlines the visual impression that PURE-LET denoising dramatically improves the quality of each individual frame of the MP time-lapse sequence of DHE-stained cells. Notably, the sum projection of the denoised image stack has also a significantly higher PSNR than the sum projection of the raw image stack. It has to be added that calculation of PSNR is a relative measure of restoration quality, since it is only valid within one image sequence [37]. Next, we compared the effect of successive frame averaging on image quality in UV-WF imaging, as described previously [137], with that in MP microscopy of DHE, the latter without or with a denoising post-processing step (Fig. 2h). As reference image, we chose the average over all frames, for both the acquired bleach-corrected UV-WF stack ($n = 28$) and the MP sequences ($n = 71$ using the same sequence). Intentionally, we chose the same image sequences as used previously allowing for direct comparison of image improvement by denoising [137]. While a single acquisition by MP microscopy without denoising was of very low quality (PSNR ~8), PURE-LET denoising of the same image sequence gave a PSNR of 22.4 already in the first acquisition (compare closed and open circles in Fig. 2h). For comparison, in case of UV-WF microscopy of DHE combined with pixel-wise bleach rate fitting and correction, already the first acquisition gives a PSNR = 34.165, while averaging of 25 bleach-corrected frames resulted in image quality exceeding that of the acquired raw MP sequences (Fig. 2h and [137]). The PSNR of total averaged bleach-corrected UV-WF and denoised MP image sequence was comparable (PSNR = 60.57 for UV-WF vs. PSNR = 59.91 for denoised MP sequence; see last data point, triangles, and open circles in Fig. 2h, respectively). Together, this analysis underlines the significant image improvement of DHE MP imaging brought by the PURE-LET denoising approach.

3.6 Tracking of Sterol Vesicles and Filopodia After Denoising of Multiphoton DHE Image Series

Tracking of intracellular vesicles requires unequivocal identification of the fluorescent particles against a relative homogenous background. Without any post-processing, single-frame acquisitions, as shown in Fig. 2a (upper row), are of too low SNR to allow for tracking vesicle positions. Thus, denoising is an essential step to identify and track sterol vesicles in successive frames. Comparing the performance of Gaussian and median-filtering with the PURE-LET denoising algorithm, we found that only the latter approach greatly improves the SNR in individual frames of an image stack, while preserving the vesicle pattern in the cells. This was not found for Gauss-filtering the data or with median-filtering the image sequence, where increasing the variance of the filter caused disruption of the specific image features (Fig. 3). Thus, PURE-LET image denoising is unique in its ability to highlight vesicular sterol patterns in the presence of high noise levels. This allowed us to track DHE-containing vesicles in CHO cells with high accuracy and to determine their dynamic properties with a time resolution of four seconds (which is given by the frame rate of the recorded time-lapse sequence). Tracking attempts of Gauss- or median-filtered image sequences looking like the examples in Fig. 3

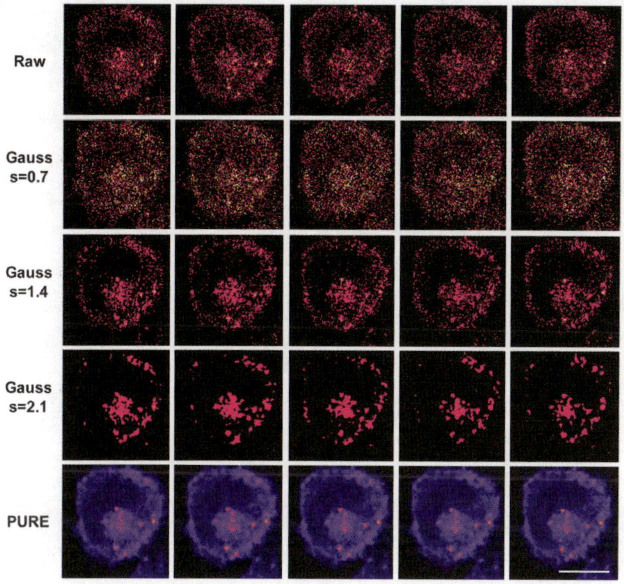

Fig. 3 Comparison of PURE-LET denoising with Gaussian filtering. The first five frames of a recorded MP time-lapse sequence of DHE-stained CHO cells are shown as raw data (*upper row*, "Raw"), after applying Gaussian filtering using tools in ImageJ with increasing variance from $\sigma = 0.7$ ("Gauss, $s = 0.7$"), $\sigma = 1.4$ ("Gauss, $s = 1.7$") to $\sigma = 2.1$ ("Gauss, $s = 2.1$") or after PURE-LET multiframe denoising (*lowest row*, "PURE"). Images are shown with identical intensity scaling in a FIRE LUT. Bar, 15 μm

were unsuccessful due to high incidence of false-linking particles in consecutive frames (not shown). A kymograph presented in Fig. 4a shows part of a cell where the red and green traces visualize the path of two different vesicles in the x- (lower frame) and y-direction (right frame) of a PURE-LET denoised image stack as a function of time. The mobility of the vesicles is low, and visual inspection of the cells indicates a highly hindered diffusion. In total, 18 vesicles were tracked over 30 to 60 frames using a dynamic programming routine, and the MSD for each trajectory was calculated [139]. Fitting a straight line to the first few points of the MSD is a standard way of obtaining diffusion constants. Doing this for the DHE-containing vesicles revealed a broad distribution of diffusion constants ranging from 3.0×10^{-5} to 5.8×10^{-4} μm^2/s with an average of $1.6 \times 10^{-4} \pm 1.4 \times 10^{-4}$ μm^2/s. Alternatively, an average MSD can be calculated for all trajectories, as shown in Fig. 4b (symbols = mean ± one standard error; line = fit of $\mathrm{MSD}(t) = 4Dt$ to initial displacement). The average MSD deviates downwards from a straight line, which is indicative of anomalous subdiffusion, in complete accordance with our previous experiments using UV-WF vesicle tracking (see above and Wüstner and Færgeman [139]). It has to be emphasized that our approach allows only for tracking sterol vesicles containing significant amounts of DHE with comparable mean intensity as in the plasma membrane. Accordingly, we

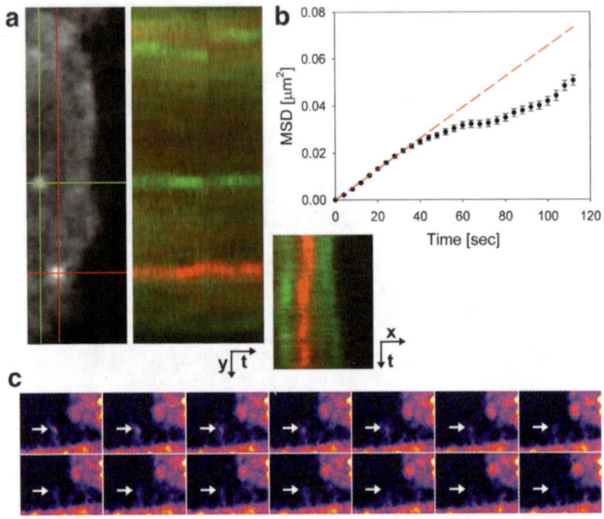

Fig. 4 Tracking of DHE-containing vesicles and filopodia in denoised MP sequences. Chinese hamster ovarian (CHO) cells were labeled with DHE from a cyclodextrin complex for 1 min at 37 °C, washed with buffer medium and chased for 30 min at 37 °C. Cells were imaged on a home-built MP microscope with an excitation wavelength of 930 nm and a pixel dwell time of 50 μs. (**a**) *upper left panel*, first denoised frame of the sequence (x, y-view in gray scale) with two vesicles tracked over 50 frames (*green* and *red cross*); *lower left panel*, kymograph (x, t-plot) and *right panel*, kymograph (y, t-plot) as color overlay with upper vesicle in green and lower vesicle in red (corresponding to the *green* and *red cross*, respectively). (**b**) Average mean square displacement of all 18 tracked vesicle trajectories (symbols = mean, error bars = SEM). A fit of the initial linear phase of the MSD until 30 s to a straight line in red is also shown. (**c**) Selected region of another cell labeled, imaged and denoised identically shows dynamic retraction of a surface protrusion over a distance of approximately 500 nm in about 25 s (*arrows*; frame rate is 4 s). See text for further information

cannot rule out the existence of much more mobile sterol vesicles in living cells which contain too little DHE to be detected by our approach. This is actually indicated by long-term tracking of another cholesterol probe, BChol, in our most recent study [71].

By PURE-LET denoising of MP images of DHE-stained cells, we can visualize small sterol-containing cell protrusions covering the whole-cell surface. Moreover, we are able to assess the dynamics of those protrusions, since individual denoised frames are of high quality. This is demonstrated for a small region of another cell, where a tiny membrane protrusion retracts over a time course of about 25 s (Fig. 4c). This conclusion could not have been inferred from the corresponding raw data set, since the surface structure can hardly be discerned in those images (not shown, but see Fig. 2). From published MP images of the weakly fluorescent DHE with comparable low quality to our single-frame images, the existence of sterol-rich domains in the plasma membrane has been postulated [150]. By applying advanced denoising routines to DHE time-lapse images, we can clearly infer that this apparently heterogeneous staining pattern is caused by small DHE-containing surface protrusions, which constantly change their shape and appearance. Sub-resolution membrane folds result in more DHE intensity per pixel, which can be misinterpreted as lateral sterol domains in a flat 2D plane. This insight was made possible with our new integrated MP denoising approach and supports our earlier spatiotemporal studies of DHE in murine J774 macrophages using UV-WF imaging combined with statistical correlation and wavelet analysis of fluorescence fluctuations [133, 139]. In fact, even in cholesterol-loaded J774 cells, where cell ruffling is strongly increased and much more sterol is in the cell membrane, DHE did not form separate domains. Staining patterns followed exactly the rough surface topography, while DHE was homogeneously distributed in cell blebs at room temperature [134]. Thus, two different imaging modalities, multicolor UV-WF imaging combined with deconvolution and MP denoising microscopy of DHE, converge to the same view that the sterol distribution in the plasma membrane is homogeneous, at least at the level of the light microscope, and apparent heterogeneity is a consequence of cell surface ruffling.

4 Conclusions and Outlook

Analysis of intracellular cholesterol transport by quantitative fluorescence microscopy of sterol analogs has made large progress over the last decade. This is the outcome of progress in various fields: fluorescence optics and detector technology, computational image analysis, biophysical characterization of the properties of sterol probes compared to cholesterol, as well as organic synthesis of new fluorescent cholesterol analogs. In this review, we have presented a particular improvement for the first time, that is, how advanced denoising of MP image sequences contributes to better imaging of DHE in living cells. It is not very likely that better analogs of cholesterol than CTL and DHE will be developed, since any further modification likely compromises the similarity to cholesterol. Further improvement in fluorescence imaging of sterols must rather come from improved detectors and even more optimization of optics and excitation schemes. To overcome the diffraction barrier and obtain insight into nanoscale sterol organization, resonance energy transfer between fluorescent sterols and suitable acceptors might be explored in the future [18, 57, 131]. Superresolution techniques like stimulated emission depletion (STED) fluorescence nanoscopy become currently adapted to conventional fluorophores, such that these methods might be useful for sterol imaging as well [85, 123].

Acknowledgements DW acknowledges funding by grants of the Lundbeck Foundation, the Novo Nordisk Foundation, the Danish Research Agency Forskningsstyrelsen, Forskningsrådet for Natur og Univers (FNU) and the Danish Research Agency Forskningsstyrelsen, Forskningsrådet for Sundhed og sygdom (FSS). We are grateful to Florian Luisier (Statistics and Information Sciences Laboratory, Harvard University, Cambridge, USA) and Daniel Sage

(Biomedical Imaging Group, Ecole Polytechnique Fédérale de Lausanne (EPFL), Lausanne, Switzerland) for their helpful comments on image denoising using the PURE-LET approach and on an early version of the manuscript.

References

1. Ariola FS, Li Z, Cornejo C, Bittman R, Heikal AA (2009) Membrane fluidity and lipid order in ternary giant unilamellar vesicles using a new bodipy-cholesterol derivative. Biophys J 96:2696–2708
2. Arthur JR, Heinecke KA, Seyfried TN (2011) Filipin recognizes both GM1 and cholesterol in GM1 gangliosidosis mouse brain. J Lipid Res 52:1345–1351
3. Axelrod D, Koppel DE, Schlessinger J, Elson E, Webb WW (1976) Mobility measurement by analysis of fluorescence photobleaching recovery kinetics. Biophys J 16:1055–1069
4. Baumgart T, Hunt G, Farkas ER, Webb WW, Feigenson GW (2007) Fluorescence probe partitioning between Lo/Ld phases in lipid membranes. Biochim Biophys Acta 1768:2182–2194
5. Beattie ME, Veatch SL, Stottrup BL, Keller SL (2005) Sterol structure determines miscibility versus melting transitions in lipid vesicles. Biophys J 89:1760–1768
6. Beaudouin J, Mora-Bermúdez F, Klee T, Daigle N, Ellenberg J (2006) Dissecting the contribution of diffusion and interactions to the mobility of nuclear proteins. Biophys J 90:1878–1894
7. Beh CT, McMaster CR, Kozminski KG, Menon AK (2012) A detour for yeast oxysterol-binding proteins. J Biol Chem 287:11481–11488
8. Behnke O, Tranum-Jensen J, van Deurs B (1984) Filipin as a cholesterol probe. I. Morphology of filipin-cholesterol interaction in lipid model systems. Eur J Cell Biol 35:189–199
9. Behnke O, Tranum-Jensen J, van Deurs B (1984) Filipin as a cholesterol probe. II. Filipin-cholesterol interaction in red blood cell membranes. Eur J Cell Biol 35:200–215
10. Bergy ME, Eble TE (1968) The filipin complex. Biochemistry 7:653–659
11. Bittman R (1978) Sterol-polyene antibiotic complexation: probe of membrane structure. Lipids 13:686–691
12. Bittman R, Chen WC, Blau L (1974) Stopped-flow kinetic and equilibrium studies of filipin III binding to sterols. Biochemistry 13:1374–1379
13. Blau L, Bittman R (1978) Cholesterol distribution between the two halves of the lipid bilayer of human erythrocyte ghost membranes. J Biol Chem 253:8366–8388
14. Borroni V, Barrantes FJ (2011) Cholesterol modulates the rate and mechanism of acetylcholine receptor internalization. J Biol Chem 286:17122–17132
15. Braga J, Desterro JM, Carmo-Fonseca M (2004) Intracellular macromolecular mobility measured by fluorescence recovery after photobleaching with confocal laser scanning microscopes. Mol Biol Cell 15:4749–4760
16. Brandt S (1999) Datenanalyse. Spektrum Verlag, Heidelberg, Berlin
17. Bridgman PC, Nakajima Y (1983) Distribution of filipin-sterol complexes on cultured muscle cells: cell-substratum contact areas associated with acetylcholine clusters. J Cell Biol 96:363–372
18. Brown AC, Towles KB, Wrenn SP (2007) Measuring raft size as a function of membrane composition in PC-based systems: Part 1–binary systems. Langmuir 23:11180–11187
19. Burger K, Gimpl G, Fahrenholz F (2000) Regulation of receptor function by cholesterol. Cell Mol Life Sci 57:1577–1592
20. Castanho MA, Coutinho A, Prieto MJ (1992) Absorption and fluorescence spectra of polyene antibiotics in the presence of cholesterol. J Biol Chem 267:204–209
21. Cedar O, Ryhage R (1964) The structure of filipin. Acta Chem Scand 18:558–560

22. Cheruku SR, Xu Z, Dutia R, Lobel P, Storch J (2006) Mechanism of cholesterol transfer from the Niemann-Pick type C2 protein to model membranes supports a role in lysosomal cholesterol transport. J Biol Chem 281:31594–31604
23. Coxey RA, Pentchev PG, Campbell G, Blanchette-Mackie EJ (1993) Differential accumulation of cholesterol in Golgi compartments of normal and Niemann-Pick type C fibroblasts incubated with LDL: a cytochemical freeze-fracture study. J Lipid Res 34:1165–1176
24. de Saint-Jean M, Delfosse V, Douguet D, Chicanne G, Payrastre B, Bourguet W, Antonny B, Drin G (2011) Osh4p exchanges sterols for phosphatidylinositol 4-phosphate between lipid bilayers. J Cell Biol 195:965–978
25. Du H, Duanmu M, Witte D, Grabowski GA (1998) Targeted disruption of the mouse lysosomal acid lipase gene: long-term survival with massive cholesteryl ester and triglyceride storage. Hum Mol Genet 7:1347–1354
26. Elson EL (2011) Fluorescence correlation spectroscopy: past, present, future. Biophys J 101:2855–2870
27. Ewers H, Smith AE, Sbalzarini IF, Lilie H, Koumoutsakos P, Helenius A (2005) Single-particle tracking of murine polyoma virus-like particles on live cells and artificial membranes. Proc Natl Acad Sci USA 102:15110–15115
28. Frolov A, Petrescu A, Atshaves BP, So PT, Gratton E, Serrero G, Schroeder F (2000) High density lipoprotein-mediated cholesterol uptake and targeting to lipid droplets in intact L-cell fibroblasts. J Biol Chem 275:12769–12780
29. Garvik O, Benediktsen P, Ipsen JH, Simonsen AC, Wüstner D (2008) The fluorescent cholesterol analog dehydroergosterol induces liquid-ordered domains in model membranes. Chem Phys Lipids 159:114–118
30. Garvik O, Benediktson P, Simonsen AC, Ipsen JH, Wüstner D (2009) The fluorescent cholesterol analog dehydroergosterol induces liquid-ordered domains in model membranes. Chem Physics Lipids 159:114–118
31. Ge L, Qi W, Miao HH, Cao J, Qu YX, Li BL, Song BL (2008) The cholesterol absorption inhibitor ezetimibe acts by blocking the sterol-induced internalization of NPC1L1. Cell Metab 7:508–519
32. Georgiev AG, Sullivan DP, Kersting MC, Dittman JS, Beh CT, Menon AK (2011) Osh proteins regulate membrane sterol organization but are not required for sterol movement between the ER and PM. Traffic 12:1341–1355
33. Gerstbrein B, Stamatas G, Kollias N, Driscoll M (2005) In vivo spectrofluorimetry reveals endogenous biomarkers that report healthspan and dietary restriction in *Caenorhabditis elegans*. Aging Cell 4:127–137
34. Ghosh RN, Mallet WG, Soe TT, McGraw TE, Maxfield FR (1998) An endocytosed TGN38 chimeric protein is delivered to the TGN after trafficking through the endocytic recycling compartment in CHO cells. J Cell Biol 142:923–936
35. Gimpl G (2010) Cholesterol-protein interaction: methods and cholesterol reporter molecules. Subcell Biochem 51:1–45
36. Gimpl G, Gehrig-Burger K (2007) Cholesterol reporter molecules. Biosci Rep 27:335–358
37. Gonzalez RC, Woods RE (2002) Digital Image Processing. Chapter 11. Prentice Hall, New Jersey, pp 675–683
38. Greenspan P, Mayer EP, Fowler SD (1985) Nile Red: a selective fluorescent stain for intracellular lipid droplets. J Cell Biol 100:965–973
39. Grosheva I, Haka AS, Qin C, Pierini LM, Maxfield FR (2009) Aggregated LDL in contact with macrophages induces local increases in free cholesterol levels that regulate local actin polymerization. Arterioscler Thromb Vasc Biol 29:1615–1621
40. Haka AS, Grosheva I, Chiang E, Buxbaum AR, Baird BA, Pierini LM, Maxfield FR (2009) Macrophages create an acidic extracellular hydrolytic compartment to digest aggregated lipoproteins. Mol Biol Cell 20:4932–4940
41. Hao M, Bogan JS (2009) Cholesterol regulates glucose-stimulated insulin secretion through phosphatidylinositol 4,5-bisphosphate. J Biol Chem 284:29489–29498

42. Hao M, Head WS, Gunawardana SC, Hasty AH, Piston DW (2007) Direct effect of cholesterol on insulin secretion: a novel mechanism for pancreatic beta-cell dysfunction. Diabetes 56:2328–2338
43. Hao M, Lin SX, Karylowski OJ, Wüstner D, McGraw TE, Maxfield FR (2002) Vesicular and non-vesicular sterol transport in living cells. The endocytic recycling compartment is a major sterol storage organelle. J Biol Chem 277:609–617
44. Hao M, Mukherjee S, Maxfield FR (2001) Cholesterol depletion induces large scale domain segregation in living cell membranes. Proc Natl Acad Sci USA 98:13072–13077
45. Hartwig Petersen N, Færgeman NJ, Yu L, Wüstner D (2008) Kinetic imaging of NPC1L1 and sterol trafficking between plasma membrane and recycling endosomes in hepatoma cells. J Lipid Res 49:2023–2037
46. Hebert B, Costantino S, Wiseman PW (2005) Spatiotemporal image correlation spectroscopy (STICS) theory, verification, and application to protein velocity mapping in living CHO cells. Biophys J 88:3601–3614
47. Henriksen J, Rowat AC, Brief E, Hsueh YW, Thewalt JL, Zuckermann MJ, Ipsen JH (2006) Universal behavior of membranes with sterols. Biophys J 90:1639–1649
48. Hoekstra M, Van Eck M, Korporaal SJ (2012) Genetic studies in mice and humans reveal new physiological roles for the high-density lipoprotein receptor scavenger receptor class B type I. Curr Opin Lipidol 23:127–132
49. Hofsass C, Lindahl E, Edholm O (2003) Molecular dynamics simulations of phospholipid bilayers with cholesterol. Biophys J 84:2192–2206
50. Horvat S, McWhir J, Rozman D (2011) Defects in cholesterol synthesis genes in mouse and in humans: lessons for drug development and safer treatments. Drug Metab Rev 43:69–90
51. IV Humphries WH, Fay NC, Payne CK (2010) Intracellular degradation of low-density lipoprotein probed with two-color fluorescence microscopy. Integr Biol (Camb) 2:536–544
52. Hölttä-Vuori M, Uronen RL, Repakova J, Salonen E, Vattulainen I, Panula P, Li Z, Bittman R, Ikonen E (2008) BODIPY-cholesterol: a new tool to visualize sterol trafficking in living cells and organisms. Traffic 9:1839–1849
53. Ipsen JH, Karlstrom G, Mouritsen OG, Wennerstrom H, Zuckermann MJ (1987) Phase equilibria in the phosphatidylcholine-cholesterol system. Biochim Biophys Acta 905:162–172
54. Jensen MO, Mouritsen OG (2004) Lipids do influence protein function-the hydrophobic matching hypothesis revisited. Biochim Biophys Acta 1666:205–226
55. Jeon JH, Tejedor V, Burov S, Barkai E, Selhuber-Unkel C, Berg-Sørensen K, Oddershede L, Metzler R (2011) In vivo anomalous diffusion and weak ergodicity breaking of lipid granules. Phys Rev Lett 106:048103
56. Jia L, Betters JL, Yu L (2011) Niemann-Pick C1-like 1 (NPC1L1) protein in intestinal and hepatic cholesterol transport. Annu Rev Physiol 73:239–259
57. John K, Kubelt J, Muller P, Wustner D, Herrmann A (2002) Rapid transbilayer movement of the fluorescent sterol dehydroergosterol in lipid membranes. Biophys J 83:1525–1534
58. Kolin DL, Wiseman PW (2007) Advances in image correlation spectroscopy: measuring number densities, aggregation states, and dynamics of fluorescently labeled macromolecules in cells. Cell Biochem Biophys 49:141–164
59. Kolovou G, Anagnostopoulou K, Mikhailidis DP, Cokkinos DV (2008) Apolipoprotein E knockout models. Curr Pharn Des 14:338–351
60. Lagane B, Mazères S, Le Grimellec C, Cézanne L, Lopez A (2002) Lateral distribution of cholesterol in membranes probed by means of a pyrene-labelled cholesterol: effects of acyl chain unsaturation. Biophys Chem 95:7–22
61. Le Guyader L, Le Roux C, Mazères S, Gaspard-Iloughmane H, Gornitzka H, Millot C, Mingotaud C, Lopez A (2007) Changes of the membrane lipid organization characterized by means of a new cholesterol-pyrene probe. Biophys J 93:4462–4473
62. Le Lay S, Ferre P, Dugail I (2004) Adipocyte cholesterol balance in obesity. Biochem Soc Trans 32:103–106
63. Lindahl E, Edholm O (2000) Spatial and energetic-entropic decomposition of surface tension in lipid bilayers from molecular dynamics simulations. J Chem Phys 113:3882–3893

64. Lippincott-Schwartz J, Altan-Bonnet N, Patterson GH (2003) Photobleaching and photoactivation: following protein dynamics in living cells. Nat Cell Biol (suppl 7):S7–14
65. Listenberger LL, Brown DA (2007) Fluorescent detection of lipid droplets and associated proteins. Curr Protoc Cell Biol 24:Unit 24.2
66. Liu JP, Tang Y, Zhou S, Toh BH, McLean C, Li H (2010) Cholesterol involvement in the pathogenesis of neurodegenerative diseases. Mol Cell Neurosci 43:33–42
67. Liu R, Lu P, Chu JW, Sharom FJ (2009) Characterization of fluorescent sterol binding to purified human NPC1. J Biol Chem 284:1840–1852
68. Loura LMS, Fedorov A, Prieto M (2001) Exclusion of a cholesterol analog from the cholesterol-rich phase in model membranes. Biochim Biophys Acta 1511:236–243
69. Luisier F, Vonesch C, Blu T, Unser M (2009) Fast Haar-wavelet denoising of multidimensional fluorescence microscopy data. Proceedings of the sixth IEEE international symposium on biomedical imaging: From nano to macro, Boston MA, USA, June 28–July 1, pp. 310–313
70. Luisier F, Vonesch C, Blu T, Unser M (2010) Fast interscale wavelet denoising of Poisson-corrupted images. Signal Processing 90:415–427
71. Lund FW, Lomholt MA, Solanko LM, Wüstner D (2012) Two-photon time-lapse microscopy of BODIPY-cholesterol reveals anomalous sterol diffusion in Chinese hamster ovary cells (Submitted for publication)
72. Masters BR, So PTC (2008) Classical and quantum theory of one-photon and multiphoton fluorescence spectroscopy. In: Masters BR, So PTC (eds) Handbook of biomedical nonlinear optical microscopy. Oxford University Press, Oxford, pp 91–152
73. Matyash V, Geier C, Henske A, Mukherjee S, Hirsh D, Thiele C, Grant B, Maxfield FR, Kurzchalia TV (2001) Distribution and transport of cholesterol in *Caenorhabditis elegans*. Mol Biol Cell 12:1725–1736
74. Maxfield FR, Menon AK (2006) Intracellular sterol transport and distribution. Curr Opin Cell Biol 18:379–385
75. Maxfield FR, Wüstner D (2002) Intracellular cholesterol transport. J Clin Invest 110:891–898
76. Maxfield FR, Wüstner D (2012) Analysis of cholesterol trafficking with fluorescent probes. Methods Cell Biol 108:367–393
77. McGookey DJ, Anderson RW (1983) Morphological characterization of the cholesteryl ester cycle in cultured mouse macrophage foam cells. J Cell Biol 97:1156–1168
78. McIntosh AL, Atshaves BP, Huang H, Gallegos AM, Kier AB, Schroeder F (2008) Fluorescence techniques using dehydroergosterol to study cholesterol trafficking. Lipids 43: 1185–1208
79. McIntosh AL, Gallegos AM, Atshaves BP, Storey SM, Kannoju D, Schroeder F (2003) Fluorescence and multiphoton imaging resolve unique structural forms of sterol in membranes of living cells. J Biol Chem 278:6384–6403
80. Mesmin B, Maxfield FR (2009) Intracellular sterol dynamics. Biochim Biophys Acta 1791:636–645
81. Mesmin B, Pipalia NH, Lund FW, Ramlall TF, Sokolov A, Eliezer D, Maxfield FR (2011) STARD4 abundance regulates sterol transport and sensing. Mol Biol Cell 22:4004–4015
82. Miao L, Nielsen M, Thewalt J, Ipsen JH, Bloom M, Zuckermann M, Mouritsen OG (2002) From lanosterol to cholesterol: structural evolution and differential effects on lipid bilayers. Biophys J 82:1429–1444
83. Mondal M, Mesmin B, Mukherjee S, Maxfield FR (2009) Sterols are mainly in the cytoplasmic leaflet of the plasma membrane and the endocytic recycling compartment in CHO cells. Mol Biol Cell 20:581–588
84. Mouritsen OG, Zuckermann MJ (2004) What's so special about cholesterol? Lipids 39:1101–1113
85. Mueller V, Ringemann C, Honigmann A, Schwarzmann G, Medda R, Leutenegger M, Polyakova S, Belov VN, Hell SW, Eggeling C (2011) STED nanoscopy reveals molecular details of cholesterol- and cytoskeleton-modulated lipid interactions in living cells. Biophys J 101:1651–1660

86. Mukherjee S, Zha X, Tabas I, Maxfield FR (1998) Cholesterol distribution in living cells: fluorescence imaging using dehydroergosterol as a fluorescent cholesterol analog. Biophys J 75:1915–1925
87. Müller F, Mazza D, Stasevich TJ, McNally JG (2010) FRAP and kinetic modeling in the analysis of nuclear protein dynamics: what do we really know? Curr Opin Cell Biol 22:403–411
88. Müller P, Herrmann A (2002) Rapid transbilayer movement of spin-labeled steroids in human erythrocytes and in liposomes. Biophys J 82:1418–1428
89. Nagao T, Qin C, Grosheva I, Maxfield FR, Pierini LM (2007) Elevated cholesterol levels in the plasma membranes of macrophages inhibit migration by disrupting RhoA regulation. Arterioscler Thromb Vasc Biol 27:1596–1602
90. Neufeld EB, Cooney AM, Pitha J, Dawidowicz EA, Dwyer NK, Pentchev PG, Blanchette-Mackie EJ (1996) Intracellular trafficking of cholesterol monitored with a cyclodextrin. J Biol Chem 271:21604–21613
91. Norman AW, Demel RA, de Kruyff B, van Deenen LLM (1972) Studies on the biological properties of polyene antibiotics. J Biol Chem 247:1918–1929
92. Petersen NO, Höddelius PL, Wiseman PW, Seger O, Magnusson KE (1993) Quantitation of membrane receptor distributions by image correlation spectroscopy: concept and application. Biophys J 65:1135–1146
93. Pipalia NH, Hao M, Mukherjee S, Maxfield FR (2006) Sterol, protein, and lipid trafficking in Chinese hamster ovary cells with Niemann-Pick type C1 defect. Traffic 8:130–141
94. Pipalia NH, Huang A, Ralph H, Rujoi M, Maxfield FR (2006) Automated microscopy screening for compounds that partially revert cholesterol accumulation in Niemann-Pick C cells. J Lipid Res 47:284–301
95. Prattes S, Horl G, Hammer A, Blaschitz A, Graier WF, Sattler W, Zechner R, Steyrer E (2000) Intracellular distribution and mobilization of unesterified cholesterol in adipocytes: triglyceride droplets are surrounded by cholesterol-rich ER-like surface layer structures. J Cell Sci 113:2977–2989
96. Puri V, Watanabe R, Singh RD, Dominguez M, Brown JC, Wheatley CL, Marks DL, Pagano RE (2001) Clathrin-dependent and -independent internalization of plasma membrane sphingolipids initiates two Golgi targeting pathways. J Cell Biol 154:535–547
97. Qian H, Sheetz MP, Elson EL (1991) Single particle tracking. Analysis of diffusion and flow in two-dimensional systems. Biophys J 60:910–921
98. Qin C, Nagao T, Grosheva I, Maxfield FR, Pierini LM (2006) Elevated plasma membrane cholesterol content alters macrophage signaling and function. Arterioscler Thromb Vasc Biol 26:372–378
99. Rodal SK, Skretting G, Garred O, Vilhardt F, van Deurs B, Sandvig K (1999) Extraction of cholesterol with methyl-beta-cyclodextrin perturbs formation of clathrin-coated endocytic vesicles. Mol Biol Cell 10:961–974
100. Rychnovsky SD, Richardson TI (1995) Relative and absolute configuration of filipin III. Angew Chem Int Ed Engl 34:1227–1230
101. Sage D, Neumann FR, Hediger F, Gasser SM, Unser M (2005) Automatic tracking of individual fluorescence particles: application to the study of chromosome dynamics. IEEE Trans Image Processing 14:1372–1383
102. Saxton MJ (1983) Lateral diffusion in an archipelago. Biophys J 64:1766–1780
103. Saxton MJ, Jacobson K (1997) Single-particle tracking: applications to membrane dynamics. Annu Rev Biophys Biomol Struct 26:373–399
104. Sbalzarini IF, Hayer A, Helenius A, Koumoutsakos P (2006) Simulations of (an)isotropic diffusion on curved biological surfaces. Biophys J 90:878–885
105. Sbalzarini IF, Koumoutsakos P (2005) Feature point tracking and trajectory analysis for video imaging in cell biology. J Struct Biol 151:182–195
106. Scheidt HA, Müller P, Herrmann A, Huster D (2003) The potential of fluorescent and spin-labeled steroid analogs to mimic natural cholesterol. J Biol Chem 278:45563–45569

107. Schrader M, Bahlmann K, Hell SW (1997) Three-photon-excitation microscopy: theory, experiment, and applications. Optik 104:116–124
108. Schroeder F, Barenholz Y, Gratton E, Thompson TE (1987) A fluorescence study of dehydroergosterol in phosphatidylcholine bilayer vesicles. Biochemistry 26:2441–2448
109. Schroeder F, Holland JF, Bieber LL (1971) Fluorometric evidence for the binding of cholesterol to the filipin complex. J Antibiot (Tokyo) 24:846–849
110. Schroeder F, Nemecz G, Gratton E, Barenholz Y, Thompson TE (1988) Fluorescence properties of cholestatrienol in phosphatidylcholine bilayer vesicles. Biophys Chem 32:57–72
111. Schütz GJ, Schindler H, Schmidt T (1997) Single-molecule microscopy on model membranes reveals anomalous diffusion. Biophys J 73:1073–1080
112. Severs NJ, Simons HL (1983) Failure of filipin to detect cholesterol-rich domains in smooth muscle plasma membrane. Nature 303:637–638
113. Shaw JE, Epand RF, Epand RM, Li Z, Bittman R, Yip CM (2006) Correlated fluorescence-atomic force microscopy of membrane domains: structure of fluorescence probes determines lipid localization. Biophys J 90:2170–2178
114. Shrivastava S, Haldar S, Gimpl G, Chattopadhyay A (2009) Orientation and dynamics of a novel fluorescent cholesterol analogue in membranes of varying phase. J Phys Chem B 113:4475–4481
115. Steck TL, Ye J, Lange Y (2002) Probing red cell membrane cholesterol movement with cyclodextrin. Biophys J 83:2118–2125
116. Steer CJ, Bisher M, Blumenthal R, Steven AC (1984) Detection of membrane cholesterol by filipin in isolated rat liver coated vesicles is dependent upon removal of the clathrin coat. J Cell Biol 99:315–319
117. Subtil A, Gaidarov I, Kobylarz K, Lampson MA, Keen JH, McGraw TE (1999) Acute cholesterol depletion inhibits clathrin-coated pit budding. Proc Nat Acad Sci U S A 96:6775–6780
118. Swedlow JR, Platani M (2002) Live cell imaging using wide-field microscopy and deconvolution. Cell Struct Funct 27:335–341
119. Tabas I, Rosoff WJ, Boykow GC (1988) Acyl coenzyme A:cholesterol acyl transferase in macrophages utilizes a cellular pool of cholesterol oxidase-accessible cholesterol as substrate. J Biol Chem 263:1266–1272
120. Tolić-Nørrelykke IM, Munteanu EL, Thon G, Oddershede L, Berg-Sørensen K (2004) Anomalous diffusion in living yeast cells. Phys Rev Lett 93:078102
121. Veatch SL, Keller SL (2002) Organization in lipid membranes containing cholesterol. Phys Rev Lett 89:268101
122. Veatch SL, Polozov IV, Gawrisch K, Keller SL (2004) Liquid domains in vesicles investigated by NMR and fluorescence microscopy. Biophys J 86:2910–2922
123. Vicidomini G, Moneron G, Han KY, Westphal V, Ta H, Reuss M, Engelhardt J, Eggeling C, Hell SW (2011) Sharper low-power STED nanoscopy by time gating. Nat Methods 8:571–573
124. Vist MR, Davis JH (1990) Phase equilibria of cholesterol/dipalmitoylphosphatidylcholine mixtures: 2 H nuclear magnetic resonance and differential scanning calorimetry. Biochemistry 29:451–464
125. Volpon L, Lancelin J-M (2000) Solution NMR structures of the polyene macrolide antibiotic filipin III. FEBS Lett 478:137–140
126. Weiss M (2004) Challenges and artefacts in quantitative photobleaching experiments. Traffic 5:662–671
127. Whitfield GB, Brock TD, Ammann A, Gottlieb D, Carter HF (1955) Filipin, an antifungal antibiotic: isolation and properties. J Am Chem Soc 77:4799–4801
128. Wiseman PW, Petersen NO (1999) Image correlation spectroscopy: II. Optimization for ultrasensitive detection of preexisting platelet-derived growth factor-beta receptor oligomers on intact cells. Biophys J 76:963–977

129. Wiseman PW, Squier JA, Ellisman MH, Wilson KR (2000) Two-photon image correlation spectroscopy and image cross-correlation spectroscopy. J Microsc 200:14–25
130. Wolf DE (2007) Fundamentals of fluorescence and fluorescence microscopy. Methods Cell Biol 81:64–91
131. Wrenn SP, Kaler EW, Lee SP (1999) A fluorescence energy transfer study of lecithin-cholesterol vesicles in the presence of phospholipase C. J Lipid Res 40:1483–1494
132. Wüstner D (2007) Fluorescent sterols as tools in membrane biophysics and cell biology. Chem Phys Lipids 146:1–25
133. Wüstner D (2007) Plasma membrane sterol distribution resembles the surface topography of living cells. Mol Biol Cell 18:211–228
134. Wüstner D (2008) Free-cholesterol loading does not trigger phase separation of the fluorescent sterol dehydroergosterol in the plasma membrane of macrophages. Chem Phys Lipids 154:129–136
135. Wüstner D (2009) Intracellular cholesterol transport. In: Ehnholm C (ed) Cellular lipid metabolism. Springer Press, Heidelberg, Germany, pp 157–190
136. Wüstner D (2012) Following intracellular cholesterol transport by linear and non-linear optical microscopy of intrinsically fluorescent sterols. Curr Pharm Biotechnol 13:303–318
137. Wüstner D, Brewer JR, Bagatolli LA, Sage D (2011) Potential of ultraviolet widefield imaging and multiphoton microscopy for analysis of dehydroergosterol in cellular membranes. Microsc Res Tech 74:92–108
138. Wüstner D, Færgeman NJ (2008) Chromatic aberration correction and deconvolution for UV sensitive imaging of fluorescent sterols in cytoplasmic lipid droplets. Cytometry A 73:727–744
139. Wüstner D, Færgeman NJ (2008) Spatiotemporal analysis of endocytosis and membrane distribution of fluorescent sterols in living cells. Histochem Cell Biol 130:891–908
140. Wüstner D, Herrmann A, Hao M, Maxfield FR (2002) Rapid nonvesicular transport of sterol between the plasma membrane domains of polarized hepatic cells. J Biol Chem 277:30325–30336
141. Wüstner D, Landt Larsen A, Færgeman NJ, Brewer JR, Sage D (2010) Selective visualization of fluorescent sterols in *Caenorhabditis elegans* by bleach-rate based image segmentation. Traffic 11:440–454
142. Wüstner D, Mondal M, Huang A, Maxfield FR (2004) Different transport routes for high density lipoprotein and its associated free sterol in polarized hepatic cells. J Lipid Res 45:427–437
143. Wüstner D, Mondal M, Tabas I, Maxfield FR (2005) Direct observation of rapid internalization and intracellular transport of sterol by macrophage foam cells. Traffic 6:396–412
144. Wüstner D, Sage D (2010) Multicolor bleach-rate imaging enlightens in vivo sterol transport. Commun Integr Biol 3:1–4
145. Wüstner D, Solanko LM, Sokol E, Lund FW, Garvik O, Li Z, Bittman R, Korte T, Herrmann A (2011) Quantitative assessment of sterol traffic in living cells by dual labeling with dehydroergosterol and BODIPY-cholesterol. Chem Phys Lipids 164:221–235
146. Xu C, Zipfel WR (2008) Multiphoton excitation of fluorescent probes. In: Masters BR, So PTC (eds) Handbook of biomedical nonlinear optical microscopy. Oxford University Press, Oxford, pp 311–333
147. Xu XX, Tabas I (1991) Lipoproteins activate acyl-coenzyme A:cholesterol acyltransferase in macrophages only after cellular cholesterol pools are expanded to a critical threshold level. J Biol Chem 266:17040–17048
148. Xu Z, Farver W, Kodukula S, Storch J (2008) Regulation of sterol transport between membranes and NPC2. Biochemistry 47:11134–11143
149. Yu L, Bharadwaj S, Brown JM, Ma Y, Du W, Davis MA, Michaely P, Liu P, Willingham MC, Rudel LL (2006) Cholesterol-regulated translocation of NPC1L1 to the cell surface facilitates free cholesterol uptake. J Biol Chem 281:6616–6624

150. Zhang W, McIntosh AL, Xu H, Wu D, Gruninger T, Atshaves B, Liu JC, Schroeder F (2005) Structural analysis of sterol distributions in the plasma membrane of living cells. Biochemistry 44:2864–2884
151. Zipfel WR, Williams RM, Webb WW (2003) Nonlinear magic: multiphoton microscopy in the biosciences. Nat Biotechnol 21:1369–1377
152. Zuckermann MJ, Ipsen JH, Miao L, Mouritsen OG, Nielsen M, Polson J, Thewalt J, Vattulainen I, Zhu H (2004) Modeling lipid-sterol bilayers: applications to structural evolution, lateral diffusion, and rafts. Methods Enzymol 383:198–229

Studying Membrane Properties Using Fluorescence Lifetime Imaging Microscopy (FLIM)

Martin T. Stöckl, Ranieri Bizzarri, and Vinod Subramaniam

Abstract Fluorescence lifetime imaging microscopy (FLIM) is a powerful tool to investigate the structure and composition of biological membranes. A wide variety of fluorescent probes suitable for FLIM experiments have been described. These compounds differ strongly in the details of their incorporation into membranes and in their responses toward changes in the membrane composition. In this chapter, we discuss and compare different classes of fluorescent membranes probes and their applications to studying biological membranes. We devote a section to a detailed description of fluorescent molecular rotors and their application to measuring local viscosity. As Förster resonance energy transfer (FRET) can be directly measured by changes in the donor fluorescence lifetime, FLIM is a very robust method to determine the distances between FRET pairs or the local concentrations of FRET-based membrane probes. Thus, we also discuss advantages and challenges of FRET-FLIM in the context of biological membranes. As biological membranes are considerably dynamic systems, imaging speed is often the limiting factor in biological FLIM experiments. Thus, novel fast imaging approaches and analysis methods to alleviate the issue of low photon statistics are also presented.

Keywords FLIM · Fluorescence Lifetime · Lipid bilayer · Membrane · Molecular Rotors

M.T. Stöckl (✉) and V. Subramaniam
Nanobiophysics, MESA+ Institute for Nanotechnology and MIRA Institute for Biomedical Technology and Technical Medicine, University of Twente, P.O. Box 217, Enschede 7500 AE, The Netherlands
e-mail: M.T.stockl@utwente.nl

R. Bizzarri
Istituto di Biofisica, Consiglio Nazionale delle Ricerche (IBF-CNR), via Moruzzi 1, Pisa 56124, Italy

Y. Mély and G. Duportail (eds.), *Fluorescent Methods to Study Biological Membranes*, Springer Ser Fluoresc (2013) 13: 215–240, DOI 10.1007/4243_2012_48,

Contents

1 Biological Membranes: A Short Introduction

Many cellular processes proceed in a localized fashion to allow for a fast and flexible regulation. Hereby, biological membranes play an important role as they on the one hand provide compartmentalization in cellular organelles and on the other hand directly offer scaffolds for membrane proteins. Therefore, regulation and signaling is often intimately associated with the structure and organization of cellular membranes. The formation of segregated lipid domains enriching selected lipid species and proteins is an especially important mechanism to form localized signaling platforms. While the exact properties and persistence times of these so-called lipid rafts is still under dispute, many lines of evidence suggest that they play a pivotal role in the regulation of numerous processes at cellular membranes [1–4]. Identifying and characterizing these entities promise to yield novel insights into the role of biological membranes in these processes. However, elucidating the impact of different factors is laborious, as an intricate network of interactions between different lipids and membrane proteins is involved. The lipid fraction alone contributes hugely to the complexity with hundreds of different types of lipids, which moreover are distributed heterogeneously, both laterally in the membrane and across the two leaflets of the bilayers. Furthermore, cell organelles also show specific lipid compositions, which may even vary within the organelle. In addition to lipids, cellular membranes also contain a wide variety of proteins, like receptors, channels, or peripheral membrane proteins. The organization of these proteins is also partly determined by their preference for certain lipid species, which guide them to specific membrane fractions. While for many peripheral membrane proteins the impact of membrane anchors or binding sites interacting with lipid head groups on the enrichment of the proteins in lipid domains or cellular trafficking has been studied, in the case of integral membrane proteins, the identification of the motifs driving their recruitment to distinct lipid domains remains elusive (for detailed reviews, see [5–8]). However, not only do lipid domains orchestrate the localization of membrane proteins, but the very presence of proteins in cellular membranes influences the structure and organization of the bilayer. Hereby presumably, the recruitment of a protein into its preferred lipid domain stabilizes the respective domain due to preferential interactions between the proteins and the lipids [9]. Also the cytoskeleton, which interacts with membrane proteins, strongly influences the properties of cellular membranes. This can be clearly seen in the case of giant plasma membrane vesicles (GPMVs), which consist of the lipids and proteins of the plasma membrane, but do not contain the actin cortex. While normally no large-scale separation of lipids into coexisting domains of micrometer size can be observed in cells, this is indeed the case for GPMVs [10, 11]. Thus, it is reasonable to conclude that, in cells, the membrane proteins together with the cytoskeleton form obstacles and corrals which influence lipid diffusion and therefore impair lipid segregation on a large scale [12].

Exploring the knots and connections of this intertwined network and identifying factors that regulate these interactions will promote our understanding of how cells communicate and interact with their environment. In the following, we discuss how fluorescence lifetime imaging microscopy (FLIM) approaches can contribute to elucidating the structure of cellular membranes and/or their interplay with membrane proteins.

2 FLIM: Advantages and Challenges

Fluorescence-based methods are ideally suited to study properties of biological membranes and their associated protein networks. The high specificity of the fluorescence signal and sensitivity of detection requires only minute amounts of fluorescent probes, minimizing the impact of the probe on the behavior of the system. The labeling technique used depends on the aim of the experiment. To study the structure and composition of cellular membranes, lipophilic probes that show a preferential enrichment into specific lipid domains can be used. In this case, the spectral properties of the probes depend on their environment. While the measurement of fluorescence spectra is straightforward in fluorescence spectrometers, in fluorescence microscopy approaches, this is not always achievable, due to the long exposure times required which lead concomitantly to increased photobleaching. This issue is especially problematic when studying membranes, as the fluorophores are localized in a strictly confined space, as opposed to being continuously distributed over a larger area. As the properties of the local (nano or micro) environment surrounding the probe most of the time also impact its fluorescence lifetime, FLIM offers a feasible approach to characterize membrane properties. Since these techniques normally use low illumination powers and, in the case of time-correlated single photon counting (TCSPC) techniques, the lifetimes of single fluorophores are summed up, photobleaching merely limits the available total photon count without affecting the measured lifetime. Interactions between proteins, or of proteins with lipid domains, can be studied using approaches relying on the transfer of energy between a donor and acceptor fluorophore in close proximity due to Förster resonance energy transfer (FRET). FRET is usually detected by measuring the fluorescence emission from the acceptor fluorophore upon the excitation of the donor (sensitized emission). Although the detection of FRET is relatively simple, the quantification of FRET by measuring the relative intensities of donor and acceptor is hindered by many factors. Here, mainly the fluorescence stemming from directly excited acceptor fluorophores and the likely inhomogeneous distribution of the interacting partners are worth mentioning. Due to the nonradiative process of energy transfer from the donor to the acceptor upon FRET, the FRET efficiency (E) is reflected in changes in the fluorescence lifetime of the donor alone ($\tau_{(D)}$) in comparison to the donor in the vicinity of an acceptor ($\tau_{(DA)}$) as in Eq. 1:

$$E = 1 - \tau_{(DA)}/\tau_{(D)} \tag{1}$$

Thus, determining the FRET efficiency requires only the measurement of the donor fluorescence lifetime in the presence and absence of the acceptor. This acceptor does not even have to be fluorescent, as it is sufficient that the acceptor molecule shows dynamic quenching of the donor fluorophore.

Moreover, the information embedded into the fluorescence lifetime can be utilized to characterize the structure of biological membranes. In order to determine whether the fluorescent probes are homogeneously distributed or enriched locally, for example in lipid domains, the fluorescence lifetimes can be analyzed, as local changes in donor or acceptor concentration change the FRET efficiency [13]. Although the overall FRET efficiency can be measured using only the fluorescence intensity, this is not feasible in inhomogeneous samples, as a measured FRET efficiency of 0.3 can arise from a homogeneous population of probes with a transfer efficiency of 0.3 or from a heterogeneous population showing fractions with varying FRET efficiencies, where the linear combination of these fractions and efficiencies averages to 0.3. Fluorescence lifetime measurements would, however, reveal in the homogeneous case a single lifetime of $\tau_{(DA)} = 0.7 * \tau_{(D)}$, while for a heterogeneous population, two or more fluorescence lifetimes would be observed.

Thus, the measurement of fluorescence lifetimes can yield additional information about the environment, the interactions of the probed molecules, or the composition of a population of molecules. However, as the determination of the fluorescence lifetimes in either the time or frequency domain requires some time, the acquisition of a single image can take up to a few minutes. This problem is especially severe if accurate lifetimes of probes embedded in lipid membranes have to be determined, as fewer pixels yielding the fluorescence signal in comparison to a continuously distributed probe are available. Therefore, longer acquisition times are required to achieve statistically sound data. This makes especially the direct visualization of lipid domains and interactions of membrane proteins with each other or with lipid domains by fluorescence lifetime contrast at least challenging, as these entities show short persistence times and rapid diffusion when not linked to the cytoskeleton or otherwise immobilized.

Furthermore, if fluorescent probes showing an environment-dependent behavior are used to determine the organization of biological membranes, first, their behavior has to be evaluated using well-defined model systems, which contain a defined lipid composition and for which the physicochemical properties like charge, degree of lipid order, or phase behavior, are known. Using these models, the basic properties of the probes, such as their response to changes in lipid composition, charge, and degree of order, can be determined. Also their preferential enrichment into certain lipid domains can be evaluated. However, care has to been taken to validate whether the results obtained in these models are also applicable for biological membranes; for instance, the differences in the degree of lipid order in model systems showing a coexistence of liquid-disordered (ld) and liquid-ordered (lo) domains are more pronounced than in biological membranes [14]. As an example, this may lead to the effect that a probe is excluded from the cholesterol rich lo domains in the model system due to the high degree of lipid order but is still found to be associated with the lipid raft fraction in a biological system [14]. In this view, GPMVs offer a better test bed, as in these vesicles, being directly derived from the plasma membrane, the biological lipid composition is preserved. However, also for this model system under certain conditions, a phase separation of different lipid domains can be observed on the μm scale, which is not present in cellular membranes [11]. At this point, the impact of the interaction between membrane proteins and the cytoskeleton becomes evident.

In summary, FLIM is well suited to characterize and visualize the localized interaction of fluorescently labeled interaction partners, as the fluorescence lifetime of the donor does not only yields information about the FRET efficiencies but also in principle reveals inhomogeneities below the optical resolution manifested as the occurrence of multiple lifetime components. Using fluorescent probes whose fluorescence lifetimes are sensitive to their environment, FLIM can also be used to spatially resolve the structure and composition of biological membranes. The challenge of using this approach, however, is that a thorough characterization and understanding of the properties of the probe are needed. In the following sections, approaches to study the structure of biological membranes as well as methods to evaluate lifetime data will be discussed in more detail.

3 Membrane Structure and Composition: Environment-Sensitive Probes

The fluorescence lifetime of a fluorescent molecule reflects the persistence of the excited state of the fluorophore, which depends on the rates at which it returns to the ground state. Thereby, the energy is lost by the emission of a photon as fluorescence, converted internally to heat or transferred in nonradiative processes to a surrounding molecule. This explains the shortening of the donor lifetime in the presence of a FRET acceptor or dynamic quencher, as these molecules can take up the energy of the excited state and therefore offer an additional route for de-excitation. In addition, the surrounding environment can influence the energy levels of ground and excited states and therefore their stability, for instance, by the formation of hydrogen bonds or dipolar interactions. Moreover, the physicochemical properties of the milieu like viscosity or polarity may influence the specific fluorescence lifetime of a fluorophore. Thus, the fluorescence lifetime of a fluorophore contains information about the lipid environment in which the probe is embedded, since different lipid domains have a specific composition and therefore varying physicochemical properties, like local charge or degree of lipid order. The characterization of the behavior of a probe to such changes in the environment has to be determined in initial experiments using model membranes with known properties. Using this approach, the fluorescence lifetimes typical for certain lipid phases and compositions can be measured, which allows the identification of analogous lipid domains in biological membranes. As this approach, unlike the detection of lipid domains by differences in the fluorescence intensity, does not depend on the preferential enrichment of the probe in a specific lipid environment, the detailed distribution behavior of the probe does not need to be determined using model systems, avoiding the pitfall of erroneous assignments of localizations due to a deviating distribution of the probe in the biological membrane compared to the model system. Moreover, in principle, a detection of lipid domains even below the resolution limit of optical microscopy should be possible, due to their characteristic lifetimes. For comparison, alternatively, the combination of fluorescence correlation spectroscopy with stimulated emission depletion (STED-FCS) is a powerful tool to study the structure of biological membranes. Using this method allowed for the identification of membrane regions in which fluorescent probes become entrapped over short periods of time [15, 16].

3.1 "Classical" Membrane Probes

The probes suitable for these types of experiments can be grouped into different classes according to their chemical structure. These compounds have in common a more or less pronounced hydrophobic moiety responsible for the integration of the probe into the hydrophobic core of the lipid bilayer. While some of the probes are derived from components present in biological membranes like cholesterol or lipids, chemically unrelated probes can also be used. An overview of probes and labeling of cells can be found in [17].

While the sterol moiety of cholesterol is ideally suited to embed into lipid bilayers due to its extensive hydrophobic ring system, this property also hinders the use of sterol analogues in the labeling of biological membranes, as the compounds normally are not readily soluble in aqueous solutions. Thus, the labeling of biological membranes often requires the use of a complex-forming substance like methyl-β-cyclodextrin (MβCD) which is able to bind sterols into its cavity. However, caution has to be exercised, as MβCD has also the ability to extract cholesterol from the plasma membrane of cells. As cholesterol is decisively involved in the formation of lipid rafts, a reduced cholesterol concentration may fundamentally change the structure of cellular membranes and impair the function of these signaling platforms. Moreover, while cholesterol itself packs well with saturated lipids and therefore is enriched in raft domains, the introduction of a fluorophore to form a fluorescent lipid probe adds a bulky moiety, which normally prevents enrichment in well-ordered lipid domains. This drawback can be clearly seen for NBD-labeled cholesterol derivatives that are excluded from lo domains. A BODIPY-labeled cholesterol compound which is able to embed also in such domains, depending on the chain length of the sphingomyelin used, has been described [18]. However, the lifetime difference of the compound localized in an ld or lo domain is quite small [19]. This point is reflected by the observation that the probe shows indistinguishable fluorescence lifetimes when incorporated in the plasma membrane or in lipid droplets [20]. In addition to these probes containing an added fluorescent moiety, also dihydroergosterol and cholestatrienol, which are intrinsically fluorescent due to the presence of three conjugated double bonds in the sterol ring system, are used to study the distribution of cholesterol in cells, as they faithfully reproduce the behavior of native cholesterol. However, due to their very low photostability and excitation wavelength in the UV, they have not been used for FLIM applications yet. For a detailed overview describing the properties of fluorescent sterol derivatives, see [21].

The most abundant building blocks of cellular membranes are phospholipids; thus, it can be construed that they offer a good basic structure for fluorescent probes for their characterization. Their main advantage is that using different acyl chains and head groups, their properties can be tailored to match the requirements of an experiment. In principle, lipids that bear the fluorophore at the head group embed into the bilayer with minimal impact on lipid packing. However, these lipids normally show a lower solubility in aqueous solutions. Additionally, the greater flexibility of the fluorophore may impair its sensitivity for the properties of the lipid bilayer. Nevertheless *N*-rhodamine-dioleoylphosphatidylethanolamine (*N*-Rho-DOPE) has been used to identify and characterize different lipid phases in lipid vesicles. For ld and gel domains, lifetimes of about 2.6–2.7 ns and for lo domains shorter lifetimes of about 1.8 ns were observed. Also changes in membrane fluidity or degree of lipid ordering, for example, caused by different fractions of cholesterol were reflected in the fluorescence lifetime of the probe [22]. In the second class of phospholipid-derived fluorescent probes, one of the fatty acids is exchanged for a fluorescent moiety. Thus, the fluorophore is embedded into the lipid bilayer or at the tightly packed interface between membrane and aqueous environment. This positioning is ideal to sense changes in lipid packing. In addition, the exchange of a hydrophobic acyl chain against a more polar structure can enhance the solubility of the probe, while retaining the preference for an enrichment into lipid bilayers, a property which facilitates the labeling of biological specimens. The bulky fluorophore however in all known cases excludes the localization of such probes to well-ordered lo or gel domains. For experiments, different BODIPY- or NBD-labeled phospholipid variants are commercially available.

BODIPY-PC embedded in GUVs composed of DOPC and DPPC showing a coexistence of ld and gel phases exhibits specific average fluorescence lifetimes of 5.2 and 3.8 ns, respectively [23]. Due to this sensitivity toward the degree of lipid order, the probe should also be usable to study the structure of biological membranes. Recently a BODIPY variant has been described whose lifetime is dependent on its subcellular localization [24]. Although the fluorescence lifetimes of the fluorescent lipid analogues N-Rho-DOPE and BODIPY-PC are sensitive to changes in the properties of lipid bilayers, the differences for different lipid phases are relatively small, even for model systems with clearly distinguished lipid domains. For biological membranes, however, a much more complex composition can be expected. Due to the high number of different lipids present, unlike in model systems, not only two or three well-defined lipid phases coexist, but it is likely that biological membranes contain a variety of lipid domains with similar properties. Thus, the differences in the fluorescence lifetimes for different lipid environments would be much less pronounced, which makes the identification of specific domains difficult. Therefore, a strong dependence of the fluorescence lifetime on the lipid environment would be desirable. This criterion is met by the fluorescent lipid analogue C6-NBD-PC. Embedded in lipid bilayers in GUV model systems, the probe showed characteristic lifetimes of about 7 ns for ld and 12 ns for lo lipid phases (Fig. 1a) [25]. The large difference between the specific fluorescence lifetimes should enable the probe to detect lipid domains which cannot be resolved by a differential fluorescence intensity, due to small size or similar fluorescence intensities. In fact, for GUVs prepared from lipid mixtures for which the coexistence of ld and lo lipid phases can be safely assumed from previous experiments, despite showing a homogeneous distribution of C6-NBD-PC on the scale of optical microscopy, the typical fluorescence lifetimes for the lipid phases can be recovered. Applying this approach to GPMVs, no coexistence of lipid domains with differing fluorescence intensities could be observed. Yet, the FLIM images revealed the coexistence of μm-sized lipid domains with different average fluorescence lifetimes. However, the corresponding fluorescence lifetime histogram showed a single broad distribution of fluorescence lifetimes. Separate data analysis of regions of interest containing only one of the domains yielded fluorescence lifetimes of about 6.7 and 9.7 ns (Fig. 1b) [25]. The difference between the fluorescence lifetimes is therefore much smaller than in the model system with well-defined ld and lo lipid phases. Accordingly in the lifetime histogram, no separate distributions are observed, as the contributions from both lifetimes strongly overlap and give rise to a single broader distribution. In line with the lack of contrast in fluorescence intensity, this suggests that properties of the lipid domains in the GPMVs exhibiting physiological lipid compositions are much more similar than those observed in GUVs prepared from simple three-component lipid mixtures. This observation renders the above-mentioned approach to detect domains below the optical resolution inapplicable for biological systems. Indeed, the fluorescence histograms recorded for the plasma membrane of cells labeled with C6-NBD-PC showed a very broad distribution of lifetimes centered around 11 ns but also including lifetimes longer than those observed for pure lo phases (Fig. 1c). This suggests that the plasma membrane of cells is composed of a variety of lipid domains with differing properties and an overall high degree of lipid order. As the fluorescence lifetimes are substantially longer than those found for GPMVs, it becomes apparent that the interactions between membrane proteins and the underlying cytoskeleton have a strong impact on the properties of the plasma membrane. With the fluorescence lifetimes reflecting the properties of the surrounding environment, changes in lipid composition and structure of the lipid bilayer should have an impact on the

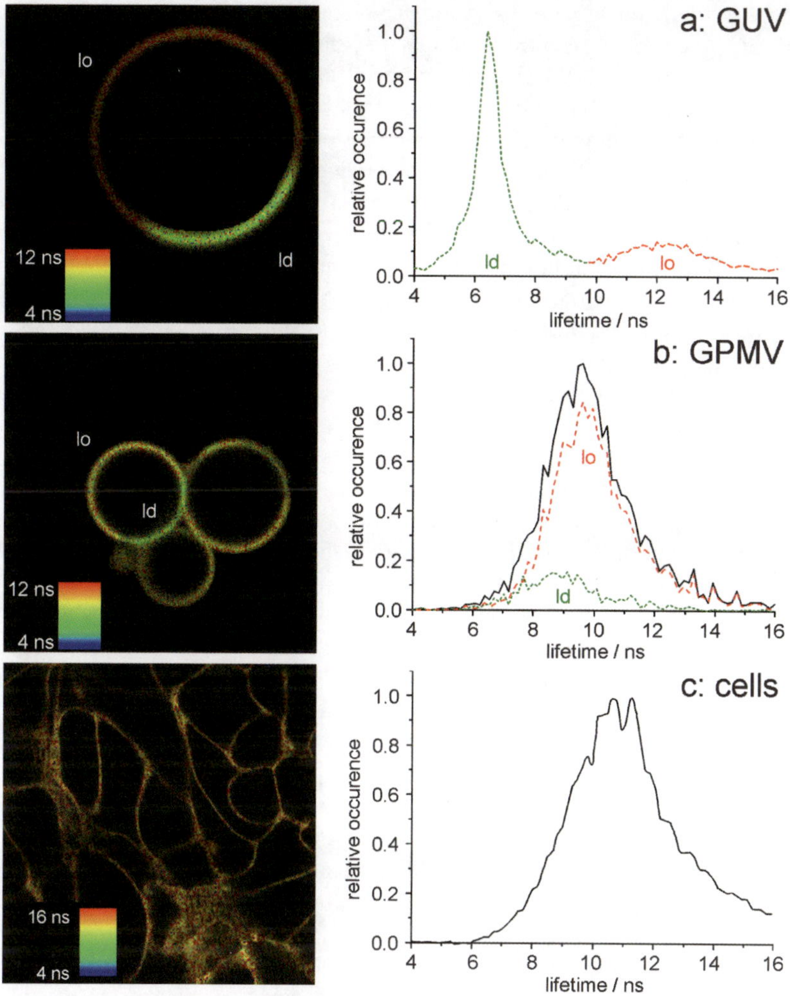

Fig. 1 Fluorescence lifetime distributions for C6-NBD-PC embedded in different lipid bilayer systems. The *left column* shows FLIM images using a pseudo-color scale based on the average fluorescence lifetime of C6-NBD-PC. Normalized lifetime histograms are shown in the *right column*. While for both GUV (DOPC/18:0-SM/Chol = 1/1/1 (**a**)) and GPMV (**b**) systems, a coexistence of liquid-disordered (*ld*) and liquid-ordered (*lo*) domains can be seen in the FLIM images, distinct lifetimes for the two phases in the total histograms can be only distinguished for the GUVs. Yet, separate analysis of ld and lo domains also yields different lifetimes in GPMV systems; however, the individual distributions overlap in the total lifetime histogram (**b**). The fluorescence lifetimes of C6-NBD-PC show a broad distribution of relatively long lifetimes when embedded in the plasma membrane of HepG2-cells (**c**), which suggests that the probe senses a heterogeneous environment with a relatively high degree of lipid order. For experimental details, refer to [25]

average fluorescence lifetimes. In this view, the lowering of the cholesterol fraction in the plasma membrane of Hela cells, due to incubation with MβCD, a cholesterol extracting compound, shifted the lifetimes recorded in the lifetime histogram to slightly shorter values. This observation can likely be attributed to the disruption of lipid rafts, therefore embedding the C6-NBD-PC in an environment with a lower degree of lipid order [25]. Using this approach, also structural rearrangements in the plasma membrane of Jurkat cells induced by cross-linking of the CD3 receptors or incubation with sphingomyelinase could be detected [26]. Although under these conditions well-ordered lipid domains, either due to the stabilization of lipid rafts or the formation of ceramides, should form, on average the fluorescence lifetime of C6-NBD-PC became shorter which is indicative of a more disordered lipid environment. Keeping in mind that the probe is equally distributed between domains with a different degree of lipid order in biological membranes (e.g., GPMVs), while it is mostly excluded from the well-ordered lo domains in model membranes, inducing such lipid domains would lead to the redistribution of C6-NBD-PC into lipid domains with a lower degree of lipid order, therefore causing the shorter lifetimes. These experiments clearly demonstrate that not only the correlation of membrane properties with fluorescence lifetimes but also a possible redistribution of the probe into different lipid domains has to be taken into account when evaluating fluorescence lifetime data.

In conclusion, fluorescent lipid analogues labeled with a NBD fluorophore at an acyl chain are well suited to study membrane properties in biological systems, as the probes are readily soluble in aqueous solutions and exhibit long fluorescence lifetimes which are very sensitive to the degree of lipid order. However, their low photostability limits the total yield of photons in TCSPC experiments and hinders long-term measurements.

Apart from fluorescent probes based on physiologically occurring components of biological membranes (cholesterol and phospholipids), also compounds with a differing chemical structure, which preferentially enrich in the hydrophobic bilayer, can be used. Prominent examples are derivatives of diphenylhexatriene (DPH). As DPH shows a very strong hydrophobicity, it embeds into the core of lipid bilayers where it is oriented in a parallel fashion along the acyl chains of the lipids. Therefore, the polarization of fluorescence is a strong indicator for the mobility and orientation of the probe and is directly related to the degree of lipid order. Additionally, such properties of the lipid bilayer are also reflected in the fluorescence lifetime. Combining both approaches, a DPH derivative has been used to analyze the structure and properties of model lipid systems using GUVs [27]. In DOPC/cholesterol mixtures in a ld phase, the ordering impact of cholesterol became apparent, leading to a stronger fluorescence polarization and a longer fluorescence lifetime. Similar behavior was observed for sphingomyelin/cholesterol mixtures in the lo phase, albeit the fluorescence lifetime values were consistently higher due to the higher degree of order. For coexisting lipid phases in ternary mixtures, the fluorescence lifetime and polarization were independently determined. The deviation from the values measured for the binary systems clearly shows that the lipids do not demix entirely but are only enriched with specific fractions in the respective domains.

Also, for an amphipathic perylene imide compound lipid domain-specific lifetimes have been determined in a model system, yielding values of 4.4 ns for the ld and 6.0 ns for the lo phase [28]. This probe also has been applied to analyze structural rearrangements of lipids in the plasma membrane of Jurkat cells. For the fluorescence lifetime, a value of 5.4 ns for untreated cells was measured. In line with our experiments using C6-NBD-PC, this value is indicative for the presence of a high degree of lipid order in the plasma membrane. In addition, also receptor cross-linking and cholesterol extraction by MβCD lead to shorter fluorescence lifetimes of 5.1 and 4.3 ns, respectively. This type of probe shows the additional advantage that the fluorophore can be modified to produce probes with different spectral characteristics [29].

A further membrane probe applicable in FLIM is di-4-ANEPPDHQ, a fluorescent compound that was initially used to measure the transmembrane potential [30]. In model lipid bilayers, the probe does not only show a shift in fluorescence emission with maxima of about 560 nm in a ld DOPC environment to about 620 nm for a lo phase (DPPC/chol = 7/3) but also exhibits lipid phase-specific fluorescence lifetimes of 1.9 ns (DOPC) and 3.5 ns (PSM/chol = 7/3) [31, 32]. This property allowed the identification of lipid domains showing different fluorescence lifetimes in the plasma membrane of HEK293-cells, which were not apparent using an intensity-based ratiometric approach. A depletion of cholesterol by MβCD or an increase in temperature, concordant with the disruption of well-ordered lipid rafts, caused a shorter lifetime component to appear. The correlation of a reduced cholesterol fraction due to MβCD treatment with shorter fluorescence lifetimes of di-4-ANEPPDHQ has also been shown for COS-cells [33]. *trans*-Parinaric acid can also be used to study the structure of biological membranes. This probe has the striking property that it shows a wide range of fluorescence lifetimes dependent on its environment [17].

A novel class of fluorophores to study the structure of lipid membranes has been recently described. Hydrophobic ruthenium complexes have been used to label cells and characterize the structure of the membrane using FLIM. These compounds are of great interest as their uptake into cells, where they enrich at cellular membranes, can be triggered by laser illumination [34]. Embedded into lipid bilayers, the compounds show very long luminescence lifetimes of about 100 ns, which depend on their environment [35, 36]. Moreover, also for gold nanoclusters coated with the lipohilic dihydrolipoic acid, an enrichment into cellular membranes has been described. These particles display environment-sensitive lifetimes with distributions around 440 ns in aqueous solution, around 660 ns in intracellular compartments, or 720 ns at the plasma membrane [37]. To study structural rearrangement of lipid domains and the IgE receptor in the plasma membrane of RBL-mast cells upon cross-linking of activated receptors, the cells were labeled with fluorescent IgE (AlexaFluor488-IgE) and the raft marker diI-C_{18}. Upon cross-linking, the receptors clustered in distinct patches which were also rich in diI-C_{18} [38, 39]. In comparison to the untreated cells, for both fluorescent probes, longer fluorescence lifetimes and stronger anisotropy of fluorescence originating from these clusters, indicative for an increased degree of lipid order, were observed. Strikingly, the changed membrane properties were not only reported by the membrane embedded diI-C_{18} but also had an impact on the fluorescence emission of the labeled IgE bound at the receptor.

3.2 Viscosity-Sensitive Probes: Fluorescent Molecular Rotors

Viscosity is one of the most relevant environmental properties that affect cell function. This importance stems from the specific role of viscosity in determining molecular diffusion rates in condensed media. Changes in mobility of chemicals within the cell influence fundamental processes such as signaling, intracellular transport, and the efficiency of bimolecular processes governed by diffusion of short-lived intermediates [40]. Owing to its large-scale molecular order, the cell plasma membrane has intrinsically high viscosity; values as high as 217 cP (water = 0.8–1 cP) have been measured [41]. Yet, membrane viscosity is concomitantly a measure of local order and a powerful modulator of the activity of membrane proteins such as receptors and ionic channels [42]. Alterations in cell membrane microviscosity have been linked to atherosclerosis [43], infarction [44], diabetes [45], Alzheimer's disease [46], and aging [47].

The remarkable role played by viscosity on cell functions (particularly at membrane level) has prompted in recent years the development of indicators that can nondestructively report on local viscosity at high spatial resolution in living specimens. Fluorescent molecular rotors (FMRs) are a class of fluorescent probes that appear particularly suitable for this challenging goal. Structurally, FMRs comprise two subunits that can rotate relative to each other around a single atomic bond. In most cases, FMRs display a common photophysical behavior that is pictorially expressed by the Jablonski diagram of Fig. 2 [48]. In the ground state, a FMR exists in a (nearly) planar conformation; photon absorption leads to an excited state that retains the same (nearly) planar conformation on account of the Franck-Condon principle. This Franck-Condon state is named locally excited state (LE). From LE, the FMR can decay back to ground state by three different pathways (Fig. 2). Two pathways entail the direct conversion of LE to the ground state, either by emission (rate constant $k_{r(LE)}$) or by nonradiative thermal conversion (rate constant $k_{nr(LE)}$). The third one entails the initial conversion of LE to give a twisted excited state (TS) at lower energy than LE (rate constant k_{rot}); TS is characterized by conformational distortion of the two FMR subunits from the planar state. Then, very fast nonradiative conversion of the twisted excited state to ground state ensues (rate constant $k_{d(TS)} \gg k_{rot}$). In some peculiar cases, also the twisted state is emissive (e.g., 1,4-dimethylamino benzonitrile or DMABN), although these molecules display a complex photophysical behavior (strong dependence of emission on environmental polarity) that hampers their use as efficient viscosity sensors [49].

The viscosity sensitivity of FMRs comes from the notable sensitivity of k_{rot} to local viscosity $\eta(x,y,z)$. Indeed, the kinetics of internal rotation (i.e., the rotation energy barrier) is strongly modulated by the viscous drag exerted by the nearby solvent molecules. Under the experimentally verified assumption that the twisting pathway is the dominant nonradiative pathway, that is. $k_{rot} \gg k_{nr(LE)}$, and considering $k_{r(LE)}$ independent of viscosity, the relationship between fluorescence quantum

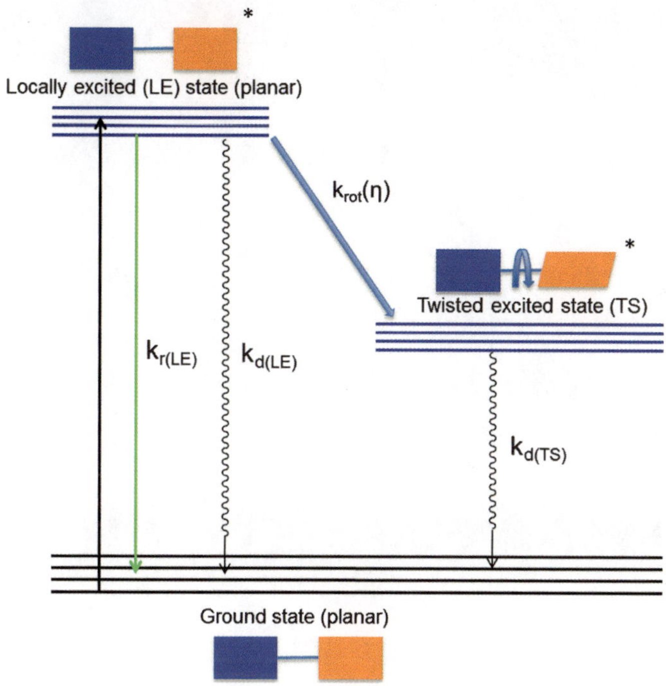

Fig. 2 Jablonski diagram for fluorescent molecular rotors. According to the Franck-Condon principle, upon excitation, the molecules transition from the ground state to the planar locally excited state, which decays radiatively ($k_{r(LE)}$) or nonradiatively ($k_{d(LE)}$) to the ground state. Additionally, a transition to a lower energy twisted state is possible; this state only decays nonradiatively ($k_{d(TS)}$). As fluorescence photons are only emitted from the locally excited state and the transition rate from the planar to the twisted state ($k_{rot(\eta)}$) is dependent on the viscosity of the environment, the local viscosity can be derived from measurements of the fluorescence lifetime of the molecular rotor

yield (Φ) or lifetime (τ) and local viscosity can be shown to be reasonably expressed by the Förster-Hoffmann equation [48, 50]:

$$\begin{aligned} \ln(\Phi) &= A + x\ln(\eta) \\ \ln(\tau) &= A' + x\ln(\eta) \end{aligned} \tag{2}$$

where $A(A')$ and x are solvent and dye-dependent parameters. The x parameter expresses the sensitivity of FMR to viscosity changes; the original Förster-Hoffmann theory sets $x = 2/3$, but experimental measurements showed values from 0.3 to 0.9 [51]. This relationship has been experimentally shown to be valid in a wide range of viscosities and in both polar and nonpolar fluids, although some deviations were sometimes described for the low-viscosity regime.

In many cases, FMRs are characterized by an electron donating subunit (EDG) conjugated, through a polyenic and/or aromatic spacer (π), to an electron accepting

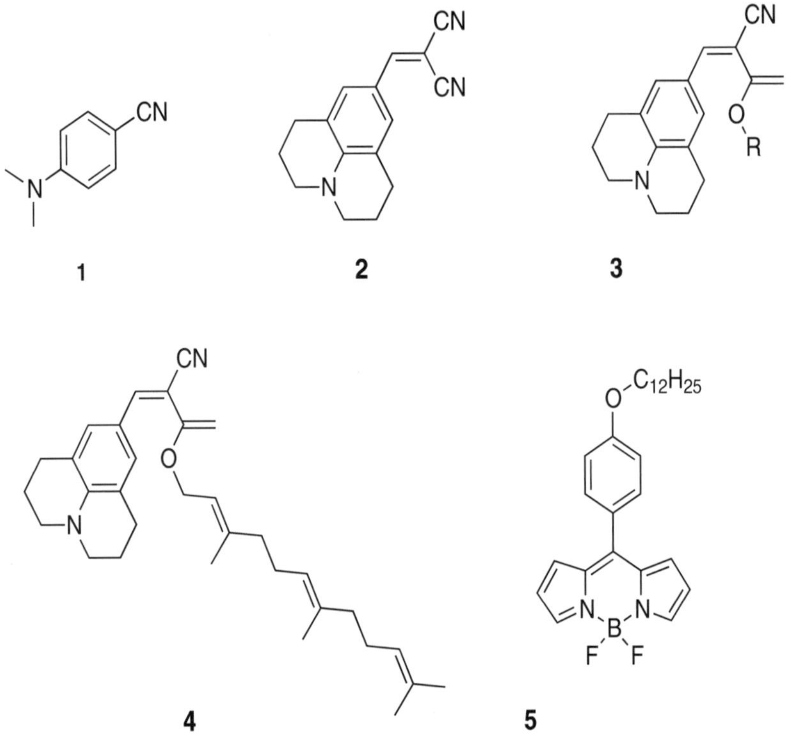

Fig. 3 Chemical structures of molecular rotors. **1**: 1,4-dimethylamino benzonitrile (DMABN) [48]; **2**: 9-(dicyanovinyl)-julolidine (DCVJ) [48]; **3**: alkyl-(2-carboxy-2-cyanovinyl)-julolidine [54]; **4**: farnesyl-(2-carboxy-2-cyanovinyl)-julolidine (FCVJ) [54, 55]; **5**: 4,4′-difluoro-4-bora-3a,4a- diaza-s-indacene [61]

group (EWG). The twisted state of these FMRs has usually a strong charge-transfer character, referred to as twisted intramolecular charge-transfer (TICT) state. Accordingly, we can group these FMRs in the TICT-FMR family. Owing to its strong polar character, the energy of TICT states is usually much lower in polar solvents than in apolar ones. This leads to the dependence of Φ and τ on solvent polarity through the parameters $A(A')$ and x of the Forster-Hoffmann equation [49].

Several FMRs have been applied to probe local viscosity in the cellular context. Aniline nitriles, *p*-(dialkylamino)benzylidene malonitriles, and stilbenes represent well-known classes of TICT-FMRs for cell imaging. 9-(dicyanovinyl)-julolidine (DCVJ) was applied to monitor changes in membrane viscosity in endothelial cells induced by external fluid shear stress, in view of elucidating the subtle role of the cell membrane in mechanotransduction [52]. Note that shear stress due to blood flow has a critical role in determining the (patho)physiology of vascular endothelial cells [53]. The same group that pioneered the use of FMRs for monitoring shear stress-induced alterations of membrane viscosity described later an improved toolbox of membrane-targeted FMRs constituted by analogues of DCVJ where one of the nitrile functionalities had been replaced by an alkyl carboxyester [54]. Among these compounds, the most promising probe for reporting on membrane viscosity was represented by (2-carboxy-2-cyanovinyl)-julolidine farnesyl ester (FCVJ). Indeed, FCVJ was shown to provide reliable and sensitive measurements of viscosity changes in both model [55] and cellular membranes [54]. These authors pursued cell membrane targeting of TICT-FMR by conjugating the julolidine system to either one hydrophobic chain or the polar head of a phospholipid. Notably, conjugation of the FMR to the polar head of the phospholipid decreased the viscosity sensitivity of the final probe, indicating the existence of specific interactions between the phospholipid scaffold and the FMR unit that decrease the effect of external viscous drag on TICT state formation [56].

In spite of these compelling results, most TICT-FMRs are characterized by several drawbacks that hamper their use as reporters of intracellular viscosity at high spatial resolution. In some cases, the sensitivity to viscosity is critically dependent on environmental polarity, preventing a universal calibration following Eq. 2 [49, 51]. Yet, the most severe problem associated to TICT-FMR probes is their failure to generate a fluorescent intensity-based signal independent of the local probe concentration, the latter being a quantity that cannot be controlled in the cellular context [50]. This prevents the determination of a spatially resolved viscosity map of the biological system under study. This issue has been addressed by conjugating a second, viscosity-independent, fluorescent unit to a classical FMR unit, making the overall system suitable for ratiometric recording [48, 57]. Ratiometric FMRs have been applied to monitor viscosity in liposome membranes [58]. This approach is synthetically demanding and increases the structural complexity of the fluorescent probe, possibly complicating some applications at the cellular level. A more elegant way to tackle this issue is to make use of either dual emission FMRs [59, 60] or to monitor viscosity by FLIM [61]. Interestingly, both approaches have been applied to FMRs whose twisted state is not characterized by a strong charge-transfer character such as some substituted boron dipyrromethene (BODIPY) derivatives and porphyrin dimers. In spite of these interesting results, to our knowledge, none of these systems has been selectively targeted to cell membranes yet.

4 Membrane Structure and Protein Interactions: FRET-FLIM

Using FLIM to characterize the structure and composition of biological membranes is based on the sensitivity of the fluorescence lifetime of the fluorophore to the surrounding environment. Determining the lifetimes specific for different lipid phases allows their identification in biological membranes. Due to the complex composition of biological membranes, this approach requires a large enough difference in the domain-specific fluorescence lifetimes and a thorough characterization of the correlation of the fluorescence lifetimes with the properties of the lipid bilayers, like degree of lipid order or fraction of charged lipids. Additionally, a differential localization of a probe between model and biological membrane may influence the apparent fluorescence lifetimes. As the properties of the lipid domains constituting biological membranes are not known, the behavior of the probes ultimately cannot be predicted, which hampers the analysis of the measured lifetime distributions. This challenge can be alleviated using an approach that is based on the transmission of energy by FRET between two probes with similar, well-defined lipid association behavior. Assuming that both probes would preferentially enrich in the same lipid domains due to similar properties, structural rearrangements which would induce such domains would increase the local concentration of both probes; therefore, the shorter average interprobe distance would increase the overall FRET efficiency, which becomes apparent in a shorter lifetime of the donor. Likewise, a disruption of such domains would lead to a decreased FRET efficiency. Yet, it has to be taken into account that changes in the properties of lipid domains, due to a changed composition or interaction with cellular proteins which change the degree of lipid order, can induce the loss of the preferential enrichment of one or both probes to that domain, also decreasing the observed FRET efficiency, although the original domains are still present. While changes in FRET efficiency can be directly related to the formation or disruption of lipid domains if the probes preferentially enrich and the domains are large enough to be resolved in the intensity images, this information is not available if structural rearrangements do not lead to the formation of such lipid domains. In any case, the presence of a FRET signal reliably indicates that two probes are in close proximity to each other.

The major advantage of FRET-based approaches is that the choice of fluorophores is not limited by the necessity that the lifetime of the fluorescent probe has to be sensitive to the different lipid phases. This allows for a wide spectrum of fluorescent molecules to be used, including but not limited to fluorescently-labeled lipids or hydrophobic molecules which embed into the lipid bilayer: fluorophores attached to various kinds of membrane anchors, for example, a GPI anchor; short peptides modified with hydrophobic moieties or transmembrane domains; and whole proteins being fluorescent due to a fusion with a fluorescent protein or due to interactions with an effector (e.g., fluorescent ligand or antibody). For many of these probes, the preferential enrichment has already been characterized, at least in GUV or GPMV model systems [11, 62, 63].

Such an approach was used to identify the interaction sites of cellular actin with the plasma membrane [64]. To this end, GFP-actin was coexpressed in mouse B16-F1 melanoma cells with RFP fused to protein domains which specifically recognize lipid domains rich in PIP2 or PIP3. PIP2 and PIP3 respectively preferentially enrich in less (H-Ras; farnesylation) or more ordered (neuromodulin, palmitoylation) microdomains. The local interaction of the GFP-actin with these fusion proteins was indicated by a shorter GFP fluorescence lifetime. These experiments suggest that the cortical actin is most intensively interacting with the plasma membrane at sites rich in PIP3. This approach is also applicable if one of the interaction partners is not a fluorescent protein, as has been shown in a study using the probe diI-C_{18}, which preferentially enriches in more ordered lipid domains, to show that the IgE-receptor (FcεRI) labeled with fluorescent antibodies preferentially enriches in cholesterol rich lipid domains [38].

In a similar fashion, in order to study the budding sites of the influenza virus, CHO cells coexpressing fusion proteins of YFP linked to transmembrane domains of influenza hemagglutinin (HA) and CFP fused with a peptide containing a GPI anchor, which preferentially enriches in lipid rafts, were used. The reduction of the CFP fluorescence lifetime due to FRET indicates that the HA transmembrane domain is a sufficient signal to mediate the localization of proteins in such domains [65]. In line with disruption of lipid rafts upon cholesterol depletion by MβCD, the decrease of overall FRET efficiency in the treated cells corroborates these findings. As the preferential enrichment to raft domains is decreased for hemagglutinin variants lacking the palmitoylation in the transmembrane domain, this suggests that the modification of the protein is decisive for its localization [65]. These findings were corroborated by experiments measuring the FRET efficiencies in cells expressing HA fused with a CFP variant and YFP fused with a double acylated peptide serving as a raft marker, which again shows a cholesterol-dependent enrichment of HA into raft domains [66]. Determining FRET efficiencies for varying concentrations of the acceptor, clustering of HA with itself and an interaction with the influenza capsid protein M2 has been shown [66, 67].

Not only the enrichment in specific lipid domains can be investigated by FRET experiments, but also the interaction of proteins in close proximity at biological membranes. In this view, the interaction of t- and v-SNARE proteins fused to fluorescent proteins has been studied. In proliferating hyphae of *Trichoderma reesei*, a filamentous fungus, complexes of t- and v-SNAREs showing differing FRET efficiencies were visualized at spatially segregated sites, suggesting that the interaction of both proteins is dependent on their localization in specific lipid domains [68]. Likewise in *Zea mays*, it has been shown that two families of plasma membrane intrinsic proteins form tight complexes in distinct lipid domains at the plasma membrane [69]. A similar approach has been used to characterize the multiprotein complex responsible for cell division in *Streptomyces* and to identify the components which interact with the plasma membrane or to characterize the interactions between Golgi-tethering factors and small GTPases [70, 71].

As an example, FRET-FLIM has also been used to investigate interactions between presenilin 1 and amyloid precursor protein (APP) variants in CHO cells. The cells were transiently transfected to express the proteins which were specifically labeled. Regions of reduced donor lifetime were identified in a patterned fashion, especially at the periphery of cells, which indicates the formation of complexes suggesting that certain lipid domains at the plasma membrane seem to be involved in the regulation of the complex formation [72]. Measuring the FRET efficiencies between eGFP attached to the membrane by a myristoyl anchor and mRFP fused to APP or presenilin 1 variants allowed the determination of the impact of point mutations related to Alzheimer's disease on the interaction of the C terminus with the membrane. For APP, corresponding results were observed if the fluorescent membrane probe CM-DiI (APP-GFP) was used. Also the impact of drugs on the orientation of C terminus of APP could be determined [73].

Signaling processes upon cell-cell interactions or plasma membrane remodulation during cell migration, endocytosis, membrane fusion, or tubule formation require a well-orchestrated interplay of many membrane proteins in a highly localized fashion. Characterizing these complexes using fusion proteins containing the proteins of interest or labeling those proteins with fluorescent antibodies allows the identification of the underlying interactions [74–76]. To this end, also probes labeling lipid domains are applied [77].

As FRET records the close proximity of two molecules, it can be used also to follow the formation of dimers or larger complexes composed of a single protein species, as long as it constitutes a mixture of proteins labeled either with a donor or acceptor fluorophore. Such an approach has been applied to compare the degree of interaction between serotonin receptors in different cell lines [78]. Additionally, the complex formation of the signal peptidase has been investigated. The patterned distribution of the dimers suggests that specific lipid domains are involved in the formation of these complexes [79].

In summary, using FLIM to detect a decrease in donor fluorescence lifetime by FRET to visualize interactions of membrane components or their enrichment in specific lipid domains is a very sensitive and robust technique, as it is largely independent of the local concentrations. Yet, the differences in the donor lifetime are often quite small, due to fact that not all donor molecules may be in close proximity to the acceptor or that the FRET pairs are aligned in an unfavorable fashion, due to limitations in orientation imposed by the lipid bilayer. This is especially severe for fluorescent proteins, as due to the large distance between the chromophores embedded in the core of these proteins the highest achievable FRET efficiency is approximately 0.5. However, fusion proteins between a fluorescent protein and the target protein still offer the least intrusive approach to label live cells. Additionally, as the environment of the fluorophores may change their fluorescence lifetimes, proper controls in which the lifetime behavior of the donor fluorophore is measured in interaction with the appropriate nonfluorescent interaction partners are necessary. In summary, while the detection of interactions is often straightforward, and even applicable in a high throughput regime, the measurement of the fraction of involved molecules and the distance between the interacting fluorophores may be tedious, as the observed average fluorescence lifetimes are influenced by both parameters [33, 80]. Additionally, the heterogeneity of the sample on a small scale is often not known, and due to the small image pixel size, the data is often noisy.

5 Challenges and Novel Evaluation Techniques

Measuring the fluorescence lifetimes of probes in biological membranes offers detailed information about the environment in which they are embedded or can be used to facilitate the determination of FRET efficiencies. As FLIM yields spatially resolved images, the structural composition of biological membranes and of the associated proteins can be studied. However, this requires the determination of the fluorescence lifetimes for each pixel of a FLIM image. In general, two kinds of approaches, based on their way to determine the lifetimes, can be discerned. In time domain methods, the fluorescence lifetimes are calculated from the decay of fluorescence after a pulsed excitation while in frequency domain methods, a continuously modulated excitation is used and fluorescence lifetimes are determined from the phase shift and amplitude demodulation of the fluorescence emission. The most common time domain technique is TCSPC, where the arrival times of single fluorescence photons after a pulsed excitation are measured, from which the decay of the fluorescence emission is rebuilt. As it has to be ensured that only single photons hit the detector in one detection cycle to prevent distortion of the decay, the achievable count rate is rather limited, leading to long acquisition times to record enough photons in each pixel, so that the often multiexponential decays can be fitted with sufficient accuracy. To alleviate the problem of fitting errors due to a low photon count, recently an algorithm to virtually increase the numbers of available photons using adaptive Monte Carlo data inflation (AMDI) has been described [81]. Instead of fitting the measured fluorescence decays with exponential decay curves, the data can also be converted into the frequency domain by calculating Fourier sine and cosine transforms [82, 83]. The lifetimes then can be visualized in polar plots, which have been originally developed for frequency domain methods (see below) [84–86]. Using this representation, it is possible in cells coexpressing CFP and eGFP variants to distinguish between both fluorescent proteins, as CFP shows a multiexponential decay, while eGFP shows a monoexponential decay, despite similar overall lifetimes for both proteins [83].

As the acquisition times of images normally are in the range of tenths of seconds to minutes for images, this precludes following fast processes in live cells. However, techniques can be applied to reduce the required frame acquisition time, for example, by binning of pixels, which sacrifices spatial resolution, or time gating, which sacrifices temporal resolution [87]. Making use of time gates, it is also possible to extract photons suitable for coherent anti-Stokes Raman scattering (CARS) microscopy from the recorded fluorescence decay [88]. A different time domain method to determine fluorescence lifetimes, which is not limited by the constraint of single photon detection, but rather records the mean delay between excitation and the fluorescence emission using an analogue amplifier with subsequent digitization, has been recently described [89, 90]. This method allowed to follow the activation of Ca channels with frame rates up to 7.7 Hz [91].

A completely different approach to measure fluorescence lifetimes is to excite the fluorophores using modulated excitation. For these approaches, normally widefield illumination in combination with a CCD-camera gated by a microchannel plate is used, allowing for short acquisition times. However, the information gained about the lifetime decay itself may be limited. In these approaches, the fluorescence lifetimes are calculated from phase shifts between excitation and emission or the demodulation of the fluorescence emission. Alternatively, the phase shift and modulation recorded for single pixels can be transformed into coordinates and analyzed graphically. Pixels showing a monoexponential lifetime thereby lie on a half circle, while pixels in which two lifetime components are present fall in the interior of the circle, on a line connecting both monoexponential components

[84–86]. This approach is especially beneficial to determine the fraction of molecules undergoing FRET in a heterogeneous population [92]. For example, using fluorescent antibodies specific for phosphotyrosines in combination with a library of proteins fused with a fluorescent protein, in a FRET-FLIM setup, polar plots also can be applied to determine the degree of phosphorylation in cell arrays, which allows to identify the proteins responsible for signal transduction [80].

While classically the excitation is sinusoidally modulated for lifetime measurements, however, also a train of picosecond pulses can be used for excitation. In combination with square-shaped modulation of the detector, this allows for a faster image acquisition and *a priori* measurement of biexponential decays [93, 94]. Prior to the detection, the fluorescence emission can be analyzed by a monochromator, which yields the spectral information for the sample. The combination of both data sets improves the separation of signals originating from donor and acceptor in a FRET pair and facilitates the exclusion of donor only constructs [95].

The widefield illumination used for the frequency domain approaches also facilitates the development of setups which allow three-dimensional sectioning of cell complexes or even whole tissue samples or organisms with a measurement of FRET, using techniques like single plane illumination microscopy or optical projection tomography [96–98].

Although the characterization of the structure and composition of biological membranes is hindered by the large numbers of different compounds from which they are formed, leading to a huge complexity, FLIM techniques have already helped to unravel some of these mysteries, as fluorescence lifetimes yield insights, for example, into heterogeneities of populations. However, the structure of the membranes is not only complex but also highly dynamic. This is especially demanding on FLIM approaches as image acquisition is time consuming, requiring several seconds up to minutes per frame. However, the development of novel techniques to alleviate this challenge will potentially allow the study of dynamics of processes in biological membranes in more detail.

References

1. Pike LJ (2006) Rafts defined: a report on the keystone symposium on lipid rafts and cell function. J Lipid Res 47(7):1597–1598. doi:E600002-JLR200, [pii] 10.1194/jlr.E600002-JLR200
2. Fessler MB, Parks JS (2011) Intracellular lipid flux and membrane microdomains as organizing principles in inflammatory cell signaling. J Immunol 187(4):1529–1535. doi:10.4049/jimmunol.1100253
3. Staubach S, Hanisch F-G (2011) Lipid rafts: signaling and sorting platforms of cells and their roles in cancer. Expert Rev Proteomics 8(2):263–277. doi:10.1586/epr.11.2
4. Simons K, Ikonen E (1997) Functional rafts in cell membranes. Nature 387(6633):569–572
5. Cho W, Stahelin RV (2005) Membrane-protein interactions in cell signaling and membrane trafficking. Annu Rev Biophys Biomol Struct 34(1):119–151. doi:10.1146/annurev.biophys.33.110502.133337
6. Resh MD (1999) Fatty acylation of proteins: new insights into membrane targeting of myristoylated and palmitoylated proteins. Biochim Biophys Acta 1451(1):1–16

7. Aicart-Ramos C, Valero RA, Rodriguez-Crespo I (2011) Protein palmitoylation and subcellular trafficking. Biochim Biophys Acta 1808(12):2981–2994. doi:10.1016/j.bbamem.2011.07.009
8. Levental I, Grzybek M, Simons K (2010) Greasing their way: lipid modifications determine protein association with membrane rafts. Biochemistry 49(30):6305–6316
9. Hancock JF (2006) Lipid rafts: contentious only from simplistic standpoints. Nat Rev Mol Cell Biol 7(6):456–462
10. Levental I, Byfield FJ, Chowdhury P, Gai F, Baumgart T, Janmey PA (2009) Cholesterol-dependent phase separation in cell-derived giant plasma-membrane vesicles. Biochem J 424 (2):163–167
11. Baumgart T, Hammond AT, Sengupta P, Hess ST, Holowka DA, Baird BA, Webb WW (2007) Large-scale fluid/fluid phase separation of proteins and lipids in giant plasma membrane vesicles. Proc Natl Acad Sci U S A 104(9):3165–3170
12. Kusumi A, Suzuki K (2005) Toward understanding the dynamics of membrane-raft-based molecular interactions. Biochim Biophys Acta 1746(3):234–251
13. Loura L, Fernandes F, Prieto M (2010) Membrane microheterogeneity: Förster resonance energy transfer characterization of lateral membrane domains. Eur Biophys J 39(4):589–607. doi:10.1007/s00249-009-0547-5
14. Sezgin E, Levental I, Grzybek M, Schwarzmann G, Mueller V, Honigmann A, Belov VN, Eggeling C, Coskun U, Simons K, Schwille P (2012) Partitioning, diffusion, and ligand binding of raft lipid analogs in model and cellular plasma membranes. Biochim Biophys Acta 1818:1777–1784. doi:10.1016/j.bbamem.2012.03.007
15. Eggeling C, Ringemann C, Medda R, Schwarzmann G, Sandhoff K, Polyakova S, Belov VN, Hein B, von Middendorff C, Schonle A, Hell SW (2009) Direct observation of the nanoscale dynamics of membrane lipids in a living cell. Nature 457(7233):1159–1162, http://www.nature.com/nature/journal/v457/n7233/suppinfo/nature07596_S1.html
16. Mueller V, Ringemann C, Honigmann A, Schwarzmann G, Medda R, Leutenegger M, Polyakova S, Belov VN, Hell SW, Eggeling C (2011) Sted nanoscopy reveals molecular details of cholesterol- and cytoskeleton-modulated lipid interactions in living cells. Biophys J 101(7):1651–1660. doi:10.1016/j.bpj.2011.09.006
17. Bastos AEP, Scolari S, Stöckl M, de Almeida RFM (2012) Applications of fluorescence lifetime spectroscopy and imaging to lipid domains in vivo. In: Conn PM (ed) Methods in enzymology, vol 504. Academic Press, Burlington, pp 57–81
18. Shaw AS (2006) Lipid rafts: now you see them, now you don't. Nat Immunol 7(11):1139–1142
19. Ariola FS, Li Z, Cornejo C, Bittman R, Heikal AA (2009) Membrane fluidity and lipid order in ternary giant unilamellar vesicles using a new bodipy-cholesterol derivative. Biophys J 96 (7):2696–2708. doi:S0006-3495(09)00390-7, [pii] 10.1016/j.bpj.2008.12.3922
20. Wüstner D, Solanko L, Sokol E, Garvik O, Li Z, Bittman R, Korte T, Herrmann A (2011) Quantitative assessment of sterol traffic in living cells by dual labeling with dehydroergosterol and bodipy-cholesterol. Chem Phys Lipids 164(3):221–235. doi:10.1016/j.chemphyslip.2011.01.004
21. Wustner D (2007) Fluorescent sterols as tools in membrane biophysics and cell biology. Chem Phys Lipids 146(1):1–25
22. de Almeida RF, Borst J, Fedorov A, Prieto M, Visser AJ (2007) Complexity of lipid domains and rafts in giant unilamellar vesicles revealed by combining imaging and microscopic and macroscopic time-resolved fluorescence. Biophys J 93(2):539–553
23. Ariola FS, Mudaliar DJ, Walvick RP, Heikal AA (2006) Dynamics imaging of lipid phases and lipid-marker interactions in model biomembranes. Phys Chem Chem Phys 8(39):4517–4529. doi:10.1039/b608629b
24. Didier P, Ulrich G, Mely Y, Ziessel R (2009) Improved push-pull-push e-bodipy fluorophores for two-photon cell-imaging. Org Biomol Chem 7(18)
25. Stöckl M, Plazzo AP, Korte T, Herrmann A (2008) Detection of lipid domains in model and cell membranes by fluorescence lifetime imaging microscopy of fluorescent lipid analogues. J Biol Chem 283(45):30828–30837

26. Zarubica A, Plazzo AP, Stockl M, Trombik T, Hamon Y, Muller P, Pomorski T, Herrmann A, Chimini G (2009) Functional implications of the influence of abca1 on lipid microenvironment at the plasma membrane: a biophysical study. FASEB J 23(6):1775–1785
27. Haluska CK, Schroder AP, Didier P, Heissler D, Duportail G, Mely Y, Marques CM (2008) Combining fluorescence lifetime and polarization microscopy to discriminate phase separated domains in giant unilamellar vesicles. Biophys J 95(12):5737–5747. doi:S0006-3495(08) 81990-X, [pii] 10.1529/biophysj.108.131490
28. Margineanu A, Hotta J, Vallee RA, Van der Auweraer M, Ameloot M, Stefan A, Beljonne D, Engelborghs Y, Herrmann A, Mullen K, De Schryver FC, Hofkens J (2007) Visualization of membrane rafts using a perylene monoimide derivative and fluorescence lifetime imaging. Biophys J 93(8):2877–2891
29. Li C, Schoneboom J, Liu Z, Pschirer NG, Erk P, Herrmann A, Mullen K (2009) Rainbow perylene monoimides: easy control of optical properties. Chemistry 15(4):878–884. doi:10.1002/chem.200802126
30. Obaid AL, Loew LM, Wuskell JP, Salzberg BM (2004) Novel naphthylstyryl-pyridium potentiometric dyes offer advantages for neural network analysis. J Neurosci Methods 134 (2):179–190. doi:10.1016/j.jneumeth.2003.11.011, S0165027003003881 [pii]
31. Jin L, Millard AC, Wuskell JP, Clark HA, Loew LM (2005) Cholesterol-enriched lipid domains can be visualized by di-4-aneppdhq with linear and nonlinear optics. Biophys J 89 (1):L04–L06. doi:S0006-3495(05)72649-7, [pii] 10.1529/biophysj.105.064816
32. Owen DM, Lanigan PM, Dunsby C, Munro I, Grant D, Neil MA, French PM, Magee AI (2006) Fluorescence lifetime imaging provides enhanced contrast when imaging the phase-sensitive dye di-4-aneppdhq in model membranes and live cells. Biophys J 90(11):L80–L82
33. Talbot CB, McGinty J, Grant DM, McGhee EJ, Owen DM, Zhang W, Bunney TD, Munro I, Isherwood B, Eagle R, Hargreaves A, Dunsby C, Neil MA, French PM (2008) High speed unsupervised fluorescence lifetime imaging confocal multiwell plate reader for high content analysis. J Biophotonics 1(6):514–521. doi:10.1002/jbio.200810054
34. Svensson FR, Matson M, Li M, Lincoln P (2010) Lipophilic ruthenium complexes with tuned cell membrane affinity and photoactivated uptake. Biophys Chem 149(3):102–106. doi:10.1016/j.bpc.2010.04.006
35. Svensson FR, Li M, Nordén B, Lincoln P (2008) Luminescent dipyridophenazine-ruthenium probes for liposome membranes. J Phys Chem B 112(35):10969–10975. doi:10.1021/jp803964x
36. Svensson FR, Abrahamsson M, Strömberg N, Ewing AG, Lincoln P (2011) Ruthenium(ii) complex enantiomers as cellular probes for diastereomeric interactions in confocal and fluorescence lifetime imaging microscopy. J Phys Chem Lett 2(5):397–401. doi:10.1021/jz101580e
37. Shang L, Azadfar N, Stockmar F, Send W, Trouillet V, Bruns M, Gerthsen D, Nienhaus GU (2011) One-pot synthesis of near-infrared fluorescent gold clusters for cellular fluorescence lifetime imaging. Small 7(18):2614–2620. doi:10.1002/smll.201100746
38. Davey AM, Krise KM, Sheets ED, Heikal AA (2008) Molecular perspective of antigen-mediated mast cell signaling. J Biol Chem 283(11):7117–7127. doi:M708879200, [pii] 10.1074/jbc.M708879200
39. Davey AM, Walvick RP, Liu Y, Heikal AA, Sheets ED (2007) Membrane order and molecular dynamics associated with IgE receptor cross-linking in mast cells. Biophys J 92(1):343–355. doi:S0006-3495(07)70832-9, [pii] 10.1529/biophysj.106.088815
40. Luby-Phelps K (2000) Cytoarchitecture and physical properties of cytoplasm: volume, viscosity, diffusion, intracellular surface area. Int Rev Cytol 192:189–221
41. Kapitulnik J, Weil E, Rabinowitz R, Krausz MM (1987) Fetal and adult human liver differ markedly in the fluidity and lipid composition of their microsomal membranes. Hepatology 7 (1):55–60. doi:S0270913987000259 [pii]
42. Emmerson PJ, Clark MJ, Medzihradsky F, Remmers AE (1999) Membrane microviscosity modulates mu-opioid receptor conformational transitions and agonist efficacy. J Neurochem 73(1):289–300

43. Thorin E, Hamilton C, Dominiczak AF, Dominiczak MH, Reid JL (1995) Oxidized-ldl induced changes in membrane physico-chemical properties and [ca2+]i of bovine aortic endothelial cells. Influence of vitamin e. Atherosclerosis 114(2):185–195. doi:002191509405482X [pii]
44. Saldanha C, Sargento L, Monteiro J, Perdigao C, Ribeiro C, Martins-Silva J (1999) Impairment of the erythrocyte membrane fluidity in survivors of acute myocardial infarction. A prospective study. Clin Hemorheol Microcirc 20(2):111–116
45. Maksimov GV, Luneva OG, Maksimova NV, Matettuchi E, Medvedev EA, Pashchenko VZ, Rubin AB (2005) Role of viscosity and permeability of the erythrocyte plasma membrane in changes in oxygen-binding properties of hemoglobin during diabetes mellitus. Bull Exp Biol Med 140(5):510–513
46. Yang X, Sheng W, He Y, Cui J, Haidekker MA, Sun GY, Lee JC (2010) Secretory phospholipase a2 type iii enhances alpha-secretase-dependent amyloid precursor protein processing through alterations in membrane fluidity. J Lipid Res 51(5):957–966. doi:jlr.M002287, [pii] 10.1194/jlr.M002287
47. Baumann M (2003) Cell ageing for 1 day alters both membrane elasticity and viscosity. Pflugers Arch 445(5):551–555. doi:10.1007/s00424-002-0979-6
48. Haidekker MA, Theodorakis EA (2010) Environment-sensitive behavior of fluorescent molecular rotors. J Biol Eng 4:11. doi:1754-1611-4-11, [pii] 10.1186/1754-1611-4-11
49. Haidekker MA, Brady TP, Lichlyter D, Theodorakis EA (2005) Effects of solvent polarity and solvent viscosity on the fluorescent properties of molecular rotors and related probes. Bioorg Chem 33(6):415–425. doi:S0045-2068(05)00076-3, [pii] 10.1016/j.bioorg.2005.07.005
50. Haidekker MA, Theodorakis EA (2007) Molecular rotors–fluorescent biosensors for viscosity and flow. Org Biomol Chem 5(11):1669–1678. doi:10.1039/b618415d
51. Zhou F, Shao J, Yang Y, Zhao J, Guo H, Li X, Ji S, Zhang Z (2011) Molecular rotors as fluorescent viscosity sensors: molecular design, polarity sensitivity, dipole moments changes, screening solvents, and deactivation channel of the excited states. Eur J Organ Chem 2011 (25):4773–4787. doi:10.1002/ejoc.201100606
52. Haidekker MA, L'Heureux N, Frangos JA (2000) Fluid shear stress increases membrane fluidity in endothelial cells: a study with DCVJ fluorescence. Am J Physiol Heart Circ Physiol 278(4):H1401–H1406
53. Ando J, Yamamoto K (2009) Vascular mechanobiology: endothelial cell responses to fluid shear stress. Circ J 73(11):1983–1992. doi:JST.JSTAGE/circj/CJ-09-0583 [pii]
54. Haidekker MA, Ling T, Anglo M, Stevens HY, Frangos JA, Theodorakis EA (2001) New fluorescent probes for the measurement of cell membrane viscosity. Chem Biol 8(2):123–131. doi:S1074-5521(00)90061-9 [pii]
55. Nipper ME, Majd S, Mayer M, Lee JC, Theodorakis EA, Haidekker MA (2008) Characterization of changes in the viscosity of lipid membranes with the molecular rotor FCVJ. Biochim Biophys Acta 1778(4):1148–1153. doi:S0005-2736(08)00021-7, [pii] 10.1016/j.bbamem.2008.01.005
56. Haidekker MA, Brady T, Wen K, Okada C, Stevens HY, Snell JM, Frangos JA, Theodorakis EA (2002) Phospholipid-bound molecular rotors: synthesis and characterization. Bioorg Med Chem 10(11):3627–3636. doi:S0968089602002407 [pii]
57. Haidekker MA, Brady TP, Lichlyter D, Theodorakis EA (2006) A ratiometric fluorescent viscosity sensor. J Am Chem Soc 128(2):398–399. doi:10.1021/ja056370a
58. Nipper ME, Dakanali M, Theodorakis E, Haidekker MA (2011) Detection of liposome membrane viscosity perturbations with ratiometric molecular rotors. Biochimie 93 (6):988–994. doi:S0300-9084(11)00058-7, [pii] 10.1016/j.biochi.2011.02.002
59. Peng X, Yang Z, Wang J, Fan J, He Y, Song F, Wang B, Sun S, Qu J, Qi J, Yan M (2011) Fluorescence ratiometry and fluorescence lifetime imaging: using a single molecular sensor for dual mode imaging of cellular viscosity. J Am Chem Soc 133(17):6626–6635. doi:10.1021/ja1104014
60. Kuimova MK, Botchway SW, Parker AW, Balaz M, Collins HA, Anderson HL, Suhling K, Ogilby PR (2009) Imaging intracellular viscosity of a single cell during photoinduced cell death. Nat Chem 1(1):69–73. doi:nchem.120, [pii] 10.1038/nchem.120

61. Kuimova MK, Yahioglu G, Levitt JA, Suhling K (2008) Molecular rotor measures viscosity of live cells via fluorescence lifetime imaging. J Am Chem Soc 130(21):6672–6673. doi:10.1021/ja800570d
62. Baumgart T, Hunt G, Farkas ER, Webb WW, Feigenson GW (2007) Fluorescence probe partitioning between lo/ld phases in lipid membranes. Biochim Biophys Acta 1768(9): 2182–2194
63. Sengupta P, Hammond A, Holowka D, Baird B (2008) Structural determinants for partitioning of lipids and proteins between coexisting fluid phases in giant plasma membrane vesicles. Biochim Biophys Acta 1778(1):20–32
64. Konig I, Schwarz JP, Anderson KI (2008) Fluorescence lifetime imaging: association of cortical actin with a pip3-rich membrane compartment. Eur J Cell Biol 87(8–9):735–741. doi:S0171-9335(08)00030-7, [pii] 10.1016/j.ejcb.2008.02.002
65. Scolari S, Engel S, Krebs N, Plazzo AP, De Almeida RF, Prieto M, Veit M, Herrmann A (2009) Lateral distribution of the transmembrane domain of influenza virus hemagglutinin revealed by time-resolved fluorescence imaging. J Biol Chem 284(23):15708–15716. doi: M900437200, [pii] 10.1074/jbc.M900437200
66. Engel S, Scolari S, Thaa B, Krebs N, Korte T, Herrmann A, Veit M (2010) Flim-fret and frap reveal association of influenza virus haemagglutinin with membrane rafts. Biochem J 425 (3):567–573. doi:10.1042/bj20091388
67. Thaa B, Herrmann A, Veit M (2010) Intrinsic cytoskeleton-dependent clustering of influenza virus m2 protein with hemagglutinin assessed by flim-fret. J Virol 84(23):12445–12449. doi:10.1128/jvi.01322-10
68. Valkonen M, Kalkman ER, Saloheimo M, Penttila M, Read ND, Duncan RR (2007) Spatially segregated snare protein interactions in living fungal cells. J Biol Chem 282(31):22775–22785
69. Zelazny E, Borst JW, Muylaert M, Batoko H, Hemminga MA, Chaumont F (2007) Fret imaging in living maize cells reveals that plasma membrane aquaporins interact to regulate their subcellular localization. Proc Natl Acad Sci U S A 104(30):12359–12364. doi:0701180104, [pii] 10.1073/pnas.0701180104
70. Willemse J, Borst JW, de Waal E, Bisseling T, van Wezel GP (2011) Positive control of cell division: Ftsz is recruited by ssgb during sporulation of streptomyces. Genes Dev 25(1):89–99. doi:10.1101/gad.600211
71. Osterrieder A, Carvalho CM, Latijnhouwers M, Johansen JN, Stubbs C, Botchway S, Hawes C (2009) Fluorescence lifetime imaging of interactions between golgi tethering factors and small gtpases in plants. Traffic 10(8):1034–1046. doi:10.1111/j.1600-0854.2009.00930.x
72. Herl L, Thomas AV, Lill CM, Banks M, Deng A, Jones PB, Spoelgen R, Hyman BT, Berezovska O (2009) Mutations in amyloid precursor protein affect its interactions with presenilin/gamma-secretase. Mol Cell Neurosci 41(2):166–174. doi:S1044-7431(09)00044-X, [pii] 10.1016/j.mcn.2009.02.008
73. Uemura K, Farner K, Nasser-Ghodsi N, Jones P, Berezovska O (2011) Reciprocal relationship between app positioning relative to the membrane and ps1 conformation. Mol Neurodegener 6(1):15
74. Martin-Villar E, Fernandez-Munoz B, Parsons M, Yurrita MM, Megias D, Perez-Gomez E, Jones GE, Quintanilla M (2010) Podoplanin associates with cd44 to promote directional cell migration. Mol Biol Cell 21(24):4387–4399. doi:E10-06-0489, [pii] 10.1091/mbc.E10-06-0489
75. Nievergall E, Janes PW, Stegmayer C, Vail ME, Haj FG, Teng SW, Neel BG, Bastiaens PI, Lackmann M (2010) Ptp1b regulates eph receptor function and trafficking. J Cell Biol 191 (6):1189–1203. doi:jcb.201005035, [pii] 10.1083/jcb.201005035
76. Bu W, Lim KB, Yu YH, Chou AM, Sudhaharan T, Ahmed S (2010) Cdc42 interaction with n-wasp and toca-1 regulates membrane tubulation, vesicle formation and vesicle motility: implications for endocytosis. PLoS One 5(8):e12153. doi:e12153, [pii] 10.1371/journal.pone.0012153

77. Dumas F, Byrne RD, Vincent B, Hobday TM, Poccia DL, Larijani B (2010) Spatial regulation of membrane fusion controlled by modification of phosphoinositides. PLoS One 5(8):e12208. doi:10.1371/journal.pone.0012208
78. Fjorback AW, Pla P, Muller HK, Wiborg O, Saudou F, Nyengaard JR (2009) Serotonin transporter oligomerization documented in rn46a cells and neurons by sensitized acceptor emission fret and fluorescence lifetime imaging microscopy. Biochem Biophys Res Commun 380(4):724–728. doi:S0006-291X(09)00108-9, [pii] 10.1016/j.bbrc.2009.01.128
79. Nyborg AC, Herl L, Berezovska O, Thomas AV, Ladd TB, Jansen K, Hyman BT, Golde TE (2006) Signal peptide peptidase (spp) dimer formation as assessed by fluorescence lifetime imaging microscopy (flim) in intact cells. Mol Neurodegener 1:16. doi:1750-1326-1-16, [pii] 10.1186/1750-1326-1-16
80. Grecco HE, Roda-Navarro P, Girod A, Hou J, Frahm T, Truxius DC, Pepperkok R, Squire A, Bastiaens PI (2010) In situ analysis of tyrosine phosphorylation networks by flim on cell arrays. Nat Methods 7(6):467–472. doi:nmeth.1458, [pii] 10.1038/nmeth.1458
81. Trinel D, Leray A, Spriet C, Usson Y, Héliot L (2011) Upgrading time domain flim using an adaptive Monte Carlo data inflation algorithm. Cytometry A 79A(7):528–537. doi:10.1002/cyto.a.21054
82. Digman MA, Caiolfa VR, Zamai M, Gratton E (2008) The phasor approach to fluorescence lifetime imaging analysis. Biophys J 94(2):L14–L16
83. Leray A, Spriet C, Trinel D, Blossey R, Usson Y, Heliot L (2011) Quantitative comparison of polar approach versus fitting method in time domain flim image analysis. Cytometry A 79 (2):149–158. doi:10.1002/cyto.a.20996
84. Jameson DM, Gratton E, Hall RD (1984) The measurement and analysis of heterogeneous emissions by multifrequency phase and modulation fluorometry. Appl Spectrosc Rev 20 (1):55–106. doi:10.1080/05704928408081716
85. Clayton AH, Hanley QS, Verveer PJ (2004) Graphical representation and multicomponent analysis of single-frequency fluorescence lifetime imaging microscopy data. J Microsc 213(Pt 1):1–5. doi:1265 [pii]
86. Redford GI, Clegg RM (2005) Polar plot representation for frequency-domain analysis of fluorescence lifetimes. J Fluoresc 15(5):805–815. doi:10.1007/s10895-005-2990-8
87. Chang CW, Mycek MA (2010) Enhancing precision in time-domain fluorescence lifetime imaging. J Biomed Opt 15(5):056013. doi:10.1117/1.3494566
88. Slepkov AD, Ridsdale A, Wan HN, Wang MH, Pegoraro AF, Moffatt DJ, Pezacki JP, Kao FJ, Stolow A (2011) Forward-collected simultaneous fluorescence lifetime imaging and coherent anti-stokes Raman scattering microscopy. J Biomed Opt 16(2):021103. doi:10.1117/1.3490641
89. Won YJ, Han WT, Kim DY (2011) Precision and accuracy of the analog mean-delay method for high-speed fluorescence lifetime measurement. J Opt Soc Am A Opt Image Sci Vis 28 (10):2026–2032. doi:10.1364/JOSAA.28.002026, 222391 [pii]
90. Moon S, Won Y, Kim DY (2009) Analog mean-delay method for high-speed fluorescence lifetime measurement. Opt Express 17(4):2834–2849. doi:176573 [pii]
91. Won Y, Moon S, Yang W, Kim D, Han WT, Kim DY (2011) High-speed confocal fluorescence lifetime imaging microscopy (flim) with the analog mean delay (amd) method. Opt Express 19(4):3396–3405. doi:10.1364/OE.19.003396, 209963 [pii]
92. Hinde E, Digman MA, Welch C, Hahn KM, Gratton E (2011) Biosensor Forster resonance energy transfer detection by the phasor approach to fluorescence lifetime imaging microscopy. Microsc Res Tech. doi:10.1002/jemt.21054
93. Elder AD, Kaminski CF, Frank JH (2009) Phi2flim: a technique for alias-free frequency domain fluorescence lifetime imaging. Opt Express 17(25):23181–23203. doi:10.1364/OE.17.023181, 192319 [pii]
94. Schlachter S, Elder AD, Esposito A, Kaminski GS, Frank JH, van Geest LK, Kaminski CF (2009) Mhflim: resolution of heterogeneous fluorescence decays in widefield lifetime microscopy. Opt Express 17(3):1557–1570. doi:176133 [pii]

95. Chen YC, Clegg RM (2011) Spectral resolution in conjunction with polar plots improves the accuracy and reliability of flim measurements and estimates of fret efficiency. J Microsc 244 (1):21–37. doi:10.1111/j.1365-2818.2011.03488.x
96. Greger K, Neetz MJ, Reynaud EG, Stelzer EH (2011) Three-dimensional fluorescence lifetime imaging with a single plane illumination microscope provides an improved signal to noise ratio. Opt Express 19(21):20743–20750. doi:10.1364/OE.19.020743, 222944 [pii]
97. McGinty J, Stuckey DW, Soloviev VY, Laine R, Wylezinska-Arridge M, Wells DJ, Arridge SR, French PM, Hajnal JV, Sardini A (2011) In vivo fluorescence lifetime tomography of a fret probe expressed in mouse. Biomed Opt Express 2(7):1907–1917. doi:10.1364/BOE.2.001907, 146182 [pii]
98. McGinty J, Taylor HB, Chen L, Bugeon L, Lamb JR, Dallman MJ, French PM (2011) In vivo fluorescence lifetime optical projection tomography. Biomed Opt Express 2(5):1340–1350. doi:10.1364/BOE.2.001340, 14367 [pii]

Fluorescence Correlation Spectroscopy to Study Membrane Organization and Interactions

Monika Zelman-Femiak, Yamunadevi Subburaj, and Ana J. García-Sáez

Abstract This chapter describes the application of fluorescence correlation spectroscopy (FCS) as a powerful technique for the study of membrane organization and interactions. Monitoring the fluorescence signal fluctuations allows resolving concentrations, diffusion coefficients, and binding of several membrane components in experiments *in vitro* as well as *in vivo*.

We discuss the basic principles of FCS and explain novel implementations of FCS introduced to overcome the technical difficulties present in the standard version of fluorescence correlation spectroscopy. Finally, we report several examples of studies with the application of FCS on both model and biological membranes to obtain interesting insight in the topic of lateral membrane organization and membrane interactions.

Keywords Chemical relaxation · Diffusion · Fluctuations · Fluorescence · Fluorescence correlation spectroscopy · Interactions · Membranes · Triplet state

Contents

M. Zelman-Femiak, Y. Subburaj and A.J. García-Sáez (✉)
Max Planck Institute for Intelligent Systems and German Cancer Research Center, Bioquant, Im Neuenheimer Feld 267, Heidelberg 69120, Germany
e-mail: ana.garcia@mf.mpg.de

Y. Mély and G. Duportail (eds.), *Fluorescent Methods to Study Biological Membranes*, Springer Ser Fluoresc (2013) 13: 241–270, DOI 10.1007/4243_2012_49,

1 Introduction

Cellular membrane biology attracts currently intense scientific interest and research. Comprehending the complex organization of the membrane together with decoding the function of its particular constituents could bring us much closer to decipher the subsistence of a cell, a fundamental unit of life.

Many questions, however, remain still unanswered including the structure of membrane proteins that consist of small amount of all resolved structures, the mechanisms piloting membrane organization and curvature, the mechanisms of organelle biogenesis and the transport between organelles, and how membranes undergo fusion and fission. It has already been shown in several studies that membrane proteins can serve as receptors, ion channels, or transporters, while the lipids provide not only a barrier between aqueous compartments but also act as signals. It is commonly known that lipids and proteins blend together to constitute and maintain membrane composition and specific membrane domains; however, the precept of molecular intertwining has not been fully understood yet.

To gain deeper insight into membrane organization and the rules governing it, as well as molecular interactions within membranes, very sensitive techniques are emerging. Precise methods which would allow for the characterization of molecular and cellular dynamics and which would deliver pivotal parameters like local concentrations, mobility coefficients, and characteristic rate constants of inter- or intramolecular reactions and through this could address so far unresolved issues.

Fluorescence correlation spectroscopy (FCS) is one of those powerful techniques that provide high spatiotemporal resolution and detection of biomolecules at extremely low concentrations. Evolved in the early 1970s as a special case of relaxation analysis, fluorescence correlation spectroscopy bases on the minuscule spontaneous fluctuations of the signal detected from fluorophores in a very small volume (fL). FCS measures intensity fluctuations originating from any deviations of the observed specimen like those occurring due to diffusion of the molecules, physical or chemical reactions, or aggregation. Such fluctuations continuously exist at ambient temperatures and are depicted as noise patterns of the measured fluorescence. Intensity fluctuations are then analyzed by applying the temporal autocorrelation of the recorded intensity signal. The strength and duration of the fluctuations are juxtaposed in flow of time, providing the image of a self-similarity of the signal and therefore describing the endurance of information acquired from the autocorrelation. In a straightforward way, this leads to the delivery of crucial information regarding the processes that govern molecular dynamics and the determination of local concentrations, mobility coefficients, and reaction rate constants of fluorescently labeled biomolecules.

Signal-correlation techniques were first experimentally applied to fluorescence in 1972 by Magde, Elson, and Webb [1]. The pioneering method was further developed in several publications by various research groups demonstrating its theoretical basis and applications, like attempts to determine particle concentration and rotational and translational mobility in two or three dimensions and even in the cellular context [2–5]. Nevertheless, these early experiments endure poor signal-to-noise ratios, chiefly due to low detection efficiency, oversized ensemble, and insufficient background suppression.

Beginning in 1993, numerous improvements in the measurement techniques, especially the usage of confocal microscopy and then two-photon microscopy, to better define the measured volume and deselect background significantly enhanced the signal-to-noise ratio and reached single-molecule sensitivity [6–8]. This aroused sudden avid interest in FCS and resulted in over 3,000 papers with the application of FCS to be found in Web of Science by August 2007 [9] and since that time almost doubled, resulting in over 5,600 publications on the topic by March 2012. Additionally, there has been a flurry of activity extending FCS in different ways, for instance, to laser scanning and spinning-disk confocal microscopy, in applying cross correlation (FCCS) between two fluorescent channels instead of autocorrelation and in using Förster resonance energy transfer (FRET) instead of fluorescence.

In any subtype of FCS, the basic concept is to achieve and maintain the number of observed molecules so low that each particular contributes essentially to the measured signal. Only under such conditions credible analyses of spontaneous, noncoordinated fluctuations can be conducted.

FCS measurements function properly only if the concentration of the studied sample is restricted to very few molecules that are detected in the focal volume at the same time, which necessitates the fluorescence photon yield per single molecule. Enhancement of FCS could be provided by the usage of efficient fluorescent dyes for molecule labeling, strong and stable light sources like lasers, and ultrasensitive detectors with single-photon sensitivity. The final breakthrough combined the FCS technique with confocal detection causing that the incoming laser light is strongly focused by a high numerical aperture objective to a diffraction-limited spot [10]. Solely the few fluorophores within the illuminated region of interest are detected after excitation. To reduce the detection volume in axial direction, a pinhole has been introduced in the image plane blocking all light not originating from the femtoliter-sized focal volume.

Most common FCS measurements are conducted on fluorescently labeled molecules diffusing in aqueous buffer solution into and out of the focal volume and delivering the signal fluctuations large enough to yield good signal-to-noise ratios. Obviously, in addition to the time a molecule spends in the focus, the alteration of the emission characteristics of the fluorophore due to chemical or photophysical reactions or conformational changes introduces the additional fluctuations in the detected signal.

Although FCS is an established technique to study diffusion in solution, several additional problems connected with slow dynamics and two-dimensional geometry strongly limit the accuracy of confocal FCS on membranes [11].

In confocal FCS, the detection volume has to be positioned in the membrane with a vertical accuracy of approximately 100 nm. Otherwise, the divergence of the laser would cause an unwanted enlargement of the effective detection area. Additionally, various optical artifacts (e.g., saturation, cover-slide thickness local variations, refractive index mismatch, astigmatism [12]) lead to deformation of the focal volume, impeding the exact determination of the detection area, which is necessary for quantitative measurements. Therefore, FCS techniques that do not rely on the calibration of the detection area are recommended for quantitative and reliable measurements on membranes.

The diffusion in lipid membranes can be several orders of magnitude slower than that of fluorophores freely diffusing in solution. To average over a significant number of independent events, appropriately long measurement times, at least 10^4 times the diffusion time, are necessary [11]. During this time, the position of the detection volume with respect to the membrane has to stay stable; otherwise, additional changes in the intensity trace are introduced with consequent distortion of the autocorrelation curve. Even minor instabilities might result in an apparent second component or non-Brownian diffusion. To measure slow diffusion, one can either use techniques that allow the correction for instabilities [13] or techniques where parallel acquisition decreases the measurement times [14,15].

Another consequence of the slow diffusion observed in membranes is the long residence time of the fluorophores in the detection volume. This causes strong photobleaching, leading to an apparent reduction of the measured diffusion times. Therefore, low laser powers have to be applied for membrane FCS. Unfortunately, low excitation power produces a weak signal, easily concealed by stray light or detector dark counts, and thus requires long measurement times.

On membranes, the two-dimensional geometry makes it difficult for an efficient replenishment of bleached fluorophores, which are depleted in the detection area. Initially, the intensity trace quickly decays with time, resulting in distorted correlation curves. Starting the measurement after the system has reached a quasi-steady state lets one to omit distortions but causes an underestimation of the concentration. Depletion is especially problematic if the reservoir of fluorophores is limited and a quasi-steady state cannot be reached. This is the case for small domains in phase-separating membranes or for closed bilayer and cell membranes. Depletion due to photobleaching can be corrected at the level of the intensity trace [14].

Even moderate excitation laser powers can induce nonradiative energy dissipation phenomena, like electrochemical alteration of the lipids or heat production. This may also lead to significant changes of the physical properties of the membrane [16].

In many experiments, a fluorescent background in the solution around the membrane cannot be avoided. Measurements in cellular membranes are precluded by intracellular fluorescence and, especially, by fluorescent vesicles in the cytosol, which have a similar diffusion time compared to molecules diffusing in the more viscous membrane. When studying a molecule with a low binding affinity to the membrane, the high concentration in solution required for significant binding can completely mask the signal originating in the membrane. To reduce the fluorescent background, very flat detection volumes that are provided by FCS with total internal reflection excitation (TIRF-FCS) [17] or with FCS using a supercritical angle objective [18] have to be employed.

Even more problematic than FCS measurements in the membranes are the ones carried out in cellular processes that confine the detection volume and the standard FCS model leads to erroneous results. In cells, due to the complicated microenvironment, larger viscosity, molecular crowding effects, and multiple interaction partners, the standard FCS approach is not very effective. Here, dual-color cross correlation spectroscopy (FCCS) is superior, provided that spectral cross talk and global movements of the cellular structures are minimized.

In this chapter, after describing the basic principles of FCS, we present and discuss recent work involving practical application of FCS for the study of model and biological membranes in the context of complex dynamics observed for membrane components in both model systems and living cells.

2 Fluorescence Correlation Spectroscopy: Brief Method Description

2.1 Typical FCS Setup

The typical FCS setup consists of a laser line with wavelengths usually ranging from 405 to 633 nm (continuous wave) or from 690 to 1,100 nm (pulsed), which is reflected into a microscope objective by a dichroic mirror. The laser beam is focused in the specimen containing fluorescent molecules in such an extremely high dilution that only a few are found within the focal spot. Usually the number of fluorescently tagged particles is between 1 and 100 per volume of 1 fL. When the molecules diffuse across the focal volume, they fluoresce. This light is collected by the same objective, and because it is red shifted with respect to the excitation light, it passes the dichroic mirror reaching a detector, typically an avalanche photodiode detector. The resulting electronic signal can be stored either directly as the intensity versus time trace to be analyzed at a later point or computed to generate the autocorrelation directly, which necessitates an online correlator. The FCS or autocorrelation curve by itself only demonstrates fluctuations in the fluorescence intensity signal originating from the labeled molecules over time, and application of appropriate models to these data leads to further conclusions on physical phenomena and interpretation of the experiment. The parameters of interest are found after fitting the autocorrelation curve to modeled functional forms [19].

2.2 Autocorrelation Function

The temporal autocorrelation function is the correlation of a signal in a time series with itself shifted by time τ, as a function of τ:

$$G(\tau) = \frac{\langle \delta I(t)\delta I(t+\tau)\rangle}{\langle I(t)\rangle^2} = \frac{\langle I(t)I(t+\tau)\rangle}{\langle I(t)\rangle^2} - 1 \tag{1}$$

where

$$\delta I(t) = I(t) - \langle I(t)\rangle \tag{2}$$

is the deviation from the mean intensity. The normalization here is the most commonly used for FCS, because then the correlation at $\tau = 0$, called $G(0)$, is related to the average number of particles in the measurement volume.

To extract the values for the parameters of interest, the autocorrelation data can be fitted, most commonly using a nonlinear least-squares algorithm. The functional form of the fit depends on the type of dynamics and the optical geometry in question.

2.3 The Measurement Volume

The measurement volume depends on illumination (excitation) and detection geometries, which results from the optical elements that are involved. The focal volume is described mathematically by the point spread function (PSF), which essentially is the image of a point source. The PSF is often presented as an ellipsoid with diffuse boundaries of few hundred nanometers in focus diameter and almost 1 μm along the optical axis. The shape of the PSF differs significantly depending on the quality of the optical elements of the microscopic system and has a large impact on the shape of FCS curves. It is crucial that, before acquiring any data, astigmatism is avoided and the real shape of the PSF is checked on the particular instrument. In the case of confocal microscopy, and for small pinholes like those around one airy unit, the PSF is well approximated by Gaussians:

$$PSF(r,z) = I_0 e^{-2r^2/\omega_{xy}^2} e^{-2z^2/\omega_z^2} \tag{3}$$

where I_0 is the peak intensity, r and z are the radial and axial positions, and ω_{xy} and ω_z are the radial and axial radii and $\omega_z > \omega_{xy}$. This Gaussian form is assumed in deriving the mathematical model to fit the autocorrelation curve.

Typically, ω_{xy} is 200–300 nm, and ω_z is 2–6 times larger [20]. One common way of calibrating the measurement volume parameters is to perform FCS on a species with known diffusion coefficient under the same experimental conditions of the measurements. The Gaussian approximation works to varying degrees depending on the optical details, and corrections can sometimes be applied to offset the errors in approximation [21].

2.4 Diffusion

2.4.1 Normal Diffusion in 3D

The experimental autocorrelation curve is fitted with a mathematical model function to extract the parameters of interest, such as the diffusion coefficient D or the concentration C. For measurements in solution, the detection volume is usually approximated by a three-dimensional Gaussian profile. The correlation function describing three-dimensional Brownian diffusion through such a profile is

$$G(\tau) = G(0)\frac{1}{(1+(\tau/\tau_D))(1+a^{-2}(\tau/\tau_D))^{1/2}} + G(\infty) \tag{4}$$

where $a = \omega_z/\omega_{xy}$ is the ratio of axial to radial e^{-2} radii of the measurement volume and τ_D is the characteristic residence time. This form was derived assuming a Gaussian measurement volume. Typically, the fit would have three free parameters: G(0), G(∞), and τ_D – from which the diffusion coefficient and fluorophore concentration can be calculated.

With the normalization used in the previous section, $G(0)$ gives the mean number of diffusing molecules in the volume $< N >$ or equivalently with knowledge of the observation volume size – the mean concentration:

$$G(0) = \frac{1}{\langle N \rangle} = \frac{1}{V_{\text{eff}}\langle C \rangle}, \tag{5}$$

where the effective volume corresponds to the detectable emission intensity distribution determined by the Gaussian laser beam profile and the pinhole used [6] and is given by

$$V_{\text{eff}} = \pi^{3/2}\omega_{xy}^2\omega_z. \tag{6}$$

τ_D gives the diffusion coefficient:

$$D = \omega_{xy}^2/4\tau_D \tag{7}$$

where ω_{xy} is the radius of confocal detection volume. The diffusion coefficient is in turn determined by two properties: the viscosity of the sample η and the hydrodynamic radius of the particle R_h:

$$D = \frac{k_B T}{6\pi\eta R_h} \tag{8}$$

where k_b is the Boltzmann constant and T is temperature.

2.4.2 Lateral Diffusion in Membranes

For measurements on membranes, the intersection of the laser with the membrane defines the two-dimensional Gaussian detection area. The corresponding correlation function is

$$G(\tau) = G(0)\,\frac{1}{(1 + (\tau/\tau_D))} \tag{9}$$

where $G(0)$ gives the mean number of diffusing molecules in the volume $< N >$ like previously described in Eq. 5.

The characterization of diffusion processes in biological systems is crucial to the understanding of the molecular interactions underlying membrane organization. Nevertheless, the information provided by studies in model membranes has always varied from the diffusion in the plasma membrane, which seems to be orders of magnitude slower than that observed in simple artificial bilayers [22,23]. Furthermore, the mobility of components in the plasma membrane shows a deviation from the standard diffusion equation, which would imply a linear growth of the mean square displacement (MSD) with time. In this case, the MSD is often found to vary with a fractional power of time (i.e., MSDμt^{α}, $0 < \alpha < 1$), and the exponent α is called anomalous exponent [24,25]. This phenomenon, referred to as anomalous diffusion or subdiffusion [26], was observed experimentally in SPT [27,28] and FRAP [29,30] measurements. More in detail, the work by Feder et al. [31] presented a direct connection between the immobile fraction reported in recovery experiments [32] and the anomaly in lipid diffusive behavior. In the context of the debate regarding the physical origins of anomalous diffusion in the plasma membrane, Nicolau et al. [33] have contributed with a study worthy of note. Using Monte Carlo simulations, three different possible sources for anomalous dynamics were analyzed: immobile randomly distributed obstacles [34], raft-like domains [27], and regular networks of immobile obstacles (i.e., cytoskeleton-anchored picket model) [23]. Although not explicitly considering the possibility of direct interactions with the underlying cytoskeleton (i.e., not mediated by membrane components; "membrane-skeleton fence model" for large proteins [22]), this work probes a wide spectrum of possibilities. Nicolau et al. concluded that tightly packed lipid domains and fixed random obstacles, in absence of other interactions, are the mechanisms most likely inducing anomalous diffusion and values of α significantly lower than unity. For these reasons, it appears clear that free-standing model membranes might not be the best choice to gain further insight into the complex dynamics characterizing the cellular membranes. Rather, direct *in vivo* measurements would be more appropriate, but unfortunately, the precise evaluation of the anomalous diffusion parameter α by means of FCS measurements on cellular membranes might be very challenging [24]. Although a more complex mathematical analysis may result very useful [35], the suboptimal experimental conditions often encountered in cellular measurements do not allow ruling out simple Brownian diffusion models [11]. For this reason, it is possible to find in the scientific literature sound works reporting a wide variety of diffusive behavior for plasma membrane components, ranging from anomalous diffusion of Golgi resident membrane proteins [36] or myelin oligodendrocyte glycoproteins [36] to simple Brownian diffusion of lipids [37,38], integral membrane proteins [13], and GPI-anchored proteins [14]. A very promising alternative approach that uses FCS to probe the details of membrane organization is that proposed by Wawrezinieck et al. [39]. Monitoring the change in diffusion time τ of fluorescent lipids or proteins as function of FCS detection area A, it is possible to characterize the submicroscopical heterogeneities of the bilayer. For free Brownian diffusion, the diffusion time is simply proportional to the detection area, while for hindered diffusion with obstacles much smaller than optical resolution, an offset τ_0 is needed to fit the experimental data points:

$$\tau = \text{Const}^{*}\text{A} + \tau_0 \tag{10}$$

Whether this offset is positive or negative depends on the exact diffusion mode: dynamic partition into microdomains results in positive τ_0 values, while confinement in cytoskeleton-dependent meshworks results in negative τ_0 values. The size of the detection area A can be varied either by using a diaphragm or by means of focusing on different z-positions around the plasma membrane [40]. Exceptional reduction of the detection volume, needed to characterize quantitatively the size of membrane heterogeneities, can be achieved using zero-mode waveguides [41]. By means of this approach, Lenne et al. [42] succeeded to determine the diffusive behavior of several membrane components, like GPI-anchored proteins or the transferrin receptor, elucidating the role of both lipid domains and cytoskeleton-mediated meshwork.

2.5 *Chemical Relaxation*

A wide range of possible FCS experiments involves chemical reactions that continually fluctuate from equilibrium because of thermal motions and then "relax." In contrast to diffusion, which is also a relaxation process, the fluctuations cause changes between states of different energies. One very simple system presenting chemical relaxation would be a stationary binding site in the measurement volume, where particles only produce signal when bound, e.g., by FRET, or if the diffusion time is much faster than the sampling interval. In this case, the autocorrelation is

$$G(\tau) = G(0)\exp(-\tau/\tau_B) + G(\infty) \tag{11}$$

where

$$\tau_B = (k_{\mathrm{on}} + k_{\mathrm{off}})^{-1} \tag{12}$$

is the relaxation time and depends on the reaction kinetics, on and off rates, and

$$G(0) = \frac{1}{\langle N \rangle}\frac{k_{on}}{k_{off}} = \frac{1}{\langle N \rangle}K \tag{13}$$

is related to the equilibrium constant K.

Most systems with chemical relaxation also show measureable diffusion, and the autocorrelation function will depend on the details of the system. If the diffusion and chemical reaction are decoupled, the combined autocorrelation is the product of the chemical and diffusive autocorrelations.

2.6 Triplet State Correction

The autocorrelations mentioned above assume that the fluctuations are not due to changes in the fluorescent properties of the particles. However, for the majority of bioorganic fluorophores like green fluorescent protein, rhodamine, Cy3, and Alexa Fluor dyes, some fraction of illuminated particles are excited to a triplet state or other nonradiative decaying states, and then they do not emit photons for a characteristic relaxation time τ_F. Typically, τ_F is on the order of microseconds, which is usually smaller than the dynamics of interest (e.g., τ_D), but large enough to be measured. A multiplicative term is added to the autocorrelation form to account for the triplet state. For normal diffusion, it looks like

$$G(\tau) = G(0)\frac{1 - F + Fe^{-\tau/\tau F}}{1 - F}\frac{1}{(1 + (\tau/\tau_D))(1 + a^{-2}(\tau/\tau_D))^{1/2}} + G(\infty) \tag{14}$$

where F is the fraction of particles that have entered the triplet state and τ_F is the corresponding triplet state relaxation time. If the dynamics of interest are much slower than the triplet state relaxation as usually in membranes, the short-time component of the autocorrelation can simply be truncated, and the triplet term is unnecessary.

2.7 *Variations of FCS*

FCS almost always refers to the single-point, single-channel, temporal autocorrelation measurements, although the term "fluorescence correlation spectroscopy" out of its historical scientific context implies no such restriction. FCS has been extended in a number of variations by different researchers, with each extension generating as well new extended name of the method.

2.7.1 Fluorescence Cross Correlation Spectroscopy (FCCS)

Fluorescence correlation spectroscopy is sometimes applied to study molecular interactions by detecting changes in the diffusion times, e.g., the product of an association reaction will be larger and thus have larger diffusion times than the reactants individually; however, FCS is relatively insensitive to molecular mass. FCS data can report correlation time τ_D in both autocorrelation and cross correlation mode. Correlation time is related to the translational diffusion coefficient D, described by the relation in Eq. 7.

Dual-color fluorescence cross correlation spectroscopy (FCCS) measures interactions by cross correlating two or more fluorescent channels (one channel for each reactant), which distinguishes interactions more sensitively than FCS, particularly when the mass changed in the reaction is small. Fluorescence cross correlation spectroscopy (FCCS) overcomes the weak dependence of the diffusion rate on the molecular mass by looking at dynamic multicolor coincidence [11,43–45]. In dual-color cross correlation, intensity traces from two spectral channels, each detecting the emission of one fluorophore species, are cross correlated. Solely if the two species interact and diffuse as an entity through the detection volume, there is a correlation between the two intensity traces which gives rise to a positive cross correlation. Dual-color cross correlation, therefore, measures dynamic colocalization and is a powerful tool to study binding or dissociation. However, both partners should be completely labeled, since nonfluorescent competitors reduce the cross correlation amplitude, as do free fluorophores in solution. I_r(t) and I_g(t) are the intensity traces of the long- and short-wavelength channel, respectively. The spectral cross correlation curve is then defined as follows:

$$G_{\mathrm{rg}}(\tau) = \frac{\langle \delta \mathrm{I_r}(t) \cdot \delta \mathrm{I_g}(t+\tau) \rangle}{\langle \mathrm{I_r}(t) \rangle \langle \mathrm{I_g}(t) \rangle} \tag{15}$$

In case of a complete overlap of the detection areas, the concentrations of the bound molecules C_{rg} can be determined from the amplitudes of the auto- and cross correlation curves:

$$\frac{C_{\mathrm{rg}}}{C_{\mathrm{rg}} + C_{\mathrm{g}}} = \frac{G_{\mathrm{rg}}(0)}{G_{\mathrm{r}}(0)}, \frac{C_{\mathrm{rg}}}{C_{\mathrm{rg}} + C_{\mathrm{r}}} = \frac{G_{\mathrm{rg}}(0)}{G_{\mathrm{g}}(0)} \tag{16}$$

C_r and C_g are the concentrations of the unbound molecules. Generally, the detection areas might not overlap completely. This can be mirrored in a loss in cross correlation. A cross correlation standard, containing only dimers of both colors, can be used to quantify the maximal cross correlation, then used as a reference. If the two detection areas are well aligned, but differ in size due to their different wavelengths, the maximal cross correlation amplitude for complete double labeling is between the autocorrelation amplitudes:

$$G_{rg}(0) = 2\left(G_r(0)^{-1} + G_g(0)^{-1}\right)^{-1} \tag{17}$$

With a two-photon laser, the two dyes might often be simultaneously excited, resulting in a perfect overlap of the excitation profiles. Another artifact, which leads to false-positive cross correlation, is spectral cross talk: if there is an overlap in the emission spectra of the two fluorophores, some fluorescence from the green fluorophores can leak into the red channel. Especially with the usage of fluorescent proteins, spectral cross talk is common. An elegant but experimentally complex way to avoid spectral cross talk is the application of alternating excitation or pulsed interleaved excitation. Repeatedly, only the green laser is switched on, and fluorescence is collected in the green channel, then only the red laser is switched on, and fluorescence is gathered in the red channel. Thereby, contributions from the different molecules can be completely separated.

The repetition rate defines the time resolution of correlation curve and has to be much faster than the timescales of interest (i.e., diffusion time). In most commercial setups, alternating excitation is not implemented. Fortunately, cross talk can be corrected during the data analysis. In dual-color FCS, mechanical instabilities and membrane movements result in a false-positive cross correlation. This artifact is difficult to quantify and, contrarily to spectral cross talk, difficult to take into account and should therefore be avoided.

2.7.2 FRET-FCS

Another FCS-based approach to study molecular interactions uses fluorescence resonance energy transfer (FRET) instead of fluorescence and is called FRET-FCS [46]. With FRET, there are two types of probes, as with FCCS; however, there is only one channel, and light is solely then detected when the two probes are very close – close enough to ensure an interaction. The FRET signal is weaker than with fluorescence but has the advantage that the signal is observed only during a reaction, aside from autofluorescence.

2.7.3 Scanning FCS

In scanning fluorescence correlation spectroscopy (sFCS), the measurement volume is moved across the specimen in a defined way. The introduction of scanning is

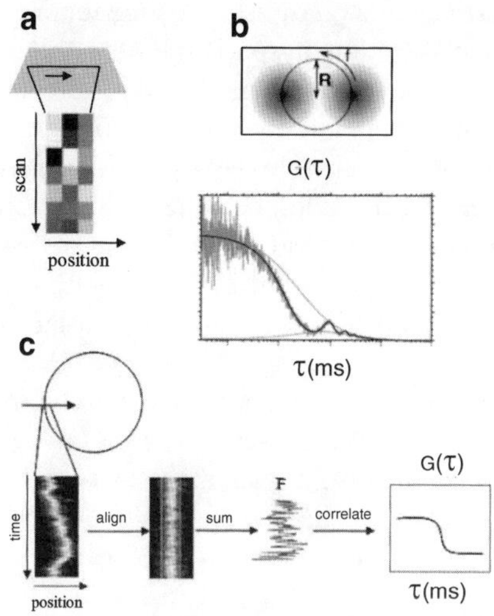

Fig. 1 Scanning FCS. (**a**) In line-scan FCS, the detection volume is scanned in a linear manner on the membrane. Cross correlation of pixels within one line and among different lines leads to the full spatiotemporal correlation curves from which calibration-free concentrations and diffusion coefficients can be determined. (**b**) The circular scan path in small-circle-scanning FCS causes oscillations in the autocorrelation curve, permitting calibration-free concentration and diffusion measurements. (**c**) Perpendicular scanning FCS. The detection volume is scanned in a linear fashion perpendicularly through a vertical membrane. When the individual lines are arranged vertically, the membrane is clearly visible. Membrane movements can be corrected for by shifting each scan. Membrane contributions are added up to result in the fluctuation intensity trace from which correlation curves can be calculated

motivated by its ability to alleviate or remove several distinct problems like photobleaching, not enough signal-to-noise ratio often encountered in standard FCS, and, thus, to extend the range of applicability of fluorescence correlation methods in biological systems [47]. Scanning FCS is a powerful technique for accurate measurements of diffusion, even extremely slow one, like in cortex or in membranes. Scanning FCS can be circular or linear [14,48,49] (Fig. 1). Two-focus spatial scanning cross correlation measures diffusion without the necessity for calibrating the detection volume and therefore greatly improves the accuracy compared to the traditional FCS. The simple implementation in a commercial laser scanning microscope should allow establishing scanning FCS as a standard method for membrane studies.

More data are obtained by performing the experiment in multiple spatial points in parallel using a laser scanning confocal microscope than getting adequate statistics from a single-point FCS experiment, which on membranes characterized by the slow diffusion may take a prohibitively long time. This approach has been called image correlation spectroscopy (ICS) [50]. Another variation of ICS performs a spatial autocorrelation on images, which gives information about the concentration of particles [51]. The correlation is then averaged in time. A natural extension of the temporal and spatial correlation versions is spatiotemporal ICS (STICS) [52]. In STICS there is no explicit averaging in space or time, only the averaging inherent in correlation. In systems with nonisotropic motion, e.g., directed flow and asymmetric diffusion, STICS can extract the directional information. A variation that is closely related to STICS by the Fourier transform is *k*-space image correlation spectroscopy (kICS) [53]. There are cross correlation versions of ICS as well [50].

Image correlation spectroscopy and its variations all were implemented on a scanning confocal or scanning two-photon microscope. Extending the dimensions to scanning whole images instead of just along a line, raster image correlation spectroscopy (RICS) [54] allows constructing diffusion and concentration maps of whole cells and bridges the gap between FCS and image correlation spectroscopy (ICS).

Raster ICS (RICS) [54] and position-sensitive FCS (PSFCS) [55] incorporate the time delay between parts of the image scan into the analysis. Also, low dimensional scans, e.g., a circular ring [56], can access timescales between single-point and full-image measurements. The scanning path has also been made to adaptively follow particles [57].

One of scanning FCS schemes, which is ready to be used with a commercial laser scanning microscope, is line-scan FCS. The detection volume is scanned repeatedly on the membrane. Each line can be used to calculate an autocorrelation curve. Since the detection volume is at different positions at various times, this corresponds to a spatial correlation of the signal. Additionally, the cross correlation of each line with every other line can be calculated. The length of the scan path and therefore the velocity can be easily determined in line-scan FCS; thus, detection volume does not require calibration, which makes the method very convenient. Line-scan FCS measures diffusion coefficients and concentrations within minimal measurement times due to the acquisition along the whole line, corresponding to a simultaneous measurement with many confocal detection volumes. The usage of line-scan FCS allows for the determination of the temperature dependence of diffusion coefficients in membranes, what is not accomplishable by static confocal FCS due to temperature-induced aberrations and therefore changes in the size of the detection volume [14].

In contrast to scanning on a straight line, movements along a circular scan path in the circle-scanning FCS benefit from avoiding any hysteresis effects or inhomogeneities because of changing direction (Fig. 1). Circular scans can be much faster than line scans, allowing measurements of faster diffusion in solution [48]. Additionally, the precise knowledge of the radius of the scan path enables calibration-free diffusion measurements [49]. Synchronization of scanning and data acquisition allows for acquiring even the flow directions [55]. Unfortunately, the circular scan path is not yet implemented in most commercial systems. The scan radius for small-circle-scanning FCS is a few 100 nm and thus of the order of the size of the detection volume itself. The scanning frequency should be chosen as high as technically possible and preferably larger than 1 kHz. Determination of the radius of the circle can be achieved with high precision by scanning over a grid with affixed line spacing. Fitting the experimental correlation curve with the proper model leads therefore to directly resolving the diffusion coefficients, not influenced by optical artifacts and without calibration of the detection volume.

Another version of scanning FCS technique is scanning FCS perpendicular to the membrane plane. Measuring the diffusion in free-standing membranes is important to clarify the surface contributions in mica- or glass-supported lipid bilayers. Commonly used as examples of free-standing membranes are large vesicles, such as GUVs. Ries et al. reported linear scanning FCS through an equatorial plane of the vesicle to measure membrane diffusion in the presence of excess free dye in solution, which usually masks less abundant slow components in the correlation curves [14]. Correlating the corresponding time windows of a linear scan resulted in a series of correlation curves. Due to the continuous scan path intersecting the surface of the GUV, the membrane contributions from each scan are added up, thus allowing the correction of membrane movements. Very long measurement times are in this case suitable, permitting the measurement of very slow diffusion as found for proteins in yeast cell membranes. In addition, the autocorrelation curves are not influenced by membrane vibrations and movements any more. Perpendicular scanning FCS can be implemented with a commercial laser scanning microscope making use of multiple colors spanning the whole visible range [13]. Scanning dual-color cross correlation enables quantitative binding measurements within membranes. Asynchronous scanning with two parallel lines allows also to perform dual-focus scanning FCS for calibration-free diffusion measurements in unstable membranes; however, the distance between the scan paths needs to be determined. This is especially useful for FCS in deeper tissue, where the refractive index mismatch causes a distortion of the detection volume that depends on the depth of focus [58].

2.7.4 Spinning-Disk FCS and Spatial Mapping

Any of the image correlation spectroscopy methods can be performed on a spinning-disk confocal microscope, which in practice allows to achieve faster imaging speeds compared to a laser scanning confocal microscope. This approach has recently been applied to diffusion in a spatially varying complex environment, producing a pixel resolution map of a diffusion coefficient [59]. The spatial mapping of diffusion with FCS has subsequently been extended to the TIRF system [60]. Spatial mapping of dynamics using correlation techniques had been applied before but only at sparse points [61] or at coarse resolution [62]. Recently, spatially resolved quantitative measurements of protein abundance, mobilities, and interactions to study diffusion processes and local dynamic equilibria inside mammalian cells have been performed with a microscope based on light-sheet illumination that allows massively parallel fluorescence correlation spectroscopy technique [63].

2.7.5 Total Internal Reflection FCS

Total internal reflection fluorescence (TIRF) is a microscopy approach that is only sensitive to a thin layer near the surface of a coverslip, which greatly minimizes background fluorescence. FCS has been extended to that type of microscope and is called TIR-FCS [64]. Because the fluorescence intensity in TIRF falls off exponentially with distance from the coverslip instead of as a Gaussian with a confocal, the autocorrelation function is different.

Comparing to confocal FCS, the lateral extension of the TIR-excitation profile is, at several micrometers, pretty large. To obtain a detection volume small enough for FCS, a pinhole in the image plane is applied for lateral confinement. For diffusion measurements in membranes, a tight lateral confinement is especially important to reach reasonable diffusion times. The drawback of this approach is significant out of focus photobleaching. Many fluorophores will get bleached in the large excitation spot before they enter the detection area. As this can be a serious limitation in the study of membrane diffusion, TIR-FCS is well suited to study binding kinetics [64].

2.7.6 Stimulated Emission Depletion FCS

Stimulated emission depletion (STED) microscopy uses a confocal detection volume and a second laser with a beam profile in the shape of a donut [65]. This second light pattern suppresses fluorescence emission due to depopulation of the excited state via stimulated emission. Since it features zero intensity in the center, fluorescence is only emitted from a central region of the detection volume of 20–50 nm in diameter. This approach has been used to realize far-field FCS with a subdiffraction detection volume, [66] particularly in two-dimensional systems such as membranes. Due to the small size of the detection volume in STED-FCS, it is optimally suited to study small spatial heterogeneities, such as lipid rafts. Eggeling et al. could show that certain raft-associated lipids, such as the ganglioside GM1, are transiently trapped in domains of $<$20 nm, whereas nonraft lipids show free diffusion even on these length scales [67].

2.7.7 Z-Scan FCS

Z-scan fluorescence correlation spectroscopy uses multiple focus planes to determine diffusion coefficients in membranes or supported phospholipid bilayers (SPBs) without the need for extrinsic calibration. Focusing on different z-positions around the plasma membrane allows the size of the detection area to be changed, what was mentioned previously in Sect. 2.4.2 [40].

The approach can be used to discriminate diffusion modes. Z-scan FCS was applied to study the diffusion of the fluorescent lipid analog DiI–C_{18} in the top membrane of giant unilamellar vesicles (GUVs) in the electroformation chamber. In contrast to SPBs, the intrinsic estimate of the radial waist might differ from the one obtained by extrinsic calibration. The validity of the Z-scan approach under these conditions was here described [68].

3 Applications: FCS on Membranes

Since a couple of decades [69], it is widely accepted that the plasma membrane of cells is not uniform but highly organized and dynamic, consisting of a multitude of interacting subdomains within the lipid membrane [70]. The lateral heterogeneities of the membrane have gathered the attention of the researchers, resulting in many models of nonrandom molecular distribution with different names, e.g., lipid rafts, microdomains, and confinement zones [70–76]. The membrane domains are considered highly dynamical structures known to be involved in signaling pathways and other cellular processes [77–79]. FCS has proved to be a powerful technique for examining the dynamic properties of lipid membranes. With single-molecule sensitivity, FCS can also be used to measure diffusion coefficients, concentration of fluorophores, size of particles, conformational changes, and association and disassociation processes [80,81]. In this section, we briefly describe some of the biological applications of FCS in lipid membranes, which are classified according to the type of system studied.

3.1 Model Membranes

According to the raft hypothesis, cell membranes have lipid and protein nanodomains enriched in sphingomyelin and cholesterol. Interestingly, model membranes with a certain composition of saturated and unsaturated lipids and cholesterol exhibit coexistence of liquid-disordered (L_d) and liquid-ordered (L_o) lipid phases. Both L_d and L_o phases are fluid, but they do differ in their organization: L_o phases are usually formed by saturated lipids and cholesterol and exhibit tightly packed acyl chains, while lipids in L_d phase are usually unsaturated and exhibit disordered acyl chains [72,82]. Due to the limitations to investigate cellular rafts, model membranes exhibiting lateral segregation have been very helpful for the study of the physical-chemical principles that govern cell membrane organization. The two excellent model membrane systems normally used for FCS measurements are giant unilamellar vesicles and supported lipid bilayers.

3.1.1 Giant Unilamellar Vesicles (GUVs)

Giant unilamellar vesicles (GUVs) are free-standing bilayers with diameters ranging from 10 to 100 μm suitable for optical microscopy. They provide a single spherical closed bilayer and with a cell-like curvature. As they provide a flat, free-standing membrane, the focal volume could be either placed on the top or scanned periodically through equatorial plane. GUVs prepared from ternary mixtures of dioleoylphosphatidylcholine/sphingomyelin/cholesterol provide a very good model for investigating the lipid spatial and dynamic organization using FCS. Kahya et al.

[83] observed formation of domains with raft-like properties for a certain range of cholesterol concentrations. The authors used FCS to analyze lipid mobility and identify the effect of cholesterol in the lipid phases. In addition, it was shown that the lipid mobility in sphingomyelin-enriched regions significantly increases at higher cholesterol concentration. The results pinpoint the key role played by cholesterol in tuning lipid dynamics in membranes.

Having known the importance of cholesterol in tuning the lipid dynamics in membranes, a new Bodipy-labeled cholesterol (Bdp-Chol) [84] derivative has been used to investigate membrane fluidity, lipid order, and partitioning in various lipid phases in GUVs as a model system. This cholesterol analog adopts a similar conformation to the parent cholesterol molecule itself, thereby partitioning into lipid phases in a comparable manner. The translational diffusion coefficient of Bdp-Chol measured using fluorescence correlation spectroscopy is $7.4 \pm 0.3 \times 10^{-8}$ cm^2/s and $5.0 \pm 0.2 \times 10^{-8}$ cm^2/s in the L_d and L_o phases, respectively. Recently, Chiantia et al. [85] used FCS to measure the diffusion coefficient of the fluorescent probes in asymmetric GUVs (with an inner leaflet composed of dioleoylphosphatidylcholine (DOPC) and an outer leaflet composed of DOPC and brain SM (bSM) using an mßCD-mediated lipid exchange). Another application by Kubiak et al. [86] on bacterial model membranes indicated the presence of gel-like domains, and the size of these domains depends on the chemical structure and concentration of lipopolysaccharide. In order to analyze the biophysical properties of myelin membranes, Yurlova et al. [87] applied FCS on model membranes. They found that lipids are tightly organized and highly ordered in myelin isolated from wild-type mice, but not from *shiverer* and ceramide synthase 2 null mice.

The membrane fluidity is controlled by the lipid composition, which is susceptible to oxidative attack by reactive oxygen species. FCS along with Raman spectroscopy has been successfully employed to characterize the lateral fluidity of membranes and the structure of lipids in model membranes under oxidative attack [88]. Hydroxyl radicals caused significantly increased membrane fluidity. The ability of vitamins C and E to inhibit ROS-induced alteration of membrane fluidity was also demonstrated with this study.

Furthermore, phase-separating GUVs have been used as a raft-mimicking system to study protein behavior. Kahya et al. [89] reconstituted a GPI-anchored protein, human placental alkaline phosphatase (PLAP), into domain-exhibiting GUVs. With a combination of confocal microscopy and FCS, the authors characterized the domain morphology, protein phase partitioning, and dynamic behavior. The results showed that PLAP has a great affinity for raft-like domains and it depends on subtle changes in lipid-lipid and lipid-protein interaction energies. DPIDA is a protein-binding lipid designed to partition into ordered membrane phases, and this can target proteins to liquid-ordered domains in lipid membranes. FCS has been successfully applied to determine the diffusion rates of DPIDA/DPPC/cholesterol containing domains verifying their liquid state. There is an equal division of vesicles into bright and dark domains [90].

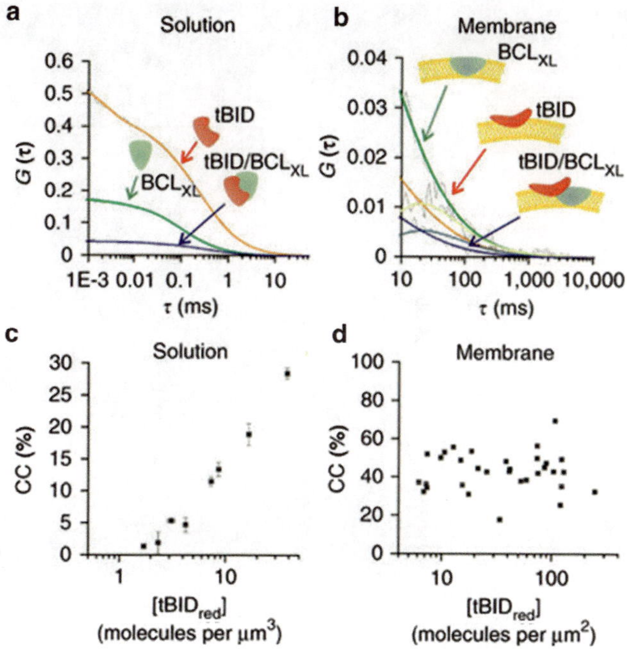

Fig. 2 (**a**) Auto- and cross correlation analysis of protein/protein interactions in solution. The fitted autocorrelation curves of $tBID_{red}$ and $BCL_{XL}\Delta Ct_{green}$ are shown in *orange* and *green*, respectively. The amplitude of the fitted cross correlation curve (*blue*) indicates formation of tBID-$BCL_{XL}\Delta Ct$ complexes. (**b**) Auto- and cross correlation global analysis of protein/protein interactions within lipid membranes using two-focus scanning FCCS measurements in GUVs. The *orange* and *green lines* correspond to the fitted autocorrelation curves of $tBID_{red}$ and $BCL_{XL}\Delta Ct_{green}$, respectively. The *light blue* and *yellow lines* show the two-focus analysis for $tBID_{red}$ and $BCL_{XL}\Delta Ct_{green}$, respectively. Interactions between tBID and $BCL_{XL}\Delta Ct$ in the membrane are demonstrated by the amplitude of the CC curve (*dark blue line*). *Thin black lines* depict raw data. (**c, d**) Quantification of tBID-$BCL_{XL}\Delta Ct$ complexes, expressed as a cross correlation (CC) percentage, as a function of tBID concentration in solution (**c**) and in the membrane (**d**)

Garcia et al. [91] performed FCCS to study protein-protein interactions within model membranes. With this technique, the authors were able to study the binding affinities of apoptotic proteins tBID and BCL_{XL} in aqueous solution as well as within the lipid bilayer using GUVs. From these studies, they found that tBID and BCL_{XL} interact in both environments, but their affinity is greatly enhanced by the membrane environment (Fig. 2). FCCS also helped to characterize the oligomerization of VDAC in GUVs containing CL, PG, and two other anionic lipids [92]. The authors concluded that the anionic lipids CL and PG play a role in the regulation of VDAC oligomerization, which could play a role in apoptosis. The functional oligomeric state of SecYEG has also been studied using the above-mentioned technique. Fluorescently labeled SecYEG complexes were reconstituted into

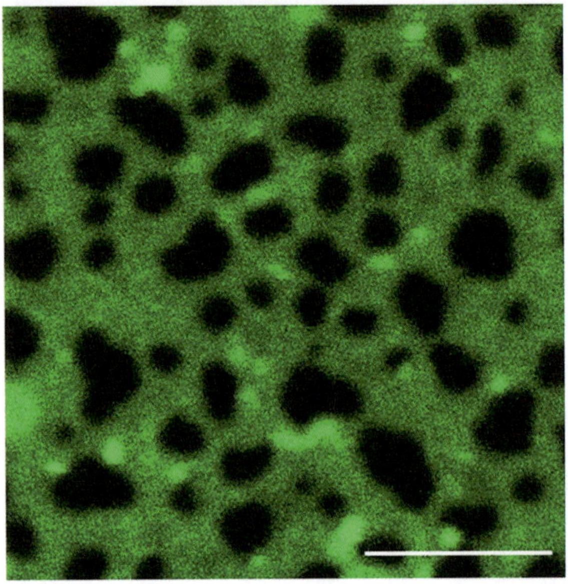

Fig. 3 Confocal image of a supported lipid bilayer with lipid composition of dioleoylphosphatidylcholine: sphingomyelin:cholesterol (2:2:1) and exhibiting phase coexistence. The liquid-disordered lipid phase is labeled with DiO (*green*). Scale bar: 5 μm

GUVs at low density. The results showed that SecYEG can form oligomers in the presence of ATP and SecA [93].

3.1.2 Supported Lipid Bilayers

Supported lipid bilayers can be formed by thermal fusion of small unilamellar vesicles (SUVs) or proteo-SUVs onto a mica support in the presence of calcium (Fig. 3) [94]. The procedure is more protein friendly that in the case of GUVs, but there might be unwanted interactions with the substrate that could hinder protein mobility, leading to artifacts. FCS has been successfully applied in this model membrane for the study of lipid and protein organization and their dynamics.

Ries et al. developed line-scan FCS to investigate domain-exhibiting supported bilayers [95]. The authors investigated the diffusion and partition behavior of fluorescent lipids in three commonly used phase-separating bilayers and to probe the temperature dependence of diffusion in several model membranes and in cellular plasma membrane patches. Their studies showed that the physical properties of the liquid-ordered phase as a model for cellular rafts are strongly influenced by (a) the specific chemical composition of the bilayer and (b) the temperature, due to the high activation energy of this lipid phase.

A combined approach of atomic force microscopy (AFM), fluorescence imaging, and two-focus sFCS was used to obtain structural and dynamical information of the supported lipid bilayers. Partition and diffusion coefficients in the different phases were measured with two-focus sFCS [96]. The same approach was used to elucidate the effects of ceramides on structure and dynamic organization of bilayers

[94]. While AFM provides structural information based on topographical variations between different lipid phases in supported lipid bilayers, FCS provides structural information based on data on lipid dynamics that are acquired on a much faster timescale. The results of this study show that the presence of ceramide affects the lipid spatial organization and that ceramide displaces cholesterol from the liquid-ordered phase [94,97].

Lieto et al. used a wide-field TIR-FCS approach to investigate the binding characteristics of IgG antibody molecules to the Fcg receptor-2 bound to a glass-supported lipid bilayer. Upon binding to the receptor, the autocorrelation curve changed to a bimodal distribution where the quantification of the bound versus unbound fractions as well as the Kd of the interaction (2.4–0.4 mM) could be determined [64].

Dual-focus FCS was used to systematically study lipid diffusion in black lipid membranes (BLM) as a function of ion valence and concentration in the surrounding buffer. From this study, it was found that divalent Ca^{2+} ions slowed down the lipid diffusion within charged membranes at high ionic strength [98].

Machán et al. [99] employed Z-scan FCS to characterize the interaction between arenicin-1 and supported lipid bilayers (SLBs) of different compositions. Z-scan diffusion law analysis indicated the presence of microdomains with less mobile lipid and peptide molecules in negatively charged SLBs. Additionally, the FCS results implied that arenicin-1 exists in the form of oligomers and/or aggregates when interacting with membranes of both compositions. Przybylo et al. [100] employed the similar method to compare the diffusion of BODIPY-tail-labeled lipid in GUVs and SLBs. The results revealed that the lipid mobility in GUVs ($D = 7.8 \pm 0.8\ \mu m^2 s^{-1}$) is significantly higher than in the bilayer created on the solid support ($D = 3.1 \pm 0.3\ \mu m^2 s^{-1}$).

3.2 Cells

FCS has been successfully applied on the membranes of living cells mainly in order to (1) characterize the lateral organization of the plasma membrane [42], (2) study receptor dynamics in the plasma membrane [101], (3) investigate receptor-ligand interactions [102–104], and (4) study oligomerization [105–107].

In an approach to identify the nanoscale membrane organization of live cells, Lasserre et al. [108] used FCS and reported that highly dynamic nanodomains exist in both the inner and the outer leaflets of the plasma membrane. They also showed that these nanodomains play a critical role in the phosphatidylinositol-3 kinase/AKT signaling pathway. Golebiewska et al. [109] used FCS to investigate the occurrence of phosphatidylinositol 4,5-bisphosphate (PIP2) on the inner leaflet of the plasma membrane. They microinjected micelles of Bodipy-TMR-PIP2 into cells and measured its diffusion coefficient on the inner leaflet of the plasma membrane of fibroblasts and epithelial cells by using FCS. Their results showed that the diffusion coefficient of fluorescent PIP2 on the inner leaflet of fibroblast plasma membranes is significantly (two- to threefold) lower than that of PIP2 in cellular blebs or in the outer leaflet of the plasma membrane and that of the diffusion of other lipids (rhodamine-PE and DiD) in the inner leaflet.

With the application of Z-scan FCS, serotonin (1A) receptor has been shown to exhibit confinement in cell membranes. The confinement could be a result of the interaction of the receptor with the actin cytoskeleton. Depletion of cholesterol caused cytoskeletal destabilization and reduced the receptor confinement [110].

As reported in many signaling pathways, ligand binding can induce receptor oligomerization. One example of these is somatostatin receptors (SSTRs). Normally these receptors exist as monomers, but ligand binding can induce homo- and hetero-oligomerization. FCS was used to demonstrate that ligand binding triggers receptor clustering, leading to formation of dimers and higher-order oligomers [102]. Another example of this kind is complex formation by the interleukin-4 receptor subunits [103]. Few more applications of FCS on ligand-receptor-binding studies include (1) the quantification of A_1-adenosine receptor-antagonist complexes in single living cells [104], (2) the characterization of the lateral mobility of receptor-ligand complex in cell membranes as in case of $GABA_A$ receptors [101], and (3) insight into the dynamics of receptor-mediated signaling processes in living cells [111].

Philip et al. [112], using single-point and scanning FCS, showed that the bradykinin receptor can diffuse with mobility corresponding to dimers/small oligomers as well as with mobility corresponding to higher-order aggregates. This suggested that these proteins exist as dimers and further associate to form higher-order signaling domains on the plasma membrane. FCS was also used in studies of dysferlin homodimerization [105] and homo- and heterodimerization of epidermal growth factor (EGFR) [106]. Garcia et al. [107] applied FCCS to study the oligomerization of Equinatoxin II, which leads to reorganization of the plasma membrane. The authors performed two-color, two-focus scanning FCS on the membrane of the membrane blebs induced by Equinatoxin II (Fig. 4).

Recently FCS combined with far-field stimulated emission depletion (STED) nanoscopy [113] was used to have detailed understanding of molecular characteristics of nanoscale lipid trapping in the plasma membrane of living cells.

3.3 *Organisms*

A combination of static-volume, two-focus, and dual-color scanning fluorescence correlation spectroscopy was used to analyze receptor-ligand interactions even in the complex biological environment of fish embryos. Reis et al. [58] quantified the mobility of fibroblast growth factor receptors Fgfr1 and Fgfr4 in cell membranes of living zebra fish embryos and determined the *in vivo* binding affinities to their ligand Fgf8.

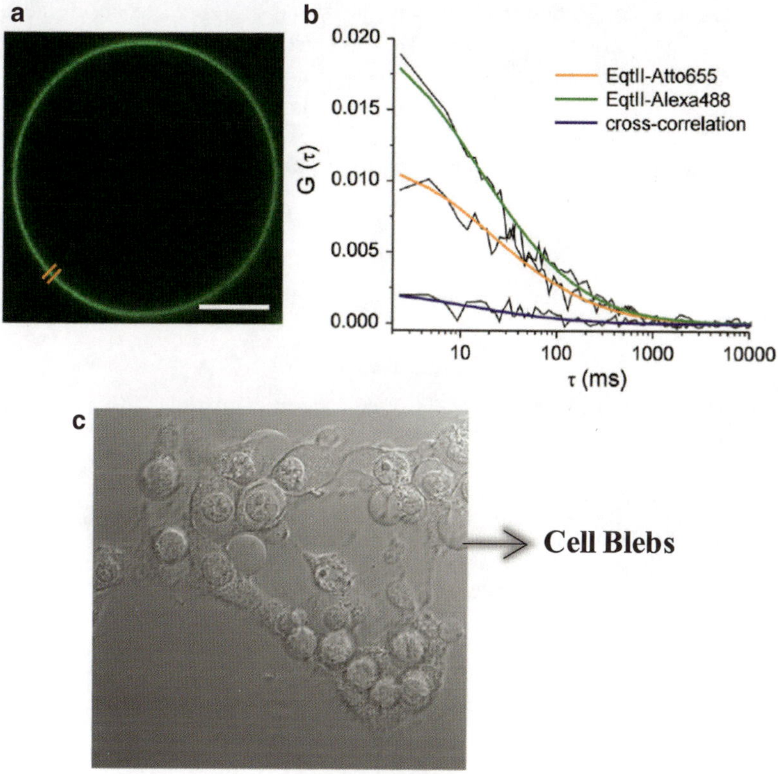

Fig. 4 Two-color, two-focus scanning FCS to investigate EqtII oligomerization in cells. COS-7 cells treated with EqtII-Al488 and EqtII-Atto655 and incubated for 1 h at room temperature. (**a**) The two foci of scanning FCS are depicted as parallel lines on one of the blebs measured. (**b**) Auto- and cross correlation fitted curves of EqtII-Al488 and EqtII-Atto655. The positive amplitude of the cross correlation curve (*blue line*) indicates direct EqtII/EqtII interactions. Raw data are shown in *black lines*. FCS was measured on the membrane of cell blebs (**c**) induced by the toxin. Scale bar: 15 μm

4 Conclusions

In this chapter, we discussed the principle, various experimental realizations, and applications of fluorescence correlation spectroscopy to study organization and interaction of membranes. FCS has proven very useful to determine concentrations, diffusion coefficients, and binding parameters of biomolecules in artificial and cellular membranes. We presented advanced implementations of FCS that greatly increase the versatility and accuracy of the technique in biological membranes and push the application to systems previously inaccessible.

Due to the fact that currently the field is very dynamic, novel variations of FCS are continuously being developed. Also common problems like the photostability of the fluorophores and the signal-to-noise ratio during the measurements have been addressed for improvements. Novel strategies in data analysis tend to extract the maximum of information from the single data set with low risk of possible artifacts.

Fluorescence correlation spectroscopy has been presented to be the adequate technique for investigation of membrane organization. FCS allows not only for the determination of molecular interactions occurring within the scope of the membrane but also every single experiment bring us closer to understand and decode the function of its particular constituents. Gathering the whole package of information together will let us depict a more clear, fundamental unit of life, a cell, and decipher its substance.

References

1. Magde D, Elson EL, Webb WW (1972) Thermodynamic fluctuations in a reacting system: measurement by fluorescence correlation spectroscopy. Phys Rev Lett 29:705–708
2. Ehrenberg M, Rigler R (1974) Rotational Brownian motion and fluorescence intensity fluctuations. Chem Phys 4:390–401
3. Elson EL, Magde D, Elson EL, Magde D (1974) Fluorescence correlation spectroscopy I. Conceptual basis and theory. Biopolymers 13:1–27
4. Magde D, Elson EL, Webb WW (1974) Fluorescence correlation spectroscopy II. An experimental realization. Biopolymers 13:29–61
5. Thompson NL (1991) In: Lakowicz JR (ed) Topics in fluorescence spectroscopy techniques, vol 1. Plenum, New York, pp 337–378
6. Rigler R, Mets U, Widengren J, Kask P (1993) Fluorescence correlation spectroscopy with high count rate and low background: analysis of translational diffusion. Eur Biophys J 22 (3):159
7. Eigen M, Rigler M (1994) Sorting single molecules: application to diagnostics and evolutionary biotechnology. Proc Natl Acad Sci USA 91:5740–5747
8. Rigler M (1995) Fluorescence correlations, single molecule detection and large number screening. Applications in biotechnology. J Biotechnol 41:177–186
9. Krichevsky O, Bonnet G (2002) Fluorescence correlation spectroscopy: the technique and its applications. Rep Prog Phys 65:251–297
10. Rigler R, Pramanik A, Jonasson P, Kratz G, Jansson OT, Nygren PA, Stahl S, Ekberg K, Johansson BL, Uhlen S, Uhlen M, Jornvall H, Wahren J (1999) Proc Natl Acad Sci USA 96:13318
11. Ries J, Schwille P (2008) New concepts for fluorescence correlation spectroscopy on membranes. Phys Chem Chem Phys 10(24):3487–3497
12. Enderlein J, Gregor I, Patra D, Dertinger T, Kaupp UB (2005) Performance of fluorescence correlation spectroscopy for measuring diffusion and concentration. Chemphyschem 6:2324–2336
13. Ries J, Schwille P (2006) Studying slow membrane dynamics with continuous wave scanning fluorescence correlation spectroscopy. Biophys J 91:1915–1924
14. Ries J, Chiantia S, Schwille P (2009) Accurate determination of membrane dynamics with line scan FCS. Biophys J 96:1999–2008
15. Hebert B, Costantino S, Wiseman PW (2005) Spatiotemporal image correlation spectroscopy (STICS) theory, verification, and application to protein velocity mapping in living CHO cells. Biophys J 88:3601–3614
16. Ayuyan AG, Cohen FS (2006) Lipid peroxides promote large rafts: effects of excitation of probes in fluorescence microscopy and electrochemical reactions during vesicle formation. Biophys J 91:2172–2183
17. Thompson NL, Steele BL (2007) Total internal reflection with fluorescence correlation spectroscopy. Nat Protoc 2:878–890

18. Ries J, Ruckstuhl T, Verdes D, Schwille P (2008) Supercritical angle fluorescence correlation spectroscopy. Biophys J 94:221–229
19. Medina MA, Schwille P (2002) Fluorescence correlation spectroscopy for the detection and study of single molecules in biology. Bioessays 24:758–764
20. Mayboroda OA, van Remoortere A, Tanke HJ, Hokke CH, Deelder AM (2003) A new approach for fluorescence correlation spectroscopy (FCS) based immunoassays. J Biotechnol 107:185–192
21. Hess ST, Webb WW (2002) Focal volume optics and experimental artifacts in confocal fluorescence correlation spectroscopy. Biophys J 83:2300–2317
22. Kusumi A, Nakada C, Ritchie K, Murase K, Suzuki K, Murakoshi H, Kasai RS, Kondo J, Fujiwara T (2005) Paradigm shift of the plasma membrane concept from the two-dimensional continuum fluid to the partitioned fluid: high-speed single-molecule tracking of membrane molecules. Annu Rev Biophys Biomol Struct 34:351–378
23. Fujiwara T, Ritchie K, Murakoshi H, Jacobson K, Kusumi A (2002) Phospholipids undergo hop diffusion in compartmentalized cell membrane. J Cell Biol 157:1071–1081
24. Schwille P, Korlach J, Webb WW (1999) Fluorescence correlation spectroscopy with single-molecule sensitivity on cell and model membranes. Cytometry 36:176–182
25. Bouchaud JP, Georges A (1990) Anomalous diffusion in disordered media – statistical mechanisms, models and physical applications. Phys Rep Rev Sect Phys Lett 195:127–293
26. Saxton MJ (1994) Anomalous diffusion due to obstacles – a Monte-Carlo study. Biophys J 66:394–401
27. Schutz GJ, Schindler H, Schmidt T (1997) Single-molecule microscopy on model membranes reveals anomalous diffusion. Biophys J 73:1073–1080
28. Kusumi A, Ike H, Nakada C, Murase K, Fujiwara T (2005) Single-molecule tracking of membrane molecules: plasma membrane compartmentalization and dynamic assembly of raft-philic signaling molecules. Semin Immunol 17:3–21
29. BrustMascher I, Feder TJ, Slattery JP, Baird B, Webb WW (1993) FPR data on mobility of cell-surface proteins reevaluated in terms of temporally constrained molecular motions. Biophys J 64:A354
30. Kenworthy AK, Nichols BJ, Remmert CL, Hendrix GM, Kumar M, Zimmerberg J, Lippincott-Schwartz J (2004) Dynamics of putative raft-associated proteins at the cell surface. J Cell Biol 165:735–746
31. Feder TJ, BrustMascher I, Slattery JP, Baird B, Webb WW (1996) Constrained diffusion or immobile fraction on cell surfaces: a new interpretation. Biophys J 70:2767–2773
32. Tamm LK (1988) Lateral diffusion and fluorescence microscope studies on a monoclonal-antibody specifically bound to supported phospholipid-bilayers. Biochemistry 27:1450–1457
33. Nicolau DV, Hancock J, Burrage K (2008) Sources of anomalous diffusion on cell membranes: a Monte Carlo study. Biophys J 92:1975–1987
34. Berry H (2002) Monte Carlo simulations of enzyme reactions in two dimensions: fractal kinetics and spatial segregation. Biophys J 83:1891–1901
35. Weiss M, Hashimoto H, Nilsson T (2003) Anomalous protein diffusion in living cells as seen by fluorescence correlation spectroscopy. Biophys J 84:4043–4052
36. Gielen E, Vercammen J, Sykora J, Humpolickova J, Vandeven M, Benda A, Hellings N, Hof M, Hof M, Engelborghs Y, Steels P, Ameloot M (2005) Diffusion of sphingomyelin and myelin oligodendrocyte glycoprotein in the membrane of OLN-93 oligodendroglial cells studied by fluorescence correlation spectroscopy. C R Biol 328:1057–1064
37. Bacia K, Scherfeld D, Kahya N, Schwille P (2004) Fluorescence correlation spectroscopy relates rafts in model and native membranes. Biophys J 87:1034–1043
38. Gombos I, Steinbach GB, Pomozi I, Balogh A, Vamosi G, Gansen A, Laszlo G, Garab G, Matko J (2008) Some new faces of membrane microdomains: a complex confocal fluorescence, differential polarization, and FCS imaging study on live immune cells. Cytometry A 73A:220–229

39. Wawrezinieck L, Rigneault H, Marguet D, Lenne PF (2005) Fluorescence correlation spectroscopy diffusion laws to probe the submicron cell membrane organization. Biophys J 89:4029–4042
40. Humpolickova J, Gielen E, Benda A, Fagulova V, Vercammen J, Vandeven M, Hof M, Ameloot M, Engelborghs Y (2006) Probing diffusion laws within cellular membranes by Z-scan fluorescence correlation spectroscopy. Biophys J 91:L23–L25
41. Wenger J, Conchonaud F, Dintinger J, Wawrezinieck L, Ebbesen TW, Rigneault H, Marguet D, Lenne PF (2007) Diffusion analysis within single nanometric apertures reveals the ultrafine cell membrane organization. Biophys J 92:913–919
42. Lenne PF, Wawrezinieck L, Conchonaud F, Wurtz O, Boned A, Guo XJ, Rigneault H, He HT, Marguet D (2006) Dynamic molecular confinement in the plasma membrane by microdomains and the cytoskeleton meshwork. EMBO J 25:3245–3256
43. Schwille P (2001) Cross-correlation analysis in FCS. In: Elson EL, Rigler R, Elson EL, Rigler R (eds) Fluorescence correlation spectroscopy. Theory and applications. Springer, Berlin/New York, pp 360–378
44. Bacia K, Schwille P (2007) Practical guidelines for dual-color fluorescence cross-correlation spectroscopy. Nat Protoc 2(11):2842–2856
45. Bacia K, Kim S, Schwille P (2006) Fluorescence cross-correlation spectroscopy in living cells. Nat Methods 3:83–89
46. Remaut K, Lucas B, Braeckmans K, Sanders NN, De Smedt SC, Demeester J (2005) FRET-FCS as a tool to evaluate the stability of oligonucleotide drugs after intracellular delivery. J Contr Release 103(1):259–271
47. Mashaghi A et al (2008) Characterization of protein dynamics in asymmetric cell division by scanning fluorescence correlation spectroscopy. Biophys J 95(11):5476–5486
48. Berland KM, So PT, Chen Y, Mantulin WW, Gratton E (1996) Scanning two-photon fluctuation correlation spectroscopy: particle counting measurements for detection of molecular aggregation. Biophys J 71:410–420
49. Petrasek Z, Schwille P (2008) Precise measurement of diffusion coefficients using scanning fluorescence correlation spectroscopy. Biophys J 94:1437–1448
50. Wiseman PW, Squier JA, Ellisman MH, Wilson KR (2000) Two-photon video rate image correlation spectroscopy (ICS) and image cross-correlation spectroscopy (ICCS). J Microsc 200:14–25
51. Petersen NO, Höddelius PL, Wiseman PW, Seger O, Magnusson KE (1993) Quantitation of membrane receptor distributions by image correlation spectroscopy: concept and application. Biophys J 65:1135–1146
52. Hebert B, Constantino S, Wiseman PW (2005) Spatio-temporal image correlation spectroscopy (STICS): theory, verification and application to protein velocity mapping in living CHO cells. Biophys J 88:3601–3614
53. Kolin DL, Ronis D, Wiseman PW (2006) *k*-Space image correlation spectroscopy: a method for accurate transport measurements independent of fluorophore photophysics. Biophys J 91(8):3061–3075
54. Digman MA, Sengupta P, Wiseman PW, Brown CM, Horwitz AR, Gratton E (2005) Fluctuation correlation spectroscopy with a laser-scanning microscope: exploiting the hidden time structure. Biophys J 88(5):L33–L36
55. Skinner JP, Chen Y, Mueller JD (2005) Position-sensitive scanning fluorescence correlation spectroscopy. Biophys J 89(2):1288–1301
56. Ruan Q, Cheng MA, Levi M, Gratton E, Mantulin WW (2004) Spatial-temporal studies of membrane dynamics: scanning fluorescence correlation spectroscopy (SFCS). Biophys J 87:1260–1267
57. Berglund A, Mabuchi H (2005) Tracking-FCS: fluorescence correlation spectroscopy of individual particles. Opt Express 13:8069–8082
58. Ries J, Yu SR, Burkhardt M, Brand M, Schwille P (2009) Modular scanning FCS quantifies receptor-ligand interactions in living multicellular organisms. Nat Methods 6(9):643–645

59. Sisan DR, Arevalo R, Graves C, McAllister R, Urbach JS (2006) Spatially resolved fluorescence correlation spectroscopy using a spinning disk confocal microscope. Biophys J 91 (11):4241–4252
60. Kannan B, Guo L, Sudhaharan T, Ahmed S, Maruyama I, Wohland T (2007) Spatially resolved total internal reflection fluorescence correlation microscopy using an electron multiplying charge-coupled device camera. Anal Chem 79(12):4463–4470
61. Wachsmuth M, Waldeck W, Langowski J (2000) Anomalous diffusion of fluorescent probes inside living cell nuclei investigated by spatially resolved fluorescence correlation spectroscopy. J Mol Biol 298(4):677–689
62. Lieto AM, Thompson NL (2004) Total internal reflection with fluorescence correlation spectroscopy: nonfluorescent competitors. Biophys J 87(2):1268–1278
63. Capoulade J, Wachsmuth M, Hufnagel L, Knop M (2011) Quantitative fluorescence imaging of protein diffusion and interaction in living cells. Nat Biotechnol 29(9):835–839
64. Lieto AM, Cush RC, Thompson NL (2003) Ligand-receptor kinetics measured by total internal reflection with fluorescence correlation spectroscopy. Biophys J 85:3294–3302
65. Klar TA, Jakobs S, Dyba M, Egner A, Hell SW (2000) Fluorescence microscopy with diffraction resolution barrier broken by stimulated emission. Proc Natl Acad Sci USA 97:8206–8210
66. Kastrup L, Blom H, Eggeling C, Hell SW (2005) Fluorescence fluctuation spectroscopy in subdiffraction focal volumes. Phys Rev Lett 94:178104
67. Eggeling C, Ringemann C, Medda R, Schwarzmann G, Sandhoff K, Polyakova S, Belov VN, Hein B, von Middendorff C, Schonle A, Hell SW (2009) Direct observation of the nanoscale dynamics of membrane lipids in a living cell. Nature 457:1159–1162
68. Gielen E, van de Ven M, Margineanu A, Dedecker P, Van der Auweraer M, Engelborghs Y, Hofkens J, Ameloot M (2009) On the use of Z-scan fluorescence correlation experiments on giant unilamellar vesicles. Chem Phys Lett 469(1–3):110–114
69. Singer SJ, Nicolson GL (1972) Fluid mosaic model of structure of cell-membranes. Science 175:720–721
70. Vereb G et al (2003) Dynamic, yet structured: the cell membrane three decades after the Singer–Nicolson model. Proc Natl Acad Sci USA 100:8053–8058
71. Thompson TE, Tillack TW (1985) Organization of glycosphingolipids in bilayers and plasma-membranes of mammalian-cells. Annu Rev Biophys Biophys Chem 14:361–386
72. Simons K, Ikonen E (1997) Functional rafts in cell membranes. Nature 387:569–572
73. Sharma P, Varma R, Sarasij RC, Ira, Gousset K, Krishnamoorthy G, Rao M, Mayor S (2004) Nanoscale organization of multiple GPI-anchored proteins in living cell membranes. Cell 116:577–589
74. Kiessling V, Wan C, Tamm LK (2009) Domain coupling in asymmetric lipid bilayers. Biochim Biophys Acta Biomembr 1788:64–71
75. Ramstedt B, Slotte JP (2006) Sphingolipids and the formation of sterol-enriched ordered membrane domains. Biochim Biophys Acta Biomembr 1758:1945–1956
76. Vigh L, Escriba PV, Sonnleitner A, Sonnleitner M, Piotto S, Maresca B, Horvath I, Harwood JL (2005) The significance of lipid composition for membrane activity: new concepts and ways of assessing function. Prog Lipid Res 44:303–344
77. Marguet D, Lenne PF, Rigneault H, He HT (2006) Dynamics in the plasma membrane: how to combine fluidity and order. EMBO J 25:3446–3457
78. Blanchette CD, Lin WC, Ratto TV, Longo ML (2006) Galactosylceramide domain microstructure: impact of cholesterol and nucleation/growth conditions. Biophys J 90:4466–4478
79. Brown DA, London E (1998) Functions of lipid rafts in biological membranes. Annu Rev Cell Dev Biol 14:111–136
80. García-Sáez AJ, Carrer DC, Schwille P (2010) Fluorescence correlation spectroscopy for the study of membrane dynamics and organization in giant unilamellar vesicles. Methods Mol Biol 606:493–508

81. García-Sáez AJ, Schwille P (2008) Fluorescence correlation spectroscopy for the study of membrane dynamics and protein/lipid interactions. Methods 46(2):116–122
82. Simons K, Vaz WLC (2004) Model systems, lipid rafts, and cell membranes. Annu Rev Biophys Biomol Struct 33:269–295
83. Kahya N, Scherfeld D, Bacia K, Poolman B, Schwille P (2003) Probing lipid mobility of raft-exhibiting model membranes by fluorescence correlation spectroscopy. J Biol Chem 278 (30):28109–28115
84. Ariola FS, Li Z, Cornejo C, Bittman R, Heikal AA (2009) Membrane fluidity and lipid order in ternary giant unilamellar vesicles using a new bodipy-cholesterol derivative. Biophys J 96 (7):2696–2708
85. Chiantia S, Schwille P, Klymchenko AS, London E (2011) Asymmetric GUVs prepared by MβCD-mediated lipid exchange: an FCS study. Biophys J 100(1):L1–L3
86. Kubiak J, Brewer J, Hansen S, Bagatolli LA (2011) Lipid lateral organization on giant unilamellar vesicles containing lipopolysaccharides. Biophys J 100(4):978–986
87. Yurlova L et al (2011) Self-segregation of myelin membrane lipids in model membranes. Biophys J 101(11):2713–2720
88. Tai WY et al (2010) Interplay between structure and fluidity of model lipid membranes under oxidative attack. J Phys Chem B 114(47):15642–15649
89. Kahya N, Brown DA, Schwille P (2005) Raft partitioning and dynamic behavior of human placental alkaline phosphatase in giant unilamellar vesicles. Biochemistry 44(20):7479–7489
90. Stachowiak JC et al (2011) Targeting proteins to liquid-ordered domains in lipid membranes. Langmuir 27(4):1457–1462
91. García-Sáez AJ, Ries J, Orzáez M, Pérez-Payà E, Schwille P (2009) Membrane promotes tBID interaction with BCLXL. Nat Struct Mol Biol 16:1178–1185
92. Betaneli V, Petrov EP, Schwille P (2012) The role of lipids in VDAC oligomerization. Biophys J 102(3):523–531
93. Kedrov A et al (2011) A single copy of SecYEG is sufficient for preprotein translocation. EMBO J 30:4387–4397
94. Chiantia S, Kahya N, Ries J, Schwille P (2006) Effects of ceramide on liquid-ordered domains investigated by simultaneous AFM and FCS. Biophys J 90:4500–4508
95. Jonas R et al (2009) Accurate determination of membrane dynamics with Line-Scan FCS. Biophys J 96:1999–2008
96. Chiantia S, Ries J, Kahya N, Schwille P (2006) Combined AFM and two-focus SFCS study of raft-exhibiting model membranes. Chemphyschem 7:2409–2418
97. Chiantia S, Kahya N, Schwille P (2007) Raft domain reorganization driven by short – and long-chain ceramide: a combined AFM and FCS study. Langmuir 23:7659–7665
98. Weiß K, Enderlein J (2012) Lipid diffusion within black lipid membranes measured with dual-focus fluorescence correlation spectroscopy. Chemphyschem 13:990–1000
99. Macháň R et al (2011) Formation of arenicin-1 microdomains in bilayers and their specific lipid interaction revealed by Z-scan FCS. Anal Bioanal Chem 399(10):3547–3554
100. Przybylo M et al (2006) Lipid diffusion in giant unilamellar vesicles is more than 2 times faster than in supported phospholipid bilayers under identical conditions. Langmuir 22 (22):9096–9099
101. Meissner O, Häberlein H (2003) Lateral mobility and specific binding to GABA(A) receptors on hippocampal neurons monitored by fluorescence correlation spectroscopy. Biochemistry 42(6):1667–1672
102. Patel RC et al (2002) Ligand binding to somatostatin receptors induces receptor-specific oligomer formation in live cells. Proc Natl Acad Sci USA 99(5):3294–3299
103. Weidemann T et al (2011) Single cell analysis of ligand binding and complex formation of interleukin-4 receptor subunits. Biophys J 101(10):2360–2369
104. Briddon SJ et al (2004) Quantitative analysis of the formation and diffusion of A1-adenosinereceptor-antagonist complexes in single living cells. Proc Natl Acad Sci USA 101(13):4673–4678

105. Xu L, Pallikkuth S, Hou Z, Mignery GA, Robia SL, Han R (2011) Dysferlin forms a dimer mediated by the C2 domains and the transmembrane domain in vitro and in living cells. PLoS One 6(11)
106. Liu P et al (2007) Investigation of the dimerization of proteins from the epidermal growth factor receptor family by single wavelength fluorescence cross-correlation spectroscopy. Biophys J 93(2):684–698
107. García-Sáez AJ, Buschhorn SB, Keller H, Anderluh G, Simons K, Schwille P (2011) Oligomerization and pore formation by equinatoxin II inhibit endocytosis and lead to plasma membrane reorganization. J Biol Chem 286(43):37768–37777
108. Lasserre R et al (2008) Raft nanodomains contribute to Akt/PKB plasma membrane recruitment and activation. Nat Chem Biol 4(9):538–547
109. Golebiewska U, Nyako M, Woturski W, Zaitseva I, McLaughlin S (2008) Diffusion coefficient of fluorescent phosphatidylinositol 4,5-bisphosphate in the plasma membrane of cells. Mol Biol Cell 19(4):1663–1669
110. Ganguly S, Chattopadhyay A (2010) Cholesterol depletion mimics the effect of cytoskeletal destabilization on membrane dynamics of the serotonin1A receptor: a zFCS study. Biophys J 99(5):1397–1407
111. Larson DR, Gosse JA, Holowka DA, Baird BA, Webb WW (2005) Temporally resolved interactions between antigen-stimulated IgE receptors and Lyn kinase on living cells. J Cell Biol 171(3):527–536
112. Philip F, Sengupta P, Scarlata S (2007) Signaling through a G protein-coupled receptor and its corresponding G protein follows a stoichiometrically limited model. J Biol Chem 282 (26):19203–19216
113. Mueller V et al (2011) STED nanoscopy reveals molecular details of cholesterol- and cytoskeleton-modulated lipid interactions in living cells. Biophys J 101(7):1651–1660

Deciphering Cell Membrane Organization Based on Lateral Diffusion Measurements by Fluorescence Correlation Spectroscopy at Different Length Scales

Vincent Rouger, Cyrille Billaudeau, Tomasz Trombik, Sébastien Mailfert, Yannick Hamon, Hai-Tao He, and Didier Marguet

Abstract The plasma membrane delineating the cell is a complex multicomponent assembly of lipids and proteins. Deciphering the lateral organization of this supramolecular complex on appropriate length and temporal scales is necessary to unravel its implication in biological function. Here, we show how measurements of diffusion may be taken to shed light on membrane function. We first describe the current methods used to report lateral diffusion of membrane components. We then focus on one fluorescent correlation spectroscopy (FCS)-based method, namely, the spot variation FCS (svFCS), which allows the characterization of the modes of molecular confinement within the plasma membrane of living cells. We next illustrate with different biological systems the progress made toward improving our understanding of cell membrane function. We also discuss the findings with regard to the current view of nanoscale domains/assemblies as a general membrane-organizing principle.

Keywords Confinement · Fluorescence correlation spectroscopy · Lateral diffusion · Nanodomains · Plasma membrane

---

V. Rouger, C. Billaudeau, T. Trombik, S. Mailfert, Y. Hamon and H.-T. He
Centre d'Immunologie de Marseille-Luminy (CIML), Aix-Marseille University, UM2 Marseille, France

Institut National de la Santé et de la Recherche Médicale (Inserm), U1104, Marseille, France

Centre National de la Recherche Scientifique (CNRS), UMR 7280 Marseille, France

D. Marguet (✉)
Centre d'Immunologie de Marseille-Luminy (CIML), Aix-Marseille University, UM2 Marseille, France

Institut National de la Santé et de la Recherche Médicale (Inserm), U1104, Marseille, France

Centre National de la Recherche Scientifique (CNRS), UMR 7280 Marseille, France

CIML, Parc Scientifique de Luminy, Case 906, F-13288 Marseille cedex 09, France
e-mail: marguet@ciml.univ-mrs.fr

Y. Mély and G. Duportail (eds.), *Fluorescent Methods to Study Biological Membranes*, Springer Ser Fluoresc (2013) 13: 271–290, DOI 10.1007/4243_2012_56,

Contents

1 Introduction

Biomembranes are complex systems which have been the focus of intensive research. This is mainly because of their molecular composition and physicochemical properties. For biologists, this supramolecular aggregate plays a key role as an active barrier filtering the information that regulates the cell fate. For physicists, the biomembranes are viewed as entities that are made of soft condensed matter capable of building collective cohesion forces using solely weak molecular interactions.

The Singer and Nicolson fluid mosaic model [1, 2], the central paradigm of membrane biology for decades, emphasized greatly the fluid nature of the membrane as the result of homogenizing driving force by Brownian molecular motion and as such failed to perceive local inhomogeneities. Indeed, the ability to achieve the formation of small domains by selective molecular interactions was underestimated, associated for a long time with a lack of proper measurements at the membranes of living cells.

In fact, our understanding of the cell membrane over the years has evolved from it playing a passive role as a barrier surrounding cells to it having a more active function integrating to some extent the chemical specificity of lipids into the biomembrane lateral organization. Considering its two-dimensional fluid nature, molecular motion would cover a large range of time and distance scales: from atomistic scales ranging around tens of angstroms and with characteristic times of picoseconds to very long spatiotemporal scales corresponding to membrane flow and recycling or cell locomotion.

Within this wide scale range, the events occurring on micrometer and millisecond scales are the focus of this review. In many cases, the mobility of membrane components is under some form of biological control external to the membrane. For example, protein motion may be confined to domains by interactions with peripheral structures such as the cytoskeleton or extracellular matrices. However, molecular motion is primarily influenced by the intrinsic properties of the molecule itself and of the membrane bilayer solvent: lateral diffusion relates to the membrane viscosity and will ultimately influence the dynamics of the molecular interactions taking place within or in the vicinity of the cell membrane. Therefore, lateral diffusion and molecular interaction are intrinsically linked. As a consequence, the membrane environment plays a key role in finely tuning biochemical reactions.

From a strict physicochemical point of view, membrane inhomogeneity can be considered as the direct consequence of a collective cohesion generated by weak molecular interactions (or repulsions) between an extraordinary diversity of lipid and protein components. Indeed, the amphiphilic structure of lipids, which constitutes the matrix of any cell membrane, can itself form the basis of diversity allowing over a thousand different species [3]. This results from the numerous possible combinations of different polar head groups with two different hydrophobic tails of variable chain length and level of unsaturation of carbon-carbon bonds. The molecular complexity of proteins is even more prominent with the associated and intrinsic membrane proteins representing up to 30 % of the cellular proteins. At the plasma membrane, most of them form multimeric complexes with large domains extending out of the membrane and accounting for a significantly larger area than that occupied by the transmembrane domains alone [4–6]. Moreover, a number of permanent or reversible posttranslational modifications are made during the protein maturation pathway. Therefore, the spatiotemporal posttranslational status of a protein will delineate another level of inhomogeneity. Finally, a significant level of asymmetry exists between the inner and outer membrane leaflets as a result of the protein insertion within membranes and of the asymmetric distribution of the different classes of lipids. From a biological point of view, investigating the dynamics of plasma membrane organization with appropriate spatiotemporal resolution and noninvasive approaches allowing the characterization of local inhomogeneity in living cells has proven to be and still remains a challenge.

The key issue is determining on which spatiotemporal scales membrane inhomogeneities take place which will allow the investigation of a possible role for such local organization on the regulation of biological functions such as signal transduction. The lipid raft concept introduced in the 1990s has stimulated the field of biomembrane research and has functionally implicated lipids in the regulation of receptor functions (for review, see [7, 8]). Much controversy had emerged as to the existence of such rafts, due mainly to the weak capability of investigating the cell membrane with appropriate methodologies [9]. Today, advanced technologies have produced compelling data depicting a lipid-based inhomogeneity within cell membranes, the extent and duration of which vary according to the dynamic association of membrane components [10].

In the present review, we will focus on the fluorescence-based approaches which have provided significant insight into the lateral organization of the cell membrane and more particularly on spot variation fluorescence correlation spectroscopy (svFCS). The rationale for recording FCS measurements at different spatial scales of observation will be detailed. We will summarize the main results obtained in the characterization of local inhomogeneity. In the last part, we will illustrate the functional implications of such nanodomain organization on different signal transduction mechanisms.

2 Lateral Diffusion in Cell Membranes by Fluorescence-Based Methods

Microscopic techniques have been adapted to study dynamic processes in living cells and more specifically to investigate membrane dynamics. Among them, the most commonly used methods aim at recording signals from fluorescently tagged proteins or lipids. Two classes of methods have been set up: those resolving single-molecule motion with appropriate spatiotemporal resolution (single-particle tracking, SPT) and those based on analyzing the collective behavior of molecules (fluorescence recovery after photobleaching (FRAP) and FCS). While the latter methods do take an average of many events, specific experimental strategies have been implemented which allow the resolution of potential inhomogeneity in the diffusion characteristics of complex molecular events. Here, we aim to sum up the principles behind these strategies and to discuss their advantages and caveats. Our review will not however include discussion on the methods yielding quantitative information on specific membrane environments which are based on biophotonic microscopy in combination with specific fluorescent lipophilic tracers (e.g., Laurdan).

2.1 Single-Particle Tracking (SPT) and Related Methods

The SPT and related methods aim at tracking single objects over long periods of time in order to investigate the diffusion characteristics at a single-molecular level. Initially, SPT was based on recording on a wide-field microscope, and at video rate, images of the light diffracted by submicrometer-sized particles (latex or polypropylene beads, gold nanoparticles) attached to a molecule of interest via specific interaction. The principle was to localize the particles in each image with very high spatial resolution and, therefore, to build individual trajectories by following their motion over time [11]. With the development of sensitive cameras (i.e., electron-multiplying charge-coupled device camera, EMCCD), it has been possible to extend the method to track fluorescently labeled molecules in living cells [12, 13]. This methodological extension circumvents one major drawback of the SPT method – the attachment of large objects on the molecule of interest.

The commonly used fluorescent microspheres or quantum dots still have some limitations relating to their size or weak determination of valence for the tagged molecules potentially influencing the mobility of the molecules. Where possible, one should privilege the use of minimally invasive single dye-tagged molecules. Nowadays, SPT can be easily set up on regular microscopes in epi-, confocal, or total internal reflection (TIR) fluorescence modes. With fluorescent tags, the spatial resolution relates directly to the signal-to-noise ratio: the use of an EMCCD camera in association with adapted image processing allows the localization of molecules significantly below the diffraction limit imposed by classical optical physics (i.e., at a few tens of a nanometer in the *xy* plane).

The data analysis follows three steps: (1) the localization of single molecules within each frame, (2) the tracking of them over the stack of images via the reconstruction of trajectories, and (3) the analysis of their diffusion based on their trajectories. This image processing is computed by different algorithms [14, 15]. The localization of the molecule is commonly done by fitting a Gaussian function to the intensity distribution of the fluorescent signal. An alternative method has been proposed by designing a pattern recognition algorithm which improves the precision when tracking particles of finite size but only for two-dimensional motion [16]. The trajectory reconstruction frequently uses robust nearest-neighbor reconnection algorithms. Trajectories are then usually analyzed by calculating the mean square displacement (MSD) which represents the average distance that a molecule travels during the lag time. The biological significance of the shape of the MSD as a function of the lag time has to be compared with theoretical models taking into account different geometries for two-dimensional diffusion in plasma membrane (for review, see [17]). Tracing of a large number of targets expands the potential of single-molecule approaches as well as the statistical relevance, with more observations. Among the methods developed to challenge the high density, the multiple-target tracing (MTT) algorithm gives accurate localization and exhaustive detection, combined with integrated historical information of trajectories, in order to provide efficient multiple-target reconnection [18]. Robustness was achieved by minimum *a priori* hypotheses on the underlying diffusion phenomena and a minimal set of parameters to adjust. Reconnection in turn allowed *a posteriori* validation of detected particles.

In conclusion, the SPT approaches have attracted a tremendous amount of interest mainly due to their simple implementation on commercial microscopes and to the beauty of the reported observations reaching nanometer-scale spatial resolution and millisecond-scale temporal resolution at a single-molecular level. Several aspects need however to be considered when performing single-molecule fluorescence imaging. The molecule of interest must be detected by SPT at low concentrations in order to distinguish individual fluorescent molecules without overlap between their respective point-spread function (PSF). One should assume that the molecules of interest are detected with the same probability. The light intensity has to be maximized to achieve high SNR without photodamage. In the meantime, the integration time of an image has to be adjusted to obtain a real snapshot of the motion. This remains a strong limitation of this technique which inherently favors the examination of slowly diffusing particles. Moreover, it is necessary to record images over long periods of time to accurately describe the trajectories and identify subdiffusion on confined diffusion events. The quality of the raw data recorded using minimally invasive fluorophores is strongly impaired by photobleaching and weak SNR; this respectively limits the length of the trajectories and the precision of localization. In this specific circumstance, the MSD analysis averages multiple single-molecule trajectories. Moreover, there is a need to improve the z resolution when trajectories are recorded on a cellular scale over a long period of time. TIRF microscopy and the use of a cylindrical lens [19, 20] are current solutions to overcome this limitation.

2.2 Fluorescence Recovery After Photobleaching (FRAP)

With the regular implementation of a FRAP module on commercial microscopes, this method is certainly the most popular technique to measure lateral diffusion in cell membranes (for review, see [21]). Exploiting the intrinsic photobleaching properties of fluorophores, the method is performed in a variety of ways. In spot FRAP, the laser light focused through the objective of the microscope delineates a small volume of excitation on a region of interest (ROI). A pulse of high laser intensity photobleaches the fluorophores present in this ROI which is then observed at low light of excitation in order to record the fluorescence recovery over time. Organic fluorophores or fluorescent proteins can be used to label the molecule of interest; in any case, the level of labeling has to be sufficiently high since the quality of the fluorescence recovery will depend on the reservoir of molecules surrounding the bleached ROI. The curve of the fluorescence intensity over time has an exponential-like shape from which kinetic parameters such as the half-time of recovery and the mobile fraction are quantified using appropriate fitting models [22–25]. For membrane components, various processes may simultaneously contribute to the overall recovery kinetics such as lateral diffusion, interactions, or reactions of the surrounding fluorescent molecules. This is an ensemble averaging method that is weakly sensitive to separate subpopulations with different characteristics.

Nowadays, FRAP experiments are commonly performed on a laser scanning confocal microscope rather than using a spot FRAP setup. Although the confocal FRAP is easy to implement, it has some considerable drawbacks which have led to significant discrepancies in the literature. Indeed, fluorescence recovery is biased if the photobleaching step is slow by comparison with the lateral diffusion. This effect is even more prominent for large ROI. Moreover, the fitting model has to integrate not only the size but also the shape of the bleached ROI. Another drawback concerns the FRAP experiments in general which are often associated with undesirable effects relating to the photobleaching process locally generating chemical reactions. While these are difficult to estimate experimentally in biological systems, they might nevertheless significantly alter measurements.

2.3 Fluorescence Correlation Spectroscopy (FCS) and Related Methods

FCS-based methods are a compelling alternative to FRAP. In fact, FCS was introduced by Magde, Elson, and Webb more than 30 years ago [26, 27] but only gained a renewal of interest in the 1990s for the study of the dynamics in living systems when it became possible to perform the observations on a microscope platform [28, 29]. The principle is based on the recording of temporal fluctuations in fluorescence emission in a small volume of observation. While the average concentration of molecules is expected to remain constant for long time periods, the value spontaneously fluctuates locally within the system due to Brownian motion in space and as a Poisson process for chemical reaction. FCS has the potential to describe phenomena which occur on nanosecond to second scales and therefore can provide information about both kinetics and thermodynamic properties of fluorescent molecules. The diffusion coefficient is classically extracted by the temporal relaxation of the recorded fluorescence fluctuations.

In a system in thermodynamic equilibrium, these fluctuations are the result either of the movement of the fluorescently labeled molecules in and out of the excitation volume, due to, *e.g.*, a change in local concentration, or to photochemical processes of the fluorescent dyes, such as changes in intersystem crossing to a nonfluorescent triplet state. Usually, the fluctuations due to lateral diffusion occur on a more lengthy time scale than those due to photochemical processes, thus allowing the separation of their respective contributions by appropriate statistical analysis.

The experimental setup is quite simple. The laser beam is focalized by the objective of the microscope as a small volume of excitation (~0.3 fL) on the plasma membrane. The emitted photons are collected on a confocal mode by an avalanche photodiode, and the signal can be immediately processed. The fluorescence fluctuates due to molecular diffusion into and out of the observation volume. Therefore, the diffusion process has a direct impact on the rate at which the fluctuations dissipate within a temporal measurement, defined as a degree of correlation. Data are usually immediately computed by calculating the autocorrelation function (ACF), a statistical tool dedicated to the analysis of temporal signal fluctuations. The shape of the ACF curve provides information on the time scale of the fluctuations within a calibrated volume. It therefore allows a relationship to be drawn between τ_d and the lateral diffusion coefficient D.

The FCS-based approaches provide a suitably temporal resolution for studying plasma membrane organization in living cells (see for review [30–33]). Moreover, FCS is minimally invasive: it operates at low light intensity with low probe concentrations (~nM range). This is of particular interest to overcome experimental bias due to the overexpression of fluorescently tagged membrane components. However, performing FCS measurements on living samples requires the careful control of key parameters such as the concentration of fluorescent molecules, labeling impact, or laser power. High concentrations mask the fluctuations in the signal intensity, and the recorded autocorrelation function becomes almost impossible to fit properly due to the marginal contribution of the fluctuations. Fluorescent dyes also need to be photostable; otherwise, molecules disappear before leaving the excitation volume which can lead to an apparent reduced diffusion time. Laser power must be adjusted to minimize photobleaching but yet provide a good count rate per molecule over a long enough acquisition time.

To date, most observations on cell membranes have been made by FCS, a few of which were made as a result of fluorescence cross-correlation spectroscopy (FCCS), which extends the observation to two species to monitor interactions or dynamic colocalization by spatiotemporal coincidence [34–36]. Moreover, different FCS modalities such as scanning FCS [37–41], two-focus FCS [42, 43], or z-scan FCS [44] have been beneficial with respect to the quantification of relevant observables on cell membranes.

2.4 Conceptual Basis of Spot Variation FCS

In a complex membrane environment, establishing a diffusion coefficient value at a single spatial scale of observation is of rather limited interest. Indeed, where the lateral membrane organization is concerned, local inhomogeneity should exist as a balance between intrinsic Brownian agitation and differential interaction among the extreme diversity of membrane components. Any single molecule experiences a combination of events over time including direct or indirect cytoskeletal interactions [45, 46], molecular crowding, and confinement within lipid-dependent domains. As a result, one should predict an overlapping distribution of the instant D values. However, conventional FCS observations, i.e., at a single waist of observation, will only discriminate subsets of molecular components when their respective diffusion coefficients D differ by a factor below ~1.6 [47]. Therefore, the autocorrelation function is usually described by an anomalous diffusion model [48–50], but assessing the biological significance of this model is rather difficult.

We have proposed to overcome this limitation by varying the focal excitation volume in order to analyze the diffusion time τ_d at different spatial scales [51–53]. This approach, namely, the spot variation FCS (svFCS), relates to the classical Fick's second diffusion law from which a strict proportionality between the diffusion time τ_d and the probed area (i.e., the square of the beam waist w^2) is expected. Experimentally, we evaluate on living cells the mean diffusion times of the fluorescently tagged molecules through observation volumes of different sizes which vary between 0.2 and 0.5 μm in radius [53]. Then, we plot τ_d as a function of w^2. Two key observables are quantified: the inverse of the slope which relates to the effective diffusion coefficient D_{eff} and the intercept on the diffusion time axis, namely, t_0. Any deviation of the t_0 from the origin is indicative of a potential confinement of the molecule of interest. Therefore, the svFCS is a perfect analog to single-particle tracking in the time domain [17, 54]. The experimental methods with which to perform svFCS are detailed elsewhere [55].

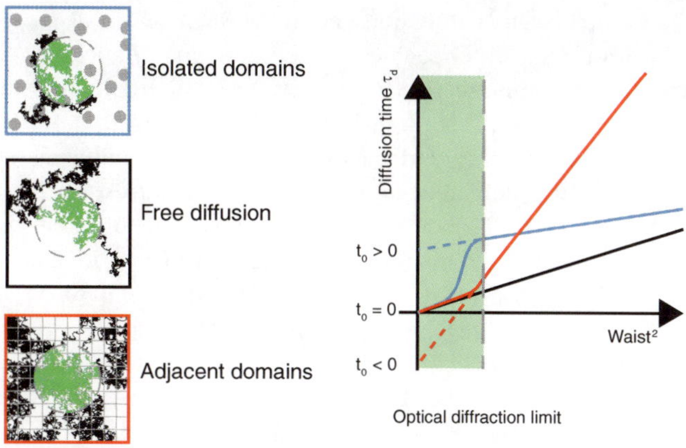

Fig. 1 *Confinement models for membrane organization probed by svFCS* – In the free diffusion model (*center panel*), molecules can freely diffuse within the plasma membrane. On the contrary, the presence of constraints modifies the diffusion of molecules. The presence of permeable barriers in the adjacent domain model prevents the free diffusion of the molecules (*bottom panel*). In the presence of isolated domains, the molecules can diffuse into and out of the domains where they are transiently trapped (*top panel*). As a consequence, the diffusion times evaluated by svFCS reveal three regimes by the plot of the diffusion time τ_d as a function of the squared waist of observation w^2. The free diffusion model exhibits a strict linearity of τ_d with w^2 with a null intercept with the time axis ($t_0 = 0$). In the presence of molecular confinement, the plot of the diffusion time (in a range accessible by the optical diffraction limit) is characterized by an affine function which exhibits an intercept with the time axis that is either negative ($t_0<0$) for adjacent domains or positive for isolated domains ($t_0>0$)

We can predict by simulations the possible effect of different geometrical membrane architectures on svFCS observations (see Fig. 1) [51]. In the absence of any constraint, the molecular motion is characterized by a constant diffusion coefficient D corresponding to the mean surface explored by a molecule per unit of time: the plot of τ_d versus w^2 is a straight line with a null value for t_0 and a slope inversely proportional to D. When adjacent permeable domains mimicking physical barriers such as those imposed by the actin-based cytoskeleton [46] are introduced, the molecules diffuse freely inside the domains (with a microscopic diffusion coefficient D_μ) before reaching the barriers which can be crossed with a given probability relating to the confinement strength. If the observation area is smaller than l, the size of the meshwork, a linear regime is observed similar to the one previously observed for free diffusion. At large observation volumes, the barriers are perceived as obstacles, increasing the diffusion time as compared to free diffusion. The plot of τ_d versus w^2 conserves the straight linearity, but the slope is now proportional to an effective diffusion coefficient D_{eff}, and the extrapolated t_0 intercept takes a negative value. The D_{eff} depends on D_μ and the ratio τ_{conf}/τ_{dom}, where τ_{conf} and τ_{dom} are the confinement time and the diffusion time into the domain, respectively. The negative t_0 intercept is mainly dominated by τ_{conf} when τ_{dom} becomes negligible. Recently, the offset of t_0 was derived as a function of the size parameter l and the effective diffusion coefficient as $t_0 \approx l^2/(36D_{eff})$ [56]. In a third model, we introduced permeable isolated domains mimicking, for instance, lipid-dependent nanodomains (see for review [32]). Particles can diffuse into and out of such domains. The transit time τ_d is the sum of the time the molecules are freely diffusing and the time spent within domains. τ_d depends on the domain density, confinement strength, and diffusion coefficients inside and outside the domains. The plot of τ_d versus w^2 reveals different regimes. For small observation areas by comparison with the size of the domains, the plot is linear such as the one observed for the free diffusion model. At larger scales of observation, the plot conserves a straight linearity, but the slope is proportional to an effective diffusion coefficient D_{eff}, and the extrapolated t_0 intercept takes a positive value. In such an instance, t_0 depends on the fraction of molecules α confined within the domains and the confinement time $t_0 \approx 2\alpha(\tau_{conf} - \tau_{dom})$, and D_{eff} is sensitive to the free diffusion constant and α by the following equation $D_{eff} \approx (1 - \alpha)D_{free}$. In a more detailed description for these parameters [57], the positive t_0 offset value is derived from general analytical expressions:

$$t_0 = \beta\tau_{trap} \exp\left(-\frac{4}{\sqrt{1-\beta}}\frac{D_{trap}}{D_{free}}\right)$$

$$D_{eff} = (1-\beta)D_{free} + \beta D_{trap}$$

where $\beta = \tau_{trap}/(\tau_{trap} + \tau_{free})$ denotes the trapped time fraction, D_{free}, D_{trap} the diffusion coefficient of free or trapped molecules, and τ_{trap} and τ_{free} the duration of free or trapped status, respectively. t_0 gives information on the confinement strength.

To recapitulate on the interpretation of the svFCS experimental data, the plot of τ_d versus w^2 is linear and intercepts the time axis origin in the case of Brownian motion. Otherwise, the plot conserves a straight linearity, but the extrapolated t_0 intercept correlates with the dominant mode of confinement: t_0 becomes negative for adjacent domains and positive for isolated domains. Although the svFCS method offers the spatiotemporal resolution necessary to characterize in living cells the inhomogeneous lateral organization of the plasma membrane, it is important to stress that the geometrical modes of confinement are not mutually exclusive: a deviation of t_0 in one direction or the other solely reveals a dominant mode of confinement [52].

Moreover, the svFCS method is limited by the classical optical diffraction limit (i.e., ~200 nm). This limit is significantly larger than the domains confining the molecules within the cell membranes; therefore, the analysis of the confinement is deduced from the t_0 value extrapolated from the plot of τ_d versus w^2. This limitation has been overcome by implementing the observations through single nanometric apertures of radii varying between 75 and 250 nm [58], near-field optical scanning microscopy [59], or more recently by combining FCS with the emerging STED microscopy [60, 61].

3 Dynamic Molecular Confinement in the Plasma Membrane Within Microdomains and by the Cytoskeleton Meshwork Revealed with svFCS Analysis

By performing svFCS analysis, we and others were able to examine the nanoscale lateral diffusion of a number of proteins and lipid analogs in the plasma membrane of different cell types (see for review [32]). The latter include both adherent (COS-7, TE671, mouse embryonic fibroblast (MEF)) and nonadherent cells (T cell lines, primary T and NK lymphocytes) [51, 52, 58, 62–69]. We observed that in general the molecular diffusion in the plasma membrane is constrained by either meshwork or domain/assembly organizations. While the first depends essentially on the actin cytoskeleton, the second can be promoted by different mechanisms even though raft domains presumably constitute the most prominently observed component (see below). These observations led to the conclusion that cytoskeleton-based and raft-dependent organizations are the two main compartmentalizing forces at work in the plasma membrane.

3.1 *svFCS Analysis Shows the Occurrence of Protein Hop Diffusion in the Plasma Membrane of Living Cells*

The motion of TfR-GFP in the plasma membrane of COS-7 fibroblasts is mainly constrained by the meshwork organization as suggested by a negative t_0 value characterized in svFCS. Following depolymerization of F-actin, TfR-GFP recovers an almost free-like motion. In addition, the t_0 is even more negative when actin polymerization is strengthened. TfR-GFP thus experiences the F-actin-based hop diffusion in these cells, as previously observed in normal rat kidney (NRK) fibroblast cells [70]. Altogether, these svFCS measurements performed with drugs acting on the cytoskeleton permitted establishing that the mesh size for the actin meshwork in COS-7 cells probed by TfR-GFP diffusion can be estimated at ~230 nm. These data appear in very good agreement with those obtained using the electron microscope and freeze-fractured specimens for actin cytoskeleton meshes in NRK cells [71]. This first set of observations has provided the proof of principle that svFCS reaches similar levels of sensitivity as SPT methods with respect to characterizing molecular confinement in living cells.

Moreover, Rose et al. [72] have found that at the surface of unstimulated primary human CD4 T lymphocytes, the IL7 receptor complex exhibits hop diffusion due to actin meshwork-dependent constraints, which could be released by depolymerizing the F-actin using cytochalasin D. Based on an approach very similar to svFCS, by producing the diffusion time τ_d plot as a function of Δz^2 using z-scan FCS (zFCS) [44], it was recently reported that the serotonin$_{1A}$ (5-HT$_{1A}$) receptor, which belongs to the G protein-coupled receptor (GPCR) superfamily, also experiences actin cytoskeleton-based hop diffusion at the surface of CHO-K1 cells and that this behavior is not altered upon serotonin binding [73].

3.2 svFCS Enables the Probing of Dynamic Membrane Nanodomains/Nanoassemblies in the Plasma Membrane of Living Cells

Identification of sphingolipid- and cholesterol-dependent nanodomains – As recapitulated previously (see section "Conceptual Basis of Spot Variation FCS"), svFCS analysis has the ability to detect dynamic molecular trapping into small membrane domains as featured by the positive deviation of the t_0 values [52]. Interestingly, some membrane domains were found to trap particular proteins and lipids previously postulated to be enriched in the so-called raft domains, the formation of which is believed to be promoted by condensation between sphingolipids, particularly sphingomyelin, and cholesterol. We thought that these entities could correspond to the long search for lipid rafts. Subsequent experiments combining svFCS analysis with pharmacological treatments of the cells have supported this hypothesis. Indeed, the membrane domains that trap fluorescently labeled ganglioside GM_1 (Bodipy-GM_1) and GFP-conjugated glycosylphosphatidylinositol (GPI)-anchored proteins (GFP-GPI) were found to be sensitive to sphingomyelinase (SMase) and cholesterol oxidase (COase) treatments. Nevertheless, SMase- and COase-mediated reactions also produce ceramide and cholestenone, respectively: these products can potentially alter raft domain organization. To further assess the involvement of sphingolipids and cholesterol in nanodomain formation, their synthesis was inhibited using myriocin and zaragozic acid, respectively. Individual treatment with myriocin or zaragozic acid had at best a partial effect on confinement of the GFP-GPI. The combined treatment, however, fully suppressed such confinement, indicating domain abolishment [66]. Thus, sphingolipids and cholesterol were found to act together and play an essential role in promoting the formation of the membrane domains, thus indicating that these entities are indeed lipid rafts.

Sensing raft nanodomains in both plasma membrane leaflets – By carrying out svFCS with GFP-conjugated proteins solely attached to either the outer or inner leaflet of the plasma membrane, lipid-dependent domains were found to be present in the two leaflets. Since sphingolipids are abundant in the outer leaflet of the plasma membrane, their interactions with cholesterol at this location presumably participate directly in this domain formation. However, the contribution of sphingolipids in the inner leaflet would be indirect considering the relative paucity of this lipid species there. It is therefore hypothesized that lipid interdigitations between the two leaflets or transmembrane proteins would contribute to this organization [74].

Of interest, not only did the proteins that are integrated into the inner leaflet of the plasma membrane exhibit raft-based constrained diffusion, the peripheral proteins bound noncovalently to the inner leaflet lipids also exhibited the same type of diffusion. Upon accumulation of phosphatidylinositol-3,4,5-triphosphate (PIP_3) in the plasma membrane inner leaflet following PI(3)K activation triggered by the surface receptor stimulation, Akt/PKB is recruited to the plasma membrane via PIP_3 binding, where it is phosphorylated and activated by membrane-localized kinases. We found that raft disruption following inhibition of cholesterol and sphingolipid biosynthesis inhibited Akt membrane association and phosphorylation, supporting the necessity for rafts in Akt signaling [66]. Furthermore, svFCS analysis allowed the direct demonstration of Akt recruitment into PIP_3-containing raft domains. This example also illustrates that raft nanodomains could play an important role in organizing signaling networks at the cytoplasmic surface of the plasma membrane.

Lipid-dependent nanodomains were initially identified by svFCS on COS-7 cells but have now been observed in many other cell types including primary mouse embryonic fibroblasts, primary mouse T lymphocytes, human and mouse T cell lines, or human NK cells [32]. However, we would like to emphasize that the svFCS characterization of confinement within nanodomains is not systematically associated with lipid-based nanodomains. For instance, for the neural cell adhesion molecule (NCAM), a positive t_0 was observed indicative of a dynamic confinement within nanodomains that is insensitive to cholesterol and sphingolipid membrane content and rather dependent on the integrity of the actin cytoskeleton [64]. In this example, the confinement is also controlled by NCAM polysialylation, the posttranslational addition of a polymer of N-acetylneuraminic acid residues on the protein known to be an important regulator of NCAM functional activity [64].

Essential characteristics of raft nanodomains – As already discussed, simulation studies have showed that in simple situations, the t_0 depends both on the partition coefficient α and the confinement time τ_{conf} within the domain, which can be expressed as $t_0 \approx 2\alpha.\tau_{conf}$. From this, it could be concluded that molecules are likely to be dynamically partitioned into rafts with a timescale of a few to tens of milliseconds. Our models postulated that the molecules are transiently trapped within discrete nanodomains and implicitly assumed that the trapping period is either shorter than or equal to the lifetime of the nanodomain [51]. Therefore, the models fitted situations involving either preexisting membrane domains or self-promoting membrane complexes. In the latter case, τ_{conf} would therefore represent the half-life of nanodomains.

Raftophilic molecules exhibiting affine diffusion regimes with positive t_0 intercepts provide an estimation of the upper limit of the size of an average domain, which is 60 nm in radius. In order to further investigate domain size, svFCS was performed in COS-7 cells using nanometric apertures (nanoholes) of different sizes, which can overcome the limit of spatial resolution set by the diffraction of light, as illustrated in Fig. 2 for GFP-GPI [58]. For both Bodipy-GM_1 and GFP-GPI, two distinct diffusion regimes were observed when aperture radius was varied. Below a characteristic radius value of ~100 nm, the transit time increased rapidly and linearly as a function of nanohole area, with a diffusion regime intercepting the origin. Above this value, transit time decreased dramatically and quickly resulted in an affine diffusion regime. COase treatment led to a free-like linear diffusion, comparable to that obtained with bodipy-coupled phosphatidylcholine (Bodipy-PC). Considering the point of transition between diffusion regimes, the average

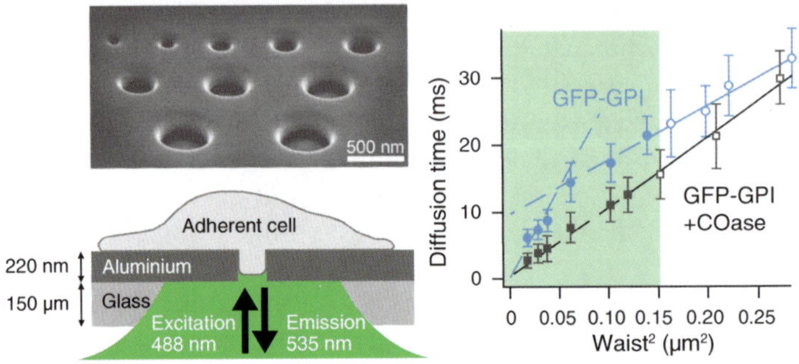

Fig. 2 *svFCS observation through nanometric apertures – Left panel*: a metallic film deposited onto a glass substrate had been drilled with nanometric apertures of radius between 75 and 250 nm. FCS was then performed by collecting the fluorescence emitted light coming from a single nanohole over the time allowing achieving observation areas below the optical diffraction limit. *Right panel*: experimental plot of diffusion time τ_d as a function of squared waist w^2 of observation for GFP-GPI molecules in COS-7 cells evaluated with classical svFCS and nanoholes. The modification of cholesterol composition of the plasma membrane by COase modifies the diffusion mode of GFP-GPI from a diffusion constrained by isolated domains enriched in cholesterol to free diffusion

confinement area that transiently trapped Bodipy-GM_1 and GFP-GPI in COS-7 cells was estimated to be 40 and 30 nm in radius, respectively.

Diverse molecular determinants are involved in the lipid-dependent nanodomain partitioning of membrane proteins – The svFCS approach was used to investigate the membrane dynamics of two members of the TNFR protein family, Fas/CD95 and Fas ligand (FasL), in relationship to their biological functions [62, 63, 65, 67, 68]. By transducing signals inside cells upon mutual binding, they are involved in various physiological and pathological mechanisms. In particular, Fas is capable of transducing signals that lead either to cell death or to growth and is thus involved in the regulation of cell development and homeostasis. We found that both Fas and FasL are dynamically partitioned into raft nanodomains, and this partitioning is required for FasL to efficiently interact with Fas and for Fas to transduce cell death signals. One of the raft-targeting signals for Fas and FasL is protein S-palmitoylation. Indeed, mutation of the cysteine residues that are responsible for the palmitoylation of human Fas and FasL abolishes their partitioning into raft nanodomains and strongly impairs their ability to elicit cell death [63, 65]. These findings also provide further evidence for the important role of raft nanodomains in the function of the Fas-FasL system. However, beside protein S-palmitoylation, other structural elements of Fas have been implicated in the trapping of Fas within rafts. These include (1) an extracellular glycosphingolipid-binding motif [67] and (2) a lysine-rich region in the cytoplasmic, membrane-proximal region [68]. Thus, the raft-partitioning of Fas could be regulated by different mechanisms, each of which may potentially be exploited to finely tune the activity of Fas. This could then lead to the discovery of interesting targets in the development of new therapeutic strategies against various diseases involving the Fas-FasL system.

3.3 Fence and Lipid-Dependent Confinements Are Two Major Constraints that Impede the Diffusion of Membrane Components

Altogether, these results indicate that membrane confinement originates from two major processes: one mediated by the cytoskeleton meshwork and the other depending on specific lipid interactions. TfR motion is mainly regulated by actin-based membrane skeleton fences whereas the molecular confinement of the lipid analogs, Bodipy-GM_1 and Bodipy-SM, and the GFP-tagged GPI-anchored proteins is mainly regulated by cholesterol and sphingomyelin. Our findings also suggest that these two confinement mechanisms are not mutually exclusive but rather simultaneously play a role in the diffusion of a given component. Very interesting features are revealed when the motion of the transmembrane proteins, such as TfR-GFP and DPP_{IV}-GFP proteins, is monitored after treatments that disrupt the actin-based cytoskeleton and/or modify the membrane cholesterol. Indeed, a single treatment abolishing one of the two major hindrance mechanisms unmasks the other. It is only following a double treatment that recovery of a free-like diffusion law is observed.

4 Conclusion and Future Outlook

While SPT does gather images of individual molecules over time without masking the molecular motion by an ensemble averaging, FRAP and FCS methods do still offer the advantage of collecting a large amount of data and allow immediate data analysis. Among the latter two methods, svFCS has proven to be one of the most efficient experimental approaches at detecting molecular confinement within the plasma membrane of living cells (Fig. 1).

Our results on the expression of TfR-GFP and DPP_{IV}-GFP proteins in COS-7 have shown that the svFCS is sensitive to meshwork confinement. The good level of agreement that we found with previous works on the lateral mobility of TfR ascertained the validity of our new experimental approach [70, 71]. Moreover, our experimental approach has provided compelling evidence for the existence of lipid rafts in living cells. Such detection ability may be directly linked to the structural features of raft nanodomains [7, 8]. The svFCS and related approaches display high temporal resolution, access spatial information on a minute scale, and are in general minimally invasive; high statistical accuracy is another prominent trait of these methods.

Acknowledgments This research project was supported by institutional grants from Inserm and CNRS and by specific grants from ANR, ARC, FRM, INCa, and CNRS. VR was awarded fellowships from the LNFCC. We thank Emily Witty (AngloScribe) for editing the English.

References

1. Singer SJ, Nicolson GL (1972) The fluid mosaic model of the structure of cell membranes. Science 175(23):720–731
2. Singer SJ (2004) Some early history of membrane molecular biology. Annu Rev Physiol 66:1–27. doi:10.1146/annurev.physiol.66.032902.131835
3. van Meer G (2005) Cellular lipidomics. EMBO J 24(18):3159–3165. doi:10.1038/sj.emboj.7600798
4. Binda C, Newton-Vinson P, Hubalek F, Edmondson DE, Mattevi A (2002) Structure of human monoamine oxidase b, a drug target for the treatment of neurological disorders. Nat Struct Biol 9(1):22–26. doi:10.1038/nsb732
5. Engelman DM (2005) Membranes are more mosaic than fluid. Nature 438(7068):578–580
6. Bracey MH, Hanson MA, Masuda KR, Stevens RC, Cravatt BF (2002) Structural adaptations in a membrane enzyme that terminates endocannabinoid signaling. Science 298(5599): 1793–1796. doi:10.1126/science.1076535
7. Lingwood D, Simons K (2010) Lipid rafts as a membrane-organizing principle. Science 327 (5961):46–50. doi:10.1126/science.1174621
8. Bagatolli LA, Ipsen JH, Simonsen A, Mouritsen OG (2010) An outlook on organization of lipids in membranes: searching for a realistic connection with the organization of biological membranes. Prog Lipid Res 49(4):378–467
9. Munro S (2003) Lipid rafts: elusive or illusive? Cell 115(4):377–388
10. Simons K, Gerl MJ (2010) Revitalizing membrane rafts: new tools and insights. Nat Rev Mol Cell Biol 11(10):688–699. doi:10.1038/nrm2977
11. Saxton MJ, Jacobson K (1997) Single-particle tracking: applications to membrane dynamics. Annu Rev Biophys Biomol Struct 26:373–399. doi:10.1146/annurev.biophys.26.1.373
12. Schutz GJ, Kada G, Pastushenko VP, Schindler H (2000) Properties of lipid microdomains in a muscle cell membrane visualized by single molecule microscopy. EMBO J 19(5):892–901. doi:10.1093/emboj/19.5.892
13. Sako Y, Minoghchi S, Yanagida T (2000) Single-molecule imaging of egfr signalling on the surface of living cells. Nat Cell Biol 2(3):168–172. doi:10.1038/35004044
14. Schmidt T, Schutz GJ, Baumgartner W, Gruber HJ, Schindler H (1996) Imaging of single molecule diffusion. Proc Natl Acad Sci USA 93(7):2926–2929
15. Kalaidzidis Y (2007) Intracellular objects tracking. Eur J Cell Biol 86(9):569–578
16. Levi V, Ruan Q, Gratton E (2005) 3-d particle tracking in a two-photon microscope: application to the study of molecular dynamics in cells. Biophys J 88(4):2919–2928. doi:10.1529/biophysj.104.044230
17. Wieser S, Schutz GJ (2008) Tracking single molecules in the live cell plasma membrane-do's and don't's. Methods 46(2):131–140. doi:10.1016/j.ymeth.2008.06.010
18. Serge A, Bertaux N, Rigneault H, Marguet D (2008) Dynamic multiple-target tracing to probe spatiotemporal cartography of cell membranes. Nat Methods 5(8):687–694. doi:10.1038/nmeth.1233
19. Kao HP, Verkman AS (1994) Tracking of single fluorescent particles in three dimensions: use of cylindrical optics to encode particle position. Biophys J 67(3):1291–1300. doi:10.1016/S0006-3495(94)80601-0
20. Holtzer L, Meckel T, Schmidt T (2007) Nanometric three-dimensional tracking of individual quantum dots in cells. Appl Phys Lett 90(5):053902–053903
21. Chen Y, Lagerholm BC, Yang B, Jacobson K (2006) Methods to measure the lateral diffusion of membrane lipids and proteins. Methods 39(2):147–153. doi:10.1016/j.ymeth.2006.05.008
22. Axelrod D, Koppel DE, Schlessinger J, Elson E, Webb WW (1976) Mobility measurement by analysis of fluorescence photobleaching recovery kinetics. Biophys J 16(9):1055–1069. doi:10.1016/S0006-3495(76)85755-4
23. Soumpasis DM (1983) Theoretical analysis of fluorescence photobleaching recovery experiments. Biophys J 41(1):95–97. doi:10.1016/S0006-3495(83)84410-5

24. Oancea E, Teruel MN, Quest AF, Meyer T (1998) Green fluorescent protein (gfp)-tagged cysteine-rich domains from protein kinase c as fluorescent indicators for diacylglycerol signaling in living cells. J Cell Biol 140(3):485–498
25. Siggia ED, Lippincott-Schwartz J, Bekiranov S (2000) Diffusion in inhomogeneous media: theory and simulations applied to whole cell photobleach recovery. Biophys J 79(4): 1761–1770
26. Elson EL, Magde D (1974) Fluorescence correlation spectroscopy. I. Conceptual basis and theory. Biopolymers 13(1):1–27
27. Magde D, Elson EL, Webb WW (1974) Fluorescence correlation spectroscopy. Ii. An experimental realization. Biopolymers 13(1):29–61
28. Chen H, Farkas ER, Webb WW (2008) Chapter 1: In vivo applications of fluorescence correlation spectroscopy. Methods Cell Biol 89:3–35. doi:10.1016/S0091-679X(08)00601-8
29. Haustein E, Schwille P (2007) Fluorescence correlation spectroscopy: novel variations of an established technique. Annu Rev Biophys Biomol Struct 36:151–169. doi:10.1146/annurev.biophys.36.040306.132612
30. Bacia K, Kim SA, Schwille P (2006) Fluorescence cross-correlation spectroscopy in living cells. Nat Methods 3(2):83–89
31. Chiantia S, Ries J, Schwille P (2009) Fluorescence correlation spectroscopy in membrane structure elucidation. Biochim Biophys Acta 1788(1):225–233. doi:10.1016/j.bbamem.2008.08.013
32. He HT, Marguet D (2011) Detecting nanodomains in living cell membrane by fluorescence correlation spectroscopy. Annu Rev Phys Chem 62:417–436. doi:10.1146/annurev-physchem-032210-103402
33. Digman MA, Gratton E (2011) Lessons in fluctuation correlation spectroscopy. Annu Rev Phys Chem 62:645–668. doi:10.1146/annurev-physchem-032210-103424
34. Schwille P, Meyer-Almes FJ, Rigler R (1997) Dual-color fluorescence cross-correlation spectroscopy for multicomponent diffusional analysis in solution. Biophys J 72(4):1878–1886
35. Larson DR, Gosse JA, Holowka DA, Baird BA, Webb WW (2005) Temporally resolved interactions between antigen-stimulated ige receptors and lyn kinase on living cells. J Cell Biol 171(3):527–536
36. Lillemeier BF, Mortelmaier MA, Forstner MB, Huppa JB, Groves JT, Davis MM (2009) Tcr and lat are expressed on separate protein islands on t cell membranes and concatenate during activation. Nat Immunol 11(1):90–96. doi:ni.1832 [pii]. 10.1038/ni.1832
37. Delon A, Derouard J, Delapierre G, Jaffiol R (2006) Measurement of surface concentration of fluorophores by fluorescence fluctuation spectroscopy. Opt Lett 31(8):1142–1144
38. Jaffiol R, Blancquaert Y, Delon A, Derouard J (2006) Spatial fluorescence cross-correlation spectroscopy. Appl Opt 45(6):1225–1235
39. Ries J, Schwille P (2006) Studying slow membrane dynamics with continuous wave scanning fluorescence correlation spectroscopy. Biophys J 91(5):1915–1924. doi:10.1529/biophysj.106.082297
40. Ruan Q, Cheng MA, Levi M, Gratton E, Mantulin WW (2004) Spatial-temporal studies of membrane dynamics: scanning fluorescence correlation spectroscopy (sfcs). Biophys J 87(2): 1260–1267
41. Petrasek Z, Schwille P (2008) Precise measurement of diffusion coefficients using scanning fluorescence correlation spectroscopy. Biophys J 94(4):1437–1448. doi:10.1529/biophysj.107.108811
42. Dertinger T, Pacheco V, von der Hocht I, Hartmann R, Gregor I, Enderlein J (2007) Two-focus fluorescence correlation spectroscopy: a new tool for accurate and absolute diffusion measurements. Chemphyschem 8(3):433–443. doi:10.1002/cphc.200600638
43. Dertinger T, Loman A, Ewers B, Muller CB, Kramer B, Enderlein J (2008) The optics and performance of dual-focus fluorescence correlation spectroscopy. Opt Express 16(19): 14353–14368. doi:171559 [pii]

44. Humpolickova J, Gielen E, Benda A, Fagulova V, Vercammen J, Vandeven M, Hof M, Ameloot M, Engelborghs Y (2006) Probing diffusion laws within cellular membranes by z-scan fluorescence correlation spectroscopy. Biophys J 91(3):L23–25. doi:10.1529/biophysj.106.089474
45. Kusumi A, Shirai YM, Koyama-Honda I, Suzuki KG, Fujiwara TK (2010) Hierarchical organization of the plasma membrane: investigations by single-molecule tracking vs fluorescence correlation spectroscopy. FEBS Lett 584(9):1814–1823. doi:10.1016/j.febslet.2010.02.047
46. Kusumi A, Nakada C, Ritchie K, Murase K, Suzuki K, Murakoshi H, Kasai RS, Kondo J, Fujiwara T (2005) Paradigm shift of the plasma membrane concept from the two-dimensional continuum fluid to the partitioned fluid: high-speed single-molecule tracking of membrane molecules. Annu Rev Biophys Biomol Struct 34:351–378
47. Meseth U, Wohland T, Rigler R, Vogel H (1999) Resolution of fluorescence correlation measurements. Biophys J 76(3):1619–1631. doi:10.1016/S0006-3495(99)77321-2
48. Bouchaud J-P, Georges A (1990) Anomalous diffusion in disordered media: statistical mechanisms, models and physical applications. Phys Reports 195(4–5):127–293
49. Saxton MJ (2007) A biological interpretation of transient anomalous subdiffusion. I. Qualitative model. Biophys J 92(4):1178–1191. doi:10.1529/biophysj.106.092619
50. Saxton MJ (2008) A biological interpretation of transient anomalous subdiffusion. Ii. Reaction kinetics. Biophys J 94(3):760–771. doi:10.1529/biophysj.107.114074
51. Wawrezinieck L, Rigneault H, Marguet D, Lenne PF (2005) Fluorescence correlation spectroscopy diffusion laws to probe the submicron cell membrane organization. Biophys J 89 (6):4029–4042. doi:10.1529/biophysj.105.067959
52. Lenne PF, Wawrezinieck L, Conchonaud F, Wurtz O, Boned A, Guo XJ, Rigneault H, He HT, Marguet D (2006) Dynamic molecular confinement in the plasma membrane by microdomains and the cytoskeleton meshwork. EMBO J 25(14):3245–3256. doi:10.1038/sj.emboj.7601214
53. Wawrezinieck L, Lenne P-F, Marguet D, Rigneault H (2004) Fluorescence correlation spectroscopy to determine diffusion laws: application to live cell membranes. Proc SPIE 5462:92–102
54. Saxton MJ (2005) New and notable: fluorescence correlation spectroscopy. Biophys J 89 (6):3678–3679
55. Billaudeau C, Mailfert S, Trombik T, Bertaux N, Rouger V, Hamon Y, He HT, Marguet D (2012) Probing the plasma membrane organization in living cells by spot variation fluorescence correlation spectroscopy. Methods Enzymol (in press)
56. Destainville N (2008) Theory of fluorescence correlation spectroscopy at variable observation area for two-dimensional diffusion on a meshgrid. Soft Matter 4(6):1288–1301
57. Ruprecht V, Wieser S, Marguet D, Schutz GJ (2011) Spot variation fluorescence correlation spectroscopy allows for superresolution chronoscopy of confinement times in membranes. Biophys J 100(11):2839–2845. doi:10.1016/j.bpj.2011.04.035
58. Wenger J, Conchonaud F, Dintinger J, Wawrezinieck L, Ebbesen TW, Rigneault H, Marguet D, Lenne PF (2007) Diffusion analysis within single nanometric apertures reveals the ultrafine cell membrane organization. Biophys J 92(3):913–919. doi:10.1529/biophysj.106.096586
59. Manzo C, van Zanten TS, Garcia-Parajo MF (2011) Nanoscale fluorescence correlation spectroscopy on intact living cell membranes with nsom probes. Biophys J 100(2):L8–10. doi:10.1016/j.bpj.2010.12.3690
60. Eggeling C, Ringemann C, Medda R, Schwarzmann G, Sandhoff K, Polyakova S, Belov VN, Hein B, von Middendorff C, Schonle A, Hell SW (2009) Direct observation of the nanoscale dynamics of membrane lipids in a living cell. Nature 457(7233):1159–1162. doi:10.1038/nature07596
61. Mueller V, Ringemann C, Honigmann A, Schwarzmann G, Medda R, Leutenegger M, Polyakova S, Belov VN, Hell SW, Eggeling C (2011) Sted nanoscopy reveals molecular details of cholesterol- and cytoskeleton-modulated lipid interactions in living cells. Biophys J 101(7):1651–1660. doi:10.1016/j.bpj.2011.09.006

62. Cahuzac N, Baum W, Kirkin V, Conchonaud F, Wawrezinieck L, Marguet D, Janssen O, Zornig M, Hueber AO (2006) Fas ligand is localized to membrane rafts, where it displays increased cell death-inducing activity. Blood 107(6):2384–2391. doi:10.1182/blood-2005-07-2883
63. Guardiola-Serrano F, Rossin A, Cahuzac N, Luckerath K, Melzer I, Mailfert S, Marguet D, Zornig M, Hueber AO (2010) Palmitoylation of human fasl modulates its cell death-inducing function. Cell Death Disease 1:e88. doi:10.1038/cddis.2010.62
64. Conchonaud F, Nicolas S, Amoureux MC, Menager C, Marguet D, Lenne PF, Rougon G, Matarazzo V (2007) Polysialylation increases lateral diffusion of neural cell adhesion molecule in the cell membrane. J Biol Chem 282(36):26266–26274. doi:10.1074/jbc.M608590200
65. Chakrabandhu K, Herincs Z, Huault S, Dost B, Peng L, Conchonaud F, Marguet D, He HT, Hueber AO (2007) Palmitoylation is required for efficient fas cell death signaling. EMBO J 26(1):209–220. doi:10.1038/sj.emboj.7601456
66. Lasserre R, Guo XJ, Conchonaud F, Hamon Y, Hawchar O, Bernard AM, Soudja SM, Lenne PF, Rigneault H, Olive D, Bismuth G, Nunes JA, Payrastre B, Marguet D, He HT (2008) Raft nanodomains contribute to akt/pkb plasma membrane recruitment and activation. Nat Chem Biol 4(9):538–547. doi:10.1038/nchembio.103
67. Chakrabandhu K, Huault S, Garmy N, Fantini J, Stebe E, Mailfert S, Marguet D, Hueber AO (2008) The extracellular glycosphingolipid-binding motif of fas defines its internalization route, mode and outcome of signals upon activation by ligand. Cell Death Differ 15(12): 1824–1837. doi:10.1038/cdd.2008.115
68. Rossin A, Kral R, Lounnas N, Chakrabandhu K, Mailfert S, Marguet D, Hueber AO (2010) Identification of a lysine-rich region of fas as a raft nanodomain targeting signal necessary for fas-mediated cell death. Exp Cell Res 316(9):1513–1522. doi:10.1016/j.yexcr.2010.03.002
69. Guia S, Jaeger BN, Piatek S, Mailfert S, Trombik T, Fenis A, Chevrier N, Walzer T, Kerdiles YM, Marguet D, Vivier E, Ugolini S (2011) Confinement of activating receptors at the plasma membrane controls natural killer cell tolerance. Sci Signal 4 (167):ra21. doi:10.1126/scisignal.2001608
70. Sako Y, Kusumi A (1995) Barriers for lateral diffusion of transferrin receptor in the plasma membrane as characterized by receptor dragging by laser tweezers: fence versus tether. J Cell Biol 129(6):1559–1574
71. Morone N, Fujiwara T, Murase K, Kasai RS, Ike H, Yuasa S, Usukura J, Kusumi A (2006) Three-dimensional reconstruction of the membrane skeleton at the plasma membrane interface by electron tomography. J Cell Biol 174(6):851–862. doi:10.1083/jcb.200606007
72. Rose T, Pillet AH, Lavergne V, Tamarit B, Lenormand P, Rousselle JC, Namane A, Theze J (2010) Interleukin-7 compartmentalizes its receptor signaling complex to initiate cd4 t lymphocyte response. J Biol Chem 285(20):14898–14908. doi:10.1074/jbc.M110.104232
73. Ganguly S, Chattopadhyay A (2010) Cholesterol depletion mimics the effect of cytoskeletal destabilization on membrane dynamics of the serotonin1a receptor: A zfcs study. Biophys J 99(5):1397–1407. doi:10.1016/j.bpj.2010.06.031
74. Kiessling V, Crane JM, Tamm LK (2006) Transbilayer effects of raft-like lipid domains in asymmetric planar bilayers measured by single molecule tracking. Biophys J 91(9): 3313–3326. doi:10.1529/biophysj.106.091421

STED-FCS Nanoscopy of Membrane Dynamics

Christian Eggeling

Abstract Lipid-lipid and lipid-protein interactions such as the formation of lipid nanodomains (often denoted "rafts") are considered to play a functional part in a whole range of membrane-associated processes. However, their direct and noninvasive observation in living cells is impeded by the resolution limit of >200 nm of a conventional far-field optical microscope. With the superior spatial resolution of STED nanoscopy, it is now possible to directly resolve nanoscale membrane organization. While direct imaging of membrane heterogeneities turns challenging due to their fast dynamics, the combination of STED nanoscopy with tools such as fluorescence correlation spectroscopy (FCS) allows the disclosure of complex nanoscopic dynamical processes. By performing FCS measurements in observation spots tuned to a diameter of down to 30 nm, new details of molecular membrane dynamics have been obtained: Unlike fluorescent phosphoglycerolipids, fluorescent sphingolipids are transiently (~10 ms) trapped on the nanoscale in often cholesterol- and cytoskeleton-assisted molecular complexes. These interactions are distinct for different lipids and may play an important role in cellular functionality. Comparison of the trapping characteristics to the organization of the different fluorescent lipid analogs in model membranes reveals details of the role of lipid "rafts". This chapter reviews how STED-FCS may shed new light on the role of lipid-protein interactions and nanodomains for membrane bioactivity.

Keywords Fluorescence correlation spectroscopy · Lipid rafts · Membrane heterogeneity · Plasma membrane · STED microscopy

C. Eggeling (✉)
Department of NanoBiophotonics, Max-Planck Institute for Biophysical Chemistry, Am Fassberg 11, Göttingen 37077, Germany
e-mail: ceggeli@gwdg.de

Y. Mély and G. Duportail (eds.), *Fluorescent Methods to Study Biological Membranes*, Springer Ser Fluoresc (2013) 13: 291–310, DOI 10.1007/4243_2012_50,

Contents

1 Introduction: Heterogeneity of Cellular Plasma Membrane

The plasma membrane plays a central role in cellular signaling. It envelopes the cell, is built up by a lipid bilayer, and is constituted by a multitude of different lipids and proteins (Fig. 1). It is well acknowledged that the plasma membrane is not just a simple fluid system, whose molecular constituents diffuse freely as introduced in 1972 by the "fluid mosaic model" [1]. The plasma membrane is rather a highly heterogeneous system with the embedded molecules usually showing highly anomalous diffusion patterns (e.g., [2–5]) (Fig. 1). Therefore, the plasma membrane is more than a cellular envelope: The different membrane molecules interact and fulfill important functions of cellular signaling. For a long time, only membrane proteins were regarded (such as receptors of extra- and inner-cellular messengers) and the surrounding lipids considered as a fluid system. In the meantime, lipids have been acknowledged as a fundamental part of the functionality and regulation of integral or associated membrane proteins (e.g., [4–6]). This follows from novel experimental techniques and the recognition that membrane functionality is governed by the extremely high structural and chemical diversity of lipids and their highly heterogeneous spatiotemporal distribution [7, 8]. As an example, specific lipid-protein interaction may result in a change of the protein's conformational state, enabling or inhibiting the protein's function (e.g., [9–13]). Lipids may on the other hand also be direct molecular receptors of, for example, viral vesicles or toxins, paving their way into or out of cells (for a review, see, e.g., [4, 5]).

Some signaling events may require a tight spatial packing of several proteins, either of the same or different types. Such signaling platforms are often referred to as membrane "rafts" or domains and may coalesce out of the above-mentioned lipid-protein interactions (e.g., [2–5, 14]) (Fig. 1). The stabilization of such platforms may occur spontaneously or be triggered by extra- or inner-cellular events. Further, due to the higher local concentration of lipids and proteins, these domains may be characterized by an increased molecular order, which may then

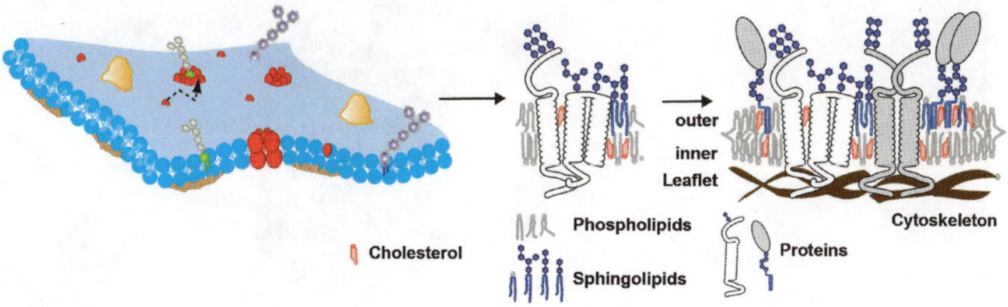

Fig. 1 *Lipids and proteins are heterogeneously distributed in the cellular plasma membrane*. This heterogeneity stems from often cholesterol-assisted lipid-protein interactions (which may be the basis for the coalescence of transient signaling platforms – denoted membrane domains or lipid "rafts," that is, spatially confined molecular assemblies of different lipids and proteins which are essential for a cellular signaling event), an asymmetric molecular distribution to the inner and outer leaflet of the bilayer, the underlying cytoskeleton (which is membrane-anchored via proteins), and from the membrane curvature and pits (Adapted from [4])

also result in changes of the involved proteins' structure and functionality. Prominent examples of such platform-triggered processes include the modulation of cell growth [15, 16], activation of lymphocytes such as T cells [17], viral uptake and budding such as of HIV [7, 18, 19], or cellular internalization of molecules during endocytosis [20–25] (for an overview, see, e.g., [4, 5]). Distinct lipid-protein interactions often play a central role in these examples. For example, cellular uptake and thus toxicity of the cholera toxin is triggered by its binding to the ganglioside lipid GM1 (e.g., [21, 25]). The same lipid acts as a receptor and thus initiator of internalization of the VP1 protein of the simian virus 40 (e.g., [24]). Thereby, the binding affinity and thus the efficiency of the cellular uptake is influenced by both the multivalency of the binding (i.e., the necessity to simultaneously bind several GM1 lipids) and the lipid structure. It has been shown that the incorporation of these virus or toxins is most effective for saturated and long-chained GM1 analogs, while short unsaturated GM1 analogs hardly facilitate this process [24]. This observation supports the assumption that such processes are realized by domains of tight molecular packing (such as "rafts"), since long saturated lipids prefer areas of higher molecular order (e.g., [26–29]). Besides these specific lipid-protein interactions, sphingolipids and membrane-associated cholesterol seem to play a central role in the formation of these signaling platforms. Quite a number of experiments showed that a lowering of the level of sphingolipids and cholesterol resulted in an interruption or disorder of cellular signaling processes (e.g., [4, 5, 30–33]).

The cortical cytoskeleton, underlying the cellular plasma membrane, plays another central role in the membrane's heterogeneous organization, dynamics, and thus biofunctionality (Fig. 1). Proteins anchor to cytoskeletal structures such as microtubule, actin, or spectrin at several points of the plasma membrane, resulting not only in the stabilization of the latter but also in an immobilization or disturbance of the mobility of membrane molecules (e.g., [34–37]). These characteristics may on one hand stabilize the mentioned domains (e.g., [35, 38])

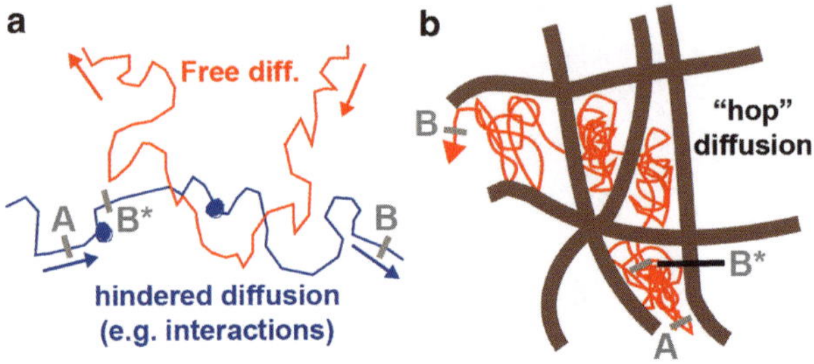

Fig. 2 *Hindered diffusion of membrane molecules.* Possible reasons are (**a**) transient molecular interactions or incorporations into domains, which leads to an interruption or slowdown of diffusion, or (**b**) compartmentalization of the membrane, which causes a "hopping" diffusion. Compartments may result from the underlying cytoskeleton along which proteins are immobilized. These picket-fence-like obstacles impede molecular diffusion from one compartment to the next. Diffusion inside the compartments is free. In the case of the "hopping" diffusion, the macroscopic diffusion from A to B is rather slow and the nanoscopic diffusion from A to B* rather fast, which is the other way around for transient molecular interactions

and on the other increase the interaction probability of less abundant molecules (e.g., [39, 40]). Other membrane characteristics further influence membrane heterogeneity and organization, such as an asymmetric distribution of lipids to the inner and outer leaflet of the plasma membrane (certain phospholipids are rather found in the inner leaflet, while sphingolipids such as gangliosides rather prefer the outer leaflet, see e.g., [41]) or the membrane curvature, which is pronounced at local caveats such as caveolae (e.g., [42–46]).

2 Diffusion Modes of Membrane Molecules

The above-mentioned heterogeneities may in one way or the other restrict the diffusion of membrane molecules. As a consequence, the diffusion of only a few molecules appears free, that is, completely follows free Brownian motion, but the mobility of most is hindered (i.e., anomalous or heterogeneous) (e.g., [34, 37, 47, 48]). For example, interactions between lipids and proteins and incorporation into putative domains of high molecular order lead to a local, transient slowdown or interruption of diffusion. As a result, the time to get from one point of the membrane to another is prolonged, especially on small spatial scales (Fig. 2a). The influence of the cytoskeleton filament can be manifold and becomes obvious when considering its position relative to the membrane. On one hand, proteins may transiently be arrested at the filament. On the other hand, proteins that are anchored along the filament may be an obstacle for other diffusing molecules. These picket-fence-like hindrances may thus split the membrane into compartments, and a molecule may have to cross several of these barriers to get from one to another point of the membrane (Fig. 2b). A logical consequence of this picket-fence model is that molecules may show a hopping diffusion (see e.g., [34, 37]) and, even if diffusion inside the compartments (i.e., on nanoscales) can be free and extremely fast [49], their diffusion may be extremely slowed down on long spatial scales [47]. Further, hindrances of molecular membrane motility may stem from other obstacles such as membrane pits induced by clathrin or caveolae (e.g., [50]). Also, molecular crowding may result in a slowdown of diffusion, since the mobility of molecules may be retarded due to the proximity to immobilized or relatively slow-moving molecules [51].

3 Detection of Membrane Heterogeneity: Optical Microscopy

The above arguments clearly indicate that a closer understanding of molecular interactions and organizations in the plasma membrane (such as the formation of signaling platforms or the influence of the cytoskeleton) may be essential for insights into cellular functionality. This demands for the development of detection technologies that can determine the spatiotemporal positions of various membrane molecules with highest possible precision. Most of the membrane heterogeneities are however highly dynamic, since the cellular plasma membrane is not in thermodynamic equilibrium and a lot of structural characteristic change over space and time. For example, most lipid-protein interactions seem to be transient, and also, the cytoskeletal structures change over time. Consequently, it is very challenging to directly map domains or protein-lipid interactions. Many experiments therefore fix the cells and observe the state of a cell at a certain point of time. Unfortunately, fixation may result in artifacts, or some membrane molecules might still be mobile after fixation [52]. A complete picture of membrane organization and dynamics therefore requires experiments on the living cell.

Optical microscopy has proven valuable for live-cell studies, since light is minimally invasive (e.g., [53]). Often, the fluorescence readout is applied, where the studied molecule (e.g., a membrane protein or a lipid) is labeled with a fluorophore (Fig. 3a) and its fluorescence emission excited by, for example, a laser and registered by a light detector (Fig. 3b). As a consequence, the position of a labeled molecule can be followed over space and time. In a confocal fluorescence microscope, a laser beam is focused to a small spot by an objective lens, which as well collects the emitted fluorescence and guides it onto a point-detector. Scanning of the focused laser spot over the sample then allows reconstructing the spatial distribution of the fluorescently tagged molecules. Similarly, in a wide-field microscopy, a larger area of the sample is illuminated at once and the spatial distribution of the fluorescence signal observed at once on a camera.

The aforementioned dynamics of the system make the direct imaging of the heterogeneous distribution of fluorescently labeled molecules difficult [54–59]. Consequently, a confocal scanning image of the distribution of the fluorescently labeled sphingolipid sphingomyelin (SM) (Fig. 3a) incorporated into the plasma membrane of a living mammalian cell renders this lipid analog homogeneously distributed (Fig. 3c) [60]. The following requirements have to be given in order to be able to image heterogeneous distributions of the lipid: (1) The image acquisition

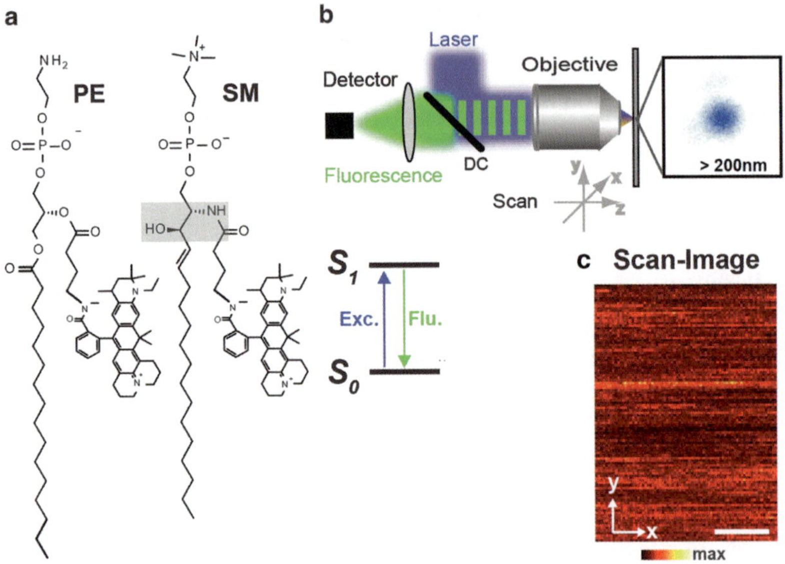

Fig. 3 *Far-field fluorescence microscopy of the distribution of lipids in the plasma membrane of living cells.* (**a**) Structures of the fluorescent lipid analogs phosphoethanolamine (*PE*) and sphingomyelin (*SM*) both tagged with the organic dye Atto647N. *Gray shaded area*: ceramide or sphingosine group of the SM lipid. (**b**) Principle of (confocal) far-field fluorescence microscopy: a laser is focused by an objective lens to a diffraction-limited (≈200 nm large) spot, where dyes are transferred from their ground S_0 to their excited state S_1 and where de-excitation results in spontaneous fluorescence emission. The fluorescence is collected by the same objective and imaged onto a detector. A dichroic mirror (*DC*) splits the fluorescence from the excitation light. Scanning of the sample relative to the laser results in an image of the spatial distribution of fluorescently tagged molecules. (**c**) Scanning fluorescence image of the SM lipid analog incorporated into the plasma membrane of live mammalian cells. Neither the temporal nor the spatial resolution of the microscope is sufficient enough to resolve heterogeneous distribution of the lipids. Scale bar: 1 μm

time, that is, the speed of the scanning process, has to be faster than the dynamics or diffusion of the lipid. (2) The spatial resolution of the image has to be larger than the size of the heterogeneity, that is, the size of, for example, putative domains should not be smaller than the resolution limit of the microscope. (3) The image will only show a contrast if the labeled molecule distributes differently between different heterogeneities: For example, more molecules have to be incorporated into domains or complexes than diffusing freely. As becomes obvious throughout this chapter, all three requirements are not fulfilled for the SM lipid when using conventional microscopy [60]. Therefore, other techniques are necessary to observe the membrane organization of this lipid with a larger contrast and a better spatiotemporal resolution.

Spectroscopic techniques such as Förster resonance energy transfer (FRET) (e.g., [6, 35, 56]) or the use of fluorescent dyes that specifically label certain areas of the plasma membrane (e.g., [61, 62]) allow an observation of membrane domains. However, such experiments may be biased since they either require an overexpression, that is, a too large concentration of the investigated molecules, or membrane incorporation of the dyes, which both may induce changes of the membrane.

4 Detection of Heterogeneous Membrane Diffusion: Single-Molecule Techniques

The direct observation of hindered diffusion of labeled molecules (instead of acquiring an instantaneous image) realizes a much better way of exploring membrane heterogeneity (see e.g., [34, 37, 63]). A prominent method is the spatiotemporal tracking of single isolated fluorescent molecules (single-molecule or single-particle tracking (SPT)). The emitted fluorescence signal is detected on a spatial sensitive detector such as a camera or several point-detectors and the molecule's position determined over time with nanometer precision (e.g., [34, 37, 39, 40, 63–66]). Figure 4a shows representative spatiotemporal diffusion trajectories of single SM lipid analogs, which reveal heterogeneous diffusion [66]. A statistical relevant evaluation of such trajectories is the mean-squared displacement (*MSD*) that the molecules on average experience during a time Δt. Diffusion is free for a linear dependence $MSD(\Delta t)$, while a deviation from this linear characteristic depicts hindered or anomalous diffusion. The $MSD(\Delta t)$ dependencies of the SM analog and of a fluorescent phosphoglycerolipid analog (phosphoethanolamine, PE, Fig. 3a) demonstrate a much more hindered diffusion of the SM compared to the PE lipids in the plasma membrane of live mammalian cells (Fig. 4b). A more detailed analysis of the single-molecule tracks revealed approximately 10-ms-long trapping events for the SM lipids (i.e., diffusion is similar to that depicted in Fig. 2a), while the diffusion of the PE lipids was interrupted for only 1 ms [66]. In the case of the SM diffusion, the trapping events were strongly dependent on the cholesterol content of the membrane: Depletion of cholesterol resulted in a weakening of the anomaly in diffusion, that is, less pronounced trapping. Unfortunately, the nonlinear $MSD(\Delta t)$ dependencies as depicted in Fig. 4b might in principle also be described by a hop-like diffusion (compare Fig. 2b) [67]. As a consequence, an accurate assignment of a diffusion mode is often not feasible. Further, either a large number of short or very long trajectories are required for a statistically relevant analysis, which usually entails long and/or extensive measurement times. On the other hand, the recording of especially long trajectories as well as a high spatiotemporal resolution demands for extremely bright and photostable fluorescent labels [68]. Therefore, SPT often employs large and bulky markers such as 20–40-nm-large gold beads or 10-nm-large quantum dots, which may themselves influence and thus bias the diffusion of the marked molecule [69].

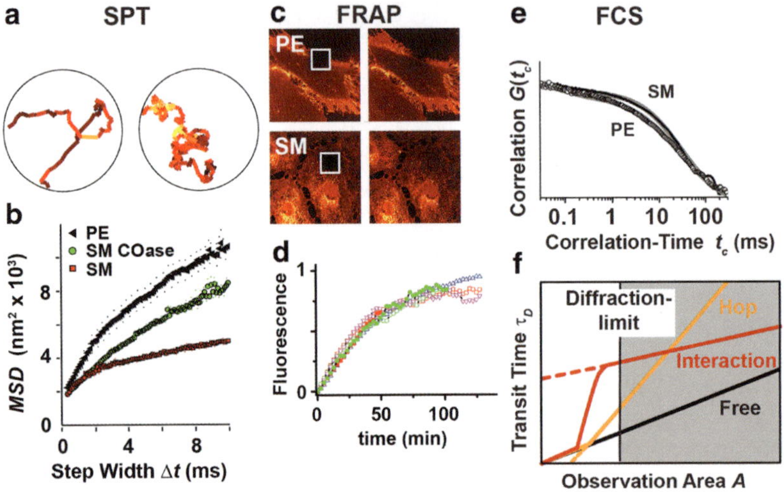

Fig. 4 *Far-field fluorescence spectroscopy of the diffusion of lipids in the plasma membrane of living cells.* (**a, b**) The observation of single fluorescent molecules over time and space (single-particle tracking, *SPT*) reveals hindered diffusion: (**a**) exemplary trajectories of the SM lipid analog through a ≈200-nm-large observation area and (**b**) the mean-squared displacement *MSD* (Δt) traveled over a time Δt for the PE (*black*), SM (*red*), and SM analog after cholesterol oxidase (*COase*) treatment for cholesterol depletion (*green*). The more $MSD(\Delta t)$ deviates from a linear dependence, the more hindered the diffusion. (**c, d**) FRAP data of the PE and SM diffusion: (**c**) confocal scanning images of the spatial distribution of the still fluorescent lipid analogs right after (*left*) and 2 h after (*right*) photobleaching the 10 × 10-μm^2-large area marked by the *white square* (the inhomogeneous distribution of fluorescence results from fluorescent lipids already incorporated into the cell). (**d**) The recovery of fluorescence results from diffusion of nonphotobleached lipid analogs into the bleached area over time, and it is almost complete and very similar for PE and SM (diffusion coefficient 0.5 μm^2) revealing no anomalous diffusion. (**e, f**) Confocal FCS data of the PE and SM diffusion through the 240-nm-large diffraction-limited observation area: (**e**) representative correlation data reveal slightly slowed down but still normal diffusion for SM in comparison to PE. (**f**) A plot of the average transit time $\tau_D(A)$ for observation areas of different size A (svFCS) can reveal more details of hindered diffusion by extrapolation to the nanoscale, such as plotted for free diffusion, diffusion with transient interactions or domain incorporation, and hopping diffusion. The diffraction limit prevents a direct observation at the relevant nanoscopic spatial scales

5 Detection of Heterogeneous Membrane Diffusion: FCS und FRAP

Methods such as fluorescence recovery after photobleaching (FRAP) [70, 71] or fluorescence correlation spectroscopy (FCS) [47, 72–74] usually require much shorter measurement times and small labels to acquire statistically relevant conclusions about the diffusion behavior of the investigated molecules. This follows from the fact that both techniques simultaneously observe the diffusion characteristics of a multitude of single molecules, which in SPT is only approximated by a large field of view and/or the use of photoswitchable fluorophores (e.g., [75]).

In FRAP, all fluorescent molecules within a micrometer-large area are photobleached (i.e., turned nonemissive) and the recovery of the fluorescence from this area detected as nonphotobleached molecules diffuse in. The recovery curve allows the determination of diffusion coefficients and fractions of immobile species. Instead of photobleaching, one may also institute photoswitchable fluorescent labels [76].

In FCS, the temporal fluctuations of the observed fluorescence signal are monitored over time as molecules diffuse in and out of the observation area or volume (e.g., given by the micrometer-large focal laser spot of a confocal microscope [77]), and the correlation function of these fluctuations is calculated. The decay time of this correlation function usually renders the average transit time of the molecules through the observation area. Hindrances in diffusion therefore either result in a shift of the correlation curve toward larger times or an anomaly or stretching of the decay.

Figure 4c–e show representative FRAP and FCS data of the SM and PE lipid analogs in the plasma membrane of live mammalian cells. In contrast to the SPT experiments, which outlined vast differences between SM and PE diffusion (Fig. 4b), the FRAP and FCS data only indicate slight differences. The FRAP recovery curves of SM and PE are basically indistinguishable, while the FCS decays are only slightly shifted: The transit time through the approximately 240-nm-large observation area is $\approx$20 ms for the PE and $\approx$30 ms for the SM lipid, but the diffusion is not anomalous for both (i.e., the decays are not stretched) [60]. These data therefore indicate that SM is just diffusing slower than PE but still homogeneous, which contradicts the results from the SPT experiments.

6 The Resolution Barrier of Optical Microscopy

The aforementioned microscopy experiments were performed in the optical far-field. This means that the focal spot of the laser is placed several micrometers away from any optical element, which allows a least invasive observation of the living cell. This is realized by using optical lenses and propagating beams. Unfortunately, since far-field optics introduce the diffraction of light, the spatial resolution of such a lens-based microscope is limited to about 200 nm [78]. A laser can therefore not be focused, or a point emitter (such as a small fluorophore) cannot be imaged to a smaller spot than these 200 nm in diameter. As a consequence, a far-field microscope cannot distinguish alike molecules that are closer together than 200 nm, and the structures below this size will appear blurred in the final image. In SPT, this issue is solved by detecting only single isolated (>200 nm apart) molecules and determining the central position of their blurred image spots. FRAP and FCS experiments are however limited by diffraction, and therefore, these measurements will average over nanoscopic hindrances in the diffusion characteristics when applied with far-field optics. For example, slowed down but free diffusion cannot be distinguished from diffusion with one or two transient traps during the transit through a >200-nm-large observation area, since in the latter case, every molecule will experience at least one trap (as for the SM diffusion [60]), that is, there will be no inhomogeneity in molecular transit times.

A way around this constraint has been suggested by spot-variation FCS (svFCS, [47, 60, 79, 80]). In svFCS correlation, data is recorded and the transit time τ_D determined for different areas A of the observation area (above the diffraction limit). The dependence $\tau_D(A)$ allows more detailed information of diffusion modes: A linear dependence is expected for free Brownian diffusion, while hindrances in diffusion (even on the nanoscale) will result in a nonlinear dependence (Fig. 4d). This nonlinearity is different for different modes of hindered diffusion as depicted for transient trapping and hopping. Using svFCS in 200-nm- to >1-μm-large observation areas, the diffusion characteristics of several different membrane lipids and proteins could be assigned to these modes [81]. Unfortunately, this assignment was only realized by an extrapolation to even smaller areas (e.g., by testing the intercepts of the $\tau_D(A)$ dependence), and further details of the molecular dynamics such as a trapping period or area could only be estimated. Further, it cannot be ruled out that the $\tau_D(A)$ dependence will be different for smaller A as approximated. Precise FCS measurements on the relevant scales, that is, with observation areas <200 nm, would provide the desired details. For example, trapping observed with small A will be much more visible: Now both the probability of being trapped during a transit would be much lower than one, and the trapping time would be much larger than the transit time of free diffusion, rendering highly heterogeneous molecular transits (some molecules would transit very fast and some very slow) and thus highly anomalous or stretched FCS data. A possible realization is the combination of FCS with sub-diffraction, super-resolution microscopy such as STED nanoscopy (STED-FCS) [82, 83].

7 Optical Microscopy with Sub-Diffraction Resolution: STED Nanoscopy

Starting in the 1990s [84], developments in optical fluorescence microscopy have realized the distinction of alike sub-diffraction features with a far-field fluorescence microscope (see e.g., [85]). The key for truly surpassing the diffraction barrier lies in the spectroscopic properties of the fluorescent markers, specifically in the possibility to modulate or reversibly inhibit their emission in space and time [86–88]. The first of such a super-resolution microscope or nanoscope was based on stimulated emission and dubbed stimulated emission depletion (STED) microscopy [84, 89]. In its usual implementation, an additional laser is added to a conventional point-scanning far-field microscope which inhibits fluorescence emission everywhere but at the center of the of the exciting laser focus (Fig. 5a). The wavelength of this second laser is tuned to the red edge of the fluorophore's emission spectrum and induces the stimulated de-excitation of the fluorophore's

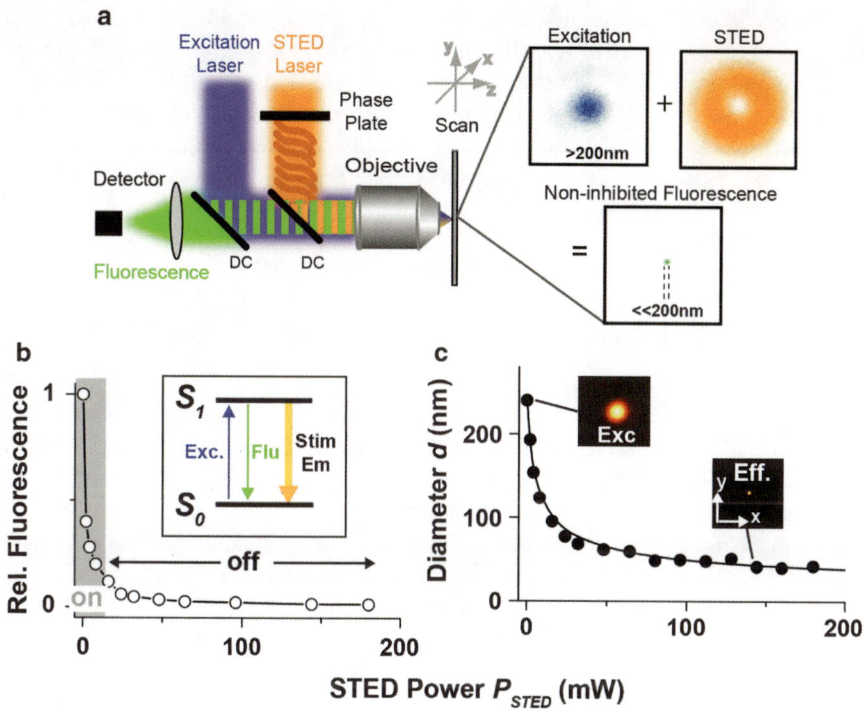

Fig. 5 *Principles of far-field fluorescence STED nanoscopy*. (**a**) Besides the fluorescence excitation laser, a second STED laser is introduced, whose wave front is altered by introducing a phase plate in such a way that, for example, a doughnutlike intensity distribution of the focused beam is realized. The overlay of the excitation and STED foci inhibits fluorescence emission everywhere but at the focal center leaving a sub-diffraction-sized area where emission is still allowed: the new observation area. (**b**) Inhibition of fluorescence emission by the STED laser is realized by stimulated emission, where the fluorophore in its excited state S_1 is de-excited to its ground state S_0 more efficiently than the spontaneous fluorescence. The efficiency of this inhibition increases with the power of the STED laser, and fluorescence inhibition is efficiently switched off above a certain power threshold. (**c**) The diameter d of the observation area scales with the power of the STED laser

excited (and fluorescent) electronic state (Fig. 5b). By detecting only the spontaneous (and not the stimulated emission), the registered signal is efficiently decreased and completely switched off by increasing the intensity of the STED laser (Fig. 5b). The introduction of a phase plate into the STED beam realizes a distortion of its wave front and an intensity distribution of the focus that features one or several local zeros, such as a doughnut-shaped intensity distribution (Fig. 5a). Consequently, increasing the intensity of the STED beams drives the area in which fluorescence emission is still allowed to smaller and smaller sub-diffraction scales (Fig. 5c). The spatial resolution of the STED microscope is therefore tuned by the intensity of the STED laser, which – as will be shown later on – is one of the central features of STED-FCS.

At first, STED microscopy was realized with both the excitation and STED laser in a pulsed mode [89, 90]. By letting the STED pulse swiftly follow the excitation laser, the fluorescence inhibition process is optimized, since the STED laser approaches the dye in the right condition: in its excited state. STED microscopy has however also been realized with continuous-wave lasers [91], an approach that has recently been significantly improved by the use of a gated detection scheme [92].

8 Detecting Heterogeneous Membrane Dynamics with STED-FCS

Figure 6a shows representative STED-FCS data of the fluorescent SM and PE lipid analogs in the plasma membrane of live mammalian cells [60]. These correlation data were recorded with an observation area of 40 nm in diameter as tuned by the STED intensity, and, in contrast to the diffraction-limited confocal data (compare Fig. 4e), show vast differences between the two lipids. While PE transits the observation area in on average 0.5 ms, the SM requires on average 3 ms. Further, the FCS data of PE can be described by normal, free diffusion, while the FCS curves of SM are more stretched and can only be described by an anomalous hindered diffusion. Following treatment of the cells with cholesterol oxidase (COase), which oxidizes and thus depletes cholesterol in the plasma membrane, diffusion of SM is very similar to PE (Fig. 6a), and one can follow that hindered diffusion of SM is cholesterol-assisted.

9 Details from the STED-FCS Experiments: Measurements at Different Sizes of the Observation Area

One can use the principle of svFCS (compare Fig. 4f) to determine the details of the hindrances in SM diffusion. Since the variation of the STED intensity allows tuning the observation area, the dependence of the transit time τ_D on the size of the observation area A can now be expanded to nanoscopic spatial scales below the diffraction barrier. Figure 6b shows the values of τ_D determined from the FCS data of the SM and PE lipid analogs in the plasma membrane of living cells recorded for different STED intensities, that is, for different diameters d = 30–200 nm of the observation area $A = \pi\,(d/2)^2$ [60]. The resulting dependencies $\tau_D(A)$ are linear for PE and nonlinear for SM, which confirms the almost free diffusion of PE and the trapped diffusion of SM. It becomes obvious that the transit times τ_D of SM do not change for small areas A with diameters d < 70–80 nm. This independency can only be explained that the transits are now ruled by transient trapping. A closer analysis of the FCS data as well as a comparison with Monte Carlo simulations

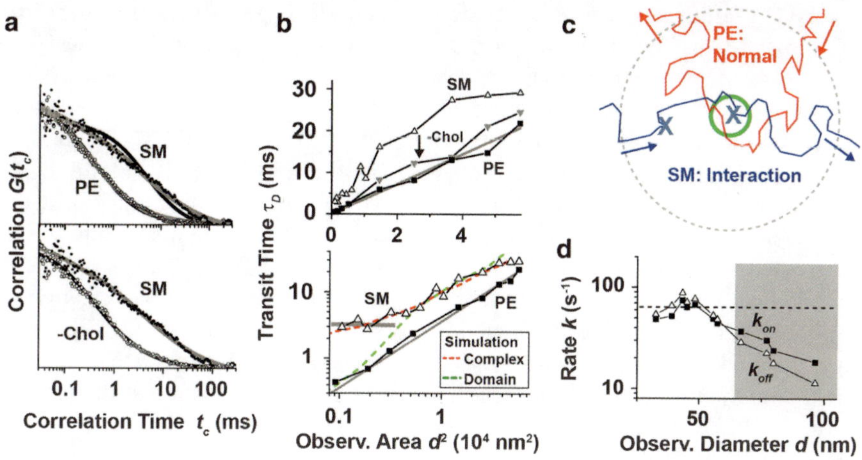

Fig. 6 *STED-FCS measurements of lipid diffusion in the plasma membrane of living cells.* (**a**) Representative STED-FCS data ($d = 40$ nm) of PE, SM, and SM after cholesterol depletion (*-Chol*). The SM data can only be described by anomalous diffusion (*black line*: fit of normal diffusion) and reveal a cholesterol-assisted hindered diffusion of the SM lipid analog. (**b**) The dependency of the transit time τ_D for different sub-diffraction-sized observation areas $A \sim d^2$ (as tuned by the STED laser power) shows an almost free diffusion (linear dependence, *gray line* diffusion coefficient 0.5 $\mu m^2/s$) for PE and SM after cholesterol depletion and a hindered diffusion (nonlinear dependence) for SM (*upper panel* linear-linear and *lower panel* log-log plot). The minimal change of τ_D for very small observation areas (*gray horizontal line*) and Monte Carlo simulations indicate that the hindrance in diffusion is caused by transient complexes with either relatively slow-moving or immobilized membrane molecules (*red dotted line*) and not by incorporation into $\geq$ 20-nm-large domains, where diffusion is slowed down (*green dotted line*). (**c**) Schematic drawing of normal PE and hindered SM diffusion (*cross*: points of interactions or complexes). The direct observation of these transient interactions is possible with the small observation area of the STED nanoscope (*green circle*) but impossible with the large diffraction-limited observation area (*gray dashed circle*). (**d**) A special analysis of the STED-FCS data results in an average effective encounter rate $k_{on} \approx 80\ s^{-1}$ and an average off-rate $k_{off} \approx 80\ s^{-1}$ of the SM interactions. An accurate determination of these rates is impossible for observation areas with diameters $d > 60$ nm (*gray shaded area*)

(Fig. 6b) showed that the SM lipids are transiently arrested for about 10 ms on average every 10 ms (with a diffusion coefficient of $\approx$0.5 $\mu m^2/s$ every 200–300 nm) and that the SM lipids hardly move during trapping (Fig. 6c) [60, 93]. Most probably, the lipids interact with other molecules such as membrane proteins, which are either relatively slow moving or even immobilized. In this respect, one can analyze the STED-FCS data with a model introducing transient binding of the diffusing lipid to a fixed or comparatively slow-moving particle with effective encounter and off-rates k_{on} and k_{off}, respectively. This model is valid if the trapping time $1/k_{off}$ is much longer than the average time the freely diffusing molecule would spend in the observation area (in this case, the diffusion is reaction-dominated and one can describe the STED-FCS data by the usual diffusion term and an added exponential term) [83, 94]. The model fitted the STED-FCS data accurately and resulted in values of k_{on} and k_{off} in the range of 80 s^{-1} for SM, that is, trapping times $1/k_{off} \approx 12.5$ ms for these complexes [83]. An appointment of these rates was impossible for diffraction-limited or even >60-nm-large observation areas (since in that case, the transit through the observation area is not reaction-dominated anymore) (Fig. 6d). The values of k_{on} and k_{off} render an equilibrium constant $K = k_{on}/k_{off} \approx 1$, that is, at a certain point of time 50 % of all SM lipids are bound (and immobilized) and 50 % are freely diffusing. As a consequence, there will be no contrast between bound and unbound, and it is impossible to image this heterogeneous distribution of lipids, even with a STED nanoscope [60]. The mobility of SM after cholesterol depletion as well as after depolymerization of the underlying cellular cytoskeleton is almost free (similar to PE, Fig. 6b), indicating on one hand that either the binding interaction or the immobilization of the binding partner is cholesterol-assisted and on the other hand that the binding partner is linked to the cytoskeleton [60, 93]. In contrast to SM, PE interacts only weakly with effective encounter and off-rates of $k_{on} \approx 190\ s^{-1}$ and $k_{off} \approx 800\ s^{-1}$, respectively, showing that PE diffuses almost freely [83].

10 Summary

The observed cholesterol- and cytoskeleton-dependent interactions of the fluorescent sphingolipids SM are well described by transient (~10 ms long) binding to relatively immobile binding partners and only undeceived by STED-FCS. The >200-nm-large diffraction-limited observation areas average over details on the nanoscale (compare Fig. 6c). The binding partners are most probably membrane proteins, whose mobility is restricted by the cytoskeleton. We can exclude that the lipids move during trapping, that is, it does not wander around inside a domain (or "raft"), where diffusion is slowed down. However, because we observed the dynamics of single molecules only, we cannot rule out the possibility that additional lipid or protein molecules were (temporarily) included in this complex. However, the complex is not of very high molecular order which is shown by a comparison with experiments on model membranes [93, 95, 96]. A model bilayer membrane of a ternary mixture of unlabeled saturated SM, unlabeled and unsaturated phosphoglycerolipids, and cholesterol separates into two different phases: a liquid-disordered and fluid Ld phase, which mainly includes the unsaturated lipids, and a liquid-ordered and less fluid Lo phase of higher molecular order, which mainly includes the saturated lipid and cholesterol (see, e.g., [26, 28, 29]). The Lo phase is often considered as a physical model system for membrane domains or "rafts" in the plasma membrane of living cells (e.g., [4, 5]). Unfortunately, the labeled SM analog does not partition into the Lo phase of this model system, but favors the Ld phase (Fig. 7a). Consequently, the lipid analog will also not be able to stain areas (or domains) of high molecular order in the plasma membrane, which therefore will be missed by the previously mentioned STED-FCS experiments. The observed strong interactions of the SM analog are therefore not driven by

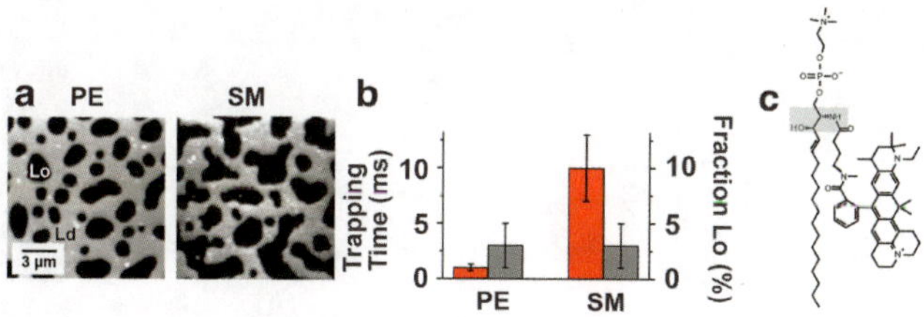

Fig. 7 *Comparison of live-cell and model membrane data.* (**a**) Phase separation into liquid-disordered (*Ld*) and liquid-ordered (*Lo*) domains of a model membrane bilayer composed of a ternary mixture [93]: both the fluorescent PE and SM lipid analogs hardly enter the Lo phase (confocal scanning fluorescence image, *black*: low signal, *white*: high signal). (**b**) Partitioning in model membranes and trapping characteristics observed by STED-FCS on live-cell plasma membranes are not correlated: trapping time (*red*, *left axis*, live-cell) and fraction of signal in Lo phase (*gray*, *right axis*, model membranes) of PE and SM. (**c**) Structure of the fluorescent SM analog: size comparison of lipid and dye. *Gray shaded area*: ceramide or sphingosine group

differences in molecular order, that is, the results of STED-FCS and phase separation experiments are not correlated. Rather, strong differences in the trapping characteristics of different saturated lipid analogs (which all hardly enter the Lo phase) have been observed [93, 96]. For example, the fluorescent PE analog partitions similarly poor into the Lo phase as the SM analog, but interacts much weaker with other constituents (Fig. 7b). Further, an elaborate study on a multitude of different lipid analogs using STED-FCS has shown that neither the properties and position of the dye label nor the partitioning characteristics of the analog but rather the chemical structure of the lipid influenced the transient trapping [93, 96]. Mainly the ceramide (i.e., the NH und OH-) unit of the lipid backbone (gray in Fig. 7c) was responsible for cholesterol- and cytoskeleton-dependent interactions, probably via hydrogen bonds. However, also the polar head groups of, for example, gangliosides lipids such as GM1 induced transient binding, however less efficient and independent of cholesterol and the cytoskeleton [93]. In general, the interaction strength, frequency, and duration seemed very lipid specific, which points to specific and functional lipid-protein bindings. Maybe, these interactions exposed by the STED-FCS experiments are those lipid-protein affinities that are the physicochemical basis of the coalescence of several molecules to the previously mentioned signaling platforms (or membrane "rafts") (Fig. 8). It is questionable, whether the investigation of this coalescence is possible with the presented fluorescent lipid analogs, since the in comparison to the lipid rather large and often charged dye (Fig. 7c) restricts the accessibility to highly ordered molecular assemblies such as the Lo phase or of the putative membrane "rafts" [96]. However, it has been shown that the order of the Lo phase generated in cellular plasma membranes (which were treated by swelling procedures generating vesicular bilayers composed of native membranes such as giant plasma membrane vesicles (GPMV) [97] or plasma membrane spheres [98]) is much lower compared to model systems [99] and that the Lo phase of the plasma membrane is thus much more efficiently penetrated by the fluorescent lipid analogs [96]. Most importantly,

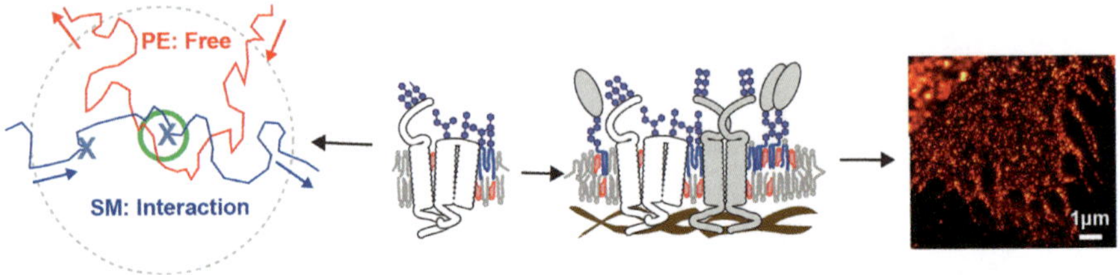

Fig. 8 *Possible interpretation of STED-FCS data*. The observed trapping of certain lipids reveals transient interactions with proteins, which might be the bases for a coalescence of several lipids and proteins into signaling platforms (or domains/"rafts") of high molecular order, as exemplified on STED images of clusters of the membrane-bound SNARE protein syntaxin-1 in fixed mammalian cells (*upper left corner*: diffraction-limited recordings not sufficiently resolving the clusters) [101]

a functional fluorescent lipid analog that does not alter phase partitioning and is compatible with STED-FCS will reveal additional details on the existence of liquid-ordered nanodomains or "rafts" in the plasma membrane of living cells (Fig. 8) [93].

In conclusion, STED-FCS is a sensitive and unique tool for studying nanoscale membrane organization and determining the cellular functions and molecular interdependencies of membrane components. More general, STED-FCS expands currently available optical microscopy and spectroscopy techniques to the nanoscale and opens up exceptional possibilities to characterize and disclose complex cellular signaling events and therefore new approaches for drug screening and development [100].

Acknowledgement This work could not have been realized without the work of the people from the department of NanoBiophotonics: Christian Ringemann, Veronika Mueller, Alf Honigmann, Rebecca Medda, Vladimir Belov, Svetlana Polyakova, Birka Llaken, Class von Middendorf und Andreas Schönle, especially by the extraordinary support and interaction with Stefan Hell, as well as by the lipid syntheses and discussions with Günter Schwarzmann (University Bonn). Further fruitful discussions with Herve Rigneault (Marseille), the Simons group (MPI Dresden), and the Schwille group (Bioquant Dresden) are greatly acknowledged.

References

1. Singer S, Nicolson GL (1972) Science 175:720–731
2. Simons K, Ikonen E (1997) Nature 387:569–572
3. Jacobson K, Mouritsen OG, Anderson GW (2007) Nat Cell Biol 9:7–14
4. Lingwood D, Simons K (2010) Science 327:46–50
5. Simons K, Gerl MJ (2010) Nat Rev Mol Cell Biol 11:688–699
6. van Meer G, Voelker DR, Feigenson GW (2008) Nat Rev Mol Cell Biol 9:112–124
7. Brügger B, Glass B, Haberkant P, Leibrecht I, Wieland FT, Kräusslich HG (2006) Proc Natl Acad Sci USA 103:2641–2646
8. Shevchenko A, Simons K (2010) Nat Rev Mol Cell Biol 11:593–598

9. Bhushan A, McNamee MG (1993) Biophys J 64:716–723
10. Mantipragada SB, Horvath LI, Arias HR, Schwarzmann G, Sandhoff K, Barrantes FJ, Marsh D (2003) Biochemistry 42:9167–9175
11. Fantini J (2003) Cell Mol Life Sci 60:1027–1032
12. Coskun Ü, Grzybek M, Drechsler D, Simons K (2011) Proc Natl Acad Sci 108:9044–9048
13. Contreras F-X, Ernst AM, Haberkant P, Björkholm P, Lindahl E, Gönen B, Tischer C, Elofsson A, von Heijne G, Thiele C, Pepperkok R, Wieland F, Brügger B (2012) Nature 481:525–529
14. Joly E (2004) BMC Cell Biol 5:3
15. Bremer EG, Schlessinger J, Hakomori S (1986) J Biol Chem 261:2434–2440
16. Kawashima N, Yoon S-J, Itoh K, Nakayama K (2009) J Biol Chem 284:6147–6155
17. Zech T, Ejsing CS, Gaus K, de Wet B, Shevchenko A, Simons K, Harder T (2009) EMBO J 28:466–476
18. Chan R, Uchil PD, Jin J, Shui G, Ott DE, Mothes W, Wenk MR (2008) J Virol 82:11228–11238
19. Waheed AA, Freed EO (2009) Virus Res 143:162–176
20. Tsai B, Gilbert JM, Stehle T, Lencer W, Benjamin TL, Rapaport TA (2003) EMBO J 22:4346–4355
21. Chinnapen DJF, Chinnapen H, Saslowsky D, Lencer W (2007) FEMS Microbiol Lett 266:129–137
22. Römer W, Berland L, Chambon V, Gaus K, Windschiegl B, Tenza D, Aly MRE, Fraisier V, Florent JC, Perrais D, Lamaze C, Raposo G, Steinem C, Sens P, Bassereau P, Johannes L (2007) Nature 450:670–675
23. Hebbar S, Lee E, Manna M, Steinert S, Kumar GS, Wenk M, Wohland T, Kraut R (2008) J Lipid Res 49:1077–1089
24. Ewers H, Römer W, Smith AE, Bacia K, Dmitrieff S, Chai W, Mancini R, Kartenbeck J, Chambon V, Berland L, Oppenheim A, Schwarzmann G, Feizi T, Schwille P, Sens P, Helenius A, Johannes L (2009) Nat Cell Biol 12:11–18
25. Wernick NLB, Chinnapen DJF, Cho JA, Lencer W (2010) Toxins 2:310–325
26. Bacia K, Scherfeld D, Kahya N, Schwille P (2004) Biophys J 87:1034–1043
27. Bagatolli LA (2006) Biochim Biophys Acta 1758:1541–1556
28. Baumgart T, Hunt G, Farkas ER, Webb WW, Feigenson GW (2007) Biochim Biophys Acta 1768:2182–2194
29. Marsh D (2009) Biochim Biophys Acta 1788:2114–2123
30. Brown DA, London E (2000) J Biol Chem 275:17221–17224
31. Hao MM, Mukherjee S, Maxfield FR (2001) Proc Natl Acad Sci 98:13072–13077
32. Pike LJ (2006) J Lipid Res 47:1597–1598
33. Wüstner D, Solanko L, Sokol E, Garvik O, Li Z, Bittman R, Korte T, Herrmann A (2011) Chem Phys Lipids 164:221–235
34. Kusumi A, Nakada C, Ritchie K, Murase K, Suzuki K, Murakoshi H, Kasai RS, Kondo J, Fujiwara T (2005) Annu Rev Biophys 34:351–378
35. Goswami D, Gowrishankar K, Bilgrami S, Ghosh S, Raghupathy R, Chadda R, Vishwakarma R, Rao M, Mayor S (2008) Cell 135:1085–1097
36. Chichili GR, Rodgers W (2009) Cell Mol Life Sci 66:2319–2328
37. Kusumi A, Shirai YM, Koyama-Honda I, Suzuki KGN, Fujiwara TK (2010) FEBS Lett 584:1814–1823
38. Machta BB, Papanikolaou S, Sethna JP, Veatch SL (2011) Biophys J 100:1668–1677
39. Andrews NL, Lidke KA, Pfeiffer JR, Burns AR, Wilson BS, Oliver JM, Lidke DS (2008) Nat Cell Biol 10:955–963
40. Jaqaman K, Kuwata H, Touret N, Collins R, Trimble WS, Danuser G, Grinstein S (2011) Cell 146:593–606
41. Mondal M, Mesmin B, Mukherjee S, Maxfield FR (2009) Mol Biol Cell 20:581–588
42. Anderson RGW, Jacobson K (2002) Science 296:1821–1825

43. Parton RG (2003) Nat Rev Mol Cell Biol 4
44. Cheng Z-J, Singh RD, Marks DL, Pagano RE (2006) Mol Membr Biol 23:101–110
45. Reynwar BJ, Illya G, Harmandaris VA, Müller MM, Kremer K, Deserno M (2007) Nature 447:461–464
46. Hatzakis NS, Bhatia VK, Larsen J, Madsen KL, Bolinger PY, Kunding AH, Castillo J, Gether U, Hedegård P, Stamou D (2009) Nat Chem Biol 5:835–841
47. Wawrezinieck L, Rigneault H, Marguet D, Lenne PF (2005) Biophys J 89:4029–4042
48. Saxton MJ (2005) Biophys J 89:3678–3679
49. Fujiwara T, Ritchie K, Murakoshi H, Jacobson K, Kusumi A (2002) J Cell Biol 157:1071–1081
50. Baumgart T, Caparo BR, Zhu C, Das SL (2011) Annu Rev Phys Chem 62:483–506
51. Niemela PS, Miettinen MS, Monticelli L, Hammaren H, Bjelkmar P, Murtola T, Lindahl E, Vattulainen I (2010) J Am Chem Soc 132:7574–7575
52. Tanaka KAK, Suzuki KGN, Shirai YM, Shibutani ST, Miyahara MSH, Tsuboi H, Yahara M, Yoshimura A, Mayor S, Fujiwara TK, Kusumi A (2010) Nat Methods 7
53. Pawley JB (2006) Handbook of biological confocal microscopy. Springer, New York
54. Munro S (2003) Cell 115:377–388
55. Lommerse PHM, Spaink HP, Schmidt T (2004) Biochim Biophys Acta 1664:119–131
56. Hancock JF (2006) Nat Rev Mol Cell Biol 7:457–462
57. Shaw AS (2006) Nat Immunol 7:1139–1142
58. Groves JT, Parthasarathy R, Forstner MB (2008) Annu Rev Biomed Eng 10:311–338
59. Pike LJ (2009) J Lipid Res 50:S323–S328
60. Eggeling C, Ringemann C, Medda R, Schwarzmann G, Sandhoff K, Polyakova S, Belov VN, Hein B, von Middendorff C, Schonle A, Hell SW (2009) Nature 457:1159–U121
61. Parasassi T, Krasnowska EK, Bagatolli L, Gratton E (1998) J Fluoresc 8:365–373
62. Kucherak OA, Oncul S, Darwich Z, Yushchenko DA, Arntz Y, Didier P, Mely Y, Klymchenko AS (2010) J Am Chem Soc 132:4907–4916
63. Lagerholm BC, Weinreb GE, Jacobson K, Thompson NL (2005) Annu Rev Phys Chem 56:309–336
64. Schutz GJ, Schindler H, Schmidt T (1997) Biophys J 73:1073–1080
65. Nishimura SY, Vrljic M, Klein LO, McConnell HM, Moerner WE (2006) Biophys J 90:927–938
66. Sahl SJ, Leutenegger M, Hilbert M, Hell SW, Eggeling C (2010) Proc Natl Acad Sci 107:6829–6834
67. Wieser S, Moertelmaier M, Fuertbauer E, Stockinger H, Schutz G (2007) Biophys J 92:3719–3728
68. Thompson RE, Larson DR, Webb WW (2002) Biophys J 82:2775–2783
69. Clausen M, Lagerholm BC (2011) Curr Protein Pept Sci 12:699–713
70. Yechiel E, Edidin M (1987) J Cell Biol 105:755–760
71. Feder TJ, Brust-Mascher I, Slattery JP, Baird BA, Webb WW (1996) Biophys J 70:2767–2773
72. Magde D, Webb WW, Elson E (1972) Phys Rev Lett 29:705–708
73. Fahey PF, Koppel DE, Barak LS, Wolf DE, Elson EL, Webb WW (1977) Science 195:305–306
74. Schwille P, Korlach J, Webb WW (1999) Cytometry 36:176–182
75. Manley S, Gillette JM, Patterson GH, Shroff H, Hess HF, Betzig E, Lippincott-Schwartz J (2008) Nat Methods 5:155–157
76. Lippincott-Schwartz J, Altan-Bonnet N, Patterson GH (2003) Nat Cell Biol 5:S7–S14
77. Widengren J, Rigler R (1990) Ultrasensitive detection of single molecules using fluorescence correlation spectroscopy. In: Klinge B, Owman C (eds) Bioscience. Lund University Press, Lund, pp 180–183
78. Abbe E (1873) Archiv für Mikroskopische Anatomie 9:413–468
79. Humpolickova J, Gielen E, Benda A, Fagulova V, Vercammen J, vandeVen M, Hof M, Ameloot M, Engelborghs Y (2006) Biophys J 91:L23–L25

80. He HT, Marguet D (2011) Annu Rev Phys Chem 62:417–436
81. Lenne PF, Wawrezinieck L, Conchonaud F, Wurtz O, Boned A, Guo XJ, Rigneault H, He HT, Marguet D (2006) EMBO J 25:3245–3256
82. Kastrup L, Blom H, Eggeling C, Hell SW (2005) Phys Rev Lett 94:178104
83. Ringemann C, Harke B, Middendorff CV, Medda R, Honigmann A, Wagner R, Leutenegger M, Schoenle A, Hell S, Eggeling C (2009) New J Phys 11:103054
84. Hell SW, Wichmann J (1994) Opt Lett 19:780–782
85. Hell SW (2007) Science 316:1153–1158
86. Hell SW, Jakobs S, Kastrup L (2003) Appl Phys A 77:859–860
87. Hell SW (2004) Phys Lett A 326:140–145
88. Hell SW (2009) Nat Methods 6:24–32
89. Klar TA, Jakobs S, Dyba M, Egner A, Hell SW (2000) Proc Natl Acad Sci 97:8206–8210
90. Donnert G, Keller J, Medda R, Andrei MA, Rizzoli SO, Lurmann R, Jahn R, Eggeling C, Hell SW (2006) Proc Natl Acad Sci 103:11440–11445
91. Willig KI, Harke B, Medda R, Hell SW (2007) Nat Methods 4:915–918
92. Vicidomini G, Moneron G, Han KY, Westphal V, Ta H, Reuss M, Engelhardt H, Eggeling C, Hell SW (2011) Nat Methods 8:571–573
93. Mueller V, Ringemann C, Honigmann A, Schwarzmann G, Medda R, Leutenegger M, Polyakova S, Belov VN, Hell SW, Eggeling C (2011) Biophys J 101:1651–1660
94. Michelman-Ribeiro A, Mazza D, Rosales T, Stasevich TJ, Boukari H, Rishi V, Vinson C, Knutson JR, McNally JG (2009) Biophys J 97:337–346
95. Honigmann A, Walter C, Erdmann F, Eggeling C, Wagner R (2010) Biophys J 98:2886–2894
96. Sezgin E, Levental I, Grzybek M, Schwarzmann G, Mueller V, Honigmann A, Belov VN, Eggeling C, Coskun Ü, Simons K, Schwille P (2012) Biochim Biophys Acta 1818:1777–1784
97. Baumgart T, Hammond AT, Sengupta P, Hess ST, Holowka DA, Baird BA, Webb WW (2007) Proc Natl Acad Sci 104:3165–3170
98. Lingwood D, Ries J, Schwille P, Simons K (2008) Proc Natl Acad Sci 105:10005–10010
99. Kaiser H-J, Lingwood D, Levental I, Sampaio JL, Kalvodova L, Simons K (2009) Proc Natl Acad Sci 106:16645–16650
100. Eggeling C, Brand L, Ullmann D, Jaeger S (2003) Drug Discov Today 8:632–641
101. Sieber JJ, Willig KI, Kutzner C, Gerding-Reimers C, Harke B, Donnert G, Rammner B, Eggeling C, Hell SW, Grubmuller H, Lang T (2007) Science 317:1072–1076

Imaging Molecular Order in Cell Membranes by Polarization-Resolved Fluorescence Microscopy

Sophie Brasselet, Patrick Ferrand, Alla Kress, Xiao Wang, Hubert Ranchon, and Alicja Gasecka

Abstract The use of light polarization properties in the analysis of fluorescence images has driven a large amount of research toward the measurement of orientational behavior of molecules in cells, in particular in their membranes. This field has been recently revisited to enlarge the possibilities of polarization-resolved fluorescence microscopy. We show that this technique allows retrieving a wealth of information on the constraints that hinder rotational mobility of lipid probes and proteins in membranes, bringing thus new insights on inter-proteins and lipid-protein interactions, on membrane morphology at the sub-diffraction length scale and on local membrane physical properties such as viscosity.

Keywords Cell membrane • fluorescence anisotropy • molecular order • polarized fluorescence microscopy • two-photon fluorescence

Contents

S. Brasselet (✉), P. Ferrand, A. Kress, X. Wang, H. Ranchon and A. Gasecka
Institut Fresnel, CNRS, Aix-Marseille Université, Ecole Centrale Marseille, Domaine Universitaire St Jérôme, Marseille 13013, France
e-mail: Sophie.brasselet@fresnel.fr

Y. Mély and G. Duportail (eds.), *Fluorescent Methods to Study Biological Membranes*, Springer Ser Fluoresc (2013) 13: 311–338, DOI 10.1007/4243_2012_51,

1 Introduction

Molecular orientation is a parameter which is important to monitor in order to understand the interactions that drive the structure and morphology of biomolecular assemblies and their consequences on related biological functions. This organization, governed by molecular interactions such as those between proteins, between lipids, or between proteins and lipids, plays an important role in biology. Proteins and lipids in cell membranes are, for instance, known to form structured assemblies (such as protein clusters or lipid domains) with collective molecular order participating in cell motility [1], vesicular trafficking [2], and signaling [3–5]. Protein interactions in supramolecular complexes such as bio-filaments are also highly driven by orientational order [6–8]. For instance, collagen and other protein filaments in the extracellular matrix undergo strong exchange forces with cell membrane proteins, which can be affected in the development of tumors. An orientational organization is quantified by the so-called molecular order, which varies between complete disorder (isotropic medium) to complete order (such as in a crystalline medium).

Imaging such information quantitatively using optical microscopy remains however a challenge. Measuring molecular orientational diffusion in solutions has been performed since the 1950s using the well-known fluorescence anisotropy technique [9], which consists in exciting the molecules by a linear polarization, and probing their fluorescence emission along two perpendicular analysis directions. This technique, based on the capacity of polarized light to photo-select only molecules which exhibit an orientation close to its polarization direction, gives access to molecular rotational diffusion time in solutions, local environment properties (temperature, viscosity) [10], but also to phenomena related to depolarization mechanisms such as fluorescence resonant energy transfer (FRET) between neighbor molecules [11]. Fluorescence anisotropy is however more delicate to implement in oriented media (e.g., membranes or fibers) where the photo-selection also depends on the orientation of the sample itself. Historically, techniques dedicated to the study of the cell membrane have been therefore essentially concentrated on measuring the spatial localization of molecules. Membranes are indeed highly heterogeneous and dynamic, constituted of different lipid types and molecules (such as cholesterol) as well as membrane proteins which are connected to the cytoskeleton in a complex way. Techniques such as fluorescence recovery after photo-bleaching (FRAP) [12], fluorescence correlation spectroscopy (FCS) [13], and single-molecule tracking (SMT) [14] have allowed a better understanding of the mobility of proteins and lipid and their perturbation by specific conditions of the cells, bringing a considerable amount of information on the nanoscale organization of the cell membrane and its associated signaling dynamics. Less attention has been directed toward protein and lipid orientations in cell membranes.

Applying fluorescence anisotropy to membrane imaging is in fact only successful under certain conditions, which limit the observation to specific parts of the membrane, since this technique only addresses two directions of detected polarization. Nevertheless, it has permitted to bring a new insight into specific molecular mechanisms in cell membranes, such as protein clustering or protein conformational changes, which cannot be only addressed by pure lateral diffusion. Such studies, which can be performed either in the time domain (time-resolved fluorescence anisotropy) or averaged over time (steady-state fluorescence anisotropy), bring new information on orientational behaviors which are complementary to other fluorescence-related imaging techniques such as lifetime imaging (FLIM) [15] and fluorescence resonant energy transfer (FRET) imaging [11], or colocalization imaging [16]. This polarization-related information is also applicable to single-molecule studies and super-resolution techniques [17].

This chapter describes the basic principles of fluorescence anisotropy imaging and polarization-resolved fluorescence imaging, in the context of steady-state measurements, in either one-photon or two-photon excitation regimes. It illustrates the use of these techniques with recent experiments performed on lipid and protein orientational behavior in cell membranes, focusing in particular on how this behavior is related to the cell lipid environment and to its mechanical properties, governed by the cytoskeleton. We introduce a more general polarization-resolved fluorescence imaging technique which can circumvent limitations of fluorescence anisotropy imaging and allow a complete mapping of molecular orientation in cell membranes. This concept can be extended to a vast variety of mechanisms in cells and tissues.

2 Defining Molecular Orientational Order in Cell Membranes

In ordered media, molecules undergo a certain degree of orientational mobility constraint, which depends on their interaction with their nearby environment. This information is formally contained in a normalized molecular angular distribution function $f(\theta,\phi)$ which defines the range of orientations (θ,ϕ) explored by the molecules after time and space averaging over the integration time of the optical measurement and the focal volume of the objective (Fig. 1a). Polarization-resolved fluorescence experiments aim at measuring accessible information on this distribution function, defined as the so-called molecular order. Molecular order, which can be used to report information on a local environment or on molecular interaction mechanisms, applies to a fluorescent "orientational probe" such as a lipid probe (for cell membrane architecture studies) or a protein label (for protein interaction studies). As in all fluorescence imaging experiments, the information retrieved is the direct signature of the behavior of this fluorescent probe, which can further be interpreted as an environment information. It is therefore important that this probe stays rigidly attached or linked to the biomolecule that it brings information on. At last, molecular orientational information as measured by polarized fluorescence is performed in microscopy where spatial resolution is limited: Therefore, the measured molecular order is always an averaged information in time (over the

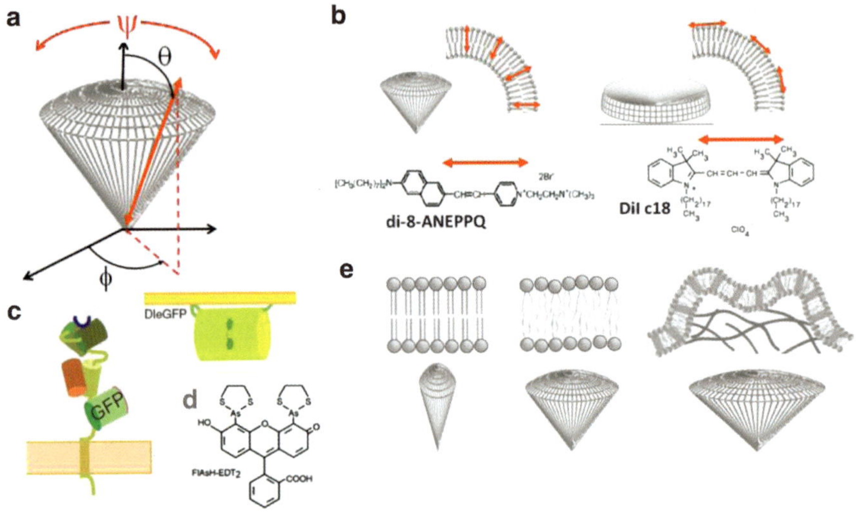

Fig. 1 Models for molecular order in membranes. (**a**) Cone aperture and molecular orientation. (**b**) Different distribution functions based on cones filled with lipid probes molecules perpendicular (*left*) or parallel (*right*) to the membrane. (**c**) Examples of protein GFP-based constructs with GFP linked in a rigid way to the protein body. Right: doubly lipidated enhanced GFP construction of a G-protein complex [18]. Left: MHC class I protein labeled with GFP rigidly attached [19–21]. (**d**) Fluorescein arsenical hairpin binder (FlAsH) [22, 23]. (**e**) Perturbations of the molecular orientational order in membranes (represented by a variation of the distribution cone aperture): local environment properties (liquid order *left*, liquid disorder *center*) or local morphology (*right*)

integration time of the detector) and in space (over the whole molecular population within the excitation volume, defined by the focal spot of an objective).

In molecular media where one-dimensional molecular probes are constraint by a known potential $U(\theta,\phi)$ at thermal equilibrium T, the orientational distribution function of the probe $f(\theta,\phi)$ follows the Boltzmann statistics:

$$f(\theta,\phi) \propto \exp\left(-\frac{U(\theta,\phi)}{k_B T}\right) \tag{1}$$

where k_B is the Boltzmann constant.

In lipid membranes, the orientation of lipids or proteins is most often defined as lying within a cone aperture (Fig. 1a) with an abrupt change of the probe potential at a defined aperture angle Ψ (Fig. 1a) [24, 25]. In such a model, the fluorophores lie statistically inside the cone aperture Ψ, meaning that they only explore a range of orientations within this cone. $f(\theta,\phi)$ is therefore defined as

$$f(\theta,\phi) = \begin{cases} \frac{1}{4\pi \sin^2(\Psi/4)} & if\ |\theta| \leq \Psi/2 \\ 0 & otherwise \end{cases} \tag{2}$$

Other smoother functions have been defined to quantify molecular order in membranes, such as based on a Gaussian distribution [26, 27] or more generally decomposed over spherical harmonic basis functions [28].

In what follows, we will focus on the cone model, which is straightforward to visualize. In a cell or lipid membrane, a lipid probe can lie either along the membrane perpendicular direction (when it is embedded in the membrane leaflets) or along its contour, depending on the molecule structure and its local interaction with lipids (Fig. 1b). In both cases, the cone distribution can be defined by two parameters: its aperture (Ψ, which is related to the "molecular order") and its average orientation in the sample plane.

In proteins, a fluorescent label is generally attached by using N- or C-terminal tagging by either a green fluorescent protein (GFP) mutant expression or by a synthetized fluorophore. In the most general case, this label is likely to rotate freely in the whole space, leaving no information on its orientation. To be able to use a protein label as a reporter of its orientation, this label needs to be at best rigidly linked to the protein body. Since a few years, a strong effort has been put on the engineering of such systems (Fig. 1c). A major histocompatibility complex class I (MHC I) construct has been obtained with a GFP probe located within the amino acid sequence of the $H2L^d$ protein, ensuring a limitation of its rotational freedom [19–21]. A G-protein complex has also been labeled in its subunits using a doubly lipidated GFP construction [18]. Recently, a strategy has also been developed based on fluorophores which can attach to proteins in several sites positions: Fluorescein arsenical hairpin binder (FlAsH) contains, for instance, a biarsenical group that can bind to a tetracysteine motif Cys-Cys-Pro-Gly-Cys-Cys in proteins [22]. This label has been used to study orientational dynamics of a G-protein-coupled receptor in cell membranes [23].

The information brought by the measurement of the molecular order Ψ can be related to local environment properties or local molecular interactions in several ways (Fig. 1e). Among the most important mechanisms which govern molecular order in cell membranes are:

- Local lipid phases of different viscosity/polarity/lipid contents. In particular, an ongoing research is dedicated to the imaging of lipid domains of different phases such as liquid order, liquid disorder, or gel phases [29–33]. These domains called rafts are of sub-resolution size, highly dynamic, and described as support platforms for proteins in different biological processes.
- Formation of proteins' clusters in cell membranes and in the cytoplasm, which are often related to mechanisms in cell signaling, cell motility, or cell trafficking [3–5].
- Local morphological changes in cell membranes due to either external interaction (with the extracellular matrix) or internal forces (due to actin, microtubule fibers, and the cytoskeleton in general), which are related to biological processes such as cell migration or adhesion [2].

3 Fluorescence and Its Relation to Light Polarization

Fluorescence is the result of two successive processes, the absorption of an incident photon and the emission of a fluorescence photon of lower frequency. These processes are separated in time due to the delay necessary for the molecule to relax from its high-energy excited state to a lower (fluorescent) excited state [34]. The steady-state fluorescence efficiency is consequently proportional to the product of two probabilities: the absorption probability between the ground and the excited state and the emission probability from the fluorescent state to the ground state. In the linear excitation regime (one-photon excitation: 1PE), the fluorescence intensity is proportional to

$$I^{1PEF} \propto P_{abs}^{1PE}.P_{em} \tag{3}$$

where the proportionality coefficient contains collection efficiencies and normalization factors. The absorption probability P_{abs}^{1PE} is proportional to

$$P_{abs}^{1PE} \propto \left| \vec{\mu}_{abs} \cdot \vec{E} \right|^2 \tag{4}$$

where $\vec{E}$ is the polarization of the incident electromagnetic field and $\vec{\mu}_{abs}$ the absorption transition dipole of the molecule.

The emission probability P_{em} depends on the analysis directions of the fluorescence emission (denoted X and Y, which define the sample plane in microscopy):

$$P_{em,X} \propto \left| \vec{\mu}_{em} \cdot \vec{X} \right|^2, \quad P_{em,Y} \propto \left| \vec{\mu}_{em} \cdot \vec{Y} \right|^2 \tag{5}$$

where $\vec{\mu}_{em}$ is the emission transition dipole of the molecule from its fluorescent state down to the ground state. The proportionality coefficient contains here the fluorescence quantum yield of the molecule and the characteristics of the radiated intensity.

Equation 5 is written assuming a planar wave approximation for the emission, which is not applicable for a high numerical aperture objective collection. In the case of a high aperture collection, a mixture of radiations along the X, Y, and Z directions occurs, and Eq. 3 has to be replaced by [24]:

$$\begin{aligned} P_{em,X} &\propto \kappa_1 \left| \vec{\mu}_{em} \cdot \vec{X} \right|^2 + \kappa_2 \left| \vec{\mu}_{em} \cdot \vec{Y} \right|^2 + \kappa_3 \left| \vec{\mu}_{em} \cdot \vec{Z} \right|^2 \\ P_{em,Y} &\propto \kappa_2 \left| \vec{\mu}_{em} \cdot \vec{X} \right|^2 + \kappa_1 \left| \vec{\mu}_{em} \cdot \vec{Y} \right|^2 + \kappa_3 \left| \vec{\mu}_{em} \cdot \vec{Z} \right|^2 \end{aligned} \tag{6}$$

where κ_1, κ_2, κ_3 are polarization mixture coefficients which depend on the numerical aperture of the objective with $\kappa_1 > (\kappa_2, \kappa_3)$ [24].

In general, $\vec{\mu}_{em}$ and $\vec{\mu}_{abs}$ have different orientations due to the fact that different states are involved in the absorption/emission processes, involving therefore different molecular conformations. For the sake of simplicity, we will assume that these dipoles are pointing along a same direction of orientation (θ,ϕ), which also defines the orientation of the molecular structure: $\vec{\mu}_{em} = \vec{\mu}_{abs} = \vec{\mu}(\theta, \phi)$. The fact that both dipoles can have different angles can be addressed by using a more complete expression accounting for an additional angle.

Finally, the one-photon fluorescence intensity, along the analysis direction X for a molecule oriented with an angle (θ,ϕ), is written in the sample frame as

$$I_X^{1PEF}(\theta, \phi) \propto \left|\vec{\mu}(\theta, \phi) \cdot \vec{E}\right|^2 \cdot J_X(\theta, \phi) \tag{7}$$

with $J_X(\theta,\phi) \propto \kappa_1\left|\vec{\mu}(\theta,\phi) \cdot \vec{X}\right|^2 + \kappa_2\left|\vec{\mu}(\theta,\phi) \cdot \vec{Y}\right|^2 + \kappa_3\left|\vec{\mu}(\theta,\phi) \cdot \vec{Z}\right|^2$.

Equation 7 can be written similarly for $I_Y^{1PEF}(\theta,\phi)$. Since polarization analysis does not depend on proportionality coefficients, they do not need to be specified here.

This formalism can be extended to multiphoton fluorescence and in particular two-photon excited fluorescence (2PEF) accounting for the two-photon excitation absorption probability P_{abs}^{2PE}, which involves a higher order nonlinear excitation [28]. Since the emission process does not change,

$$I_X^{2PEF}(\theta,\phi) \propto \left|\vec{\mu}(\theta,\phi) \cdot \vec{E}\right|^4 \cdot J_X(\theta,\phi) \tag{8}$$

and similarly for $I_Y^{2PEF}(\theta,\phi)$.

Equation 8 emphasizes the fact that the two-photon fluorescence intensity depends on the square of the incident intensity, whereas this dependence is linear in the case of one-photon fluorescence.

In both one- and two-photon excitation cases, one can see that when a linearly polarized light excites a single molecule the highest probability of absorption occurs when its transition dipole moment $\vec{\mu}_{abs}$ is oriented parallel to the incident polarization $\vec{E}$. This property, called angular photo-selection, is at the origin of the polarization dependence of the fluorescence process. Denoting Θ the angle between the absorption/emission molecular transition dipole and the exciting polarization, the one-photon angular photo-selection is proportional to $P_{abs}^{1PE}(\Theta) \propto \cos^2\Theta$, whereas in the TPEF process, this photo-selection is proportional to $P_{abs}^{2PE}(\Theta) \propto \cos^4\Theta$. This makes the two-photon photo-selection narrower than for a one-photon excitation, which allows measuring molecular orientations with a finer precision.

Finally, in a collection of molecules, one has to account for the fluorescence emitted by an ensemble of these molecules contained in the focal volume of the objective, with an orientational constraint defined by the function $f(\theta,\phi)$[28]:

$$I_X^{1PEF} \propto \int_0^{2\pi}\int_0^{\pi} \left|\vec{\mu}(\theta,\phi)\cdot\vec{E}\right|^2 \cdot J_X(\theta,\phi)\cdot f(\theta,\phi)\cdot \sin\theta d\theta d\phi \quad (9)$$

and similarly for I_Y^{1PEF}.

Equation 9 permits to calculate the dependence of the collected fluorescence as a function of the incident polarization $\vec{E}$. This expression can therefore be used to define fluorescence anisotropy (Sect. 4) or more refined polarization-resolved signals (Sect. 5). It shows also that from a polarized fluorescence measurement, if $\vec{E}$ is known, one can derive information on the orientational distribution$f(\theta, \phi)$. In the case of a cone model,$f(\theta, \phi)$is written as in Eq. 2 and therefore depends only on the molecular order cone aperture Ψ and its orientation in the macroscopic sample frame. We will show in Sect. 5 that this information can be determined experimentally without ambiguity.

4 Fluorescence Anisotropy

Fluorescence anisotropy has been widely developed since decades to study molecule rotational diffusion properties, media viscosity, and protein conformations and interactions. Here, we develop the principles of fluorescence anisotropy imaging.

4.1 *Fluorescence Anisotropy in Isotropic Media*

The idea to investigate a molecular structure by a polarization-resolved fluorescence method was introduced by Perrin in 1926 [35] and applied by Weber in 1953 to the structural study of the binding of small molecules to proteins [9]. This method is based on the fact that even though in a solution the macroscopic medium is isotropic, the fluorescence emission is polarization-dependent since the microscopic structures investigated are anisotropic. In a traditional fluorescence anisotropy setup (Fig. 2a), a solution is excited using a beam propagating along the *Z*-axis and polarized along *X*. The fluorescence is measured along the perpendicular propagation direction *Y* and analyzed along the *X* and *Z* directions using a polarizer, the measured quantity being called $I_{//}$ and $I_{\perp}$, respectively.

A ratiometric analysis of these two situations provides the so-called fluorescence anisotropy, independent on the intensity fluctuations and solely dependent on orientational properties:

$$A = \frac{I_X - I_Y}{I_X + I_Y + I_Z} = \frac{I_{//} - I_{\perp}}{I_{//} + 2I_{\perp}}. \quad (10)$$

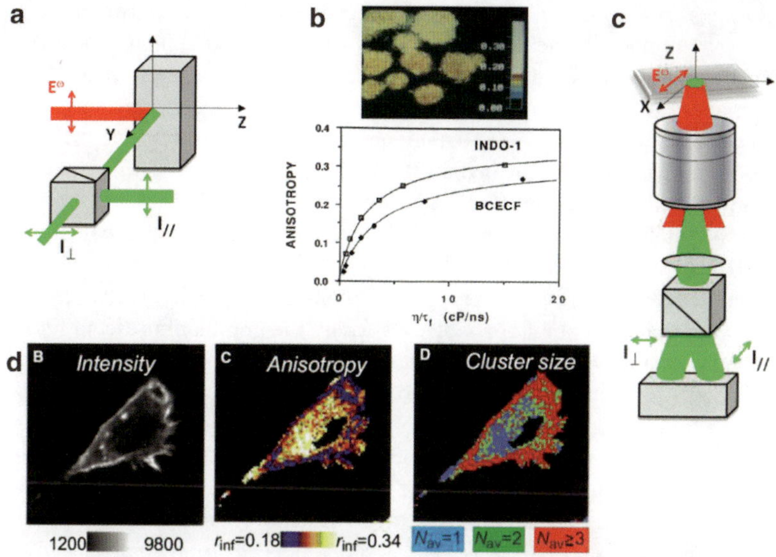

Fig. 2 Fluorescence anisotropy. (**a**) Traditional fluorescence anisotropy measurement in solution. (**b**) Typical anisotropy versus viscosity (η, normalized by the fluorescence lifetime τf) curve obtained from fluorophores Indo-1 and BCEF in solution of glycerol of increasing concentration. The values measured have been compared to anisotropy in MDCK cells (*above*) to retrieve cytoplasmic viscosity information [36]. (**c**) Fluorescence anisotropy imaging in microscopy. (**d**) Illustration of anisotropy changes in cells where protein (GPI-GFP) clusters form close to the membrane: The decrease of anisotropy is correlated with a depolarization due to energy transfer between proteins. Intensity (*left*), steady-state anisotropy (*middle*), and cluster-size deduced from anisotropy analysis (*right*). The latter two images are binned to 80 × 80 pixels [37]

The normalization denominator represents the total fluorescence intensity, summed over all possible analysis directions. The factor 2 originates from the fact that in a isotropic solution, $I_Y = I_Z = I_\perp$ [10].

Originally, the fluorescence anisotropy was used to measure rotational diffusion times of proteins in solutions. In an isotropic sample where molecules are fixed, then $f(\theta, \phi) = 1/2\pi^2$ in Eq. 9. Setting $\overrightarrow{E}//X$ as the incident polarization leads to $A = 0.4$ in a one-photon regime and $A = 0.57$ in a two-photon excitation regime. These numbers are the maximum anisotropy values possibly reachable in an isotropic medium. In practice, the measured value can be lower than this limit for three main reasons: (1) The angle between $\overrightarrow{\mu}_{abs}$ and $\overrightarrow{\mu}_{em}$ is not 0°, (2) the molecules undergo some rotational mobility, therefore having different orientations between their absorption and emission events, and (3) the molecules transfer energy between each other; therefore, the absorption step takes place for one molecular orientation and the emission step for another, uncorrelated orientation. These three processes lead to depolarization, meaning that the absorption and emission events do not take place with identical orientations. Case (2) has lead to numerous studies on proteins conformational changes and local viscosity monitoring (Fig. 2b).

Fluorescence anisotropy has been more recently applied to microscopy imaging (Fig. 2c), where fluorescence excitation is performed in a wide field geometry using an incident polarized excitation along a direction X in the sample plane (X,Y). I_X and I_Y images are formed on a CCD camera along the X and Y analysis directions, using a polarization splitter. This technique has been used to retrieve orientational diffusion rates in intracellular media [36, 38].

Among the depolarization factors widely studied in fluorescence anisotropy imaging is fluorescence resonance energy transfer between similar molecules (homo-FRET) (case (3) mentioned above). Homo-FRET is the property of an excited molecule to transfer its energy in a non-radiative way, through a dipole-dipole interaction, to a neighbor molecule which then emits fluorescent light [39]. Since the orientations of the two molecules are uncorrelated, the absorption and emission steps are decoupled in the fluorescence process, which creates depolarization and therefore a loss of orientational information [11]. This process, which occurs at molecular inter-distances of a few nanometers [40], has been used in cells to identify possible proteins clusters [5, 37, 41–43] (Fig. 2d). Homo-FRET information can be derived in a quantitative way to measure cluster sizes and is thus used to monitor oligomerization of protein constructs. This has been applied to glycosyl-phosphatidylinositol (GPI)-anchored proteins labeled with GFP, revealing that GPI forms clusters close to cell membranes with an average size of more than two subunits (Fig. 2d) [37].

4.2 *Fluorescence Anisotropy Imaging in Cell Membranes and Its Limitations*

In microscopy imaging in oriented samples such as cell membranes, the situation is different since the distribution $f(\theta, \phi)$ is no more isotropic but confined in a cone of *a priori* unknown aperture and orientation. In this situation, the geometry of a photo-selection by a linear-fixed polarization is no more adapted since the averaged orientation of the molecules in the sample is unknown. In order to avoid such photo-selection artifacts, the incident excitation is therefore often changed into a circular polarization (Fig. 3a). A new anisotropy factor is defined, which accounts for the symmetry of the setup:

$$A = \frac{I_X - I_Y}{I_X + I_Y} \tag{11}$$

with A measured for an incident circularly polarized light. The (X,Y) directions also define the sample plane (Fig. 3a). This new anisotropy factor is therefore essentially sensitive to the emission probability polarization dependence. One can deduce from Eq. 11 that A varies between -1 (for transition dipoles strictly oriented along the Y-axis) and 1 (for transition dipoles strictly oriented along the X-axis).

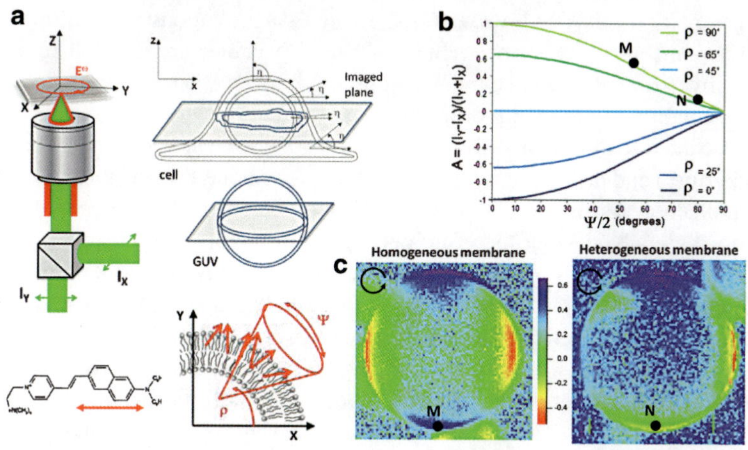

Fig. 3 Fluorescence anisotropy confocal imaging in membranes. (**a**) Experimental scheme: the I_X and I_Y analyzed intensities are recorded for an incident circular polarization (some schemes use a linear polarization). The sample is schematized as a spherical lipid membrane in which molecules are assembled within a cone aperture, normal to the membrane. All images (cell, GUVs) are performed in the equatorial plane of the membrane. (**b**) Theoretical dependence of the polarization factor A (Ψ) as a function of the cone aperture Ψ of a filled cone distribution, for several tilt angles ρ of the cone in the sample plane. (**c**) Anisotropy A measured for a giant unilamellar vesicle made of DOPC (*left*) and a mixture of DOPC, sphingomyelin, and cholesterol (1:1:1) which leads to heterogeneous lipid domain formation (*right*). The points of measurement M and N at the position $\rho = 90°$ are represented on the theoretical A (Ψ) graph, leading to an estimation of the molecular order value at this position of the sample [44]

In all intermediate situations, such as different orientation and different distribution functions, A will take intermediate values.

Figure 3b depicts the evolution of A for fluorescent molecules lying within a cone aperture Ψ, for different orientations ρ of the cone in the (X,Y) plane (the cone is supposed to lie in the sample plane; therefore, the membrane is imaged in its equatorial plane). This graph is generated using Eq. 9 in a two-photon excitation regime, assuming parallel absorption and emission dipole moments in the molecules. Similar curves are obtained for a one-photon excitation. One can immediately notice that in order to retrieve a molecular order information (a Ψ value) in a sample, the knowledge of the cone orientation in the sample plane (ρ) is required with high precision. Indeed, determining a single experimental value A does not allow retrieving more than one parameter on the sample. In addition, this measurement is only applicable when the cone lies close to the X- or Y-axes, the case $\rho = \pi/4$ leading to an indetermination ($A = 0$) of the molecular order since both X and Y projections are equivalent. Retrieving Ψ from an anisotropy measurement is nevertheless feasible in a membrane of round shape, such as in a giant unilamellar vesicle (GUV) (Fig. 3c). All points of its equator can indeed be assigned with a position angle ρ, which gives access to molecular order values even if the membrane is heterogeneous.

This method shows however its strongest limitations when applied to membranes of arbitrary shapes where the local ρ values are difficult to estimate visually from the fluorescence image. Applying fluorescence anisotropy imaging to the investigation of molecular order in cell membranes is finally only successful in cases restricted to simple distribution functions of cylindrical symmetry (such as a cone aperture) and necessitates an *a priori* knowledge of either the mean orientation of the molecular distribution or its shape [25, 44–46]. Many fluorescence anisotropy imaging studies have been therefore performed in membranes of spherical shapes such as in swelling cells (Fig. 4a) or in particular regions of the membrane where the direction ρ is identified as being along *X* or *Y* (Figs. 4b, c, 5a). A large amount of investigations have been carried out in this field since the seminal work of Axelrod in 1979 [24], in which the orientation of long chain carbocyanine dyes was studied in spherical lipid membranes of red blood cells. All further studies in lipid membranes have been limited so far to simple spherical membrane geometries such as in artificial giant unilamellar vesicles (GUVs) [25], red blood cells [47], swelling cells [21], spherical cells [26, 27], and spherical nuclear envelopes [48]. In these membrane studies, the measurement was performed on the perimeter of the spherical membrane, and the in-plane average orientation ρ of the molecular distribution function of the fluorescent probes in the lipid membrane was assumed to follow the direction normal to the membrane. Apart from membrane studies, fluorescence anisotropy imaging has been applied to the determination of the width of molecular angular distributions in biopolymers of cylindrical symmetry such as actin filaments [6], muscle fibers [45], and septin filaments [8]. In these studies, the orientation ρ of the fibers in the sample plane was visualized in the image and used as a known parameter to reduce the complexity of the studies.

Even though the limitations of fluorescence anisotropy imaging are quite constraining, this technique is still able to probe interesting mechanisms in cell membranes such as drug treatments effects or mechanical perturbations. Changes relative to lipid composition (varying the cholesterol content) have been observed on lipid probes which are seen to lose their orientational order when cholesterol is depleted from the membrane, indicating a relation between local rigidity and cholesterol content [26]. Cell membrane local morphology at a sub-resolution scale can also be probed using such technique. A recent observation has evidenced cell ruffling at the junction of two cells, demonstrated by a loss of anisotropy contrast of lipid probes in the junction [3]. We have recently demonstrated the modifications of the lipid probe's molecular order when perturbing the cell membrane tension by a mechanical perturbation (by a hypotonic shock-provoking cell swelling) or by drug treatments affecting the cytoskeleton (using latrunculin A, an actin polymerization inhibitor, or cytochalasin D, an actin depolymerization inducer) [19]. A higher orientational order is observed when the cell membrane is locally more tense due to the relaxed constraint from the cytoskeleton. This emphasizes the correlation between molecular order and local membrane morphology [19].

Of particular interest are recent studies on membrane proteins (Fig. 4). A large amount of research is indeed dedicated to label proteins in a rigid way such as the attached fluorescent probe can report the orientation of the protein itself (Fig. 1c).

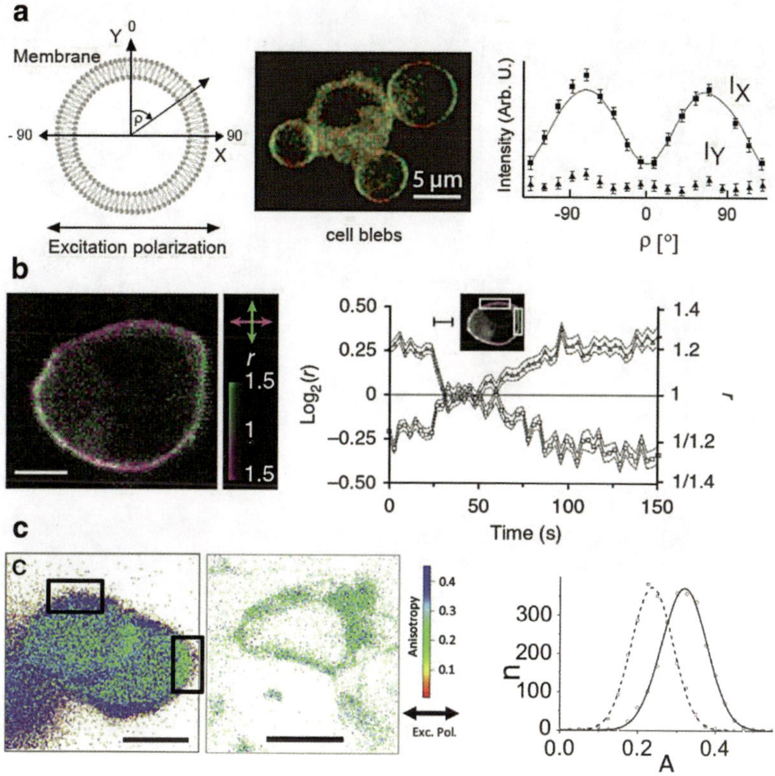

Fig. 4 Fluorescence anisotropy imaging in cell membranes (see Fig. 1c for the scheme of the molecules involved). (**a**) MHC I rigidly labeled with intra-sequence GFP in cell blebs. The anisotropy contrast shown as a function of the contour angle around the round cell membrane is a signature of rigid labeling [21]. (**b**) 2-photon fluorescence anisotropy imaging of G-protein activation. The anisotropy is followed in time for a Gα subunit labeling using doubled lipidic attachment with yellow fluorescent protein (YFP), after addition of an agonist (norepinephrine) [18] (scale bar: 5 μm). (**c**) Anisotropy images of *left image*: the intracellular loops (IL3) of a G-protein-coupled receptor (GPCR), labeled with FlAsH connecting a bridge between its N- and C-termini, *right image*: the parathyroid hormone receptor (PTHR) used as a nonrigid membrane FlAsH-binding motif where the labeling takes place at the C-terminus (scale bar: 10 μm). *Right*: The anisotropy distribution taken in a particular region of the cell membrane shows a visible contrast [23]

A major histocompatibility complex class I (MHC I) construct rigidly labeled with GFP has been shown to exhibit strong anisotropy contrast in a cell bleb membrane, indicating a degree of orientation constraint as compared to other nonrigid constructs where the GFP label is attached to the cytoplasmic tail of the protein [21] (Fig. 4a). Recent constructs applied to protein receptors have also allowed measuring, using two-photon fluorescence anisotropy, changes of proteins conformation upon G-protein activation. In particular, interactions involving the different subunits of heterotrimeric G-proteins could be visualized in the presence of

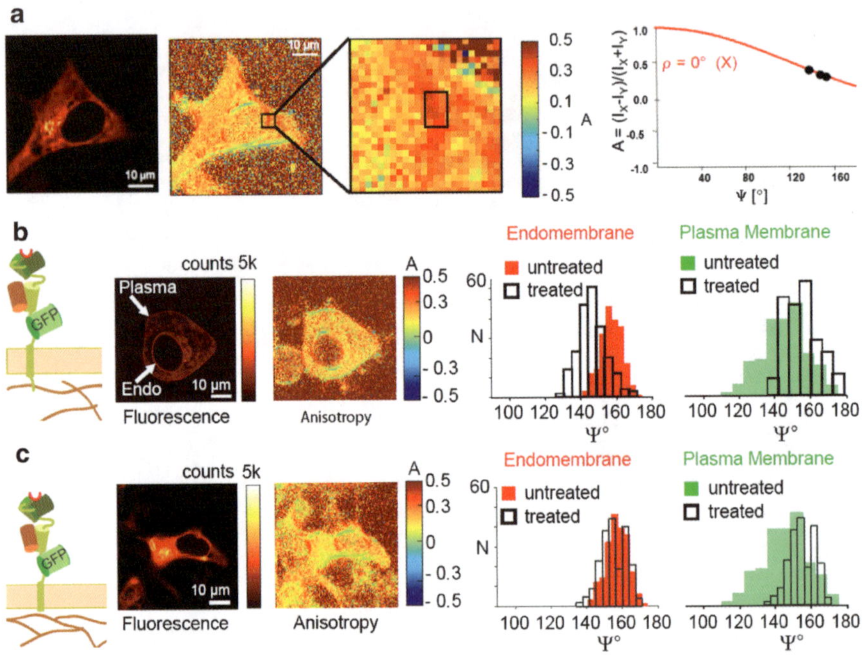

Fig. 5 Fluorescence anisotropy imaging in MHC I proteins rigidly labeled with GFP. (**a**) *Left*: fluorescence image. *Middle*: map of anisotropy and selected region of interest for structural studies. *Right*: A(Ψ) graph used to retrieve a quantitative number of the molecular order Ψ. Several points of measurements are shown. (**b, c**) Changes in molecular order Ψ in the plasma membrane and the endomembrane for (**b**) MHC I protein rigidly labeled with GFP perturbed by a hypotonic shock [19], (**c**) truncated construct where the cytoplasmic tail has been removed

a suitable G-protein-coupled receptor (GCPR), responding to the presence of a receptor agonist by a conformation change, probed by a modification of anisotropy (Fig. 4b) [18]. The decrease of anisotropy in the superior region of the cell membrane is indeed a signature of a change of relative orientation angle compared to the membrane, indicating a folding of the labeled subunit toward the membrane direction. Labeling engineering using fluorophores attached at two sites of a protein has also permitted to probe orientational order in membrane receptors (Fig. 4c) [23].

The relation between orientational properties of membrane proteins and the cell membrane mechanical properties has been further explored by applying treatments affecting the cytoskeleton (Fig. 5) ([19]). Contrary to what was observed for lipid probes, we recently observed that membrane proteins MHC I rigidly labeled with GFP in COS-7 cells undergo a loss of molecular order when the cytoskeleton is disrupted by a hypotonic shock (Fig. 5b) or drug treatments (latrunculin A, cytochalasin D). Similar effects could be observed when perturbing the microtubule network by a colcemid treatment. This shows that the orientational constraint of membrane proteins is highly governed by their mechanical or chemical link to nearby cytoskeleton molecules, as previously emphasized in other studies based, for instance, on lateral diffusion observations [49–53]. A new construct of similar MHC I rigidly labeled but with a truncated cytoplasmic chain has shown similar loss of order, confirming the role of the attachment of the protein to the cytoskeleton in its orientational behavior (Fig. 5c). The MHC I orientational behavior was also probed in the endomembrane, showing higher disorder due to the more complex endomembrane composition and morphology as compared to the plasma membrane and to the less mature stage of the protein in this location before its migration to the outer region of the cell (Fig. 5). Interestingly, the behavior of MHC I in the endomembrane can also be affected by cytoskeleton drug treatments, revealing mechanical relations between the plasma membrane and the interior of the cell [54].

Note that in these studies, it is important to ensure that homo-FRET does not take place since it would lead to depolarization and therefore to an underestimation of the anisotropy values. This characterization can be done using photo-bleaching studies, measuring possible changes of the anisotropy, while molecules are progressively removed in the samples by photo-bleaching (therefore lowering the probability of homo-FRET) [19].

At last, note that one-photon versus two-photon fluorescence leads to a slightly different behavior in the anisotropy versus molecular order dependence. As mentioned above, the photo-selection is indeed finer when using a two-photon excitation process, which leads to a more refined dependence. Two-photon fluorescence imaging is however subject to other drawbacks such as a higher probability to deteriorate the cell upon intense pulsed illumination, which is necessary to overcome the lower absorption probability cross sections of fluorophores. Among the examples found in the literature, both techniques have been used to profit from their respective advantages.

5 Polarization-Resolved Fluorescence

In general, membranes are highly heterogeneous and exhibit shapes and features that cannot be fully explored using fluorescence anisotropy due to the limitations mentioned above. In order to be able to map molecular order in the whole membrane contour, more refined techniques need to be implemented.

5.1 Principle

Polarization-resolved fluorescence has been developed in order to circumvent limitations of fluorescence anisotropy imaging. In this technique, the incident polarization is no more purely linear along one direction or circular but is rotated between 0° and 180° (relative to the X sample horizontal direction), with angle steps small enough to retrieve a complete information on molecular order (typically 20° or less).

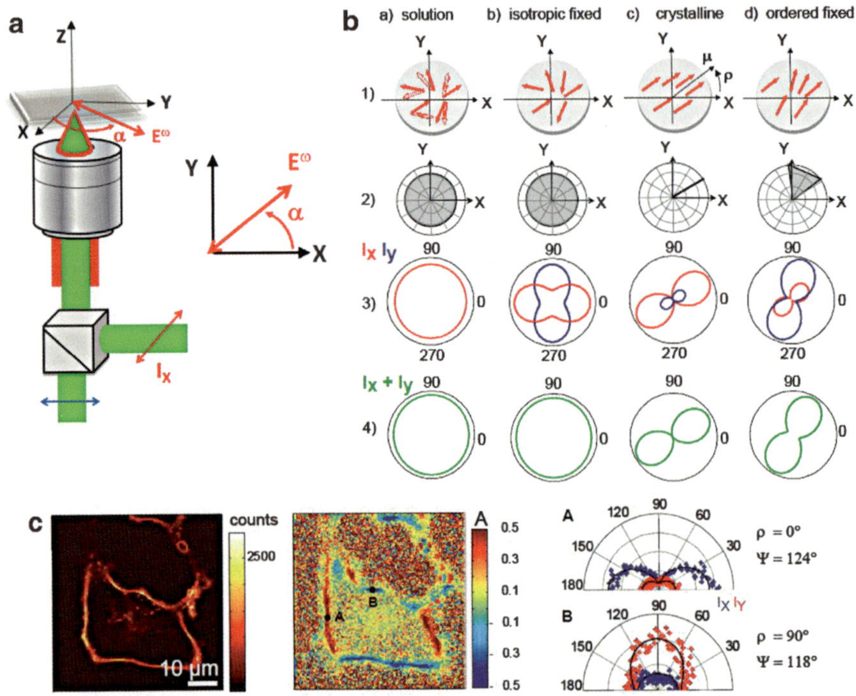

Fig. 6 Polarization-resolved fluorescence. (**a**) Setup. (**b**) Example of different fluorescence polarimetric responses depending on the sample explored, from a solution to oriented molecules in a cone aperture. From top to bottom: molecular dipoles represented as *arrows*, distribution seen in the plane of the sample, polarimetric responses in a one-photon regime $I_X(\alpha)$ and $I_Y(\alpha)$ represented as polar plots, and total polarimetric sum response $I_X(\alpha) + I_Y(\alpha)$. (**c**) Illustration in a COS-7 cell membrane labeled with di-8-ANEPPQ. *Left*: fluorescence image, *middle*: anisotropy image, *right*: polarimetric responses measured in the two points A and B visible on the anisotropy image. The distribution parameters Ψ and ρ are deduced from a fit of the polarimetric responses

Polarization-resolved fluorescence techniques (Fig. 6a) have been developed in both one-photon [19, 55] and two-photon excitation regimes [44, 56, 57]. An incident laser excitation (either continuous for one-photon or pulsed such as from a tunable Ti:sapphire laser, which delivers 150 fs pulses at a repetition rate 80 MHz, for two-photon) is used in a confocal mode. The laser beam is reflected by an appropriate dichroic mirror and focused on the sample by a microscope objective which numerical aperture (NA) can vary between 0.6 and 1.2. The use of high apertures presents the advantage of a higher optical resolution (typically 300 nm) and a more efficient signal collection in the backward epi-geometry; however, it exhibits more polarization distortion and mixing for the collected signal, which should be accounted for in the signal analysis. In an epi-geometry setup, the backward emitted signal is collected by the same objective and directed to a polarization beam splitter that separates the beam toward two detectors (avalanche photodiodes or photomultipliers). Images are performed by scanning either the sample on a piezoelectric stage or the beam position by galvanometric mirror piezoelectric scanning. In the first case, the sample is imaged and then a precise location is chosen for the measurement of the polarimetric response. The second case is more advantageous in terms of decrease of the acquisition time (typically a rate of 1 image per second is reached) and consists in recording one image per incident polarization state. In order to vary the incident linear polarization, a half waveplate, mounted on a step rotation motor, is placed before the dichroic mirror. An increase of the polarization switching speed can be obtained by the use of fast linear polarization controller, such as a Pockels cell placed before a quarter waveplate. For each value of the polarization angle α (relative to X) from 0° to 180° (the second half of the measurement being redundant), the emitted signal is recorded on the two perpendicular directions X and Y (Fig. 6a). This setup can be extended to forward detection schemes, the forward emitted signals being generally detected through a low-aperture objective (NA ~ 0.5).

The resulting data are no more one anisotropy value as described above but rather a so-called polarimetric response in both X and Y analysis directions, which will be denoted $I_X(\alpha)$ and $I_Y(\alpha)$. Examples of polarimetric responses are given in Fig. 6b as polar plots with respect to α. In isotropic media, the polarimetric responses can take different shapes depending on the degree of depolarization in the sample: For fast rotating molecules (or complete depolarization cases), both $I_X(\alpha)$ and $I_Y(\alpha)$ resemble a circle. When molecules are fixed in an isotropic distribution, $I_X(\alpha)$ and $I_Y(\alpha)$ are no more identical, even though the total polarization response $I_X(\alpha) + I_Y(\alpha)$ is a circle since the distribution is still isotropic. This is an indication that an analysis of the signal in both X and Y directions can bring considerable information on the depolarization mechanisms taking place in the sample. In oriented samples, the polarimetric response takes various two-lobe shapes in which aperture and orientation are, respectively, related to the molecular order Ψ and its average orientation ρ in the sample plane. From a fit of such polarimetric responses, it is therefore possible to retrieve a direct and unambiguous determination of both parameters (Ψ, ρ). An illustration of this is shown in Fig. 6c: Independently on the point taken around a labeled cell membrane, the simultaneous fit of polarimetric responses $I_X(\alpha)$ and $I_Y(\alpha)$ leads to the determination of both parameters of the distribution function modeled as a cone. This situation therefore completes advantageously the anisotropy image which is not exploitable in all points of the image. Here, the information can be retrieved over the whole contour of the cell equatorial plane.

The fit of polarimetric curves relies on the model developed in Eq. 9, using for $\vec{E}$ a function of α (ideally, for a perfectly controlled linear rotating polarization: $(E_X(\alpha) = \cos\alpha; E_Y(\alpha) = \sin\alpha)$), supposing $\vec{\mu}_{abs}$ and $\vec{\mu}_{em}$ parallel and taking for $f(\theta, \phi)$ the cone distribution described above with two unknown parameters (Ψ, ρ). However, one has to take care of different deviations from this perfect model in order to account for instrumental and sample artifacts:

- In a polarization-resolved fluorescence setup, controlling and correcting for polarization distortions is determining. Indeed, the polarization state of the exciting field has to be perfectly known. Calibration experiments are therefore necessary in order to introduce possible ellipticity and dichroism factors, introduced principally by the dichroic mirror which reflects the incident laser excitation [58].

- The high numerical aperture focusing induces a nonhomogeneous polarization state within the focal excitation volume. In particular, small contributions polarized in the axial Z direction appear on the edge of the focal spot [59]. Numerical analyses have shown that these distortions are not determining when molecules lie mostly in the sample plane [60].
- The assumption of parallel absorption and emission dipole moment does not apply to all fluorescent molecules. The angle between the two dipoles should be known beforehand and introduced in the model. It has been shown however that below 20°, this angle has no detrimental effect on the deformation of polarimetric responses [44]. In many probes of molecular orientation, this angle is found to lie in this range [61, 62].
- The assumption that the cone aperture lies in the sample plane relies on the fact that measurements are performed in the equator of the membrane. If the measurement is performed out of this plane, a known tilt angle has to be introduced in the model (angle η represented in Fig. 3a). Nevertheless, studies have shown that the in-plane approximation leads to only a few degrees of error on Ψ for η angles below 45° [44, 55].
- At last, this technique is sensitive to depolarization mechanisms, in particular homo-FRET between nearby molecules. When such mechanisms occur, the polarimetric response from a cone aperture loses its characteristics and tends to equalize the analyzed $I_X(\alpha)$ and $I_Y(\alpha)$ responses [44]. Similarly to anisotropy measurements, the absence of homo-FRET needs to be ascertained to avoid any bias on the retrieved molecular order angles. In Sect. 5.3, we describe a way to circumvent such bias.

5.2 *Application to Molecular Order Imaging in Heterogeneous Membranes*

The main advantage of polarization-resolved fluorescence is to provide a spatial information of molecular order independently on the membrane shape or heterogeneity. This technique has been applied to the investigation of the orientational organization in coexisting liquid phase with short-range liquid order (Lo) and liquid disordered (Ld) fluid domains of micrometric sizes in artificial membranes (giant unilamellar vesicles (GUVs)) and in cell membranes of nonspherical shapes.

Giant unilamellar vesicles (GUVs) made of lipid mixtures are considered as model systems for the investigation of lipid interactions [63–66]. They can exhibit coexisting domains with different fluidity, elasticity, and polarity properties. This phase segregation into gel, liquid ordered and disordered environments, is closely related to fundamental cell processes, in particular in cell signaling where the

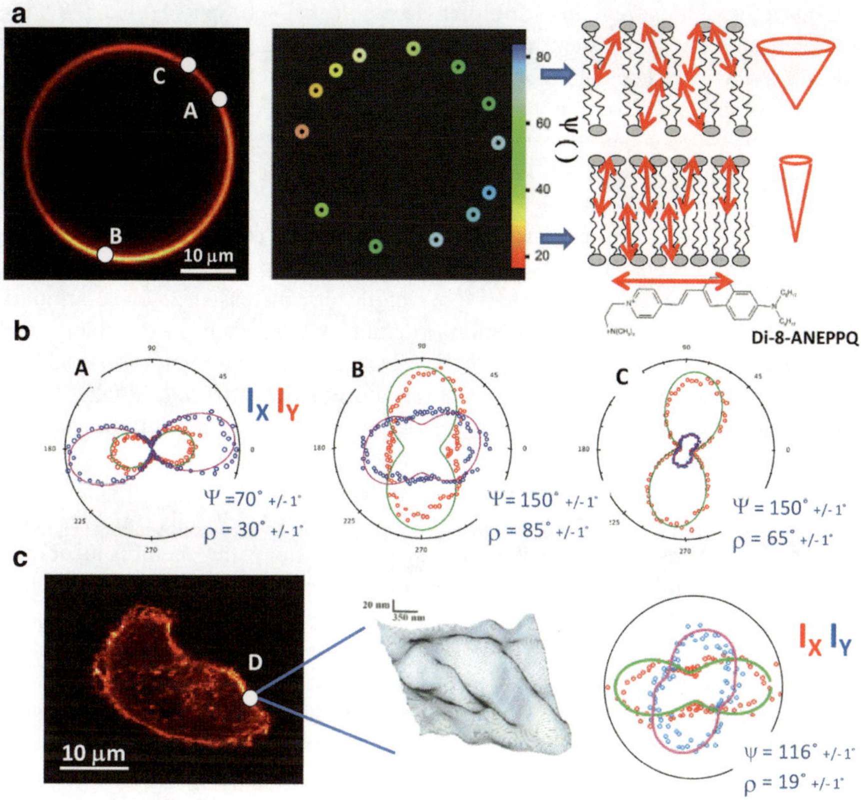

Fig. 7 (**a, b**) Polarization-resolved two-photon fluorescence in artificial lipid membranes labeled with di-8-ANEPPQ: giant unilamellar vesicles (GUVs) made of Ld and Lo phases from a DOPC: sphingomyelin:cholesterol (1:1:1) mixture. (**a**) *Left*: fluorescence image. *Right*: retrieved color-coded molecular order image showing the angle value Ψ obtained from a fit of polarimetric responses in several points of the GUV. (**b**) Typical polarimetric responses, represented in polar plots, for three points of measurements shown in (**a**). The fit (*continuous line*) of the 2PEF polarimetric data (*markers*) is performed using a filled cone model of orientation ρ in the sample plane and aperture Ψ. Two populations could be found, characteristics of ordered and disordered phases [44]. (**c**) Polarization-resolved 2PEF in a COS-7 cell labeled with di-8-ANEPPQ, which exhibits higher molecular disorder representative of local membrane morphology features (Represented schematically from [72])

existence of lipid-specific functional "rafts" platforms is still a debate [29–33]. So far, imaging lipid domains has relied on the use of dedicated fluorescent probes specifically partitioning in regions of known lipid composition or local polarity [67–69]. Such probes are also expected to undergo specific orientational order detectable by fluorescence anisotropy [67, 69, 70]. Recent studies have been dedicated to molecular order monitoring in such phases [44, 71].

Figure 7 shows typical two-photon fluorescence (2PEF) polarimetric responses of fluorescent molecules di-8-ANEPPQ in GUVs formed from a ternary mixture of the lipids sphingomyelin, 1,2-dioleoyl-sn-glycero-3-phosphocholine (DOPC), and cholesterol [64]. In such mixtures, Lo domains are known to be essentially constituted by enriched sphingomyelin and cholesterol regions, whereas DOPC is mainly present in disordered liquid regions (Ld). Although the Lo or Ld environments cannot be identified in a pure fluorescence image, the 2PEF polarimetric analysis permits to directly create an image of the spatial distribution of molecular order (Fig. 7a). Typical cone aperture values range between $30° < \Psi < 80°$ (Lo phases) and $90° < \Psi < 170°$ (Ld phases) depending on the probe molecule (Fig. 7b). In the Ld phase, where the lipid acyl chains are highly disordered, the aperture angle is therefore significantly increased. 2PEF polarimetry shows overall that molecular order information is dependent on the fluorescent probe structure, primarily because it is driven by lipid-fluorophore interactions which are influenced by the molecular head position. In particular, Ψ takes larger values when the molecule inclusion localization takes place in the periphery part of the membrane [27, 69, 73].

Similar studies have been performed on cells. One- and two-photon polarimetry measurements performed on di-8-ANEPPQ probes in COS-7 cells (Fig. 7c) show that the molecules mean orientation roughly lies along the membrane normal direction. In many points, the membrane is of complex shape, and its global orientation is difficult to visualize. The simultaneous fitting on both Ψ and ρ parameters makes it possible to avoid speculating on the local membrane contour as usually done with fluorescence anisotropy. The measured cone aperture angle on COS-7 cell membranes shows that the di-8-ANEPPQ probes behave in a slightly more disordered way than in DOPC GUV membranes. This is most probably due to a more perturbed membrane spatial morphology, originating from folding at spatial scales much below the diffraction limit (from 20 to 100 nm) [1, 2, 72, 74].

These studies show that membrane morphology features below the 300 nm optical resolution limit can be probed at any position of a membrane using polarization-resolved fluorescence: Any folding of the membrane at nanometer scales (due to ruffling, vesiculation...) [3, 75] will indeed lead to a loss of the measured molecular order.

5.3 Circumventing Depolarization Effects

In order to avoid the sensitivity of polarization-resolved fluorescence to depolarization mechanisms as mentioned above, we developed recently a technique, derived from polarization-resolved fluorescence but purely governed by photo-selection [55]. In this method, the incident polarization is varied, but the emission is not analyzed along a given direction; therefore, the sum $I(\alpha) = I_X(\alpha) + I_Y(\alpha)$ is detected. This way, the possible decorrelation between absorption and emission events is not visible in the data, which are only sensitive to the photo-selection taking place in the sample. This technique may appear less precise for the estimation of molecular order parameters since only one polarization response is recorded

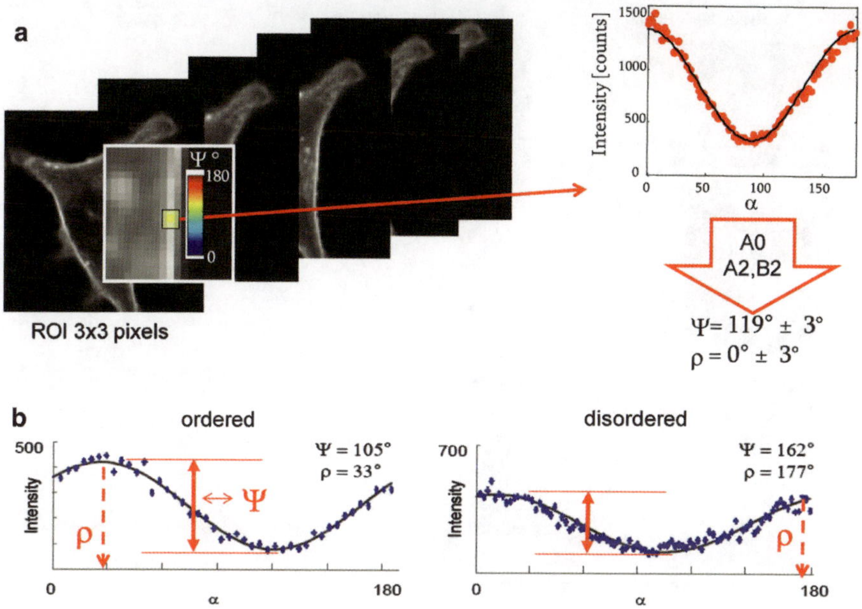

Fig. 8 Fourier decomposition of the polarization-dependent fluorescence signal and application to molecular order imaging. (**a**) A COS-7 cell doped with di-8-ANEPPQ is imaged by one-photon confocal fluorescence. A stack of images is recorded with one image per incident polarization angle. Fourier coefficients are directly deduced from the polarimetric response $I(\alpha)$ of each pixel, leading to the determination of both parameters Ψ and ρ. (**b**) Illustration of polarimetric response $I(\alpha)$ obtained from an ordered region in the cell membrane (*left*) and disordered region (*right*). The amplitude of the curves is related to the molecular order parameter, while its phase is related to the orientation ρ

instead of two; however, this information is mathematically sufficient to retrieve the parameters required for the quantification of molecular order.

The data analysis is based the Fourier analysis of the polarimetric response $I(\alpha)$ (Fig. 8a). Given the square modulus factor in Eq. 9, this response can be fully described using the finite Fourier series:

$$I(\alpha) = A_0 + A_2 \cos 2\alpha + B_2 \sin 2\alpha. \tag{12}$$

Using the initial derivation of Eq. 9 for a cone aperture model, it can be shown that the normalized A_2/A_0 and B_2/B_0 coefficients are directly related to the molecular distribution parameters Ψ and ρ, with no ambiguity of determination. This allows to build up a 2D lookup table that gives for a wide set of (A_2/A_0,B_2/A_0) couples, the values of Ψ and ρ. In particular the sinusoidal dependence of Eq. 12 has an amplitude related to the molecular order (Ψ) and a phase related to the orientation of the molecular distribution in the sample plane (ρ) (Fig. 8b). Since the coefficients A_2/A_0 and B_2/B_0, as Fourier coefficients, can be computed from the experimental data at every pixel of the image (Fig. 8), Ψ and ρ can be retrieved directly using the

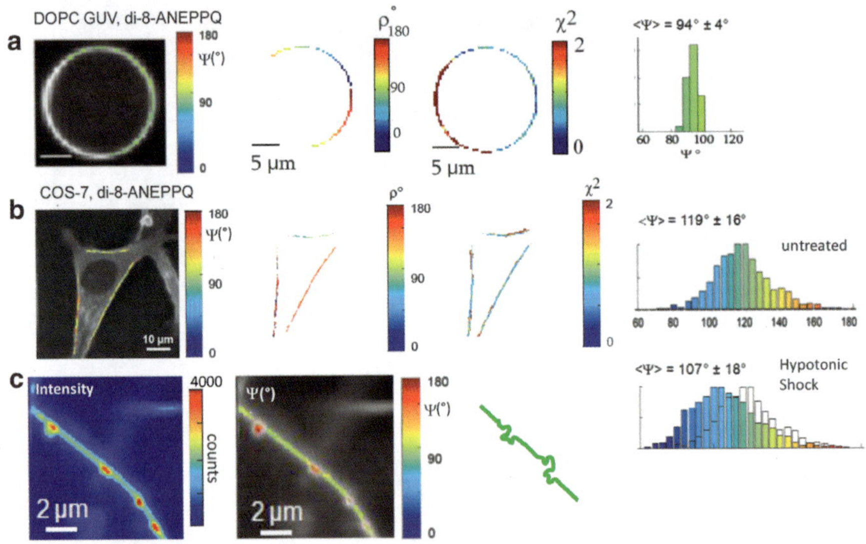

Fig. 9 Imaging molecular order Ψ by one-photon confocal polarization-resolved fluorescence around the cell contour. (**a**) DOPC GUV and (**b**) COS-7 cells labeled with di-8-ANEPPQ. From left to right: fluorescence image superimposed with its local Ψ values; ρ angle map; χ2 mean square error map (values above 5 correspond to discarded pixels). Histogram of the Ψ values obtained over the whole contour. A comparison between an untreated cell and cells treated by a hypotonic shock is shown. (**c**) Zoom on a cell membrane showing macroscopic and submicrometric size ruffling sites with high orientational disorder

lookup table. Note that the obtained values are systematically validated by performing a normalized χ^2 test that quantifies the agreement between experimental data and the expected theoretical polarimetric response. In practice, a χ^2 value larger than 5 is a good indicator of measurements that were affected by photobeaching, sample drift, etc. Such measurements are systematically discarded [55]. As compared to the previous data-fitting process, this Fourier analysis approach has permitted to considerably speed up the data processing, allowing a complete polarimetric analysis over a whole cell contour in a few seconds.

This technique has been applied to the analysis of molecular order in cell and artificial membranes labeled with di-8-ANEPPQ, using one-photon confocal polarimetry analysis (Fig. 9) [55]. Strong heterogeneities could be visualized in cells, contrary to GUV, with values confirming the ranges previously measured on a complete polarization dependence fit with two-photon fluorescence. Several conclusions were drawn from these studies:

– The GUV molecular order in a homogeneous lipid composition is highly homogeneous, with molecular order distributions widths close to the noise limit (Fig. 9a).

- In single cells, histograms of the Ψ values are much larger than GUVs, signature of a high degree of heterogeneity (Fig. 9b). The histogram over one population of 10 cells is typically of same shape as for one single cell.
- In cells, local parts of high disorder can be visualized, signature of the fact that the observed heterogeneity is spatially distributed along the cell contour (Fig. 9c).

As supported by previous measurements in fluorescence anisotropy, the high heterogeneity visualized in cells is most probably due to membrane local changes of morphology. In particular, membrane treatments provoking a perturbation of the actin cytoskeleton network (hypotonic shock, latrunculin A, cytochalasin D) induced a gain in molecular order around the whole cell contour (Fig. 9b). This confirms that the molecular order indicator Ψ is a sensitive reporter of membrane local morphology, in particular due to its nanometric-size folds which are not visible in a standard fluorescence image. Polarization-resolved fluorescence with unpolarized detection, with the additional capacity of being depolarization insensitive, therefore allows a map of membrane features which are not accessible by fluorescence anisotropy. This technique might be able to reveal new information at particular points of a cell membrane or in protein assemblies undergoing physical or chemical perturbations.

6 Summary, Conclusions, and Outlook

The polarization dependence of fluorescence signals in microscopy carries rich information on the orientational behavior of molecular probes. During the last decades, considerable progress has been done in terms of instrumentation, polarized fluorescence image analysis, and more importantly, in the development of fluorescent probe attachment strategies which now permit to reveal the orientation behavior of proteins and lipids in cell membranes with high precision.

Polarization signatures being a reporter of local environment rigidity and morphology, this technique is therefore likely to evolve toward the determination of important complementary information to already existing fluorescence-related techniques. Polarization resolution is indeed now introduced in a large scope of techniques such as super-resolution fluorescence imaging and single-molecule spectroscopy, as well as in a wide variety of contrasts such as in nonlinear optics in tissues (second-harmonic and third-harmonic generation, coherent anti-Stokes Raman scattering).

Acknowledgements The authors thank H. Rigneault, P. Réfrégier, and J. Duboisset (Institut Fresnel, Marseille, France), as well as D. Marguet, H.T. He, and T. Trombik (Centre d'Immunologie de Marseille Luminy, Marseille, France), for helpful discussions and advices. The Institut Fresnel work mentioned in this chapter was supported by CNRS, Agence Nationale de la Recherche, the region Provence Alpes Côte d'Azur.

References

1. Cheresh D, Leng J, Klemke R (1999) Regulation of cell contraction and membrane ruffling by distinct signals in migratory cells. J Cell Biol 146:1107–1116
2. Anantharam A, Onoa B et al (2010) Localized topological changes of the plasma membrane upon exocytosis visualized by polarized TIRFM. J Cell Biol 188:415–428
3. Benninger RKP, Vanherberghen B, Onfelt B (2009) Live cell linear dichroism im-aging reveals extensive membrane ruffling within the docking structure of natural killer cell immune synapses. Biophys J 96:L13–L15
4. Fooksman DR, Grönvall GK et al (2006) Clustering class I MHC modulates sensitivity of T cell recognition. J Immunol 176:6673–6680
5. Pentcheva T, Edidin M (2001) Clustering of peptide-loaded MHC class I molecules for endoplasmic reticulum export imaged by fluorescence resonance energy transfer. J Immunol 166:6625–6632
6. Borejdo J, Burlacu S (1993) Measuring orientation of actin filaments within a cell: orientation of actin in intestinal microvilli. Biophys J 65:300–309
7. Brack AS, Brandmeier BD et al (2004) Bifunctional rhodamine probes of Myosin regulatory light chain orientation in relaxed skeletal muscle fibers. Biophys J 86:2329–2341
8. Vrabioiu AM, Mitchison TJ (2006) Structural insights into yeast septin organization from polarized fluorescence microscopy. Nature 443:466–469
9. Weber G (1953) Rotational Brownian motion and polarization of the fluorescence of solutions. Adv Protein Chem 8:415–459
10. Cantor CR, Schimmel PR (1980) Biophysical chemistry, part 2: techniques for the study of biological structure and function. W. H. Freeman, San Francisco
11. Chan FTS, Kaminski CF, Kaminski Schierle GS (2011) HomoFRET fluorescence anisotropy imaging as a tool to study molecular self-assembly in live cells. Chem Phys Chem 12:500–509
12. Schlessinger J, Koppel DE et al (1976) Lateral transport on cell membranes: mobility of concanavalin A receptors on myoblasts. Proc Natl Acad Sci USA 73:2409–2413
13. Schwille P, Haupts U et al (1999) Molecular dynamics in living cells observed by fluorescence correlation spectroscopy with one- and two-photon excitation. Biophys J 77:2251–2265
14. Schütz GJ, Kada G et al (2000) Properties of lipid microdomains in a muscle cell membrane visualized by single molecule microscopy. EMBO J 19:892–901
15. Oida T, Sako Y, Kusumi A (1993) Fluorescence lifetime imaging microscopy (flimscopy). Methodology development and application to studies of endosome fusion in single cells. Biophys J 64:676–685
16. Manders EEM, Verbeek FJ, Aten JA (1993) Measurement of co-localisation of objects in dual-colour confocal images. J Microsc 169:375–382
17. Gould TJ, Gunewardene MS, Gudheti MV, Verkhusha VV, Yin SR, Gosse JA, Hess ST (2008) Nanoscale imaging of molecular positions and anisotropies. Nat Methods 5:1027–1030
18. Lazar J, Bondar A, Timr S, Firestein SJ (2011) Two-photon polarization microscopy reveals protein structure and function. Nat Methods 8:684–690
19. Kress A, Ferrand P, Rigneault H, Trombik T, He HT, Marguet D, Brasselet S (2011) Probing MHC class I protein and lipid order in cell membranes by fluorescence polarisation-resolved imaging. Biophys J 101:468–476
20. Marguet D, Spiliotis ET et al (1999) Lateral diffusion of GFP-tagged H2Ld molecules and of GFP-TAP1 reports on the assembly and retention of these molecules in the endoplasmic reticulum. Immunity 11:231–240
21. Rocheleau JV, Edidin M, Piston DW (2003) Intrasequence GFP in class I MHC molecules, a rigid probe for fluorescence anisotropy measurements of the membrane environment. Biophys J 84:4078–4086
22. Griffin BA, Adams SR, Tsien RY (1998) Specific covalent labeling of recombinant protein molecules inside live cells. Science 281:269–272

23. Spille JH, Zürn A, Hoffmann C, Lohse MJ, Harms GS (2011) Rotational diffusion of the α2a adrenergic receptor revealed by FlAsH labeling in living cells. Biophys J 100:1139–1148
24. Axelrod D (1979) Carbocyanine dye orientation in red cell membrane studied by microscopic fluorescence polarization. Biophys J 26:557–573
25. Florine-Casteel K (1990) Phospholipid order in gel- and fluid-phase cell-size liposomes measured by digitized video fluorescence polarization microscopy. Biophys J 57:1199–1215
26. Benninger RKP, Onfelt B et al (2005) Fluorescence imaging of two-photon linear dichroism: cholesterol depletion disrupts molecular orientation in cell membranes. Biophys J 88:609–622
27. Haluska CK, Schröder AP et al (2008) Combining fluorescence life-time and polarization microscopy to discriminate phase separated domains in giant unilamellar vesicles. Biophys J 95:5737–5747
28. Brasselet S (2011) Polarization resolved nonlinear microscopy: application to structural molecular and biological imaging. Adv Opt Photonics 3:205–271
29. Ikonen E (2001) Roles of lipid rafts in membrane transport. Curr Opin Cell Biol 13:470–477
30. McConnell HM, Vrljic M (2003) Liquid-liquid immiscibility in membranes. Annu Rev Biophys Biomol 32:469–492. 24
31. Munro S (2003) Lipid rafts: elusive or illusive? Cell 115:377–388. 26
32. Simons K, Ikonen E (1997) Functional rafts in cell membranes. Nature 387:569–572
33. Veatch SL, Keller SL (2002) Organization in lipid membranes containing cholesterol. Phys Rev Lett 89:268101
34. Lakowicz JR (1999) Principles of fluorescence spectroscopy. Springer, New York
35. Perrin F (1926) La fluorescence des solutions. Polarisation. Vie moyenne des molécules dans l'état excité. J de Phys 7:390–401
36. Dix JA, Verkman AS (1990) Mapping of fluorescence anisotropy in living cells by ratio imaging. Biophys J 57:231–240
37. Bader AN, Hofman EG, Voortman J, van Bergen en Henegouwen PMP, Gerritsen HC (2009) Homo-FRET imaging enables quantification of protein cluster sizes with subcellular resolution. Biophys J 97:2613–2622
38. Foster TH, Pearson BD, Mitra S, Bigelow CE (2005) Fluorescence anisotropy imaging reveals localization of meso-tetrahydroxyphenyl chlorin in the nuclear envelope. Photochem Photobiol 81:1544–1547
39. Clegg RM (1996) Fluorescence resonance energy transfer. In: Fluorescence imaging spectroscopy and microscopy, vol 137, 13th edn, Chemical analysis. Wiley, New York
40. Patterson GH, Piston DW, Barisas BG (2000) Förster distances between green fluorescent protein pairs. Anal Biochem 284:438–440
41. Blackman SM, Piston DW, Beth AH (1998) Oligomeric state of human erythrocyte band 3 measured by fluorescence resonance energy homotransfer. Biophys J 75:1117–1130
42. Gautier I, Tramier M et al (2001) Homo-FRET microscopy in living cells to measure monomer-dimer transition of GFP-tagged proteins. Biophys J 80:3000–3008
43. Varma R, Mayor S (1998) GPI-anchored proteins are organized in submicron domains at the cell surface. Nature 394:798–801
44. Gasecka A, Han T-J et al (2009) Quantitative imaging of molecular order in lipid membranes using two-photon fluorescence polarimetry. Biophys J 97:2854–2862
45. Dale RE, Hopkins SC et al (1999) Model-independent analysis of the orientation of fluorescent probes with restricted mobility in muscle fibers. Biophys J 76:1606–1618
46. Kinosita K, Kawato S, Ikegami A (1977) A theory of fluorescence polarization decay in membranes. Biophys J 20:289–305
47. Blackman SM, Cobb CE et al (1996) The orientation of eosin-5-maleimide on human erythrocyte band 3 measured by fluorescence polarization microscopy. Biophys J 71:194–208
48. Mattheyses AL, Kampmann M et al (2010) Fluorescence anisotropy reveals order and disorder of protein domains in the nuclear pore complex. Biophys J 99:1706–1717
49. Barda-Saad M, Braiman A et al (2005) Dynamic molecular interactions linking the T cell antigen receptor to the actin cytoskeleton. Nat Immunol 6:80–89

50. Bunnell SC, Kapoor V et al (2001) Dynamic actin polymerization drives T cell receptor-induced spreading: a role for the signal transduction adaptor LAT. Immunity 14:315–329
51. Edidin M, Kuo SC, Sheetz MP (1991) Lateral movements of membrane glyco-proteins restricted by dynamic cytoplasmic barriers. Science 254:1379–1382
52. Edidin M, Zuniga MC, Sheetz MP (1994) Truncation mutants define and locate cytoplasmic barriers to lateral mobility of membrane glycoproteins. Proc Natl Acad Sci USA 91:3378–3382
53. Treanor B, Depoil D et al (2010) The membrane skeleton controls diffusion dynamics and signaling through the B cell receptor. Immunity 32:187–199
54. Schauer K, Duong T et al (2010) Probabilistic density maps to study global endomembrane organization. Nat Methods 7:560–566
55. Kress A, Wang X, Ranchon H, Ferrand P, Brasselet S (2012) Revealing molecular orientation and order in cell membranes of arbitrary shape (submitted)
56. Brasselet S, LeFloc'h V, Treussart F, Roch J, Zyss J (2004) In situ diagnostics of the crystalline nature of single organic nanocrystals by nonlinear microscopy. Phys Rev Lett 92:207401
57. LeFloc'h V (2003) Monitoring of orientation in molecular ensembles by polarization sensitive nonlinear microscopy. J Phys Chem B 107:12403–12410
58. Schön P, Munhoz F et al (2008) Polarization distortion effects in polarimetric two-photon microscopy. Opt Express 16:20891–20901
59. Richards B, Wolf E (1959) Electromagnetic diffraction in optical systems. II. Structure of the image field in an aplanatic system. Proc Roy Soc Lond A 253:358–379
60. Schön P, Behrndt M, Ait-Belkacem D, Rigneault H, Brasselet S (2010) Polarization and phase pulse shaping applied to structural contrast in nonlinear microscopy imaging. Phys Rev A 81:013809
61. Moyano F, Biasutti MA, Silber JJ, Correa NM (2006) New insights on the behavior of Prodan in homogeneous media and in large unilamellar vesicles. J Phys Chem B 110:11838–11846
62. Muller JM, Harryvan DH, Verhagen JCD, van Faassen EE, van Ginkel G (1996) The orientation of the transition dipole moments of TMADPH embedded in a poly(vinyl alcohol) film. Chem Phys 211:413–420
63. Baumgart T, Hess ST, Webb WW (2003) Imaging coexisting fluid domains in biomembrane models coupling curvature and line tension. Nature 425:821–824
64. Dietrich C, Bagatolli LA, Volovyk ZN, Thompson NL, Levi M (2001) Lipid rafts reconstituted in model membranes. Biophys J 80:1417–1428
65. Samsonov AV, Mihalyov I, Cohen FS (2001) Characterization of cholesterol-sphingomyelin domains and their dynamics in bilayer membranes. Biophys J 81:1486–1500
66. Scherfeld D, Kahya N, Schwille P (2003) Lipid dynamics and domain formation in model membranes composed of ternary mixtures of unsaturated and saturated phosphatidylcholines and cholesterol. Biophys J 85:3758–3768
67. Bagatolli LA, Gratton E (1999) Two-photon fluorescence microscopy observation of shape changes at the phase transition in phospholipid giant unilamellar vesicles. Biophys J 77:2090–2101
68. Baumgart T, Hunt G, Farkas ER, Webb WW, Feigenson GW (2007) Fluorescence probe partitioning between Lo/Ld phases in lipid membranes. Biochim Biophys Acta 1768:2182–2194
69. Kim HM, Choo H-J, Jung S-Y, Ko Y-G, Park W-H (2007) A two-photon fluorescent probe for lipid raft imaging: C-laurdan. Chembiochem 8:553–559
70. Gidwani A, Holowka D, Baird B (2001) Fluorescence anisotropy measurements of lipid order in plasma membranes and lipid rafts from RBL-2 H3 mast cells. Biochemistry 40:12422–12429
71. Farkas ER, Webb WW (2010) Multiphoton polarization imaging of steady-state molecular order in ternary lipid vesicles for the purpose of lipid phase assignment. J Phys Chem B 114:15512–15522

72. van Rheenen J, Jalink K (2002) Agonist-induced PIP(2) hydrolysis inhibits cortical actin dynamics: regulation at a global but not at a micrometer scale. Mol Biol Cell 13:3257–3267
73. Chong PLG, Wong PTT (1993) Interactions of Laurdan with phosphatidylcholine liposomes: a high pressure FTIR study. Biochim Biophys Acta 1149:260–266
74. Adler J, Shevchuk AI et al (2010) Plasma membrane topography and interpretation of single-particle tracks. Nat Methods 7:170–171
75. Sund SE, Swanson JA, Axelrod D (1999) Cell membrane orientation visualized by polarized total internal reflection fluorescence. Biophys J 77:2266–2283

Near-Field Optical Nanoscopy of Biological Membranes

Thomas S. van Zanten, Carlo Manzo, and Maria F. Garcia-Parajo

Abstract The specific organization and distribution of protein receptors and lipids on the cellular plasma membrane play a crucial role for the spatiotemporal control of many different cellular processes. A great deal of novel knowledge in this area is currently being generated thanks to the advent of modern surperresolution optical techniques combined with single-molecule approaches. In this chapter, we focus on near-field nanoscopy, a technique particularly well suited for the study of biological cell surfaces at the nanometer scale. We first describe the general concept of near-field scanning optical microscopy (NSOM) and specifically focus on how NSOM is being exploited to map the spatiotemporal organization of proteins and lipids. Novel routes toward surperresolution using optical nanoantennas and first applications for cell membrane nanoimaging are discussed. The last part of the chapter describes recent technical breakthroughs that enable the application of NSOM in living cells providing detailed dynamic information on diffusion processes occurring at the nanoscale.

Keywords Cell membrane nanoscale compartmentalization · Fluorescence correlation spectroscopy in reduced volumes · Near-field scanning optical microscopy · Optical nanoantennas · Superresolution microscopy

T.S. van Zanten and C. Manzo
ICFO-Institut de Ciències Fotòniques, Mediterranean Technology Park, 08860, Castelldefels (Barcelona), Spain

M.F. Garcia-Parajo (✉)
ICFO-Institut de Ciències Fotòniques, Mediterranean Technology Park, 08860, Castelldefels (Barcelona), Spain

ICREA-Institució Catalana de Recerca i Estudis Avançats, 08010 (Barcelona), Spain
e-mail: maria.garcia-parajo@icfo.es

Y. Mély and G. Duportail (eds.), *Fluorescent Methods to Study Biological Membranes*, Springer Ser Fluoresc (2013) 13: 339–364, DOI 10.1007/4243_2012_52,

Contents

1 Introduction

Understanding the complex relationship between structure, function, and dynamics of biomolecules in the living cell is a fundamental driving force in the research field of cell biology. Propelled by the vast amount of tools that have been made available in recent years, modern molecular biology has made enormous progress. Nevertheless, observing single molecular processes in living cells has remained a major challenge. In the particular case of cellular membranes, key multimolecular interactions that define and regulate cell function occur at the nanoscale, a size regime not accessible to classical optical microscopy, owing to the diffraction of light.

For a long time, electron microscopy has been the only technique available to observe nanoscale spatial organization of proteins in the plasma membrane. Gold particles attached to membrane proteins via antibodies can be visualized with the electron microscope to estimate protein clustering and colocalization. However, harsh sample preparations and imaging modes make live cell imaging extremely challenging. Although live cell imaging with nanometer resolution can be obtained by atomic force microscopy (AFM), this technique lacks chemical specificity. Therefore, the observation of molecular processes taking place at the cell membrane has been mainly obtained by means of fluorescence-based optical techniques, which offer chemical specificity and minimal invasiveness. However, standard lens-based optical techniques are diffraction limited and do not allow optical observation beyond ~300 nm.

A fluorescence-based technique that does provide information at the nanoscale is Förster resonance energy transfer (FRET), which exploits a distance-dependent nonradiative energy transfer between two fluorophores to measure molecular distances below 10 nm [1]. Although FRET has been an important tool to demonstrate the existence of concentration-independent nanodomains on the cell membrane [2, 3] and to support the hypothesis of actively maintained lipid nanoplatforms [4], it is unable to bridge the gap between 10 and 300 nm.

Recently, valuable techniques have been devised to bridge this gap by taking advantage of the photophysical properties of fluorophores in combination with standard lens-based microscopes. For instance, stochastic optical reconstruction microscopy (STORM) or the analogous, fluorescent photoactivatable localization microscopy (FPALM/PALM) is capable of reconstructing surperresolution images of different cellular structures [5–7] and is currently being used to address the organization of the cell membrane at the nanometer scale [8–11]. Another lens-based technique, stimulated emission depletion (STED) microscopy, has also been applied to estimate protein-cluster sizes [12] and their nanoscale distribution on the membrane [13]. More recently, STED has been also combined with fluorescence correlation spectroscopy (FCS) to observe nanoscale dynamics of membrane lipids and lipid-anchored proteins in living cells [14].

Besides these techniques, nanometric optical resolution in a fluorophore-independent fashion can be obtained by near-field approaches. Indeed, the interaction of light with subwavelength apertures results in a nonpropagating electromagnetic field that can be used as a nanometric excitation light source. This concept has been exploited in near-field scanning optical microscopy (NSOM), a surperresolution technique that can reach optical resolutions between 50 and 90 nm on fully intact cell specimens. By impinging visible light on a subwavelength aperture and scanning the aperture in close proximity to the sample, NSOM allows cell membrane imaging with a lateral resolution essentially determined by the size of the aperture. In addition, the evanescent field generated at the aperture extends only a few tens of nanometer in the axial direction and, as such, inherently provides specificity to the cell membrane with high signal-to-noise ratios.

In this chapter, we describe the principles of NSOM and its implementation for biological studies. We then highlight the major contributions of the technique in providing quantitative understanding of the nanolandscape of proteins and lipids on the plasma membrane. Finally, we discuss emerging advancements on near-field nanoscopy aimed at improving both its temporal and spatial resolution.

2 Near-Field Nanoscopy: Concept and Implementation

2.1 Principles

The diffraction limit in conventional microscopy arises from the size of the spot that a light beam can be focused to with normal lens elements. In practice, the diffraction limit implies that the minimum distance Δx required to resolve independently two distinct objects is dependent on the wavelength of the light used to observe the specimen λ, and the overall objective lens system, through the numerical aperture (NA) of the objective, as $\Delta x \approx \lambda/2\text{NA}$. With modern objectives having an NA as high as ~1.4, the resolution of conventional microscopy becomes 250–300 nm in the case of visible light.

When an object, such a microscopic specimen, is illuminated with a (monochromatic) plane wave, the transmitted or reflected light is collected by a lens and projected onto a detector to form the image. Usually, the detector is placed in the far field, that is, multiple wavelengths away from the sample, so that the far-field component of the light, which propagates in an unconfined way, is the only component used to generate the image. On the other hand, the interaction between the imaging light and the specimen also generates a near-field component, which consists of a nonpropagating (evanescent) field existing only near the object at distances less than the wavelength of the light. Since the near field decays exponentially within a distance less than the wavelength, usually it cannot be collected by the lens, and it is not detected. The essence of near-field optical microscopy is to detect the near-field component before it undergoes diffraction. Alternatively, light can also impinge on a subwavelength aperture, and the evanescent field generated at the end of this aperture can be used to locally illuminate the sample, converting the evanescent components in far-field radiation that can then be detected using conventional objectives.

It is interesting to note that the idea of employing near-field optics to achieve images beyond the diffraction limit was initially formulated by Edward H. Synge already in 1928 [15]. His original suggestion was to create a minute aperture (of the order of 100 nm) in a metallic screen to create a very small light source. This tiny spot of light would then be used as local illumination source for the imaging of biological specimens. In order to achieve near-field illumination, Synge pointed out that it is necessary to keep the aperture at a distance (*d*) from the sample much shorter than the aperture dimensions ($d < a$). The light transmitted through the sample would be later recorded point-by-point with a sensitive detector to form the image. However, to be able to engage in nanoscale imaging, Synge acknowledged two crucial resolution-limiting factors that prohibited the implementation of the technique at that time, (1) the need of having well-defined subwavelength apertures and (2) the need of keeping the aperture extremely close to the sample, in the near-field region. It took nearly 50 years until Ash and Nicholls realized such a scheme operating in the microwave region, reporting a resolution of the order of one-sixtieth of the wavelength [16]. In the mid-1980s, through the invention of the scanning tunneling microscope (STM), two independent labs [17, 18] realized the original Synge's scheme with visible light. In both instruments, the key innovation was the implementation and fabrication of subwavelength optical sources with sufficient output power.

Since then, many different concepts for aperture fabrication have been explored [19]. The most used approach for biological research was introduced in 1991 by Betzig et al. [20], consisting in the use of single-mode optical fibers tapered to a tip size of about 50 nm. Nowadays, NSOM probe fibers are fabricated through heating-pulling methods [20], wet etching [21], or a combination of both [22]. Subsequent aluminum evaporation prevents light leakage from the fiber. The very end of the tip is milled open using focused ion beam (FIB) technology, resulting in apertures of 50–100 in nm diameter [23].

As already mentioned, the second and crucial requirement for near-field nanoimaging is the correct positioning of the probe close to the specimen. Indeed, since the intensity at the subwavelength aperture probe decays exponentially away from the probe, efficient excitation requires accurate control of the probe-sample distance below 10 nm. A way to control this distance in a fiber-based instrument is by relying on shear forces between the probe and the sample [24, 25]. This can be achieved using quartz tuning forks oscillating at their resonance frequency. When in proximity to the sample, shear forces dampen the amplitude and the phase of the fork oscillation, and the feedback on either of these signals is then used to keep the distance constant. In addition, the feedback signal is used to generate a topographic map of the sample surface, with resolution and sensitivity comparable to tapping-mode AFM. Other successful implementations to control the probe-sample distance make use of concepts similar to tapping-mode AFM, where the tip is bent by ~90° and used to vertically oscillate above the sample surface [22, 26]. In the last few years, different concepts have been also implemented that allow reproducible imaging under liquid conditions, enabling imaging of biological specimens in physiological settings [27–30].

The evanescent character of the near-field illumination makes NSOM a surface-sensitive technique, ideally suited for probing cell membranes. Additionally, the small illumination volume ($\sim 10^5$ nm^3) strongly reduces background components arising from cell autofluorescence and/or internalized fluorophores. Noteworthy, NSOM imaging can be obtained with any fluorophores, and the illumination geometry allows straightforward implementation of multicolor excitation. Taken together, these properties constitute an important asset of NSOM over other surperresolution techniques, enabling truly quantitative mapping of the molecular organization of the cell membrane. Although as other raster-scanning techniques, NSOM imaging is inherently slow, combination of NSOM with FCS provides measurement of dynamics on living cell membranes in ultra-confined volumes. Additionally, probe design based on optical antennas further improves lateral resolution below 30 nm. For these reasons, NSOM constitutes a valuable tool to characterize spatiotemporal details of many biological processes occurring on the cell membrane.

2.2 *Implementation for Biological Membrane Nanoimaging*

Depending on the application, NSOM can be implemented in multiple different geometries [31]. For biological studies, the most widely used NSOM configuration relies on the use of aperture probes, incorporated into an inverted optical microscope, with near-field excitation through the probe and far-field detection using a high NA objective and sensitive detectors [32, 33], as schematically depicted in Fig. 1.

For NSOM operation, the excitation light is coupled into the cleaved end of the fiber probe, reaching the sample through the nanometric aperture at the other end of

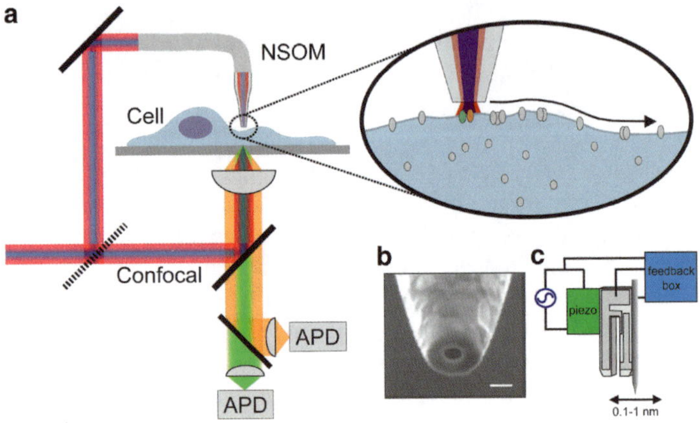

Fig. 1 (**a**) Combined confocal/NSOM setup for cell membrane studies. Excitation of the sample occurs either in confocal mode or by coupling the excitation light into the NSOM probe (zoom in). For surperresolution imaging, the NSOM probe is raster-scanned over the cell membrane, following the topographical features of the cell. (**b**) Scanning electron microscopy image of a representative NSOM probe (Scale bar: 100 nm). (**c**) Schematic view of the shear force feedback implemented using tuning forks as piezoelectric sensor elements

the fiber and exciting fluorescent markers attached to the biological molecules under investigation (proteins and/or lipids). The entire setup configuration can also include the possibility to switch to standard confocal excitation by directing the excitation light into the microscope objective. In both cases, after excitation of the sample, the fluorescent emission is detected in confocal configuration, and imaging is achieved by scanning the sample with respect to the excitation source.

By using single-photon-counting APD detectors and an optimized emission path, NSOM can allow detection of individual proteins on intact cell membranes with high signal-to-background ratios, achieving localization accuracies between 3 and 5 nm. In typical experiments, the sample is first scanned in a confocal fashion, and then NSOM is used to obtain surperresolution images of selected regions of the plasma membrane. Imaging of biological samples in liquid conditions while preserving efficient distance feedback is obtained by enclosing the tuning fork resonator into an airtight sealed diving bell [27, 34].

3 NSOM Imaging of Cell Membranes

3.1 Compartmentalization of the Cell Membrane

The fluid mosaic model, proposed in the seminal paper by Singer and Nicolson [35], describes the plasma membrane as a lipid bilayer, forming a two-dimensional fluid in which molecules are randomly distributed and diffuse freely. In contrast with this

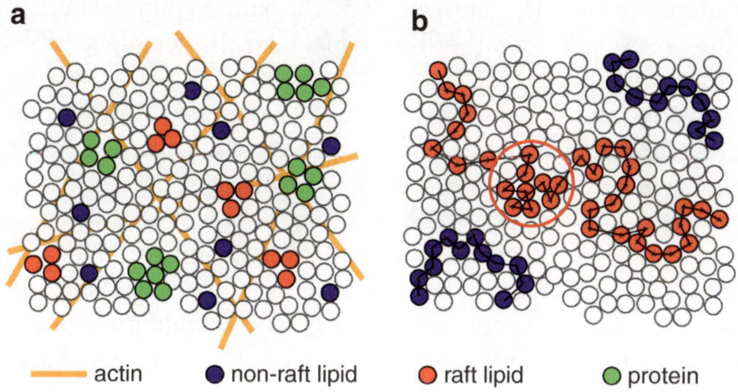

Fig. 2 Spatiotemporal heterogeneity of the cell membrane. (**a**) Different biological players contribute to the spatial compartmentalization of the cell membrane, including the actin cytoskeleton, protein-protein, and lipid-lipid interactions. (**b**) These players also compartmentalize the cell membrane in a dynamic fashion. For instance, lipids can diffuse freely on the cell membrane and get transiently confined (*red circle*) by the action of the actin network, actin-bound proteins, and/or cholesterol-enriched regions

hypothesis of simple homogeneous lipid mixing, an ever increasing number of publications appear to confirm that cell membranes are heterogeneously arranged both in the plane of the bilayer and across the two leaflets. This heterogeneity has been evidenced by the spatial and temporal confinement exhibited by proteins and lipids in defined micro- and nanoscale areas of the membrane (Fig. 2) [36, 37]. Transient associations between lipids and proteins within these microdomains can have direct impact on the biological function of these molecules and therefore on cellular processes they control. Yet, the mechanisms orchestrating the formation of membrane domains in biological membranes remain enigmatic and are thought to include protein-protein and cytoskeletal or lipid raft associations [36, 38–40].

In particular, membrane microdomains enriched in sphingolipids and cholesterol known as lipid rafts [41] have been postulated to favor segregation of specific membrane receptors and glycosylphosphatidylinositol-anchored proteins (GPI-APs). Lipid rafts have been shown to play an important role in various biological phenomena, ranging from cell adhesion [42] to pathogen binding [43], endocytosis [44], and immune cell signaling [45]. However, there is no consensus yet on the spatiotemporal regimes regulating lipid rafts in the resting state. This is mainly because their evaluation requires tools compatible with live cell imaging and capable of resolving heterogeneities at the nanoscale, a spatial regime not accessible to standard fluorescence microscopy. Nevertheless, recent advances in biophysical techniques have led to the identification of cholesterol-dependent nanoassemblies, supporting their existence in living cells [14, 46, 47].

Besides lipid-lipid interactions that serve to target proteins to lipid rafts [3, 48], protein-lipid [49] as well as protein-protein [50] interactions can have a significant impact on lipid raft localization. It should be noted that studies have also demonstrated the formation of microdomains in activated T cells exclusively created by protein-protein networks and not maintained by interactions with lipid rafts [40].

The diffusional trapping through protein-protein interactions generated microdomains that could recruit or reject specific cell membrane proteins during signal transduction.

Protein-protein interactions can be also modulated by the presence of tetraspanins, a large family of proteins that traverse the membrane four times [51]. Tetraspanins have been found to interact strongly among each other [52] as well as with other integral proteins [53], which has led to the hypothesis that tetraspanins are involved in the lateral sorting of proteins on the cell membrane. Indeed, not only can tetraspanins organize themselves in a structure also known as the tetraspanin web [54, 55], but on endothelial cells the incorporation of a multitude of cell adhesion proteins in tetraspanin-enriched areas have been identified as preorganized adhesion platforms [56].

Another family of proteins that can form networks is represented by galectins. Galectins can generate scaffolds or lattices through interactions of their carbohydrate recognition domains (CRD). These CRDs can bind glycosylated proteins at the membrane, and therefore galectins that have multiple CRDs can recruit and compartmentalize receptors. Galectin-1, for example, has been found essential for EGFR signaling through the formation and sustaining of active H-Ras nanoclusters [48, 57]. In addition, other galectins such as galectin-3 can even form links between carbohydrate and noncarbohydrate ligands [58]. As such, galectins can act as versatile molecular organizers of the cell membrane.

Aside from lipid and protein interactions, the actin cytoskeleton that lies just below the membrane is also able to compartmentalize the membrane. Transmembrane proteins anchored to the actin cytoskeleton meshwork can act as "rows of pickets" that temporarily confine diffusing lipids and proteins [59, 60]. More recently, an exact relationship between protein dynamics and actin-defined compartments has been directly visualized [61]. In this elegant paper, the authors not only showed that the diffusion of the IgE receptor (FcεRI) is confined within actin-poor areas but also demonstrated that the size and location of these actin barriers changed over time, indicating that the type of diffusion barriers formed by the actin cytoskeleton is time-dependent [61].

3.2 NSOM Contribution to Study Biological Membranes

Because of its exquisite surface sensitivity, NSOM is particularly useful for nanoscale imaging of proteins and lipids at the cell membrane. In fact, the feasibility of using NSOM to image nanoscale features on biological membranes was already demonstrated in 1992 by the pioneering work of Betzig and Trautman [62]. A few years later, Enderle et al. used for the first time dual-color NSOM to directly measure the association of host and parasite proteins in malaria (*Plasmodium falciparum*)-infected erythrocytes [63]. Since then, NSOM has been applied to the study of many different receptors and lipids on biological membranes.

3.2.1 Lipid Rafts

In the context of lipid rafts, NSOM has been the used to visualize the nanoscale organization of gangliosides GM1 and GM3, two sphingolipids thought to partition in rafts. Using NSOM in combination with quantum dot labeling, Chen and colleagues found that these two characteristic raft lipids organized in small nanodomains on the membrane of MDCK cells [64]. The nanodomains were comparable in size with diameters of 160 and 190 nm for GM1 and GM3, respectively. By taking advantage of the simultaneous topographic information provided by NSOM, the researchers found that GM3 domains were preferentially localized at the peak of microvilli-like protrusions on the cell membrane. In contrast, GM1 lipid domains were localized at the slope or valley of the membrane fluctuations, indicating that the two raft lipids do not spatially colocalize. Although such a segregation of distinct raft domains was already observed in polarized T cells and directly correlated with cell migration capacity [65], spatial segregation of different raft constituents on resting cells had not been reported yet.

As an alternative raft marker to GM1, Abulrob et al. examined the membrane partitioning of asialo-GM1, an abundant glycosphingolipid in HeLa cells [66]. NSOM showed that asialo-GM1 is localized in small nanoclusters of about 90 nm in size. In addition to clusters, a second unclustered and more diffusive population of asialo-GM1 was observed. The authors hypothesized that nanoclusters would correspond to raft lipids while the diffusive asialo-GM1 population would not be associated with rafts [66].

Recently single-molecule near-field nanoscopy has been used to visualize the nanolandscape of GM1 after binding by its ligand cholera toxin (CTxB) on intact monocyte membranes [67]. Pentavalent binding of CTxB to GM1 was sufficient to initiate a minimal raft coalescence unit, resulting in the formation of cholesterol-dependent GM1 nanodomains $<$120 nm in size. This particular arrangement appeared independent of cell type and GM1 expression level on the membrane, since GM1 nanodomains were observed both on monocytes as well as on dendritic cells. Cholesterol depletion abrogated GM1 nanoclustering demonstrating the crucial role of cholesterol for maintaining the integrity of these nanodomains. Simultaneous dual-color high-resolution images revealed that the canonical raft component CD55 (a GPI-AP) was recruited to regions proximal ($<$150 nm) to CTxB-GM1 nanodomains without physical intermixing (Fig. 3), but not the nonraft protein CD71 [67]. These results demonstrated the existence of raft-based compositional connectivity at the nanoscale crucially mediated by cholesterol. The data further suggested that such a connective condition on resting cell membranes constitutes an obligatory step toward the hierarchical evolution of large-scale raft coalescence upon cell activation. More conventional techniques such as FRET are unable to report on such a spatial proximity at spatial scales $>$10 nm. On the other extreme, diffraction-limited techniques such as confocal microscopy would misleadingly show colocalization between different components located at distances $<$300 nm (Fig. 3a). As such, NSOM is capable of bridging the gap between 10 and 300 nm providing exquisite information at these important spatial scales.

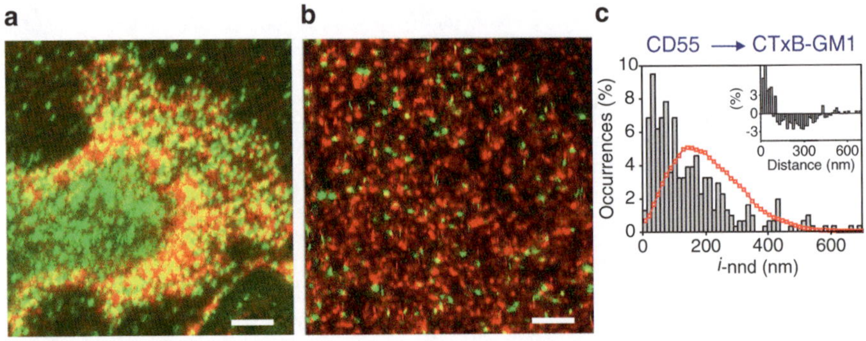

Fig. 3 Imaging nanoscale compositional connectivity with NSOM. (**a**) Confocal image of the raftophilic protein CD55 (*green*) and CTxB-GM1 (*red*) on monocytes. An artifactual high degree of colocalization (*yellow*) is observed as a result of the limited resolution in confocal (Scale bar: 5 μm). (**b**) Dual-color NSOM image of CD55 (*green*) and CTxB-GM1 nanodomains (red) on monocytes. No colocalization (lack of yellow spots) between these components is observed at the nanoscale (Scale bar: 1 μm). (**c**) Interparticle-nearest neighbor distribution (*i*-nnd) of CD55 to its closest CTxB-GM1 nanodomain (*bars*) together with simulations of random spatial distribution of proteins and CTxB-GM1 nanodomains (*red line*). Inset: Residuals between the distributions obtained by experiments and random simulations. The increased occurrence of distances <150 nm, as compared to the simulations, demonstrates a spatial proximity, or compositional connectivity, between CD55 and the CTxB-GM1 nanodomains. Adapted with permission from Ref. [67]

3.2.2 Receptor Nanoclustering

Besides the lipid-lipid interactions that define lipid raft microdomains, many membrane receptors have been seen to precluster in the resting state independent of their raftophilic character [11, 12, 42, 43, 68]. An example studied with NSOM is the receptor DC-SIGN, expressed on antigen presenting cells and involved in the recognition of several pathogens [43]. Previous studies with electron microscopy showed that although DC-SIGN partially associates with lipid rafts, depletion of cholesterol did not affect DC-SIGN capacity for forming nanodomains on the surface of immature dendritic cells (imDCs) [43]. Single-molecule NSOM confirmed that as much as 80 % of the receptors form nanoclusters of 185 nm in size randomly distributed on the membrane of imDCs [69]. Interestingly, the molecular density of these nanoclusters was extremely heterogeneous, with some nanoclusters containing only a few molecules while others contained as much as 20 DC-SIGN molecules per cluster. Together with the broad capability to bind to many different pathogens, these results suggest that the particular heterogeneous packing of DC-SIGN might serve to maximize its binding strength to a large variety of viruses and pathogens having different binding affinities to the receptor. In contrast to DC-SIGN, clusters of IL2R and IL15R (two members of the interleukin family expressed in human T lymphoma cells) were found to have a constant packing density albeit forming clusters of different sizes [70]. The linear increase in number of receptors with domain size suggested a general "building block" type of assembly for these receptors on the cell membrane.

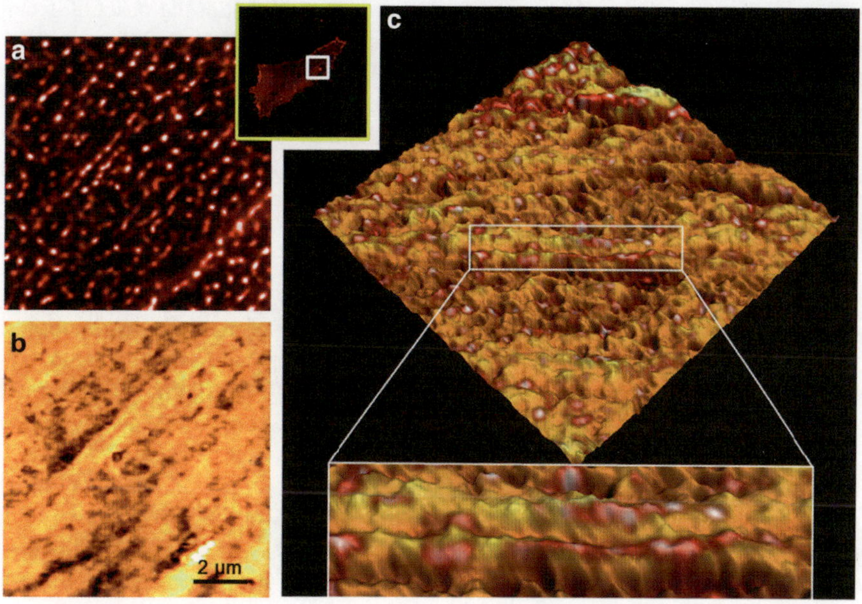

Fig. 4 NSOM images of a local area of cells stained for CD44. (**a**) NSOM fluorescence image obtained by zooming the indicated area in the inset. (**b**) Corresponding NSOM topographic image. (**c**) Composite 3D image, the 3D image displays a combination of the topographic data and fluorescence data. The inset shows CD44 molecules locate on slender, microvillus-type plasma membrane protrusions. Reproduced with permission from Ref. [73]

In another NSOM study, Vobornik et al. could address whether cluster size of a particular receptor is dependent on its expression level [71]. Although the expression levels of the β_2-adrenergic receptor (β_2AR) were increased by threefold, the average cluster size remained similar, and in the order of ~130–150 nm, suggesting that these receptors might have a biological preferred organization in nanodomains. Along these lines, the epidermal growth factor receptor (EGFR) also organizes in nanoclusters, and this organization is preserved even after ligand presentation [72].

Aside from cluster size, composition, or packing density, the specific distribution of receptor nanoclusters on the cell membrane might also reflect a different function. While in some cases receptor nanoclusters might be randomly distributed on the cell surface, in many other cases, receptors can be enriched in specific regions such as microvilli-like structures, indicating for instance a preference for high membrane curvature sites. Exploiting the simultaneous topography and fluorescence information afforded by NSOM, Chen et al. found that the hyaluronan receptor CD44 was confined in nanometer-sized domains of about 275 nm [73]. In addition, the spot density was significantly higher on the filopodia regions and membrane protrusions, as confirmed by the topography images (Fig. 4). The authors speculated that since most of the CD44 is located on filopodia, the receptor enhances its binding properties to the extracellular matrix by concentrating in these regions. Moreover, the same group recently found that at the onset of T cell activation, nanodomains of the early activation marker CD69 polarized on the peaks of membrane fluctuations [74]. In contrast, the transferrin receptor (CD71) organized in the valleys of membrane fluctuations [74]. Since CD71 is mainly involved in iron uptake via endocytosis, the authors suggested that its positioning in the valleys might correspond to sites for endocytosis.

3.2.3 Multimolecular Interactions

Cell-signaling events commonly involve a multitude of proteins and/or lipids. Although many of the molecular players involved in specific signaling processes can be readily identified by biochemical means, very little is still known about their tight spatial organization on the membrane. Zhong et al. used NSOM to investigate the nanoscale organization of CD3 and CD4 at different stages of TCR/CD3 activation on T cells [75]. In the resting state, CD3 and CD4 organized mainly as small nanoclusters made of a few proteins, with only 6–10 % of CD3 colocalizing with CD4 in 70–110-nm nanoclusters. Stimulating the T cells with anti-CD3 antibodies induced a significant reorganization of CD3 and CD4. This reorganization was not only restricted to enhanced clustering of the individual proteins. In fact, there was a significant co-clustering of the CD3 and CD4 nanoclusters into 200–500-nm-size domains on the T cell membrane [75]. The production of interleukins by the T cells due to anti-CD3 stimulation suggested that this co-clustering was sustaining TCR/CD3 signaling and activation. Interestingly, applying a different pathway for T cell activation only resulted in an enhanced cluster formation of CD3 and CD4 but not co-clustering [75]. This demonstrated that CD3 and CD4 co-clustering is an intrinsic event specific for the TCR/CD3-mediated activation pathway rather than an effect of activation-induced changes in the cortical cytoskeleton.

Dual-color NSOM has been also used to investigate the association of β_2AR and caveolae on the surface of cardiac myocytes [76]. The study showed that 15–20 % β_2AR nanoclusters colocalized in caveolae, which are small ~60-nm invaginations of the cell membrane. The lack of complete colocalization suggested that the diverse functional properties of β_2AR could arise from multiprotein complexes that are not caveolar in nature. Nevertheless, the fraction of β_2AR not colocalizing with caveolae appeared to be organized proximal to it. Interestingly, this indicates that β_2AR clusters are both preassembled in, that is, colocalized, or close to caveolae.

Recently, we investigated the spatiofunctional relationship between the integrin receptor LFA-1 involved in leukocyte adhesion and raft components (GPI-APs) [77] using dual-color NSOM. While LFA-1 formed nanoclusters of ~85 nm in size on resting monocytes, ~70 % of the GPI-APs organized as monomers, and the remaining 30 % formed small oligomers containing up to two to four molecules. Surprisingly, whereas GPI-AP monomers distributed randomly on the cell surface, the oligomers resided in regions proximal to each other (within ~250 nm). In the resting state, that is, prior to LFA-1 activation, ~50 % of the GPI-AP oligomers were found close to LFA-1 nanoclusters, whereas GPI-AP monomers exhibited no particular spatial correlation with respect to LFA-1. Ligand-mediated LFA-1 activation not only resulted in a spatial interlocking of the integrin and GPI-APs generating nascent adhesion sites but importantly resulted in an interconversion from monomers to nanodomains of GPI-APs [77]. These data demonstrated the existence of nanoplatforms composed by integrins and rafts as essential intermediates in nascent cell adhesion. Since raft association with a variety of membrane proteins other than LFA-1 has been documented, we proposed that hotspot regions enriched with raft components and functional receptors may constitute a prototype of nanoscale inter-receptor assembly and correspond to a generic mechanism to offer cells with privileged areas for rapid cellular function and responses to the outside world.

3.3 Nanoscale Fluorescence Correlation Spectroscopy on Living Cell Membranes

Dynamics of proteins and lipids can be measured by several techniques, such as fluorescence recovery after photobleaching (FRAP) [78, 79], single particle tracking (SPT) [80, 81], and fluorescence correlation spectroscopy (FCS) [82]. The use of FCS has been found advantageous to measure fast diffusion owing to the high temporal resolution, easy data collection, and analysis offered by this technique [83]. FCS is based on the measurement of fluctuations of the fluorescence intensity from a well-defined area. These fluctuations arise from fluorescent molecules diffusing in and out of the excitation volume and can be autocorrelated in time to provide information regarding the diffusion of the labeled molecules [82]. The average dwell time of fluctuations is directly related to the diffusion coefficient, while the shape of the autocorrelation function (ACF) indicates the type of diffusion [84].

In heterogeneous systems such as the cell membrane, diffusion is influenced by protein confinement and/or interactions. Since these effects might take place on short spatial and temporal scale (Fig. 2b), their detection requires an illumination area comparable to the membrane heterogeneities length scale, combined with high temporal resolution [14]. In this context, the reduced illumination size offered by NSOM probes combined with an FCS approach offers an exquisite tool to monitor single-molecule dynamics at the nanoscale, beyond the limited time resolution associated with the conventional scanning probe approach, thus obtaining high temporal resolution in ultra-confined illumination volumes.

The first demonstration of FCS using an NSOM approach was performed in 2008 by Vobornik et al. [85], by measuring the mobility of fluorescent lipids in supported lipid bilayers. More recently, the group of Naber succeeded for the first time to measure the transport kinetics at a single nuclear pore using the combination of FCS and NSOM [86]. In here, the distinct characteristic of the evanescent field was exploited to determine diffusion of ligands transported axially through the

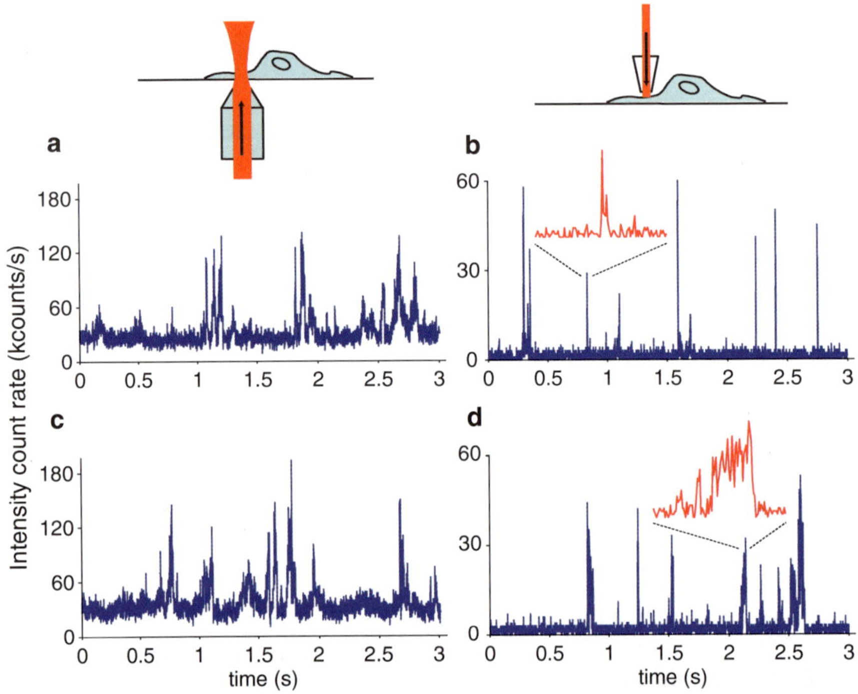

Fig. 5 Fluorescence bursts of different lipids inserted in the membrane of living CHO cells. (**a–b**) Fluorescent bursts arising from diffusing PE lipids with confocal (**a**) and NSOM (**b**) excitation obtained with an $r = 60$ nm aperture probe. Inset shows a 100-ms zoom of a typical fluorescence burst. (**c–d**) Fluorescent bursts arising from diffusing SM lipids with confocal (**c**) and NSOM (**d**) excitation. Inset shows a 100-ms zoom of a typical fluorescence burst

pore. These examples demonstrated that NSOM-FCS might indeed become a valuable tool to study single-molecule mobility both on the lateral and/or axial direction and at the nanoscale. Yet, the application of NSOM-FCS on intact living cells has been challenged by the difficulties in keeping the probe in close proximity to the soft and flickering cell membrane in a noninvasive way.

Recently, we optimized the feedback system based on tuning forks operating in a diving bell configuration to stably maintain the NSOM probe in close proximity to the cell membrane, with distance fluctuations of the order of 1 nm [87]. We then used the system to demonstrate the capability of NSOM-FCS to measure the lateral mobility of different lipids on the membrane of living CHO cells [87].

As an example, fluorescence fluctuations obtained in confocal and NSOM mode for the lipid analog PE (nonraft lipid) and SM (raft lipid) incorporated in the membrane are shown in Fig. 5a–d. In all cases, we observed the characteristic fluorescence bursts associated with fluorophores diffusing in and out of the excitation area. Comparison of confocal and NSOM fluorescence traces highlights the specificity of NSOM to the cell membrane. Since the evanescent field exiting the

NSOM probe only excites the membrane-proximal cytosolic region, we observed a five- to tenfold increase of signal-to-noise ratio with respect to confocal illumination. At the large illumination area of confocal excitation, both lipids display similar bursts. Notably, with the reduced illumination area by NSOM, the width of the fluorescence bursts for the lipid analog PE becomes narrower. In contrast, the fluorescence bursts arising from SM show a smaller width reduction with respect to confocal FCS and a wide variety of widths, implying different transit times through the NSOM illumination region (Fig. 5d). The difference in fluorescence fluctuations between PE and SM already hints at a different diffusion of SM on a spatial scale comparable to the illumination area of the NSOM probe.

Further details of the lipid diffusion can be obtained by calculation and fitting of the ACF. In general, the autocorrelation curve is fitted with a suitable function to obtain the diffusion characteristic parameters [83]. Determination of the fitting function requires the knowledge of the field intensity distribution in the illumination area. Whereas for confocal illumination the intensity has a Gaussian profile and the corresponding ACF can be analytically calculated [88], the exact intensity distribution from a circular NSOM probe is exceedingly complex so that an analytical ACF is difficult to obtain. However, Vobornik et al. [85] found that the analytical ACF obtained for an NSOM square aperture satisfactorily approximate the circular case. Generalization of this ACF to take into account anomalous diffusion provides:

$$G(\tau) = \left[erf\left(2\left(\frac{\tau_D}{\tau}\right)^{\frac{\alpha}{2}} \right) - \frac{1}{2\sqrt{\pi}} \left(\frac{\tau}{\tau_D}\right)^{\frac{\alpha}{2}} \left(1 - e^{-4\left(\frac{\tau_D}{\tau}\right)^{\alpha}} \right) \right]^2$$

where $erf(x)$ represents the error function, τ the lag time, τ_D the average time spent by the fluorescent lipid in the excitation area and α the anomalous diffusion exponent. For free diffusion, this anomalous exponent is equal to 1. Diffusion coefficients are calculated as $D = A/(4\tau_D)$, where A is the effective illumination area. In the confocal case, $A = d^2/(2\ln 2)$ for a Gaussian profile with full-width-at-half-maximum $d = 250$ nm. For NSOM-FCS, $A = \pi r^2$, where r is the radius of the aperture of the NSOM probe [85].

The fitting of ACF curves obtained for the lipid analog PE in confocal FCS and NSOM-FCS ($r = 60$ nm and 90 nm) provided $\alpha = 1$, consistent with a free diffusion model. The diffusion time showed a linear dependence with the excitation area (Fig. 6a), confirming Brownian diffusion and providing a diffusion coefficient value $D = 0.5\ \mu m^2/s$ [87]. The linearity of the diffusion time and the consistency of the diffusion coefficient with previously published values [14, 89] ensured the robustness and the reproducibility of the NSOM-based approach. Contrarily to PE, the correlation curves for SM show a markedly different behavior. Although when measured with confocal FCS, SM showed a similar behavior as PE [87], at the nanometric area of the NSOM probe, the correlation time shows a smaller reduction compared to the one-order-of-magnitude change observed for PE. Moreover, the NSOM correlation curve of SM also displays a slower decay associated with a

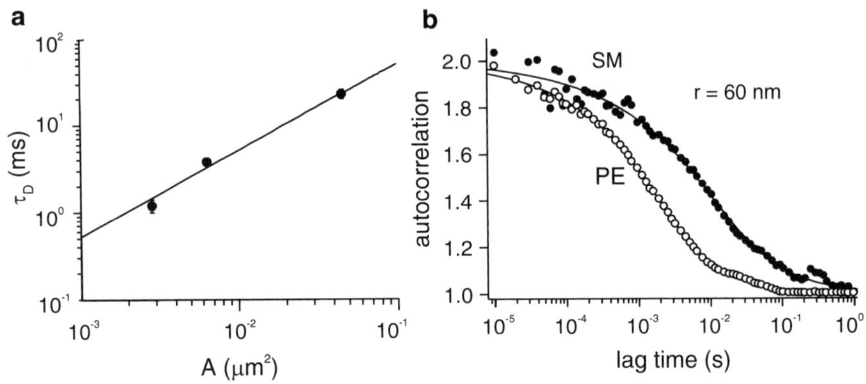

Fig. 6 NSOM-FCS on living cell membranes. (**a**) Dependence of the average time spent by the lipid analog PE in the illumination area as the illumination area decreases. (**b**) Normalized correlation curves for PE (*open circles*) and SM (*filled circles*) obtained with NSOM illumination through an $r = 60$ nm aperture probe. Lines over data denote curve fitting as explained in the text. Adapted with permission from Ref. [87]

lower α value, typical of confined diffusion (Fig. 6b, $\alpha = 0.79 \pm 0.07$) [87]. The anomalous behavior of SM is entirely consistent with a cholesterol-induced confinement of sphingolipids [14, 90, 91]. Since this trapping occurs in nanometric membrane domains, its effect on the diffusion dynamics could only be revealed by the ultrasmall illumination volume provided by NSOM.

There is a large potential for the use of NSOM-FCS since it combines membrane (proximal) specificity and high signal-to-noise ratios. The high temporal resolution together with the reduced excitation area allowed a dynamic inquiry of lipid behavior at the nanoscale. Further application of the FCS scaling law [90] should lead to more information regarding trapping times, the exact size of the nanoscale heterogeneities that hinder SM diffusion as well as its true free diffusion coefficient. Since multiple wavelengths exciting the NSOM probe overlap at the nanoscale, an exciting path lies ahead, where dual-color cross correlation might be able to address questions of inner-outer leaflet coupling, signaling complex formation, and the intimate relation between rafts and the actin cytoskeleton. Critical for this progress will be the fabrication of bright probes that confine the excitation light to truly nanoscale dimensions.

3.4 *Superresolution Imaging of Cell Membranes Using Optical Antennas*

Despite the significant gain in resolution using conventional NSOM probes with respect to standard fluorescent techniques, there are still open challenges regarding its spatial resolution. Indeed, valuable information at the nanometric scale might still be obscured if the resolution cannot be pushed toward the ultimate level of single proteins or lipids, below 10 nm.

The challenge regarding optical resolution using probe-based NSOM is largely due to the intensity of the field exiting the subwavelength aperture. Light throughput (measured in the far field) for 100-nm-sized NSOM probes drops to about 10^{-4}–10^{-6} of the incoming input. This, together with a maximum power input related to the probe damage threshold, limits the practical resolution to about 50–70 nm for single-molecule applications. The reason for the intensity drop is that there is a cutoff region in the tapered fiber where the diameter is $\sim\lambda$, after which light propagation is not possible and the intensity decreases exponentially until the aperture. In order to increase the light throughput of standard optical fibers, one can approach the original idea of Synge where a metallic plate containing a subwavelength aperture is presented in front of a propagating wave. Cutting the fiber before the cutoff region and reevaporating a thin metal layer at this end-facet with a nanohole in the center indeed improve throughput by 100-fold as well as the damage threshold by 40-fold [92]. Although this type of probe fabrication allows one to work with smaller apertures [92], the maximum resolution is still limited by the finite skin depth of the metal. In addition, the large end-facet of these probes complicates measurements on cell membranes where smaller structures might force the aperture to move out of the near-field region.

An alternative approach is to use optical antennas where surperresolution at the nanoscale is accompanied by electric field enhancement, thus providing brighter illumination sources. The main idea of optical antennas is to localize and enhance the optical radiation into a nanometric region, similar to electromagnetic antennas, which convert propagating radiation into a confined zone. Research into photonic antenna structures is still at the fundamental level, and studies so far have mainly concentrated on, for example, the influence of photonic antennas on dipole emission [93] or how the generated near-field can influence the photophysical properties of dyes [94]. Nevertheless, the first applications in the field of biology are emerging, as for instance the imaging of single Ca^{2+} channels on erythrocyte plasma membranes at 50-nm optical resolution in aqueous conditions [29]. In these experiments, single gold particles positioned at the end of NSOM probes have been used as optical antennas to localize the incident far-field laser irradiation proximal to erythrocyte plasma membranes with fluorescently labeled Ca^{2+} pumps. However, large background due to the far-field irradiation required signal-to-noise improvement using modulation methods [95]. A different excitation scheme that suppresses background illumination was first proposed by Frey et al. [96] and more recently, refined by Taminiau et al. [97]. In these tip-on-aperture antennas, the local illumination properties of aperture-type NSOM are used to drive the antenna to resonance.

The antenna probes are fabricated in a way similar as to conventional NSOM probes, but FIB milling is used to create the antenna on the flat end-face next to the circular aperture. The length, width, and radius of curvature of the antennas can be carefully controlled by the FIB milling process. To drive the antenna at its resonance, it is necessary to apply a field along the axis of the antenna, that is, in the z

direction. For an aperture probe, it has been demonstrated experimentally [23, 98] and theoretically [99, 100] that this component is present at the aperture edges in the direction of the incoming polarization. Therefore, during fabrication, the antenna is positioned at the edge of the aperture, that is, at the interface between the glass and the aluminum. Using this configuration, single-molecule detection with 30-nm resolution and virtually no background has been demonstrated [97]. This approach allowed to obtain promising results for nanoscale imaging on cell membranes [101].

Although there is a strong analogy between optical antennas and their radio frequency and microwave counterparts, there are some crucial differences. These differences mainly arise from the fact that metals are not perfect conductors at optical frequencies. This affects the scaling properties of the antenna in relation to resonance. In order to relate the antenna geometry to the resonance conditions, the effective wavelength for resonance can be estimated using [102]:

$$\lambda_{eff} = \frac{\lambda}{\sqrt{\varepsilon_S}}\sqrt{\frac{x(\lambda)}{1+x(\lambda)}} - 4R$$

where λ is the incident wavelength, ε_S the dielectric constant of the medium, R the antenna radius, and $x(\lambda)$ is given by:

$$x(\lambda) = 4\pi^2\varepsilon_S\left(\frac{R^2}{\lambda^2}\right)\left(a_1 + a_2\frac{\lambda}{\lambda_p}\right)^2$$

where λ_p is the plasma wavelength of the metal and a_1 and a_2 are coefficients that depend on the metal properties [102]. For a monopole antenna, resonance occurs at $L_{res} = \lambda_{eff}/4$. In the case of Al-antennas, resonant at $\lambda = 633$ nm in liquid conditions, realistic designs using FIB technology consist on antennas of R ~ 25 nm and length ~80 nm. Interestingly, while on conventional NSOM probes the reduction of the aperture size dramatically decreases the total field intensity, a reduction in the diameter of the antenna not only confines the field to a smaller area but additionally increases the power density at the antenna apex.

Since antenna probes are fully compatible with standard NSOM setups, they can be readily used as nanosources for the imaging of biological specimens. Figure 7a shows a confocal image of fluorescently labeled antibodies attached to a glass surface, while Fig. 7b displays the counterpart image as obtained using an antenna probe with $L = 110$ nm and $R = 25$ nm, as shown in the inset. The exceptional increase in resolution is clearly apparent by comparing both images. Remarkable from the cross sections shown in Fig. 7c–d is not only the increase in resolution but also the excellent signal-to-background (S/B) ratio obtained. This S/B ratio of ~30 results from the highly localized electric field at the antenna apex while keeping the background contribution negligible. As comparison, conventional aperture probes with similar incoupling conditions at the back-end of the probe result in S/B ~10. Hence, the accuracy on the determination of the center-of-mass of the fluorescence

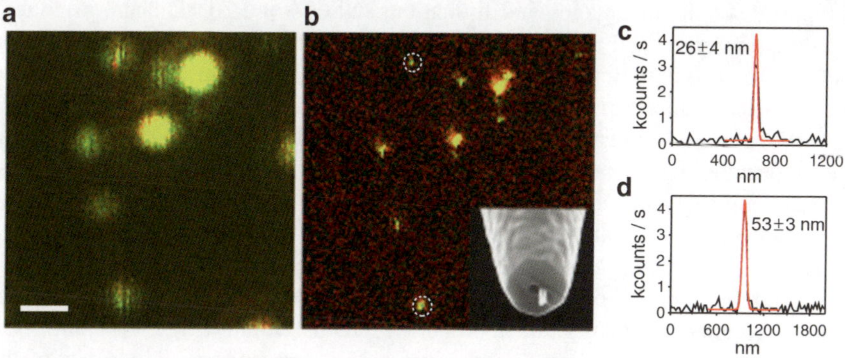

Fig. 7 Increased resolution and field confinement of optical antennas for nanoimaging. (**a**) Fluorescently labeled antibodies imaged by standard confocal microscopy. Scale bar: 500 nm. (**b**) Imaging the same area as in (**a**) using a probe-based optical antenna excited at 633 nm. Inset shows the optical antenna carved at the end of a standard NSOM probe. (**c**, **d**) Cross sections of the intensity profile for two different fluorescence features as encircled in (**b**) and corresponding Gaussian fits (*red line*) rendering widths of 26 ± 4 nm in (**c**) and 53 ± 3 nm in (**d**). Adapted with permission from Ref. [104]

spots (given by ~FHWM/(S/B) where FHWM is the full-width-at-half-maximum of the fluorescence spot, [103]) can be as good as ~1 nm, demonstrating the high degree of field localization afforded by probe-based antennas. More recently, we have been using these antenna geometries to study the nanoscale organization of adhesion receptors on the membrane of intact monocytes, obtaining a superb resolution of 30 nm [104].

4 Summary

In this chapter, we surveyed recent technological developments in near-field nanoscopy and discussed recent studies in which NSOM has been applied to address key questions in the field of cell membrane biology. Owing to its surface sensitivity and excellent localization accuracy, NSOM has been able to map molecular interactions and their spatial nanoscale proximity on intact mammalian cell membranes in relation to function. We have highlighted first exciting results where optical antennas have been already extended to biological applications to quantitatively image, with exceptional resolution, intact cell membranes in physiological conditions. Concepts from FCS implemented in ultra-confined volumes as afforded by NSOM probes allow nowadays measurements of ultrafast dynamics at the nanoscale in living cell membranes.

Within the context of nanoscale imaging of biological samples, it is important to realize that aside from NSOM, a palette of new far-field surperresolution approaches are also becoming available to the biophysical community. The various techniques such as NSOM, STED, (F-)PALM, and STORM are complementary to each other in terms of specific advantages and limitations. NSOM and STED share a reduced excitation volume. In the case of NSOM, the evanescent character of the excitation field in the axial direction grants exquisite sensitivity for cell membrane studies but also restricts its application toward intracellular nanoimaging. In contrast, STED can take advantage of standard confocal geometries and therefore allow optical sectioning at the nanoscale and even provide video-rate imaging. However, STED requires fluorophores that can cycle many times between dark and bright states. In addition, since the two beams that reshape the excitation field should be aligned properly in time and space, the optical design is rather complex and multicolor imaging challenging. NSOM on the other hand does not rely on the photophysical properties of the fluorescent markers (organic fluorophores, QDots, autofluorescent proteins, etc.) and provides multicolor imaging in a straightforward manner. In contrast to the complex setups of NSOM and STED, PALM and STORM require relatively simple wide-field setups. To reconstruct high-quality surperresolution images, a single-molecule-sensitive camera should detect all emitters labeling a specific structure with sufficient number of photons to get position accuracy down to tens of nm. This can require long measurement times and therefore restricts imaging to slow-mobile or static structures. Still, using sufficient fluorophores in order to fulfill the Nyquist requirements, it is possible to readily obtain impressive reconstructed surperresolution images. Further improvement of far-field methods, such as STED/(F) PALM/STORM, and their application to tackle complex biological questions crucially depend on the photophysical properties of the dyes used.

For the near-field technique discussed in this chapter, improvement of methods or geometries that can generate high electromagnetic fields at the nanometer scale will become crucial for successful applications in biology. In this sense, concepts from the booming field of optical antennas to quantitatively image with unprecedented resolution and/or concepts from FCS to allow the measurements of ultrafast dynamics at the nanoscale should enrich the palette of near-field biophysical methods to gain understanding of fundamental processes that occur on the cell membrane. Other possibilities lie in the use of different shapes such as nanostars that generate highly localized plasmonic hotspots [105] or the use of large-aperture nanostructured NSOM probes to increase throughput but maintain field confinement at the nanoscale [106]. We await an exciting time where nanofabrication and progress in nanophotonics will allow us to have a closer look at the fundamental properties of cellular membrane organization and function.

Bibliography

1. Jares-Erijman EA, Jovin TM (2003) FRET imaging. Nat Biotechnol 21(11):1387–1395
2. Varma R, Mayor S (1998) GPI-anchored proteins are organized in submicron domains at the cell surface. Nature 394:798–801
3. Sharma P, Varma R, Sarasij RC, Ira GK, Krishnamoorthy G, Rao M, Mayor S (2004) Nanoscale organization of multiple GPI-anchored proteins in living cell membranes. Cell 116:577–589

4. Mayor S, Rao M (2004) Rafts: scale-dependent, active lipid organization at the cell surface. Traffic 5:231–240
5. Betzig E, Patterson GH, Sougrat R, Lindwasser OW, Olenych S, Bonifacino JS, Davidson MW, Lippincott-Schwartz J, Hess HF (2006) Imaging intracellular fluorescent proteins at nanometer resolution. Science 313:1642–1645
6. Hess ST, Girirajan TPK, Mason MD (2006) Ultra-high resolution imaging by fluorescence photoactivation localization microscopy. Biophys J 91:4258–4272
7. Rust MJ, Bates M, Zhuang X (2006) Sub-diffraction-limit imaging by stochastic optical reconstruction microscopy (STORM). Nat Methods 3:793–795
8. Lillemeier BF, Mörtelmaier MA, Forstner MB, Huppa JB, Groves JT, Davis MM (2010) TCR and Lat are expressed on separate protein islands on T cell membranes and concatenate during activation. Nat Immunol 11:90–96
9. Sherman E, Barr V, Manley S, Patterson GH, Balagopalan L, Akpan I, Regan CK, Merrill RK, Sommers CL, Lippincott-Schwartz J, Samelson LE (2011) Functional nanoscale organization of signaling molecules downstream of the T cell antigen receptor. Immunity 35(5):705–720
10. Sengupta P, Jovanovic-Talisman T, Skoko D, Renz M, Veatch SL, Lippincott-Schwartz J (2011) Probing protein heterogeneity in the plasma membrane using PALM and pair correlation analysis. Nat Methods 8(11):969–975
11. Williamson DJ, Owen DM, Rossy J, Magenau A, Wehrmann M, Gooding JJ, Gaus K (2011) Pre-existing clusters of the adaptor Lat do not participate in early T cell signaling events. Nat Immunol 12:655–662
12. Sieber JJ, Willig KI, Kutzner C, Gerding-Reimers C, Harke B, Donnert G, Rammner B, Eggeling C, Hell SW, Grubmüller H, Lang T (2007) Anatomy and dynamics of a supramolecular membrane protein cluster. Science 317:1072–1076
13. Kellner RR, Baier CJ, Willig KI, Hell SW, Barrantes FJ (2007) Nanoscale organization of nicotinic acetylcholine receptors revealed by stimulated emission depletion microscopy. Neuroscience 144:135–143
14. Eggeling C, Ringemann C, Medda R, Schwarzmann G, Sandhoff K, Polyakova S, Belov VN, Hein B, von Middendorff C, Schönle A, Hell SW (2009) Direct observation of the nanoscale dynamics of membrane lipids in a living cell. Nature 457:1159–1162
15. Synge EA (1928) A suggested method for extending microscopic resolution into the ultramicroscopic region. Philos Mag 6:356–362.
16. Ash EA, Nicholls G (1972) Super-resolution aperture scanning microscope. Nature 237 (5357):510–512
17. Pohl DW, Denk W, Lanz M (1984) Optical stethoscopy – image recording with resolution lambda/20. Appl Phys Lett 44(7):651–653
18. Lewis A, Isaacson M, Harootunian A, Muray A (1984) Development of a 500-a spatial-resolution light-microscope. 1. Light is efficiently transmitted through gamma-16 diameter apertures. Ultramicroscopy 13(3):227–231
19. Peasler MA, Moyer PJ (1996) Near-field optics: theory, instrumentation and applications. Wiley, New York
20. Betzig E, Trautman JK, Harris TD, Weiner JS, Kostelak RL (1991) Breaking the diffraction barrier: optical microscopy on a nanometric scale. Science 251:1468–1470
21. Stockle R, Fokas C, Deckert V, Zenobi R, Sick B, Hecht B, Wild UP (1999) High-quality near-field optical probes by tube etching. Appl Phys Lett 75(2):160–162
22. Burgos P, Lu Z, Ianoul A, Hnatovsky C, Viriot M-L, Johnston LJ, Taylor RS (2003) Near-field scanning optical microscopy probes: a comparison of pulled and double-etched bent NSOM probes for fluorescence imaging of biological samples. J Microscopy 211:37–47
23. Veerman JA, Otter AM, Kuipers L, van Hulst NF (1998) High definition aperture probes for near-field optical microscopy fabricated by focused ion beam milling. Appl Phys Lett 72:3115–3117
24. Betzig E, Finn PL, Weiner JS (1992) Combined shear force and near-field scanning optical microscopy. Appl Phys Lett 60:2484–2486

25. Toledo-Crow R, Yang PC, Chen Y, Vaez-Iravani M (1992) Near-field differential scanning optical microscope with atomic force regulation. Appl Phys Lett 60:2957–2959
26. Lieberman K, BenAmi N, Lewis A (1996) Fully integrated near-field optical, far-field optical, and normal-force scanned probe microscope. Rev Sci Instrum 67(10):3567–3572
27. Koopman M, de Bakker BI, Garcia-Parajo MF, van Hulst NF (2003) Shear force imaging of soft samples in liquid using a diving bell concept. Appl Phys Lett 83:5083–5085
28. Taylor RS, Vobornik D, Lu Z, Chisholm RA, Johnston LJ (2010) Damping behavior of bent fiber NSOM probes in water. J Appl Phys 107(4):043526
29. Höppener C, Novotny L (2008) Antenna-based optical imaging of single Ca^{2+} transmembrane proteins in liquids. Nano Lett 8:642–646
30. Höppener C, Siebrasse JP, Peters R, Kubitscheck U, Naber A (2005) High-resolution near-field optical imaging of single nuclear pore complexes under physiological conditions. Biophys J 88:3681–3688
31. Heinzelmann H, Pohl DW (1994) Scanning near-field optical microscopy. Appl Phys A 59(2):89–101
32. Garcia-Parajo MF, de Bakker BI, Koopman M, Cambi A, de Lange F, Figdor CG, van Hulst NF (2005) Near-field fluorescence microscopy: an optical nanotool to study protein organization at the cell membrane. Nanobiotechnology 1:113–120
33. de Lange F, Cambi A, Huijbens R, de Bakker BI, Rensen WHJ, Garcia-Parajo MF, van Hulst NF, Figdor CG (2001) Cell biology beyond the diffraction limit: near-field scanning optical microscopy. J Cell Sci 114:4153–4160
34. Koopman M, Cambi A, de Bakker BI, Joosten B, Figdor CG, van Hulst NF, Garcia-Parajo MF (2004) Near-field scanning optical microscopy in liquid for high resolution single molecule detection on dendritic cells. FEBS Lett 573:6–10
35. Singer SJ, Nicolson GL (1972) The fluid mosaic model of the structure of cell membranes. Science 175(4023):720–731
36. Jacobson K, Sheets ED, Simson R (1995) Revisiting the fluid mosaic model of membranes. Science 268(5216):1441–1442
37. Maxfield FR (2002) Plasma membrane microdomains. Curr Opin Cell Biol 14:483–487
38. Simons M, Friedrichson T, Schulz JB, Pitto M, Masserini M, Kurzchalia TV (1999) Exogenous administration of gangliosides displaces GPI-anchored proteins from lipid microdomains in living cells. Mol Biol Cell 10(10):3187–3196
39. Kusumi A, Suzuki KGN (2005) Toward understanding the dynamics of membrane-raft-based molecular interactions. Biochim Biophys Acta 1746(3):234–251
40. Douglass AD, Vale RD (2005) Single-molecule microscopy reveals plasma membrane microdomains created by protein-protein networks that exclude or trap signaling molecules in T cells. Cell 121:937–950
41. Lingwood D, Simons K (2010) Lipid rafts as a membrane-organizing principle. Science 327:46–50
42. Cambi A, Joosten B, Koopman M, de Lange F, Beeren I, Torensma R, Fransen JA, Garcia-Parajo MF, van Leeuwen FN, Figdor CG (2006) Organization of the integrin LFA-1 in nanoclusters regulates its activity. Mol Biol Cell 17:4270–4281
43. Cambi A, de Lange F, van Maarseveen NM, Nijhuis M, Joosten B, van Dijk EMHP, de Bakker BI, Fransen JA, Bovee-Geurts PHM, van Leeuwen FN, van Hulst NF, Figdor CG (2004) Microdomains of the C-type lectin DC-SIGN are portals for virus entry into dendritic cells. J Cell Biol 164:145–155
44. Mayor S, Pagano RE (2007) Pathways of clathrin-independent endocytosis. Nat Rev Mol Cell Biol 8:603–612
45. Mañes S, Viola A (2006) Lipid rafts in lymphocyte activation and migration. Mol Membr Biol 23:59–69
46. Suzuki KGN, Fujiwara TK, Sanematsu F, Iino R, Edidin MA, Kusumi A (2007) GPI-anchored receptor clusters transiently recruit Lyn and Gα for temporary cluster immobilization and Lyn activation: single-molecule tracking study 1. J Cell Biol 177:717–730

47. Lasserre R, Guo X-J, Conchonaud F, Hamon Y, Hawchar O, Bernard A-M, Soudja SMH, Lenne P-F, Rigneault H, Olive D, Bismuth G, Nunès JA, Payrastre B, Marguet D, He H-T (2008) Raft nanodomains contribute to Akt/PKB plasma membrane recruitment and activation. Nat Chem Biol 4:538–547
48. Prior IA, Muncke C, Parton RG, Hancock JF (2003) Direct visualization of Ras proteins in spatially distinct cell surface microdomains. J Cell Biol 160:165–170
49. Anderson RGW, Jacobson K (2002) A role for lipid shells in targeting proteins to caveolae, rafts, and other lipid domains. Science 296(5574):1821–1825
50. Cherukuri A, Carter RH, Brooks S, Bornmann W, Finn R, Dowd CS, Pierce SK (2004) B cell signaling is regulated by induced palmitoylation of CD81. J Biol Chem 279:31973–31982
51. Hemler ME (2005) Tetraspanin functions and associated microdomains. Nat Rev Mol Cell Biol 6:801–811
52. Maecker HT, Todd SC, Levy S (1997) The tetraspanin superfamily: molecular facilitators. FASEB J 11:428–442
53. Berditchevski F (2001) Complexes of tetraspanins with integrins: more than meets the eye. J Cell Sci 114:4143–4151
54. Charrin S, le Naour F, Silvie O, Milhiet P-E, Boucheix C, Rubinstein E (2009) Lateral organization of membrane proteins: tetraspanins spin their web. Biochem J 420:133–154
55. Yanez-Mo M, Barreiro O, Gordon-Alonso M, Sala-Valdés M, Sanchez-Madrid F (2009) Tetraspanin-enriched microdomains: a functional unit in cell plasma membranes. Trends Cell Biol 19:434–446
56. Barreiro O, Zamai M, Yanez-Mo M, Tejera E, López-Romero P, Monk PN, Gratton E, Caiolfa VR, Sanchez-Madrid F (2008) Endothelial adhesion receptors are recruited to adherent leukocytes by inclusion in preformed tetraspanin nanoplatforms. J Cell Biol 183:527–542
57. Belanis L, Plowman SJ, Rotblat B, Hancock JF, Kloog Y (2008) Galectin-1 is a novel structural component and a major regulator of H-ras nanoclusters. Mol Biol Cell 19(4):1404–1414
58. Nieminen J, Kuno A, Hirabayashi J, Sato S (2007) Visualization of galectin-3 oligomerization on the surface of neutrophils and endothelial cells using fluorescence resonance energy transfer. J Biol Chem 282(2):1374–1383
59. Kusumi A, Sako Y (1996) Cell surface organization by the membrane skeleton. Curr Opin Cell Biol 8:566–574
60. Kusumi A, Nakada C, Ritchie K, Murase K, Suzuki KGN, Murakoshi H, Kasai RS, Kondo J, Fujiwara TK (2005) Paradigm shift of the plasma membrane concept from the two-dimensional continuum fluid to the partitioned fluid: high-speed single-molecule tracking of membrane molecules. Annu Rev Biophys Biomol Struct 34:351–378
61. Andrews NL, Lidke KA, Pfeiffer JR, Burns AR, Wilson BS, Oliver JM, Lidke DS (2008) Actin restricts FcεRI diffusion and facilitates antigen-induced receptor immobilization. Nat Cell Biol 10(8):955–963
62. Betzig E, Trautman JK (1992) Near-field optics: microscopy, spectroscopy, and surface modification beyond the diffraction limit. Science 257:189–195
63. Enderle T, Ha T, Ogletree DF, Chemla DS, Magowan C, Weiss S (1997) Membrane specific mapping and colocalization of malarial and host skeletal proteins in the *Plasmodium falciparum* infected erythrocyte by dual-color near-field scanning optical microscopy. Proc Natl Acad Sci 94(2):520–525
64. Chen Y, Qin J, Chen Z W (2008) Fluorescence-topographic NSOM directly visualizes peak-valley polarities of GM1/GM3 rafts in cell membrane fluctuations. J Lipid Res 49 (10):2268–2275
65. Gomez-Mouton C, Abad JL, Mira E, Lacalle RA, Gallardo E, Jimenez-Baranda S, Illa I, Bernad A, Mañes S, Martinez-A C (2001) Segregation of leading-edge and uropod components into specific lipid rafts during T cell polarization. Proc Natl Acad Sci 98 (17):9642–9647

66. Abulrob A, Lu Z, Brunette E, Pulla D, Stanimirovic D, Johnston LJ (2008) Near-field scanning optical microscopy detects nanoscale glycolipid domains in the plasma membrane. J Microscopy 232:225–234
67. van Zanten TS, Gómez J, Manzo C, Cambi A, Buceta J, Reigada R, Garcia-Parajo MF (2010) Direct mapping of nanoscale compositional connectivity on intact cell membranes. Proc Natl Acad Sci 107:15437–15442
68. Lillemeier BF, Pfeiffer JR, Surviladze Z, Wilson BS, Davis MM (2006) Plasma membrane-associated proteins are clustered into islands attached to the cytoskeleton. Proc Natl Acad Sci 103:18992–18997
69. de Bakker BI, de Lange F, Cambi A, Korterik JP, van Dijk EMHP, van Hulst NF, Figdor CG, Garcia-Parajo MF (2007) Nanoscale organization of the pathogen receptor DC-SIGN mapped by single-molecule high-resolution fluorescence microscopy. ChemPhysChem 8:1473–1480
70. de Bakker BI, Bodnár A, van Dijk EMHP, Vámosi G, Damjanovich S, Waldmann TA, van Hulst NF, Jenei A, Garcia-Parajo MF (2008) Nanometer-scale organization of the alpha subunits of the receptors for IL2 and IL15 in human T lymphoma cells. J Cell Sci 121:627–633.
71. Vobornik D, Rouleau Y, Haley J, Bani-Yaghoub M, Taylor RS, Johnston LJ, Pezacki JP (2009) Nanoscale organization of β2-adrenergic receptor-Venus fusion protein domains on the surface of mammalian cells. Biochem Biophys Res Commun 382:85–90
72. Abulrob A, Lu Z, Baumann E, Vobornik D, Taylor RS, Stanimirovic D, Johnston LJ (2010) Nanoscale imaging of epidermal growth factor receptor clustering: effects of inhibitors. J Biol Chem 285:3145–3156
73. Chen J, Pei Y, Chen ZW, Cai J (2010) Quantum dot labeling based on near-field optical imaging of CD44 molecules. Micron 41(3):198–202
74. Zhong L, Zhang Z, Lu X, Huang D, Chen CY, Wang RC, Chen ZW (2011) NSOM/QD-based fluorescence and topographic image fusion directly reveals nano-spatial peak and valley polarities of CD69 and CD71 activation molecules on cell-membrane fluctuations during T-cell activation. Immunol Lett 140(1–2):44–51
75. Zhong L, Zeng G, Lu X, Wang RC, Gong G, Yan L, Huang D, Chen ZW (2009) NSOM/QD-based direct visualization of CD3-induced and CD28-enhanced nanospatial coclustering of TCR and coreceptor in nanodomains in T cell activation. PloS one 4:e5945
76. Ianoul A, Grant DD, Rouleau Y, Bani-Yaghoub M, Johnston LJ, Pezacki JP (2005) Imaging nanometer domains of beta-adrenergic receptor complexes on the surface of cardiac myocytes. Nat Chem Biol 1:196–202
77. van Zanten TS, Cambi A, Koopman M, Joosten B, Figdor CG, Garcia-Parajo MF (2009) Hotspots of GPI-anchored proteins and integrin nanoclusters function as nucleation sites for cell adhesion. Proc Natl Acad Sci 106:18557–18562
78. Meder D, Moreno MJ, Verkade P, Vaz WLC, Simons K (2006) Phase coexistence and connectivity in the apical membrane of polarized epithelial cells. Proc Natl Acad Sci 103:329–334
79. Kenworthy AK, Nichols BJ, Remmert CL, Hendrix GM, Kumar M, Zimmerberg J, Lippincott-Schwartz J (2004) Dynamics of putative raft-associated proteins at the cell surface. J Cell Biol 165:735–746
80. Saxton MJ, Jacobson K (1997) Single-particle tracking: applications to membrane dynamics. Annu Rev Biophys Biomol Struct 26:373–399
81. Schmidt T, Schütz GJ, Baumgartner W, Gruber HJ, Schindler H (1996) Imaging of single molecule diffusion. Proc Natl Acad Sci 93:2926–2929
82. Schwille P, Haupts U, Maiti S, Webb WW (1999) Molecular dynamics in living cells observed by fluorescence correlation spectroscopy with one- and two-photon excitation. Biophys J 77(4):2251–2265
83. Kim SA, Heinze KG, Schwille P (2007) Fluorescence correlation spectroscopy in living cells. Nat Methods 4:963–973

84. Banks DS, Fradin C (2005) Anomalous diffusion of proteins due to molecular crowding. Biophys J 89:2960–2971
85. Vobornik D, Banks DS, Lu Z, Fradin C, Taylor RS, Johnston LJ (2008) Fluorescence correlation spectroscopy with sub-diffraction-limited resolution using near-field optical probes. Appl Phys Lett 93:163904
86. Herrmann M, Neuberth N, Wissler J, Pérez J, Gradl D, Naber A (2009) Near-field optical study of protein transport kinetics at a single nuclear pore. Nano Lett 9:3330–3336
87. Manzo C, van Zanten TS, Garcia-Parajo MF (2011) Nanoscale fluorescence correlation spectroscopy on intact living cell membranes with NSOM probes. Biophys J 100:L8–L10
88. Schwille P, Haustein E (2009) Fluorescence correlation spectroscopy an introduction to its concepts and applications. Anal Chem 94(22):1–33
89. Sahl SJ, Leutenegger M, Hilbert M, Hell SW, Eggeling C (2010) Fast molecular tracking maps nanoscale dynamics of plasma membrane lipids. Proc Natl Acad Sci 107:6829–6834
90. Wawrezinieck L, Rigneault H, Marguet D, Lenne P-F (2005) Fluorescence correlation spectroscopy diffusion laws to probe submicron membrane organization. Biophys J 89:4029–4042
91. Wenger J, Conchonaud F, Dintinger J, Wawrezinieck L, Ebbesen TW, Rigneault H, Marguet D, Lenne P-F (2007) Diffusion analysis within single nanometric apertures reveals the ultrafine cell membrane organization. Biophys J 92:913–919
92. Neumann L, Pang Y, Houyou A, Juan ML, Gordon R, van Hulst NF (2011) Extraordinary optical transmission brightens near-field fiber probe. Nano Lett 11:355–360
93. Taminiau TH, Stefani FD, Segerink FB, van Hulst NF (2008) Optical antennas direct single-molecule emission. Nat Photonics 2:234–237
94. Kinkhabwala A, Yu Z, Fan S, Avlasevich Y, Müllen K, Moerner WE (2009) Large single-molecule fluorescence enhancements produced by a bowtie nanoantenna. Nat Photonics 3:654–657
95. Höppener C, Beams R, Novotny L (2009) Background suppression in near-field optical imaging. Nano Lett 9:903–908
96. Frey HG, Witt S, Felderer K, Guckenberger R (2004) High-resolution imaging of single fluorescent molecules with the optical near-field of a metal tip. Phys Rev Lett 93:200801
97. Taminiau TH, Moerland RJ, Segerink FB, Kuipers L, van Hulst NF (2007) $\lambda/4$ resonance of an optical monopole antenna probed by single molecule fluorescence. Nano Lett 7:28–33
98. Betzig E, Chichester RJ (1993) Single molecules observed by near-field scanning optical microscopy. Science 262:1422–1425
99. Bethe HA (1944) Theory of diffraction by small holes. Phys Rev 66:163–182
100. Bouwkamp CJ (1950) On Bethe's theory of diffraction by small holes. Philips Res Rep 5:321–332
101. Garcia-Parajo MF (2008) Optical antennas focus in on biology. Nat Photonics 2:201–203
102. Novotny L (2007) Effective wavelength scaling for optical antennas. Phys Rev Lett 98:266802
103. Thompson RE, Larson DR, Webb WW (2002) Precise nanometer localization analysis for individual fluorescent probes. Biophys J 82:2775–2783
104. van Zanten TS, Lopez-Bosque MJ, Garcia-Parajo MF (2010) Imaging individual proteins and nanodomains on intact cell membranes with a probe-based optical antenna. Small 6:270–275
105. Hrelescu C, Sau TK, Rogach AL, Jäckel F, Laurent G, Douillard L, Charra F (2011) Selective excitation of individual plasmonic hotspots at the tips of single gold nanostars. Nano Lett 11:402–407
106. Mivelle M, Ibrahim IA, Baida F, Burr GW, Nedeljkovic D, Charraut D, Rauch JY, Salut R, Grosjean T (2010) Bowtie nano-aperture as interface between near-fields and a single-mode fiber. Opt Express 18(15):15964–15974

Part III
Characterization of Membrane Proteins and Receptors by Advanced Fluorescence-Based Imaging Techniques

Unveiling Biophysical and Biological Properties of a Hypothetical Membrane Receptor by Exploiting Recent Imaging Advances

Pauline Gonnord and Rajat Varma

Abstract Fluorescence microscopy is indispensable in the study of biological systems at various length scales. This rapidly evolving field continues to offer researchers cutting-edge techniques that enhance spatial and temporal resolution, especially with the invention of superresolution methodologies. In this chapter, we focus on techniques that have aided in the understanding of various cell biological phenomena. Each technique has certain boundaries of spatial and temporal resolution, and fluorophore density in a particular biological sample may limit the applicability of some techniques. We discuss strengths and weaknesses of many such techniques by considering their use in understanding the biological function of a hypothetical membrane receptor. We conclude that a combination of techniques is required to fully understand any cell biological process.

Keywords Fluorescence · Light microscopy · Molecular Interactions · Spectroscopy · Superresolution microscopy

Contents

P. Gonnord and R. Varma (✉)
Laboratory of Systems Biology, National Institute of Allergy and Infectious Diseases, National Institutes of Health, 4 Center Drive, Bethesda 20892, MD, USA
e-mail: varmarajat@niaid.nih.gov

Y. Mély and G. Duportail (eds.), *Fluorescent Methods to Study Biological Membranes*, Springer Ser Fluoresc (2013) 13: 367–388, DOI 10.1007/4243_2012_53,

1 Introduction

Genetics offers us the power of identifying the molecular players in any biological process, and biochemical tools give us invaluable information about how these players interact with each other and in what sequence. These tools, however, become limiting in understanding how these molecular players interact with each other in cells and living organisms. The reason for this limitation is that components of biological systems are compartmentalized in space and time. Optical microscopy offers us the opportunity to understand this biological organization by observing biological systems and processes with high spatial and temporal resolution. Over the last four decades, optical microscopy has revolutionized our understanding of various cell biological processes such as endocytosis, cell adhesion and migration, signal transduction, cell division, cell energetics, cell death, and protein synthesis and has uncovered a plethora of principles governing the compartmentalization of cellular components. Optical microscopy is a rapidly evolving field that is constantly breaking barriers of spatial and temporal resolution. Today, a toolbox of optical techniques is available to cell biologists, and the challenge lies in effectively using these techniques to uncover the mechanism of action of a group of proteins or a particular biological process.

Before choosing any imaging technique to answer a biological question, one needs to ask three questions: (1) what is the spatial resolution desired? (2) What is the temporal resolution desired or on what time scale is the phenomenon occurring? (3) What is the light output that will be obtained given the knowledge of the expression level of the molecule and its labeling efficiency? One can make a decision about which imaging technique to use based on these criteria if one is aware of the limitations of each technique. Usually, implementing a combination of tools helps overcome the limitations of a single technique.

Here, we shall give examples of how to understand the biology of the FFP (fantastic fake proteins) family of proteins by reviewing the different techniques that can allow us to study the protein at different spatial and temporal resolution.

This hypothetical protein is a transmembrane receptor. FFP1 is expressed in motile cells such as lymphocytes that migrate in response to various stimuli. FFP proteins are found in specialized structures involved in motility such as filopodia at the leading edge of migrating cells. FFP proteins can also form homo- and hetero-oligomeric complexes that have been implicated in signaling through these receptors. Downregulation of signaling is associated with endocytosis of these receptors.

Imaging of membrane proteins can be challenging. For example, using wide-field and confocal microscopy, it can be difficult to distinguish between a surface localization of receptors from a distribution of proteins in membrane proximal vesicles. Membrane dynamics and cell motility can cause loss of resolution when imaging live cells. In this chapter, we discuss a variety of techniques that allow us to examine different phenomena at high spatial and temporal resolution.

2 Taking a Glance at FFP1 Localization by Confocal Microscopy

2.1 Cellular Distribution of FFP1 by Laser Scanning Confocal Microscopy

For a recently cloned family of proteins such as FFPs that is expressed in lymphocytes as transmembrane receptors, a cellular biologist would like to know its subcellular localization.

One of the most widely available imaging techniques is laser scanning confocal microscopy (LSCM). Compared to epifluorescence microscopy, LSCM has better lateral and axial resolution. This is achieved by pinholes in front of the detectors that filter out-of-focus fluorescence. For a 100× magnification and a 1.4 numerical aperture (NA) objective, the lateral resolution will vary between 200 and 250 nm, and the axial resolution will be greater than 600 nm depending on the wavelength of light [1]. LSCM is an easily accessible primary approach to study the three-dimensional distribution of FFP1 in cells.

The limits of LSCM are longer times needed for scanning the sample leading to excessive light exposure and possible photobleaching and photodamage. Additionally, the imaging of moving cells becomes challenging due to this slow speed. We next discuss technologies that improve the speed of confocal microscopy.

2.2 Imaging Distribution of Molecules in Motile Cells by High-Speed Confocal Microscopy

Several approaches have been developed to improve the speed of LSCM. These include the use of resonant scanners in LSCM and the development of the slit confocal microscope and spinning-disk confocal microscope (SDCM).

The speed of LSCM has been improved with the use of resonant scanners. This scanning method uses resonant galvanometers that scan at a high speed of 8 kHz and achieve video-rate imaging [2]. While this method achieves video-rate imaging, it is at the cost of noisy images.

The second technology that has improved the speed of confocal imaging is the development of the slit confocal microscope. In this system, the pinhole is replaced by a slit, which allows acquisition of a single line at a time. The increase in acquisition speed comes, however, at the expense of a decrease in image resolution and optical sectioning because of more out-of-focus light coming through the slit [3, 4].

The third technology that has improved the speed of confocal microscopy is spinning-disk confocal microscopy [4]. As opposed to LSCM where the pinhole allows the confocal imaging of one spot of the sample at a time, in SDCM, several confocal spots are generated through the use of an aperture disk with many pinholes (called the Nipkow disk). Each pinhole works as both source and detector, and the rotation of the aperture disk at high speeds allows fast confocal detection of all the points in the object [5]. A significant improvement came about by the design of a second disk containing micro-lenses, aligned with the pinholes on the Nipkow disk that allow focusing of the incoming excitation light and improved illumination of the sample [6, 7]. The fast scan of the entire field of view not only improves the frame acquisition rate (up to 2,000 frames/s in theory) but also reduces the intensity of excitation light since scanning is faster, decreasing photobleaching and photodamage defects [8]. Unlike LSCM, here, the entire field of view can be recorded by a camera, such as a charge-coupled device (CCD) camera, which has better quantum efficiency than a PMT, thus increasing the signal-to-noise ratio of the images. A side-by-side comparison of SDCM with LSCM has been made, and it was found that SDCM performed better in most cases [9]. Currently, users have options for using electron multiplication CCD (EMCCD) camera, complementary metal oxide semiconductor (cMOS) cameras, or intensified cameras that provide the speed and sensitivity required for SDCM.

The pinhole size in SDCM is fixed and has been optimized for use with high NA objectives. When used with lower NA objectives for imaging larger fields of view, there will be loss in fluorescence signal. And as with other confocal systems, SDCM suffers from low axial resolution, which can reduce the quality of 3D image reconstruction [10].

To analyze FFP1 localization in motile cells or motile organelles of the cell, the fast acquisition rate of SDCM could be useful. FFP1 localization in small structures such as filopodia can be imaged as described in another example where filopodia in interleukin-2-inducible T cell kinase (Itk)-deficient T cells were observed by SDCM [11].

3 Analyzing FFP1 Clustering by Total Internal Reflection Fluorescence Microscopy (TIRFM)

To dissect further the biological significance of the localization of FFP1 observed by confocal microscopy, one needs to image the protein at higher lateral and axial resolution.

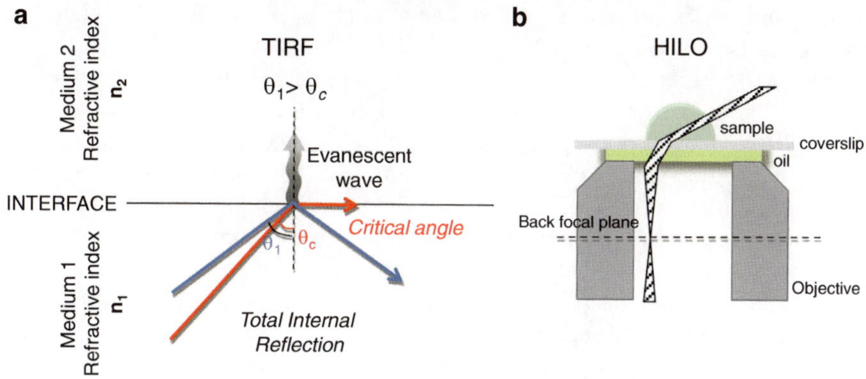

Fig. 1 (**a**) Principle of Total Internal Reflection. When light travels between two media with different refractive indices (n_1 and n_2) at a critical incident angle θ_c, it is refracted along the interface. At angles greater than the critical angle the beam undergoes total internal reflection and an evanescent wave is created in the medium with the lower refractive index. (**b**) HILO microscopy. In HILO microscopy, the laser beam is highly inclined by refraction and is grazing the sample at very low angles

TIRFM, a near-field imaging technique, achieves a narrow depth of field at the glass-water interface compared to confocal imaging (from 500–600 nm for confocal to 100–200 nm for TIRFM). It also has an improved signal-to-noise ratio of images due to acquisition of fluorescence emitted only by molecules localized near the coverslip [12].

The principle of TIRFM relies on the generation of an evanescent wave at the interface between media of different refractive indices upon total internal reflection (Fig. 1a). When light travels from a medium of high refractive index (such as glass) to a medium with lower refractive index (such as water or cell cytosol), it undergoes total internal reflection at angles greater than the critical angle. Under these conditions, an evanescent wave is generated in the medium with the lower refractive index. The intensity of this wave along the z axis, $I_{(z)}$, decays exponentially according to the equation

$$I_{(z)} = I_0 e^{-\frac{z}{dp}}$$

$$d_p = \frac{\lambda_0}{4\pi\left(n_1^2 \sin^2\theta - n_2^2\right)^{\frac{1}{2}}}$$

where d_p, the penetration depth, is a function of the incident wavelength λ_0, the refractive indices of the two media n_1(liquid) and n_2 (solid), and the angle of incidence θ [13].

There are two different ways of doing TIRFM experiments. One is through the objective, and the other is using a prism in which the laser beam is incident at an angle outside the objective. When doing objective-based TIRFM, high NA objectives are needed; otherwise, the critical angle is not achieved [13].

TIRFM is especially suited to image the plasma membrane and membrane proximal vesicles with high signal-to-noise ratio, making it an ideal tool to study endocytosis. One can study the kinetics of redistribution of FFP1 in clathrin-coated pits by dual-color TIRFM of red fluorescent protein (RFP)-tagged FFP1 and green fluorescent protein (GFP)-tagged clathrin. This experimental design has been used to study endocytosis of integrins at focal adhesions in migrating cells [14] or to look at how clathrin-coated pit dynamics are affected by the cargo [15]. One could extend this study and image other molecules of the endocytosis machinery such as dynamin [16] (see [17] for a review).

Imaging vesicles as they bud from the plasma membrane can be problematic as the vesicles move away from the membrane; hence, switching between TIRFM and epifluorescence settings can be a solution to image deeper in the sample [18]. The drawback, however, is that the high signal-to-noise ratio achieved through TIRFM is lost with epi-illumination. Recently, a technique based on the optics used to achieve TIRFM has been adapted to image deeper in the sample with a better signal-to-noise ratio than epifluorescence. In the technique known as highly inclined and laminated optical sheets (HILO) microscopy (not to be mistaken with HiLo which is another technique based on structured illumination), the laser beam is highly inclined by high refraction and collimated as an optical sheet through the sample side (Fig. 1b). This setting is similar to that observed just before achieving TIRF illumination when the beam has not undergone total internal reflection but is grazing the sample at very low angles. The signal-to-noise ratio is improved by up to 3.5-fold as the out-of-focus light is lowered by reduction of the illuminated section [19]. Switching between TIRF and HILO modes could be a useful implementation to follow endosome trafficking after internalization to recycling or degradation compartments.

TIRFM is also well suited to study diffusion and dynamics of proteins in the membrane with high resolution. If FFP1 undergoes clustering upon ligand engagement, it can ideally be observed by TIRFM as has been observed for T cell receptor (TCR) microclusters and B cell receptor (BCR) microclusters [20, 21]. The high signal-to-noise ratio achieved in TIRFM has also made it an ideal choice for single-molecule tracking experiments. It has been used extensively to study the dynamics of single motor proteins [22] and to measure diffusion coefficients of lipids and proteins [23, 24] (for a complete review, see [25] and papers cited in it). TIRFM is also a very suitable technique to study signaling in response to ligand engagement. Ligand-engaged receptors undergo phosphorylation by membrane-associated kinases and often recruit other cytoplasmic and/or membrane-associated signaling adaptors or kinases. The kinetics of such signaling processes has been extensively studied using TIRFM, as the imaging geometry is ideally suited for such applications. Douglass and Vale studied the interaction of Lck and CD2 with TCR microclusters using TIRFM [26]. Several groups have studied signaling in T cells, B cells, mast cells, and other cell types (for a complete review, see [27] and papers cited in it).

In the above section, we discussed how TIRFM can be used to track single particles and measure their diffusion coefficients. Another powerful technique that can measure diffusion of molecules and interaction between different molecules is fluorescence correlation spectroscopy (FCS) [28]. FCS measurements can be performed in solution, in the cytoplasm, or in the membranes of living cells [29, 30]. FCS is performed by recording fluorescence from a small confocal volume at time resolutions faster than the diffusion of molecules. The density of fluorophores needs to be adjusted so that at any given time, only a few fluorophores are in the confocal volume. The time series from such an observation gives rise to fluctuating intensities as only a few molecules are diffusing in and out of the confocal volume. These fluctuations in intensities would be statistically correlated over time scales (τ) that correspond to diffusion of the fluorophore through the confocal volume. A statistical analysis of these fluctuations yields a correlation function as a function ofτ. Fitting the correlation curve to a model of diffusion in either solution or in the membrane yields diffusion coefficients [31]. The technique has been extended to observing the fluctuations of two fluorophores within the same confocal volume, known as fluorescence cross-correlation spectroscopy (FCCS) [32]. FCCS can be used to study the interaction between two molecules. When molecules are interacting, they will co-diffuse through the confocal volume, and the diffusion coefficient of the interacting pair can be obtained. The cross-correlation amplitude can yield the fraction of interacting molecules, and from the concentrations of the individual molecules, the dissociation constant can be determined [33].

Kolin and colleagues developed another technique that falls in the category of correlation spectroscopy called k-space image correlation spectroscopy (KICS) [34]. The advantage of this technique over FCS is that one does not have to measure fluctuations in fluorescence from a single spot. Instead, a time series of images are analyzed by converting them into Fourier space (k-space). The Fourier transform of the images contains information on spatial variation of densities of fluorophores in the membrane. The analysis extracts number densities, diffusion coefficients, and velocities of molecules moving in the membrane. This method is less sensitive to the density of fluorophores in the membrane than FCS.

4 Investigating the Distribution of FFP1 in Cells Using Superresolution Techniques

4.1 Localizing FFP1 in the Cellular Architecture by Structured Illumination Microscopy (SIM)

Most techniques discussed so far have imaged the distribution of FFP1 at the diffraction-limited resolution of light microscopy. The field of fluorescence microscopy has undergone a revolution with the development of many techniques that circumvent the diffraction barrier, offering higher resolution. One of those techniques is structured illumination microscopy (SIM). SIM can improve the resolution of light microscopy by at least a factor of 2.

The principle relies on illumination of the sample with excitation light that is structured in space (e.g., a spatially varying sinusoidal pattern). When two such patterns are overlaid, what one observes is an interference pattern known as "moire fringes." If the second pattern contains spatial frequencies (fluorophore distribution) below the resolution of light, an interference pattern is still observed. When imaging a biological sample, the fluorophore distribution is the unknown pattern, which is excited with a known pattern of light. In practice, a series of phase-shifted images with the same patterned illumination is acquired and processed to extract information that will allow reconstruction of higher resolution images [35]. Phase shifts are implemented by rotating the patterned illumination with respect to the sample plane.

The main advantage of SIM is that the microscope setup is simple and can be adapted on a wide-field microscope. Since the technology does not involve a pinhole, light detection is efficient and sensitive.

Gustafsson and colleagues have implemented a 3D version of SIM by using three mutually coherent beams to generate the interference images. In this manner, they could achieve resolution doubling in all three dimensions [36]. This technique has been used to observe endosomal sorting complex required for transport (ESCRT) protein complexes during cytokinesis [37, 38].

The limitation of SIM and its 3D version is the time needed for acquisition since multiple frames need to be captured for each z plane. Thus, dynamic process can suffer from reconstruction artifacts if the events recorded are moving faster than the time needed to acquire all the phase-shifted images. Very thick and highly scattering samples will also suffer from out-of-focus fluorescence interfering with the pattern and the reconstruction.

4.2 Distribution of FFP1 in Filopodia Investigated with Stimulated Emission Depletion (STED) Microscopy

Based on confocal microscopy analysis that showed the localization of FFP1 toward the leading edge of migrating cells, its role in cell migration is evident. The leading edge of the migrating cell is a complex structure consisting of the lamellipodium but can also have filopodia emerging from it to make new adhesion contacts or contain interleaved filopodia [39]. To gain further insight into the role of FFP1 in cell migration, a higher resolution distribution of the protein at the leading edge is required.

Several techniques now circumvent the diffraction-limited resolution of light microscopy (200–300 nm) and allow resolution below 100 nm. Stimulated emission depletion (STED) microscopy improves resolution by shaping the point-spread function (PSF) using physical characteristics of the fluorescent probe [40].

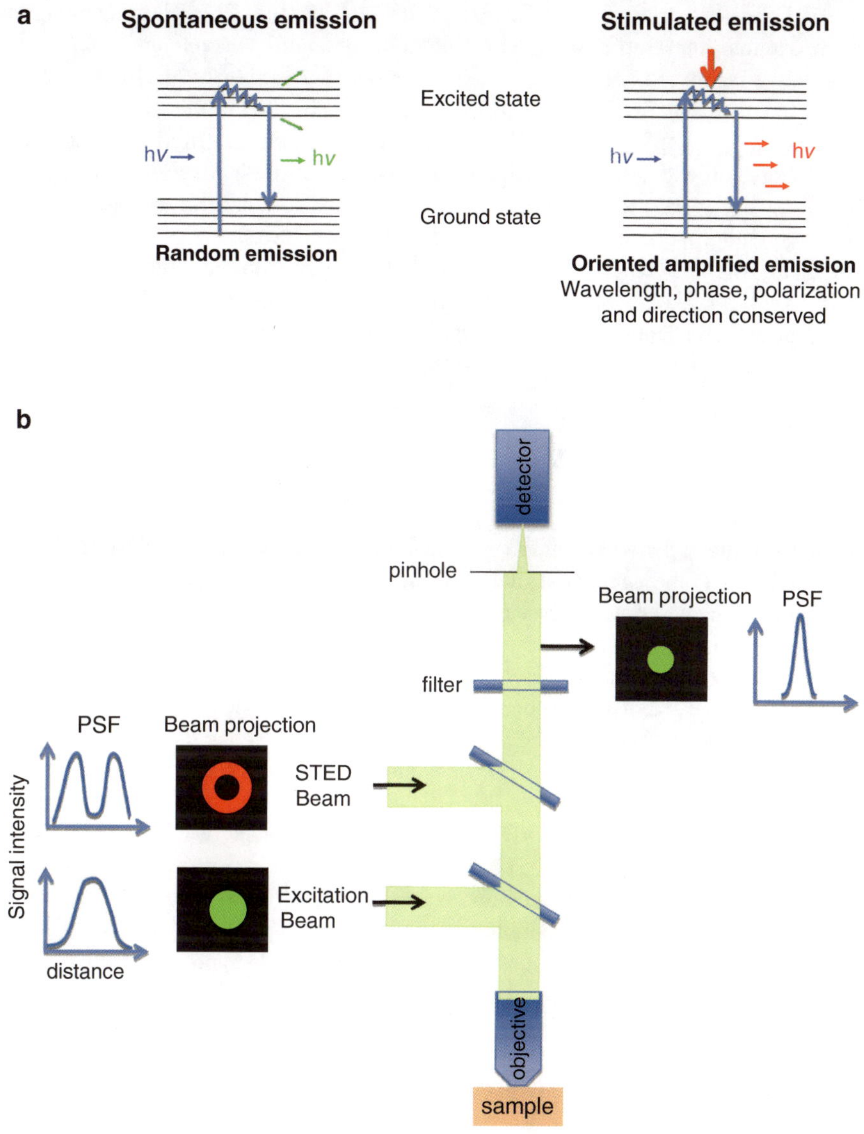

Fig. 2 (**a**) Stimulated Emission principle. Photons are mitted with random phase and direction when a fluorescent molecule transitions from an excited state to a ground state. When a fluorophore in its excited state is stimulated with a red-shifted laser, it can be forced to go to the ground state and emit a photon of the same phase, wavelength, direction, and polarization as the stimulating beam. (**b**) Stimulated Emission Depletion Microscopy. STED uses two laser beams to excite the sample. The first beam excites the sample (green) with its rather widespread point-spread function (PSF). A second doughnut-shaped red-shifted laser beam excites the sample before fluorescence emission can occur from the first laser excitation. Fluorescence quenching occurs at the periphery of the illumination spot, which shapes the PSF and lead to enhanced resolution

STED uses two laser beams to excite the sample (Fig. 2b). The first focused beam excites the sample at the corresponding excitation wavelength. A second doughnut-shaped red-shifted laser beam excites the sample before fluorescence emission can occur from the first laser excitation. This "STED beam," exploiting the phenomenon of stimulated emission, depletes the excited state of the fluorophores in the doughnut-shaped area. During stimulated emission, excited molecules are forced to go back to their ground state by emitting photons of the same wavelength, phase, polarization, and direction as the excitation laser (Fig. 2a). Thus, only the molecules in the doughnut hole will emit fluorescence that will be recorded [41, 42]. Fluorescence quenching occurs at the periphery of the illumination spot, which shapes the PSF and leads to enhanced resolution.

The resolution (R) in STED microscopy is defined as

$$R_{STED} = \frac{\lambda}{2NA} \times \frac{1}{\sqrt{\frac{I}{I_s} + 1}}$$

where λ is the light wavelength, NA the numerical aperture, I the STED beam intensity, and I_s the saturation intensity. Increased resolution has been achieved using STED microscopy by increasing the intensity of the STED beam [43]. Lipid nanodomains in a living cell membrane were observed at resolution of 30 nm [44].

STED microscopy could be useful to determine the number of FFP1 molecules in the clusters observed at the tip of filopodia, as it has been determined for syntaxin clusters [45]. All members of the FFP family could be resolved at the tips of filopodia as was shown for human voltage-dependent anion channel (hVDAC) in the membrane of mitochondria. Using two-color STED, the authors could resolve the distribution of the three different isoforms of hVDAC in the outer membrane of mitochondria [46]. And more recently, the power of STED microscopy was demonstrated by studying the dynamics of dendrites remodeling in live cells [47, 48] and also to image actin dynamics inside those structures [49].

One of the main disadvantages of STED is that multi-fluorophore imaging is difficult since, for each fluorophore, both an excitation channel and a STED channel have to be available, and implementing such a setup is a significant challenge. The acquisition times for STED microscopy are generally slow; however, high-speed imaging for small fields of view has been achieved [50]. Theoretical calculations show that the resolution in STED, microscopy depends on the fluorophore density, fluorophore lifetime, and also the diffusion of fluorophores in the sample [51].

4.3 *Dissection of Protein Complexes by Pointillism Microscopy (STORM, PALM, iPALM)*

Another class of microscopy techniques that circumvents the diffraction-limited resolution of light microscopy is known as pointillist techniques [40] because they are based on accurately determining the coordinates of single fluorophores. These pointillist superresolution techniques are photoactivated localization microscopy (PALM) and stochastic optical reconstruction microscopy (STORM). They have been implemented in three dimensions as well and offer superior resolution improvement.

In contrast to STED microscopy, PALM and STORM techniques are more flexible for multiple protein labeling and are thus more suitable for characterization of multimolecular complexes.

PALM and STORM techniques take advantage of the fact that the center of the PSF of a single molecule can be determined with high precision of 20 nms [52]. This methodology is already used in single-particle tracking experiments. If the centers of all the molecules in the field of view can be determined, then the image can be reconstructed with this information.

To allow single-molecule visualization, PALM uses photoswitchable or photoactivable fluorescent proteins [53], and STORM exploits the photoswitching capacities of fluorescent dyes that are enhanced in the presence of buffers containing oxygen scavenging system and reducing agents [54, 55] to control in both cases the density of fluorescent molecules that is recorded at a given time [56] (for a review, see [10]).

Photoactivation switches the fluorescent protein between a dark and fluorescent state, while photoswitching changes the spectral characteristics of the fluorescent protein, for example, by converting it into a red fluorophore from a green one by the process of photoconversion. Some photoactivable or photoswitchable proteins are also reversible. Several variants of photoswitchable proteins with better quantum efficiency have been developed [57]. These have been excellently reviewed in Reference [58]. The challenges in developing these probes are to achieve a good separation between the excitation and photoconverting wavelengths and also to achieve better separation between the photoswitched wavelengths to avoid making Forster resonance energy transfer (FRET) pairs [59].

Imaging is based on repeated cycles of photoactivation or photoswitching, excitation, and photobleaching. In a single cycle, a small percentage of fluorophores are either photoactivated or photoswitched, and multiple images of these activated fluorophores are collected until all of them are photobleached. From these set of images, sub-diffraction localization for the activated fluorophores that are separated enough from each other for the algorithm to distinguish them is calculated [60]. As molecules are activated randomly, unlike in scanning-based techniques, several cycles are necessary to reconstruct the whole field of view, leading to slow acquisition speeds; hence, most applications of PALM and STORM have been limited to fixed samples. Future developments of algorithms that allow determination of centers of point-spread functions from fewer images would speed up the technique permitting imaging of live samples [61].

Pointillist techniques could be really valuable to analyze the distribution of FFP1 during its intracellular trafficking. Wu and colleagues have analyzed membrane traffic reactions on membrane sheets using STORM analysis. They could describe the coating of tubular structures emerging from clathrin-coated buds by a Bin

Amphiphysin Rvs (BAR)-domain-containing protein, Fes/CIP4 homology Bar (F-BAR) 17. BAR-domain-containing proteins have the capacity to bind phospholipids and bend membranes and thus are responsible for membrane curvature [62]. In the field of TCR signaling, several studies using PALM have described the nanoscale organization of the membrane adapter linker for activated T cells (LAT) in the plasma membrane and in endocytic vesicles. Lillemeier and colleagues first demonstrated that LAT and TCR were pre-clustered in the plasma membrane and that upon signaling, these pre-clustered islands concatenate [63]. Williamson et al. used novel data analysis and proposed that signaling primarily occurred in vesicles containing LAT whereas Sherman et al. showed that signaling occurred in LAT-containing membrane nanoclusters [64, 65]. Multicolor STORM provided insights on the molecular architecture of different types of synapses in the brain [66].

Several advances have been made to improve the resolution in the axial direction with 3D implementation of pointillist techniques. In a variation called biplane FPALM, the light detection path was modified to allow simultaneous detection of two axially different planes without axial scanning. A 50/50 beam splitter split the emission such that the two images were recombined onto different regions of the same camera. The transmitted beam had a slightly shorter path than the reflected beam thus allowing imaging from two different axial planes. The signals recorded from those two planes were sufficient to obtain the 3D PSF. The x, y, and z coordinates of the molecules in the sample were then deduced from fitting, and an xyz resolution of 30 nm × 30 nm × 75 nm was obtained [67]. In another study, the astigmatism imaging method was adapted to STORM. A cylindrical lens was used in the light path that caused the PSF shape to become elliptical in the x-y direction as the z position changed. A calibration curve relating the ellipticity of the PSF to the z position allowed determination of the precise z position for all fluorescent molecules identified by STORM analysis. High-resolution 3D images of clathrin-coated pit structures were acquired by this 3D-STORM implementation [68].

4.4 *Protein-Protein Interactions Studied with FRET-Based Techniques*

4.4.1 Homo-FRET Measurement to Investigate FFP1 Oligomerization

Before the development of superresolution techniques, FRET-based techniques were used to dissect molecular interactions that could not be resolved by light microscopy due to the diffraction limit. FRET is a phenomenon in which energy is transferred nonradiatively from an excited donor fluorophore to a nearby acceptor fluorophore. The energy transfer is a function of the distance between the two fluorophores, the orientation of their dipole moments, and the spectral overlap between the donor emission and acceptor absorption spectra. As the distance dependency is in the scale of 1 to 10 nm (100 Å), this technique has been widely used to examine protein-protein interactions that could not have been resolved by the diffraction-limited resolution of light microscopy [69].

The energy transfer efficiency (*E*) varies inversely with the sixth power of the distance between the donor and the acceptor according to

$$E = \frac{1}{1 + (r/R_0)^6}$$

where r is the distance between the donor and the acceptor and R_o is the "Förster radius."

Homo-FRET refers to FRET between identical fluorophores and occurs because there is a spectral overlap between excitation and emission spectrum of most fluorophores. FRET also depends on the relative orientation between the two fluorophores. This dependence causes the light emitted from the acceptor molecule to be depolarized. As a result, homo-FRET measurements can be made by determining the loss in anisotropy of fluorescence emission after a FRET event. Fluorophores are excited by plane-polarized light, and fluorescence emission is recorded using polarizers in the two orthogonal directions (parallel and perpendicular with respect to excitation light plane of polarization). Fluorescence emission anisotropy (r) is calculated according to

$$r = \frac{I_{||} - I_{\perp}}{I_{||} + 2I_{\perp}}$$

where $I_{//}$ is the fluorescence emission intensity recorded in the parallel direction and $I_{\perp}$ is the fluorescence emission intensity recorded in the perpendicular direction [70]. Homo-FRET efficiency will be calculated as

$$E = 1 - \frac{r_{fret}}{r_{basal}}$$

The anisotropy of fluorescence also depends on the rate of rotational diffusion, which is related to the viscosity of the medium. For homo-FRET measurements, the rate of rotational diffusion has to be slower than the lifetime of the fluorophore; otherwise, the fluorescence emission will be depolarized. If fluorophores are highly oriented, one may not observe depolarization by FRET; however, this is not a concern for proteins in biological membranes that do undergo rotational diffusion. Time-resolved anisotropy measurements are particularly powerful as they provide information about both rotational diffusion and homo-FRET as the two phenomena occur at different time scales [71].

One can use homo-FRET measurements to determine the oligomeric state of FFP1 as was described for glycosylphosphatidylinositol (GPI)-anchored proteins [72].

One of the features of anisotropy measurements is that high NA objectives caused depolarization of fluorescence and prevented users from obtaining high-resolution anisotropy images. Vogel and colleagues elegantly demonstrated that the depolarization by high NA objectives was such that relative changes between samples were still maintained [71]. Anisotropy imaging using TIRF illumination, two photon excitation, and high-resolution PALM imaging (see later) is also possible allowing versatile applications [70].

4.4.2 FLIM to Investigate Protein-Protein Interactions Among FFPs

FRET measurements between different fluorophores are useful approach to study protein-protein interactions, and fluorescence lifetime imaging microscopy (FLIM) is a powerful technique to measure hetero-FRET.

The lifetime of a fluorophore is the amount of time the fluorophore spends in the excited state. When an ensemble of fluorophores is excited by a pulse of light, the decay from this excited state to the ground state is exponential and defined by the equation

$$I_{(t)} = I_0 e^{-\frac{t}{\tau}}$$

where τ is the fluorescence lifetime defined by

$$\tau = \frac{1}{(k_r + k_{nr})}$$

where k_r is the radiative rate constant and k_{nr} is the nonradiative rate constant.

The intrinsic lifetime of a fluorophore is influenced by many factors in its environment such as pH, temperature, oxygenation, and FRET. When a donor fluorophore transfers its energy to an acceptor fluorophore, its fluorescence is quenched leading to a reduction in its lifetime.

There are in general two methods used to measure fluorophore lifetimes. One is in the frequency domain, and the other is in time domain. In the frequency domain, the excitation source is modulated to have a sinusoidal temporally varying intensity. This results in a phase-shifted sinusoidal fluorescence emission. The phase shift is related to the lifetime of the fluorophore [73]. Frequency domain methods may have some limitations, and a comparison between time domain and frequency domain methods is discussed in [74].

Perhaps the most popular time-domain method to measure fluorescence lifetime is time-correlated single photon counting (TCSPC). In this technique, the sample is excited using a pulsed laser, and the fluorescence emission is measured using a detector capable of detecting single photons. The arrival time of each detected photon relative to the excitation pulse is recorded. A histogram plot of these arrival times gives the fluorescence delay [75].

These days, FLIM is typically implemented on 2-photon microscopes allowing lifetime measurements on a pixel by pixel basis as described [76]. One of the drawbacks of lifetime measurements is that acquisition times can be slow. When using donor lifetime measurements to quantify FRET, the major problem of bleed-through of donor emission into the acceptor "FRET detection" channel is circumvented. In some cases, fluorescence lifetime is independent of probe concentration and excitation intensity, which makes it more reliable to compare measurements [77]. Increasing fluorophore density leads to a concentration-dependent homo-FRET. This can have an impact on fluorophore lifetime. As discussed above, FRET leads to a decrease in donor lifetime, but consequently, it also leads to an increase in acceptor lifetime. For some fluorophores, these changes cancel each other out, and one does not observe a net change in fluorophore lifetime with increasing concentration (measured for the fluorescent protein Venus, personal communication Steven Vogel (2012) and [78]), even though homo-FRET occurs. But for many fluorophores, this may not be the case. A shortening of fluorophore lifetime has been observed as a function of concentration due to homo-FRET [79, 80].

FRET-FLIM can be used to investigate the interaction between FFP1 and another FFP family member. Oligomerization of proteins has been assessed by FLIM. Storez et al. demonstrated the homo-oligomerization of beta-arrestins [81]. The interaction of the different Rat sarcoma (Ras) protein isoforms (H-Ras, N-Ras, and K-Ras) with c-Raf has been studied by FLIM. The authors were able to show that the domain in Ras recognizing c-Raf had different orientations between the different isoforms of Ras [82]. FLIM was also used to study signaling in NK cells by looking at FRET between a GFP-tagged NK inhibitory receptor (KIR2DL1) and a Cy3-labeled anti-phosphotyrosine antibody at the contact zone between an NK and a target cell [83].

5 Superresolution Microscopy in Three Dimensions

5.1 Observation of Filopodial Dynamics Using Light-Sheet Fluorescence Microscopy and Bessel Beam Illumination

As previously mentioned, superresolution techniques come at the cost of speed of acquisition with the exception of STED imaging that can be fast but only for a small field of view.

A recent development in the field of light microscopy is light-sheet fluorescence microscopy (LSFM). This technique has offered significant advances in speed for imaging in three dimensions, thus making observation of highly dynamic processes feasible. This technique is based on illuminating an entire plane of a sample using a sheet of light generated using optics positioned perpendicular to the objective. Detection of fluorescence occurs in a wide-field configuration [84] (Fig. 3a).

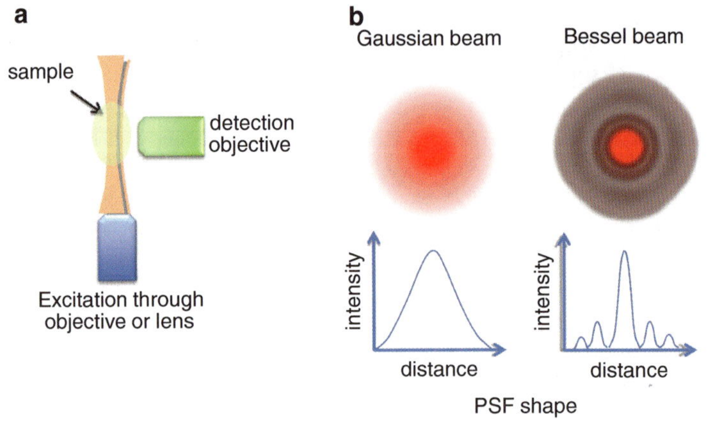

Fig. 3 (**a**) Light-Sheet Microscopy. Excitation is provided by an objective with low numerical aperture (NA) or a lens to create a sheet of light illuminating one focal plane of the sample. Emitted light is collected by a second objective with high NA positioned perpendicular to the excitation one. (**b**) Difference between Gaussian and Bessel beams. Gaussian beams are inherently diffractive and in LSFM settings they will exhibit the characteristic widening on the edges. The beam waist will determine the depth of field. Bessel beams on the contrary, are nondiffractive and hence the thickness of the central beam does not change as it is focused through the objective to make a light sheet. A pure Bessel beam does not exist but some techniques allow to create approximations of Bessel beams whose diffraction is limited enough to create large depth of field for the objective creating the light sheet and small depth of field for the objective imaging the sample

This technique has proven to be really useful for in vivo imaging, particularly to look at thick samples or imaging the development of whole embryos such as *C. elegans* [85].

The main problem of LSFM is that the axial resolution is even lower than in confocal microscopy and lateral resolution is the same. A recent implementation has allowed the achievement of better z resolution by creating the light sheet with a Bessel beam (Fig. 3b) [86]. In a study comparing light-sheet microscopy with Gaussian and Bessel beams, the authors show improvements in penetration, directional propagation stability, and depth of field. These physical characteristics of Bessel beams improved the signal-to-noise ratio and resolution of images [87]. Bessel beam illumination was combined with structured illumination or 2-photon excitation to achieve axial resolution of .27 and .5 um, respectively. This is a cutting-edge and rapidly evolving area of fluorescence microscopy and in coming years will revolutionize how three-dimensional imaging is done.

One can take advantage of the high-speed image acquisition of light-sheet microscopy and its Bessel beam implementation to observe the highly dynamic remodeling of filopodia as observed on the surface of Hela cells [86]. LSFM has proven to be suitable also to study larger scale cellular structures such as neuronal networks in the mouse olfactory system [88]. Thus, the lower scattering effect of

Bessel beams could also be used to image lymphocyte migration inside the lymph nodes upon FFP1 stimulation using an FFP-GFP transgenic mouse.

5.2 Imaging of Three-Dimensional Molecular Organization In Situ Using Interferometric PALM

The highest three-dimensional resolution achieved so far used interferometric PALM (iPALM). In this technique, the lateral resolution is similar to 2D PALM, but the axial resolution is improved by using a dual objective setup as in the 4Pi microscope configuration. Two objectives are positioned on opposite sides of the sample, and photons emitted by the sample are propagated through the top and bottom objectives simultaneously and recombined by a beam splitter. The principle of iPALM is that the path length of the photon collected by the top objective will be different from the path length of the photon collected by the bottom objective for a molecule located at a specific z location (unless it is absolutely in the center of the two objectives). The light-path differences for each photon emitted and their phase changes are recorded by CCD cameras and analyzed to calculate the z position of the molecule. The authors could resolve axial coordinates at 10-nm resolution [89]. Molecular organization of focal adhesion complexes was imaged using this technique and resolved at 10-nm resolution [90]. The instrumentation required for PALM- and STORM-based techniques is of high complexity, and some commercial solutions are available; nevertheless, these techniques are the future of microscopy as they provide significantly higher resolution.

6 Conclusion

We have made an attempt to take the reader on a journey of exploration of different optical techniques that are available for researchers to study the cell biological properties of membrane receptors. Earlier, all techniques operated within the boundaries of the diffraction-limited resolution of light microscopy. Molecular interactions had to be studied using spectroscopic techniques such as FRET and FCS. With the advent of superresolution microscopy, researchers now have more options. Biological systems are incredibly complex, and so a single technique will always be insufficient to address all questions related to the system or a molecule in particular. The choice of microscopy technique to image a biological system at a desired spatial and temporal resolution is dictated by fluorophore density and photon flux. Here, we propose that using a combination of different techniques always gives more insight.

Acknowledgments We would like to thank Steve Vogel for valuable discussions and Travis Crites for critical reading of the manuscript.

References

1. Conchello JA, Lichtman JW (2005) Optical sectioning microscopy. Nat Methods 2 (12):920–931. doi:10.1038/nmeth815
2. Sanderson MJ, Parker I (2003) Video-rate confocal microscopy. Methods Enzymol 360:447–481. doi:10.1016/S0076-6879(03)60123-0
3. Botcherby EJ, Booth MJ, Juskaitis R, Wilson T (2009) Real-time slit scanning microscopy in the meridional plane. Opt Lett 34(10):1504–1506. doi:10.1364/OL.34.001504
4. Toomre D, Pawley JB (2006) Disk-scanning confocal microscopy, vol chapter 10. In: Handbook of biological confocal microscopy. Springer, US. doi:10.1007/978-0-387-45524-2_10
5. Wilson T (2010) Spinning-disk microscopy systems. Cold Spring Harb Protoc 2010 (11):pdb top88. doi:10.1101/pdb.top88
6. Ichihara A, Tanaami T, Isozaki K, Sugiyama Y, Kosugi Y, Mikuriya K, Abe M, Uemura I (1996) High-speed confocal fluorescence microscopy using a nipkow scanner with microlenses - for 3-d imaging of fluorescent molecule in real-time. Bioimages 4:57–62
7. Stehbens S, Pemble H, Murrow L, Wittmann T (2012) Imaging intracellular protein dynamics by spinning disk confocal microscopy. Methods Enzymol 504:293–313. doi:10.1016/B978-0-12-391857-4.00015-X
8. Nakano A (2002) Spinning-disk confocal microscopy – a cutting-edge tool for imaging of membrane traffic. Cell Struct Funct 27(5):349–355. doi:10.1247/csf.27.349
9. Wang E, Babbey CM, Dunn KW (2005) Performance comparison between the high-speed yokogawa spinning disc confocal system and single-point scanning confocal systems. J Microsc 218(Pt 2):148–159. doi:10.1111/j.1365-2818.2005.01473.x
10. Fischer RS, Wu Y, Kanchanawong P, Shroff H, Waterman CM (2011) Microscopy in 3d: a biologist's toolbox. Trends Cell Biol 21(12):682–691. doi:10.1016/j.tcb.2011.09.008
11. Carrizosa E, Gomez TS, Labno CM, Klos Dehring DA, Liu X, Freedman BD, Billadeau DD, Burkhardt JK (2009) Hematopoietic lineage cell-specific protein 1 is recruited to the immunological synapse by il-2-inducible t cell kinase and regulates phospholipase cgamma1 microcluster dynamics during t cell spreading. J Immunol 183(11):7352–7361. doi:10.4049/jimmunol.0900973
12. Mattheyses AL, Simon SM, Rappoport JZ (2010) Imaging with total internal reflection fluorescence microscopy for the cell biologist. J Cell Sci 123(Pt 21):3621–3628. doi:10.1242/jcs.056218
13. Axelrod D (2008) Chapter 7: Total internal reflection fluorescence microscopy. Methods Cell Biol 89:169–221. doi:10.1016/S0091-679X(08)00607-9
14. Ezratty EJ, Bertaux C, Marcantonio EE, Gundersen GG (2009) Clathrin mediates integrin endocytosis for focal adhesion disassembly in migrating cells. J Cell Biol 187(5):733–747. doi:10.1083/jcb.200904054
15. Puthenveedu MA, von Zastrow M (2006) Cargo regulates clathrin-coated pit dynamics. Cell 127(1):113–124. doi:10.1016/j.cell.2006.08.035
16. Merrifield CJ, Feldman ME, Wan L, Almers W (2002) Imaging actin and dynamin recruitment during invagination of single clathrin-coated pits. Nat Cell Biol 4(9):691–698. doi:10.1038/ncb837
17. Rappoport JZ, Simon SM, Benmerah A (2004) Understanding living clathrin-coated pits. Traffic 5(5):327–337. doi:10.1111/j.1600-0854.2004.00187.x
18. Leonard D, Hayakawa A, Lawe D, Lambright D, Bellve KD, Standley C, Lifshitz LM, Fogarty KE, Corvera S (2008) Sorting of egf and transferrin at the plasma membrane and by cargo-specific signaling to eea1-enriched endosomes. J Cell Sci 121(Pt 20):3445–3458. doi:10.1242/jcs.031484
19. Tokunaga M, Imamoto N, Sakata-Sogawa K (2008) Highly inclined thin illumination enables clear single-molecule imaging in cells. Nat Methods 5(2):159–161. doi:10.1038/nmeth1171

20. Campi G, Varma R, Dustin ML (2005) Actin and agonist mhc-peptide complex-dependent t cell receptor microclusters as scaffolds for signaling. J Exp Med 202(8):1031–1036. doi:10.1084/jem.20051182
21. Carrasco YR, Fleire SJ, Cameron T, Dustin ML, Batista FD (2004) Lfa-1/icam-1 interaction lowers the threshold of b cell activation by facilitating b cell adhesion and synapse formation. Immunity 20(5):589–599. doi:10.1016/S1074-7613(04)00105-0
22. Reck-Peterson SL, Yildiz A, Carter AP, Gennerich A, Zhang N, Vale RD (2006) Single-molecule analysis of dynein processivity and stepping behavior. Cell 126(2):335–348. doi:10.1016/j.cell.2006.05.046
23. Murase K, Fujiwara T, Umemura Y, Suzuki K, Iino R, Yamashita H, Saito M, Murakoshi H, Ritchie K, Kusumi A (2004) Ultrafine membrane compartments for molecular diffusion as revealed by single molecule techniques. Biophys J 86(6):4075–4093. doi:10.1529/biophysj.103.035717
24. Umemura YM, Vrljic M, Nishimura SY, Fujiwara TK, Suzuki KG, Kusumi A (2008) Both mhc class II and its gpi-anchored form undergo hop diffusion as observed by single-molecule tracking. Biophys J 95(1):435–450. doi:10.1529/biophysj.107.123018
25. Reck-Peterson SL, Derr ND, Stuurman N (2010) Imaging single molecules using total internal reflection fluorescence microscopy (tirfm). Cold Spring Harb Protoc 2010 (3):pdb top73. doi:10.1101/pdb.top73
26. Douglass AD, Vale RD (2005) Single-molecule microscopy reveals plasma membrane microdomains created by protein-protein networks that exclude or trap signaling molecules in t cells. Cell 121(6):937–950. doi:10.1016/j.cell.2005.04.009
27. Dustin ML, Depoil D (2011) New insights into the t cell synapse from single molecule techniques. Nat Rev Immunol 11(10):672–684. doi:10.1038/nri3066
28. Maiti S, Haupts U, Webb WW (1997) Fluorescence correlation spectroscopy: diagnostics for sparse molecules. Proc Natl Acad Sci USA 94(22):11753–11757. doi:10.1073/pnas.94.22.11753
29. Chiantia S, Ries J, Schwille P (2009) Fluorescence correlation spectroscopy in membrane structure elucidation. Biochim Biophys Acta 1788(1):225–233. doi:10.1016/j.bbamem.2008.08.013
30. Garcia-Saez AJ, Schwille P (2007) Single molecule techniques for the study of membrane proteins. Appl Microbiol Biotechnol 76(2):257–266. doi:10.1007/s00253-007-1007-8
31. Sengupta P, Balaji J, Maiti S (2002) Measuring diffusion in cell membranes by fluorescence correlation spectroscopy. Methods 27(4):374–387. doi:10.1016/S1046-2023(02)00096-8
32. Bacia K, Schwille P (2007) Practical guidelines for dual-color fluorescence cross-correlation spectroscopy. Nat Protoc 2(11):2842–2856. doi:10.1038/nprot.2007.410
33. Rarbach M, Kettling U, Koltermann A, Eigen M (2001) Dual-color fluorescence cross-correlation spectroscopy for monitoring the kinetics of enzyme-catalyzed reactions. Methods 24(2):104–116. doi:10.1006/meth.2001.1172
34. Kolin DL, Ronis D, Wiseman PW (2006) K-space image correlation spectroscopy: a method for accurate transport measurements independent of fluorophore photophysics. Biophys J 91(8):3061–3075. doi:10.1529/biophysj.106.082768
35. Gustafsson MG (2000) Surpassing the lateral resolution limit by a factor of two using structured illumination microscopy. J Microsc 198(Pt 2):82–87. doi:10.1046/j.1365-2818.2000.00710.x
36. Gustafsson MG, Shao L, Carlton PM, Wang CJ, Golubovskaya IN, Cande WZ, Agard DA, Sedat JW (2008) Three-dimensional resolution doubling in wide-field fluorescence microscopy by structured illumination. Biophys J 94(12):4957–4970. doi:10.1529/biophysj.107.120345
37. Elia N, Sougrat R, Spurlin TA, Hurley JH, Lippincott-Schwartz J (2011) Dynamics of endosomal sorting complex required for transport (escrt) machinery during cytokinesis and its role in abscission. Proc Natl Acad Sci USA 108(12):4846–4851. doi:10.1073/pnas.1102714108

38. Guizetti J, Schermelleh L, Mantler J, Maar S, Poser I, Leonhardt H, Muller-Reichert T, Gerlich DW (2011) Cortical constriction during abscission involves helices of escrt-III-dependent filaments. Science 331(6024):1616–1620. doi:10.1126/science.1201847
39. Mattila PK, Lappalainen P (2008) Filopodia: molecular architecture and cellular functions. Nat Rev Mol Cell Biol 9(6):446–454. doi:10.1038/nrm2406
40. Toomre D, Bewersdorf J (2010) A new wave of cellular imaging. Annu Rev Cell Dev Biol 26:285–314. doi:10.1146/annurev-cellbio-100109-104048
41. Hell SW, Wichmann J (1994) Breaking the diffraction resolution limit by stimulated emission: stimulated-emission-depletion fluorescence microscopy. Opt Lett 19(11):780–782. doi:10.1364/OL.19.000780
42. Klar TA, Jakobs S, Dyba M, Egner A, Hell SW (2000) Fluorescence microscopy with diffraction resolution barrier broken by stimulated emission. Proc Natl Acad Sci USA 97 (15):8206–8210. doi:10.1073/pnas.97.15.8206
43. Hell SW (2009) Microscopy and its focal switch. Nat Methods 6(1):24–32. doi:10.1038/nmeth.1291
44. Eggeling C, Ringemann C, Medda R, Schwarzmann G, Sandhoff K, Polyakova S, Belov VN, Hein B, von Middendorff C, Schonle A, Hell SW (2009) Direct observation of the nanoscale dynamics of membrane lipids in a living cell. Nature 457(7233):1159–1162. doi:10.1038/nature07596
45. Sieber JJ, Willig KI, Kutzner C, Gerding-Reimers C, Harke B, Donnert G, Rammner B, Eggeling C, Hell SW, Grubmuller H, Lang T (2007) Anatomy and dynamics of a supramolecular membrane protein cluster. Science 317(5841):1072–1076. doi:10.1126/science.1141727
46. Neumann D, Buckers J, Kastrup L, Hell SW, Jakobs S (2010) Two-color sted microscopy reveals different degrees of colocalization between hexokinase-i and the three human vdac isoforms. PMC Biophys 3(1):4. doi:10.1186/1757-5036-3-4
47. Berning S, Willig KI, Steffens H, Dibaj P, Hell SW (2012) Nanoscopy in a living mouse brain. Science 335(6068):551. doi:10.1126/science.1215369
48. Tonnesen J, Nadrigny F, Willig KI, Wedlich-Soldner R, Nagerl UV (2011) Two-color sted microscopy of living synapses using a single laser-beam pair. Biophys J 101(10):2545–2552. doi:10.1016/j.bpj.2011.10.011
49. Urban NT, Willig KI, Hell SW, Nagerl UV (2011) Sted nanoscopy of actin dynamics in synapses deep inside living brain slices. Biophys J 101(5):1277–1284. doi:10.1016/j.bpj.2011.07.027
50. Westphal V, Rizzoli SO, Lauterbach MA, Kamin D, Jahn R, Hell SW (2008) Video-rate far-field optical nanoscopy dissects synaptic vesicle movement. Science 320(5873):246–249. doi:10.1126/science.1154228
51. Lee CJ, Boller KJ (2012) The noise-limited-resolution for stimulated emission depletion microscopy of diffusing particles. Opt Express 20(12):12793–12798. doi:10.1364/OE.20.012793
52. Bates M, Huang B, Zhuang X (2008) Super-resolution microscopy by nanoscale localization of photo-switchable fluorescent probes. Curr Opin Chem Biol 12(5):505–514. doi:10.1016/j.cbpa.2008.08.008
53. Betzig E, Patterson GH, Sougrat R, Lindwasser OW, Olenych S, Bonifacino JS, Davidson MW, Lippincott-Schwartz J, Hess HF (2006) Imaging intracellular fluorescent proteins at nanometer resolution. Science 313(5793):1642–1645. doi:10.1126/science.1127344
54. Aitken CE, Marshall RA, Puglisi JD (2008) An oxygen scavenging system for improvement of dye stability in single-molecule fluorescence experiments. Biophys J 94(5):1826–1835. doi:10.1529/biophysj.107.117689
55. Bates M, Blosser TR, Zhuang X (2005) Short-range spectroscopic ruler based on a single-molecule optical switch. Phys Rev Lett 94(10):108101
56. Rust MJ, Bates M, Zhuang X (2006) Sub-diffraction-limit imaging by stochastic optical reconstruction microscopy (storm). Nat Methods 3(10):793–795. doi:10.1038/nmeth929

57. Wiedenmann J, Gayda S, Adam V, Oswald F, Nienhaus K, Bourgeois D, Nienhaus GU (2011) From eosfp to mirisfp: structure-based development of advanced photoactivatable marker proteins of the gfp-family. J Biophotonics 4(6):377–390. doi:10.1002/jbio.201000122
58. Lippincott-Schwartz J, Patterson GH (2009) Photoactivatable fluorescent proteins for diffraction-limited and super-resolution imaging. Trends Cell Biol 19(11):555–565. doi:10.1016/j.tcb.2009.09.003
59. Brakemann T, Stiel AC, Weber G, Andresen M, Testa I, Grotjohann T, Leutenegger M, Plessmann U, Urlaub H, Eggeling C, Wahl MC, Hell SW, Jakobs S (2011) A reversibly photoswitchable gfp-like protein with fluorescence excitation decoupled from switching. Nat Biotechnol 29(10):942–947. doi:10.1038/nbt.1952
60. Gould TJ, Verkhusha VV, Hess ST (2009) Imaging biological structures with fluorescence photoactivation localization microscopy. Nat Protoc 4(3):291–308. doi:10.1038/nprot.2008.246
61. Shroff H, Galbraith CG, Galbraith JA, Betzig E (2008) Live-cell photoactivated localization microscopy of nanoscale adhesion dynamics. Nat Methods 5(5):417–423. doi:10.1038/nmeth.1202
62. Wu M, Huang B, Graham M, Raimondi A, Heuser JE, Zhuang X, De Camilli P (2010) Coupling between clathrin-dependent endocytic budding and f-bar-dependent tubulation in a cell-free system. Nat Cell Biol 12(9):902–908. doi:10.1038/ncb2094
63. Lillemeier BF, Mortelmaier MA, Forstner MB, Huppa JB, Groves JT, Davis MM (2010) Tcr and lat are expressed on separate protein islands on t cell membranes and concatenate during activation. Nat Immunol 11(1):90–96. doi:10.1038/ni.1832
64. Williamson DJ, Owen DM, Rossy J, Magenau A, Wehrmann M, Gooding JJ, Gaus K (2011) Pre-existing clusters of the adaptor lat do not participate in early t cell signaling events. Nat Immunol 12(7):655–662. doi:10.1038/ni.2049
65. Sherman E, Barr V, Manley S, Patterson G, Balagopalan L, Akpan I, Regan CK, Merrill RK, Sommers CL, Lippincott-Schwartz J, Samelson LE (2011) Functional nanoscale organization of signaling molecules downstream of the t cell antigen receptor. Immunity 35(5):705–720. doi:10.1016/j.immuni.2011.10.004
66. Dani A, Huang B, Bergan J, Dulac C, Zhuang X (2010) Superresolution imaging of chemical synapses in the brain. Neuron 68(5):843–856. doi:10.1016/j.neuron.2010.11.021
67. Juette MF, Gould TJ, Lessard MD, Mlodzianoski MJ, Nagpure BS, Bennett BT, Hess ST, Bewersdorf J (2008) Three-dimensional sub-100 nm resolution fluorescence microscopy of thick samples. Nat Methods 5(6):527–529. doi:10.1038/nmeth.1211
68. Huang B, Wang W, Bates M, Zhuang X (2008) Three-dimensional super-resolution imaging by stochastic optical reconstruction microscopy. Science 319(5864):810–813. doi:10.1126/science.1153529
69. Jares-Erijman EA, Jovin TM (2006) Imaging molecular interactions in living cells by fret microscopy. Curr Opin Chem Biol 10(5):409–416. doi:10.1016/j.cbpa.2006.08.021
70. Ghosh S, Saha S, Goswami D, Bilgrami S, Mayor S (2012) Dynamic imaging of homo-fret in live cells by fluorescence anisotropy microscopy. Methods Enzymol 505:291–327. doi:10.1016/B978-0-12-388448-0.00024-3
71. Vogel SS, Thaler C, Blank PS, Koushik S (2009) Time resolved fluorescence anisotropy vol chapter 10. In: Flim microscopy in biology and medicine. CRC Press Taylor & Francis Group, London
72. Sharma P, Varma R, Sarasij RC, Ira GK, Krishnamoorthy G, Rao M, Mayor S (2004) Nanoscale organization of multiple gpi-anchored proteins in living cell membranes. Cell 116 (4):577–589. doi:10.1016/S0092-8674(04)00167-9
73. Sun Y, Hays NM, Periasamy A, Davidson MW, Day RN (2012) Monitoring protein interactions in living cells with fluorescence lifetime imaging microscopy. Methods Enzymol 504:371–391. doi:10.1016/B978-0-12-391857-4.00019-7
74. Rosso L, Fernicola VC (2006) Time- and frequency- domain analyses of fluorescnece lifetime for temperature sensing. Rev Scientific Instrumentation 77:034901. doi:10.1063/1.2176085

75. Stockl MT, Herrmann A (2010) Detection of lipid domains in model and cell membranes by fluorescence lifetime imaging microscopy. Biochim Biophys Acta 1798(7):1444–1456. doi:10.1016/j.bbamem.2009.12.015
76. Duncan RR, Bergmann A, Cousin MA, Apps DK, Shipston MJ (2004) Multi-dimensional time-correlated single photon counting (tcspc) fluorescence lifetime imaging microscopy (flim) to detect fret in cells. J Microsc 215(Pt 1):1–12. doi:10.1111/j.0022-2720.2004.01343.x
77. Wallrabe H, Periasamy A (2005) Imaging protein molecules using fret and flim microscopy. Curr Opin Biotechnol 16(1):19–27. doi:10.1016/j.copbio.2004.12.002
78. Nguyen TA, Sarkar P, Veetil JV, Koushik SV, Vogel SS (2012) Fluorescence polarization and fluctuation analysis monitors subunit proximity, stoichiometry, and protein complex hydrodynamics. PLoS One 7(5):e38209. doi:10.1371/journal.pone.0038209
79. Koushik SV, Vogel SS (2008) Energy migration alters the fluorescence lifetime of cerulean: implications for fluorescence lifetime imaging forster resonance energy transfer measurements. J Biomedical Optics 13(3):031204. doi:10.1117/1.2940367
80. Luchowski R, Matveeva EG, Gryczynski I, Terpetschnig EA, Patsenker L, Laczko G, Borejdo J, Gryczynski Z (2008) Single molecule studies of multiple-fluorophore labeled antibodies. Effect of homo-fret on the number of photons available before photobleaching. Curr Pharmaceutical Biotechnol 9(5):411–420
81. Storez H, Scott MG, Issafras H, Burtey A, Benmerah A, Muntaner O, Piolot T, Tramier M, Coppey-Moisan M, Bouvier M, Labbe-Jullie C, Marullo S (2005) Homo- and hetero-oligomerization of beta-arrestins in living cells. J Biol Chem 280(48):40210–40215. doi:10.1074/jbc.M508001200
82. Abankwa D, Gorfe AA, Inder K, Hancock JF (2010) Ras membrane orientation and nanodomain localization generate isoform diversity. Proc Natl Acad Sci USA 107 (3):1130–1135. doi:10.1073/pnas.0903907107
83. Treanor B, Lanigan PM, Kumar S, Dunsby C, Munro I, Auksorius E, Culley FJ, Purbhoo MA, Phillips D, Neil MA, Burshtyn DN, French PM, Davis DM (2006) Microclusters of inhibitory killer immunoglobulin-like receptor signaling at natural killer cell immunological synapses. J Cell Biol 174(1):153–161. doi:10.1083/jcb.200601108
84. Keller PJ, Stelzer EH (2010) Digital scanned laser light sheet fluorescence microscopy. Cold Spring Harb Protoc (5):pdb top78. doi:10.1101/pdb.top78
85. Tomer R, Khairy K, Keller PJ (2011) Shedding light on the system: studying embryonic development with light sheet microscopy. Curr Opin Genet Dev 21(5):558–565. doi:10.1016/j.gde.2011.07.003
86. Planchon TA, Gao L, Milkie DE, Davidson MW, Galbraith JA, Galbraith CG, Betzig E (2011) Rapid three-dimensional isotropic imaging of living cells using Bessel beam plane illumination. Nat Methods 8(5):417–423. doi:10.1038/nmeth.1586
87. Fahrbach FO, Rohrbach A (2012) Propagation stability of self-reconstructing bessel beams enables contrast-enhanced imaging in thick media. Nat Commun 3:632. doi:10.1038/ncomms1646
88. Holekamp TF, Turaga D, Holy TE (2008) Fast three-dimensional fluorescence imaging of activity in neural populations by objective-coupled planar illumination microscopy. Neuron 57(5):661–672. doi:10.1016/j.neuron.2008.01.011
89. Shtengel G, Galbraith JA, Galbraith CG, Lippincott-Schwartz J, Gillette JM, Manley S, Sougrat R, Waterman CM, Kanchanawong P, Davidson MW, Fetter RD, Hess HF (2009) Interferometric fluorescent super-resolution microscopy resolves 3d cellular ultrastructure. Proc Natl Acad Sci USA 106(9):3125–3130. doi:10.1073/pnas.0813131106
90. Kanchanawong P, Shtengel G, Pasapera AM, Ramko EB, Davidson MW, Hess HF, Waterman CM (2010) Nanoscale architecture of integrin-based cell adhesions. Nature 468 (7323):580–584. doi:10.1038/nature09621

New Fluorescent Strategies Shine Light on the Evolving Concept of GPCR Oligomerization

Martin Cottet, Orestis Faklaris, Eric Trinquet, Jean-Philippe Pin, and Thierry Durroux

Abstract GPCR oligomerization has been a matter of intense research these last years. FRET and BRET methods have paved the way to a generalized concept of potential GPCR oligomerization in artificial systems (transfected cell lines). More recently, the use of fluorescent ligands compatible with time-resolved FRET studies has opened the possibility of GPCR oligomerization study in their native context and brought evidence of their existence. Furthermore, recent applications of original fluorescence techniques are unveiling new information on the dynamics that govern these complexes and are changing the way we see GPCR oligomeric structures.

Keywords BRET • FCS • Fluorescent ligand • Fluorescent protein • FRET • G-protein-coupled receptors • Lanthanide cryptate • Oligomerization • Single-particle tracking • Time-resolved FRET

Contents

M. Cottet (✉), O. Faklaris, J.-P. Pin and T. Durroux (✉)
Institut de Génomique Fonctionnelle, Centre National de la Recherche Scientifique, Montpellier, France

Institut National de la Santé et de la Recherche Médicale, Montpellier, France

Université Montpellier 1 and 2, Montpellier, France
e-mail: martin.cottet@igf.cnrs.fr; thierry.durroux@igf.cnrs.fr

E. Trinquet
Cisbio Bioassays, Bagnols sur Cèze, France

Y. Mély and G. Duportail (eds.), *Fluorescent Methods to Study Biological Membranes*, Springer Ser Fluoresc (2013) 13: 389–416, DOI 10.1007/4243_2012_54,

Abbreviations

ACP	Acyl carrier protein
BiFC	Bimolecular fluorescence complementation
BRET	Bioluminescence resonance energy transfer
CFP	Cyan fluorescent protein
co-IP	Co-immunoprecipitation
FCS	Fluorescence correlation spectroscopy
FP	Fluorescent protein
FRET	Fluorescence resonance energy transfer
GABA	γ-Aminobutyric acid
GFP	Green fluorescent protein
GPCR	G-protein-coupled receptor
hAGT	O^6-Alkylguanine-DNA alkyltransferase
HTS	High-throughput screening
pbFRET	Photobleaching FRET
RET	Resonance energy transfer
Rluc	*Renilla luciferase*
RTK	Receptor tyrosine kinase
smFRET	Single-molecule FRET
SPT	Single-particle tracking
TIRF	Total internal reflection fluorescence
TR-FRET	Time-resolved FRET
YFP	Yellow fluorescent protein

1 Introduction to GPCR Oligomerization

Many cell membrane proteins have been known to interact with each other and to form oligomers [1]. Protein oligomerization has been shown to be fundamental for their proper function, as in the case of multimeric ion channels [2]. It can also be involved in ligand binding and transmembrane signaling as for receptor tyrosine kinases (RTKs) [3]. The concept of G-protein-coupled receptor (GPCR) oligomerization arose, explaining various anomalies observed in pharmacological studies [4–10]. However, strong biochemical evidence of receptor dimerization could only be found for a few GPCRs. Such is the case of metabotropic glutamate receptors (mGluRs) as a covalent disulfide bridge connects the extracellular domains of two receptors [11]. For other GPCRs, the characterization of oligomers remains much more elusive and subject to controversy [12], and numerous results have been compiled in a dedicated database ([13, 14]). This emphasizes the necessity of new techniques to definitely characterize these complexes.

This controversy also stems from a fundamental aspect regarding GPCR oligomerization that remains unanswered. It concerns the possible physiological role of these complexes. Some hypotheses have considered that the various subunits within an oligomer could play different roles. This has been well illustrated by the GABAB obligatory heterodimer model. The $GABA_{B1}$ subunit is responsible for binding γ-aminobutyric acid (GABA) whereas the $GABA_{B2}$ subunit is able to activate the G protein [15–17]. Whether such a cross talk can be generalized to other heteromer models (oligomers of different kinds of receptors) remains to be established. In the case of homomers (oligomers of identical receptors), their role is not as clear. A few hypothesis and models have shown the possibilities offered by such structures [18–20]. However, their relevance in vivo still remains to be established as homomer formation could impair GPCR function [21–23], although the opposite has also been established [24]. Identifying oligomeric structures and understanding how oligomerization can modify receptor signaling are crucial steps in pharmacology and drug discovery as they can provide unique targets and new ways to specifically address pathologies [25].

Therefore, we aim at reviewing different fluorescence techniques that have been used to study GPCR oligomerization and how these various techniques have gradually brought answers to some of the most burning questions:

- *Are GPCRs in direct interaction?* Various data provided by binding and biochemical experiments have suggested that GPCRs can constitute complexes by interacting with each other. However, GPCR oligomerization involves direct contact between receptors. Sufficient resolution is required to distinguish direct contact from integration in larger protein complexes or accumulation in cellular or membrane subcompartments.
- *Is this interaction observed for physiological expression levels?* It must be verified that this direct interaction is not merely an artifact of receptor overexpression in artificial models (i.e., transfected cell lines). This implies the development of models mimicking as closely as possible physiological conditions as well as increased sensitivity for oligomer detection.
- *What is the size of these oligomers?* Detection of direct interaction does not allow the distinction between dimers, tetramers, and higher-order oligomers. Defining the size of the oligomers can be an important step to understanding their signaling mechanisms, for example. Additionally, the existence of multiple

Table 1 Overview of fluorescent techniques used to study GPCR oligomerization

Techniques		Advantages	Drawbacks	Results for GPCRs
Preliminary	Fluorescence colocalization	Compatible with direct or indirect fluorescent labeling of GPCRs	Diffraction-limited resolution unable to characterize direct interactions	Imaging of receptor colocalization (~250 nm)
		Renewed interest with super-resolution microscopy		
	Co-immunoprecipitation	Detect and identify receptor interactions in native tissue	No characterization of direct interactions between receptors	Receptor interactions in larger protein complexes
FRET-based	FRET with fluorescent proteins (FPs)	Direct labeling with FPs Detect direct interactions (<10 nm)	No direct measure of FRET, use of indirect measures (pbFRET)	GPCR oligomerization Receptor conformational changes
	BRET	Higher signal specificity through bioluminescence	No distinction between surface and intracellular receptor interactions	GPCR oligomerization Signaling protein recruitment
	Time-resolved FRET	Simple measure of TR-FRET Study of native receptors with fluorescent ligands	Indirect labeling step No information concerning the proportion of dimeric species	GPCR oligomerization: with covalent labeling ex vivo with fluorescent ligands
Complementary	FRAP	Image receptor diffusion in live cells	Measurement averaged over numerous fluorescent receptors	Study of oligomer stability
	Single-particle tracking (SPT)	Tracking of individual receptor diffusion	Adapt fluorescent labeling/density	Surface receptors (co-) diffusion
		Improved spatial resolution	Resolution still unable to characterize direct interactions (~30 nm)	Conclude to dimerization Measure monomer-dimer equilibrium
	smFRET	Characterize direct interactions for single complexes	Not yet implemented for GPCR oligomerization	Potentially confirm direct interactions indicated by SPT
	FCS	Additional information brought by cross correlation calculations	Diffraction-limited resolution unable to characterize direct interactions	Receptor diffusion, complex stoichiometry, effect of ligand binding

FRET Fluorescence resonance energy transfer, *FP* Fluorescent protein, *pbFRET* Photobleaching FRET, *BRET* Bioluminescence resonance energy transfer, *TR-FRET* Time-resolved FRET, *FRAP* Fluorescence recovery after photobleaching, *SPT* Single-particle tracking, *smFRET* Single-molecule FRET, *FCS* Fluorescence correlation spectroscopy

orthosteric binding sites in an oligomer can be at the origin of ligand positive or negative cooperativity.

- *Do GPCRs form stable oligomers, or is there a dynamic equilibrium between monomers and oligomers?* First considered as monomers and then as oligomers, both structures could coexist at the cell surface. Whether an equilibrium exists between these multiple forms remains to be firmly established. This also raises the question of which parameters, such as ligand binding, receptor signaling, and receptor localization, could influence this equilibrium.
- *Which role and localization for oligomers?* According to their localization, the role of oligomers can be different: indeed, they can play key roles in receptor trafficking (targeting to the cell surface or internalization) or receptor signaling (ligand binding or receptor activation). Therefore, defining their potential role requires the distinction of oligomers present at the cell surface and thus able to interact with extracellular stimuli from those trapped inside the cell.
- *Physiological relevance of in vitro results?* Design and validation of experimental models are systematically carried out on heterologous systems. Results have consequently accumulated concerning GPCR oligomerization in vitro. However, the possibilities to reproduce these experiments on physiologically relevant systems such as native tissues are much more limited. Nonetheless, this remains one of the most fundamental questions: do GPCR oligomers exist in their native context?

These are examples of issues that have been raised over these last two decades as light has been shown on the existence and the nature of GPCR oligomers. In an effort to summarize the contribution of the various techniques to this topic, a table has been provided in this chapter (see Table 1), proposing an overview on the evolution of GPCR oligomerization concepts.

2 GPCRs Oligomerization Revealed by FRET

Quite rapidly, it has been accepted that fluorescence techniques such as traditional fluorescence microscopy are not powerful enough to characterize the close interaction that occurs between receptors composing an oligomer. Indeed, colocalization with a typical resolution of approximately 250 nm is not conclusive: two contiguous GPCRs need to be resolved with a precision of at least 10 nm to characterize a dimeric structure. On the other hand, this nanometric scale is particularly well adapted to resonance energy transfer (RET) techniques. Indeed, energy transfer occurs between fluorescent probes when they are in close proximity.

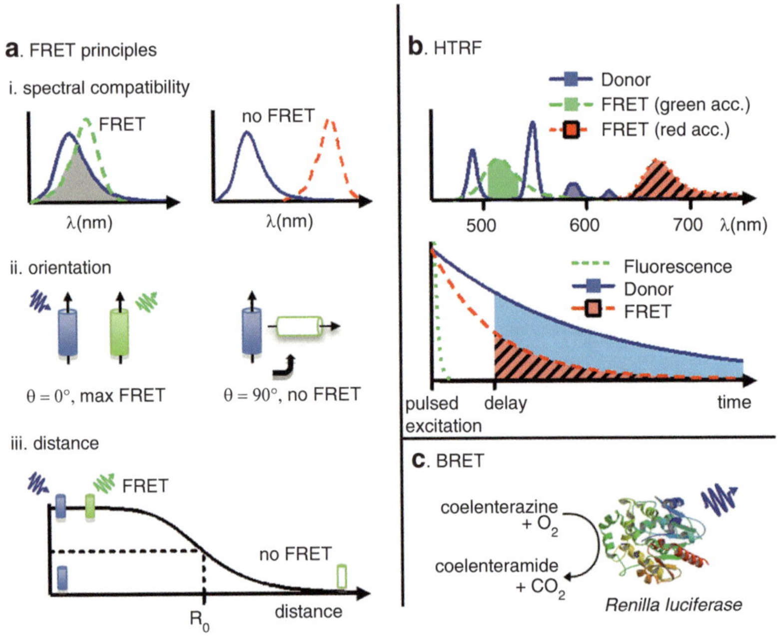

Fig. 1 *Overview of FRET, TR-FRET, and BRET principles*. (**a**) Resonance energy transfer (RET) depends (1) on the overlap (*shaded area*) of the donor emission spectrum (*full line*) and the acceptor absorption spectrum (*dotted line*). It also depends (2) on the relative orientation and (3) on the distance of the fluorophore dipoles. (**b**) Use of rare-earth cryptates as donor, such as Lumi4-Tb, produces atypical donor emission spectra with multiple emission wavelengths, allowing the use of various acceptors (here with fluorescein-like (*green* acc.) or cy5-like (*red* acc.) fluorophores as acceptors). The long fluorescence lifetimes of cryptates also offer temporal selectivity. After a pulsed excitation, short-lived autofluorescence emission is not recorded (*green dotted line*), an electronic delay of several microseconds on the detector allowing a better measure of specific donor (*blue line*) or FRET (*red dotted line*) longer-lasting fluorescence emission. (**c**) For BRET, the donor fluorophore is replaced with a bioluminescent protein, (here *Renilla luciferase*, Rluc). In the presence of oxygen, Rluc can metabolize coelenterazine to emit blue light centered around 475 nm

2.1 *Transfer of Energy: Spectral Compatibility, Orientation, and Distance*

Resonance energy transfer (RET) was studied by Theodor Förster in the middle of the twentieth century [26] and has proven an invaluable tool for biologists studying protein-protein interactions. This nonradiative dipole-dipole energy transfer is based on three conditions [27, 28] (Fig. 1a):

- *Spectral compatibility*. The two species involved in the energy transfer, called the donor and the acceptor, must present spectral compatibilities, that is, the emission spectrum of the donor must overlap with the excitation spectrum of the acceptor.

- *Transition dipoles' relative orientation.* The efficacy of the energy transfer depends on the relative orientation of the transition dipole of the molecules. It is null if the dipoles are perpendicular and maximal if they are parallel.
- *Distance.* The efficacy of energy transfer is inversely proportional to the sixth power of the distance between the donor and the acceptor:

$$E = \frac{R_0^6}{R_0^6 + r^6}$$

where R_0 is the distance corresponding to 50 % energy transfer efficiency. This distance depends on the spectral compatibility of the two species as well as their orientation and is therefore specific to every donor-acceptor pair. A typical R_0 of 5 nm therefore results in FRET efficiency below 1 % for distances over 10 nm.

RET is compatible with various labels: fluorescent or bioluminescent proteins as well as synthetic dyes such as organic fluorophores, quantum dots, or rare-earth cryptates, as will be presented further on. The principles of RET and its resulting spatial resolution make this technique particularly well suited to study close interactions.

2.2 *Direct GPCR Labeling with Fluorescent Proteins and FRET Between Receptors*

Over the last 20 years, numerous fluorescent proteins (FPs) have been derived from the original green fluorescent protein found in the jellyfish *Aequorea victoria* [29, 30], found in other species [31], and extensively used to label protein of interest in live cells. Indeed, the FP is genetically fused to the protein of interest that will be labeled as it is synthesized in the endoplasmic reticulum. Thus, no complementary labeling steps are needed, and the activity of the protein of interest can be directly monitored. Further development has generated new and improved fluorescent proteins with improved photostability, brighter, and more varied fluorescence emission spectra. This method is therefore widely used for labeling receptors and has proven its usefulness in numerous fluorescence techniques [7, 32]. Recently, super-resolution imaging has also gained from these improvements with photoactivatable [33], photoswitchable [34], and photoconvertible probes [35, 36] allowing single-molecule imaging [37].

When used with GPCRs, the 30-kDa size of the fluorescent protein must however be taken into account, and the fusion of this protein at the intracellular C-terminus of receptors (Fig. 2b) can possibly interfere with its coupling to G proteins [38]. Second, the stability of fluorescent proteins can be sometimes questionable. For example, photoconversion of YFP into CFP-like species during photobleaching protocols has been reported [39], and pH sensitivity of CFP and

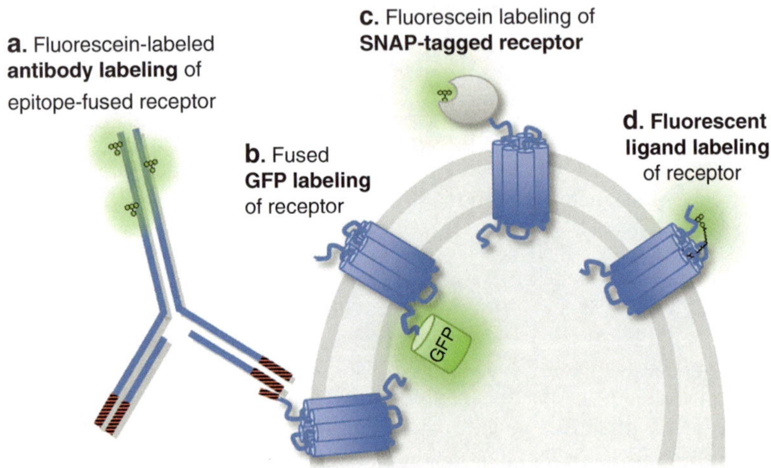

Fig. 2 *Comparing sizes of various tools to fluorescently label GPCRs*. (**a**) 150-kDa fluorescently labeled antibody directed against fused extracellular epitopes (cMyc, Flag, etc.). (**b**) Direct labeling with 30-kDa fluorescent (CFP, GFP, YFP, etc.) or luminescent proteins (Rluc). (**c**) Extracellular labeling with fusion proteins, here the 20-kDa SNAP-tag. (**d**) Use of fluorescently labeled ligands, ranging from 0.5 kDa for small molecules to 3 kDa for larger peptidic ligands

YFP has also been described [40]. Finally, the tendency of certain earlier FPs to constitutively for dimers can be a drawback when studying GPCR dimers [41].

This simple and complete labeling of receptors is however adapted for FRET experiments. Indeed, various pairs of fluorescent proteins are compatible to perform FRET. For example, BFP/GFP [42] or CFP/Venus [43] have been used to study molecular interaction but, to our knowledge, have not been adapted to the detection of GPCR oligomerization. By contrast, CFP/YFP pairs have been frequently used although they present some disadvantages due to their broad fluorescence spectra. Indeed, CFP and YFP exhibit overlapping excitation spectra and emission spectra (Fig 1a). This complicates the specific excitation of only one fluorophore and the separate detection of fluorescence emitted by the donor and by the acceptor [44, 45]. Multiple strategies have therefore been developed to estimate the FRET signal. A first method consists in measuring the fluorescence intensity at the acceptor emission wavelength. Subsequently, a ratiometric analysis is performed in order to subtract fluorescence intensities due to the emission of the donor or to the direct excitation of the acceptor [46]. Others strategies based on photobleaching FRET (pbFRET) have also been used. They consist in selectively photobleaching the donor or the acceptor and in measuring, respectively, the decrease of the acceptor fluorescence emission or the increase of the donor emission [47, 48]. These techniques however significantly reduce the sensitivity of the method. Indeed, most of the fluorescence detected must be either subtracted or bleached to ascertain the proportion corresponding to FRET and therefore resulting from direct interaction between receptors.

2.2.1 Dimerization Characterized by BRET

Bioluminescence resonance energy transfer (BRET) assays have been implemented in order to improve sensitivity [49, 50]. BRET similarly to FRET relies on a transfer of energy, the donor probe being replaced by a complex formed by a bioluminescent enzyme, usually *Renilla luciferase* (Rluc), and its substrate coelenterazine (clz). Donor excitation is achieved with a chemical substrate instead of light excitation, therefore preventing any undue excitation of the acceptor. BRET was initially used with the complex Rluc/clz as donor (emission peak at 480 nm) and YFP as acceptor (emission peak at 525 nm) [51]. Because of the significant overlap of Rluc and YFP emission spectra, BRET intensity is determined by calculating the BRET ratio, a mathematical calculation allowing the subtraction of donor emission at the acceptor wavelength (see [52] for further details). Optimization of the enzyme/substrate complexes and use of new acceptors to reduce emission spectra overlap have been carried out, and variants of BRET have been implemented [53–56]: BRET1 (Rluc or Rluc8/clz/YFP), BRET2 (Rluc or Rluc8/clz400/GFP2), and BRET3 (Rluc8/clz/mOrange) are the most frequently used variants.

This has allowed the development of suitable assays to detect GPCR oligomerization [57, 58]. Furthermore, due to the bioluminescent nature of this method, any autofluorescence that could interfere with classic fluorescence measurements is avoided, an issue particularly important in FRET studies when fluorescence signals can be weak. Once again, these assays have allowed biologists to confirm the possible oligomerization of GPCRs [59]. However, the physiological relevance of these results is still a matter of controversy as it has been shown that BRET assays must be performed under close monitoring of labeled receptor expression levels [60, 61]. Furthermore, typical energy transfer assays allow us to conclude on the interactions between at least two partners, but they do not provide any direct information concerning the implication of more than two partners. Alternative techniques have therefore been developed to investigate the possible distinction between strict dimers and higher-order oligomers.

2.2.2 Oligomerization Characterized by BiFC

Bimolecular fluorescence complementation (BiFC) is based on the principle that complementary fragments of fluorescent or luminescent proteins have the ability to reform a fluorescent (or luminescent) unit when they are brought in close proximity [62–67]. This technique has been adapted to the study of GPCR interactions and, coupled with RET, has provided evidence for the possible assembly of receptors in higher-order oligomers [68, 69]. Urizar and colleagues have also successfully used this technique to show heteromer-specific interactions between GPCR oligomers and G proteins [70]. The coupling of BiFC and RET techniques provides means to study the assembly of multiple partners [71] and can therefore bring valuable insight into the possible structure of GPCR oligomers and allow a better understanding of their coupling to different partners.

Taken together, methods using fluorescent or bioluminescent proteins described above have brought useful information about the existence of potential direct interactions between multiple GPCRs. However, the signal measured corresponds to an average signal of receptors targeted to the cell surface as well as receptors internalized or immature receptors trapped in subcellular compartments. Indeed, GPCR oligomers may assemble early during receptor synthesis in the endoplasmic reticulum, as shown by BRET experiments on vasopressin and oxytocin receptors [58]. Cell surface oligomer identification would therefore require more complex microscopy approaches [72, 73]. A BRET microscopy technique has been recently implemented [74, 75] but offers limited resolution and sensitivity.

2.3 Indirect GPCR Labeling via External Tags

A simple way to selectively monitor interactions occurring at the cell surface is to specifically label membrane receptors with nonpermeable probes. Indirect labeling presents significant advantages as it offers the possibility to label the same receptor with a vast range of synthetic fluorophores. Synthetic dyes can themselves be fine-tuned to present specificity adapted to different assays [76–79] and, for example, be engineered to emit more stable and brighter fluorescence and be better adapted for FRET studies. An example is the recently developed time-resolved FRET technique (TR-FRET) [80–82].

This technique is based on the use of lanthanides complexed either by chelates or cryptates. The role of the cage is crucial since chelates or cryptates allow lanthanides to be linked to the protein of interest. They also prevent quenching of lanthanides, by water, for example, and finally, they impact their fluorescent properties by modifying absorbance and emission fluorescent spectra and by increasing lanthanide brightness. The most commonly used lanthanide cryptate, Lumi4-Tb, displays an absorbance peak around 340 nm and four emission peaks around 490, 550, 585, and 620 nm. Therefore, Lumi4-Tb is capable of transferring its energy onto fluorescein-like "green" acceptor fluorophores (absorbance and emission peak around 490 and 530 nm, respectively) or cy5-like "red" acceptor fluorophores (absorbance and emission peak around 645 and 670 nm, respectively) [83] (Fig. 1b). UV excitation of Lumi4-Tb (under 350 nm) leads to a negligible direct excitation of the acceptor. Moreover, emission of the donor is reduced at the emission wavelength of the acceptor. Both factors contribute to an optimal spectral selectivity.

Lanthanide cryptates also present longer fluorescence lifetimes that span over several hundred microseconds, compared to the nanosecond fluorescence lifetime of classic fluorophores. This allows a simple temporal selection of the fluorescence of interest, achieved by applying a time delay (typically 50 μs) between the excitation of the sample and the measurement of the resulting longer-lasting fluorescence (donor fluorescence or FRET) (Fig. 1b). This characteristic results in an additional temporal selectivity that further improves the signal-to-noise ratio. Consequently, these new fluorescent donor species available for labeling extracellular proteins offer a promise for assays with improved sensitivity provided by optimal spectral and temporal selectivity.

2.3.1 Antibody Labeling

Following the tradition of immunohistology labeling, first approaches to monitor GPCRs expressed at the cell membrane were based on the use of fluorescently labeled antibodies. Due to the noncovalent labeling of receptors, these methods rely on the availability of high-affinity antibodies in order to achieve a specific and complete labeling. The difficulty to obtain such antibodies [84–89] led to the use of standard chimeric epitopes fused to receptors (e.g., cMyc, Flag, 6-His). Using antibodies directed against these epitopes, fluorescently labeled GPCRs present at the cell surface can be monitored [90–93]. However, considering the short distances involved in oligomerization, the considerable size of antibodies (150 kDa), and their bivalent nature that could potentially drive protein association, they are not ideally suited for oligomerization studies (Fig. 2a).

2.3.2 Chemical Labeling with Smaller Tags

Alternative solutions have been developed and are particularly effective for the labeling of GPCRs. Short peptides or proteins are fused to the receptor of interest. These proteins will afterward be recognized as the specific site for receptor labeling. As reviewed by O'Hare et al. [94], some of these solutions rely on self-labeling tags (i.e., SNAP-tag [95]), on enzyme-catalyzed fluorescent labeling (i.e., ACP tag [96]), or on simple chelation of fluorescent substrates (i.e., FlAsH labeling [97]).

Self-labeling tags, for example, derived from the human O^6-alkylguanine-DNA alkyltransferase (hAGT, 21 kDa), have extensively been used these last years to label GPCRs (Fig. 2c). As different techniques (SNAP-tag, CLIP-tag, HaloTag) have emerged with the engineering of new proteins, they enable stable, covalent, and specific labeling of membrane receptors [94, 95, 98, 99]. The possibility to add various fluorophores to their specific substrates opens a wide range of possible labeling. Indeed, combined with the use of lanthanide cryptates, these labeling techniques allowed the stable and efficient detection of cell surface GPCR oligomers, such as mGluR strict dimers and $GABA_B$ tetramers [100, 101], and the screening of heteromer populations [102].

Necessitating enzyme-catalyzed fluorescent labeling, smaller tags such as the acyl carrier protein (ACP, 9 kDa) are also available [96]. ACP tags were used to study GPCR oligomerization [103], providing a less cumbersome labeling. Indeed, fusing tags to the N-terminus of GPCRs can possibly alter pharmacological properties of the receptor or hinder the access to the binding site but can also generate additional interaction interfaces between receptors. Smaller tags that maintain a high labeling specificity while reducing their steric hindrance can therefore allow many novel studies regarding GPCR mechanisms. Ultimate reduction of tags can be achieved with tetracysteine-based approaches. In this case, the simple fusion of a six-amino acid motif, CCPGCC, allows the covalent chelation of fluorescent substrates [97, 104]. However, this approach only permits the use of a limited number of fluorescent substrates (green FlAsH or red ReAsH), reducing the possibilities to study GPCR oligomerization.

Indirect labeling approaches have gradually started to confirm oligomerization observations obtained with direct labeling using genetically encoded probes. However, the continued use of artificial systems raises the issue of verifying these results in a native context [105]. The use of chimeric receptors as presented above is not adapted to perform FRET between receptors in native tissue without resorting to genetically modified animals expressing two tagged GPCRs. To our knowledge, such animals have not yet been developed even though knockin mice expressing GFP-tagged GPCRs [106] can constitute a first step toward the development of such a technique.

3 Oligomerization in Native Tissues Studied with TR-FRET

Very little is known regarding oligomerization in native tissues. As mentioned before, resolutions displayed by conventional microscopy or co-IP techniques are not high enough to conclude to direct interactions between proteins colocalized or belonging to the same complex. Tools have therefore been developed to adapt FRET strategies and more specifically a TR-FRET approach to study native tissues.

3.1 Fluorescent Ligand-Based Labeling

The use of fluorescent ligands seems particularly well suited for these types of studies. Fluorescent ligands have been developed as tools to replace their radioactive counterparts in the case of binding assays [107, 108]. As has been reviewed by Cottet et al. [109], the use of fluorescence-based techniques presents numerous advantages, reducing the cost of these assays and the logistical complexities inherent to the use of radioactivity. In the case of FRET-based studies, the interest lies in the compatibility of some of these ligands with time-resolved FRET techniques. For binding studies, this has been assessed with donor-labeled receptors transferring their fluorescence to the acceptor-labeled ligand and providing a tool to discriminate effectively between free ligands and the bound fraction, engaged in FRET with the receptor [107].

Another aspect of ligand-based methods derives from the nature of the ligands, which are naturally designed to discriminate between receptor subtypes to bind specifically to a receptor. Because GPCRs present high variations in their ligand-binding sites [110], these tools are therefore particularly powerful to achieve receptor-specific labeling.

An obstacle for the use of fluorescent ligands is the difficulty to fluorescently label them without altering their pharmacological properties of binding and function. This issue becomes even more critical when considering smaller molecules such as nonpeptidic ligands (Fig. 2d). If labeling a 150-kDa antibody with multiple fluorophores can have limited impact on its binding properties, labeled ligands have to be systematically verified, as fluorophores may significantly modify their binding affinity or binding mechanism. Therefore, the design of a fluorescent ligand must systematically be supplemented with robust verifications to verify the affinity and the receptor selectivity of the generated probe [111, 112]. However, this strategy is not limited by the size of the ligand, whether it is a peptide or a small organic molecule, and numerous functional fluorescent ligands displaying high affinities are now available [107, 108, 113–115].

3.2 *Evidence for GPCR Oligomerization in Native Tissues*

The foremost advantage of fluorescent ligand-based methods is that they are automatically applicable to receptors, requiring no genetic modification. This makes them particularly well suited for labeling endogenous receptors and their study in a native context. Two aspects are frequently overlooked in GPCR oligomerization studies. The first is the control of the expression level. Secondly, if the overall expression level in a population of transfected cells can be adjusted to mimic that of a global tissue, it becomes more difficult to address concerns regarding localized expression patterns between cells or between membrane microdomains, or cell-type-dependent expression specificities. To address these issues, conclusive data can only be obtained directly in native tissues. Using time-resolved FRET technology combined with fluorescent ligands, Albizu et al. were able to bring evidence of GPCR oligomerization in a native context and more precisely of the oxytocin receptor in the mammary gland of the lactating rat [116].

This strategy, based on the use of fluorescent ligands, presents both advantages and drawbacks. It depends directly on the availability of high-affinity fluorescent ligands as it relies on pharmacological equilibrium for the ligands to bind to both receptors and therefore induces an uncertainty in the stoichiometry of receptor labeling (Fig. 3a). Adding an excess of unlabeled ligand reproduces the conditions of a competition experiment and offers the opportunity to verify the specificity of the labeling.

This method can also reveal additional oligomer properties. Indeed, ligand binding and therefore receptor labeling can profoundly be modified by cooperativity phenomena observed for GPCRs. According to mathematical models

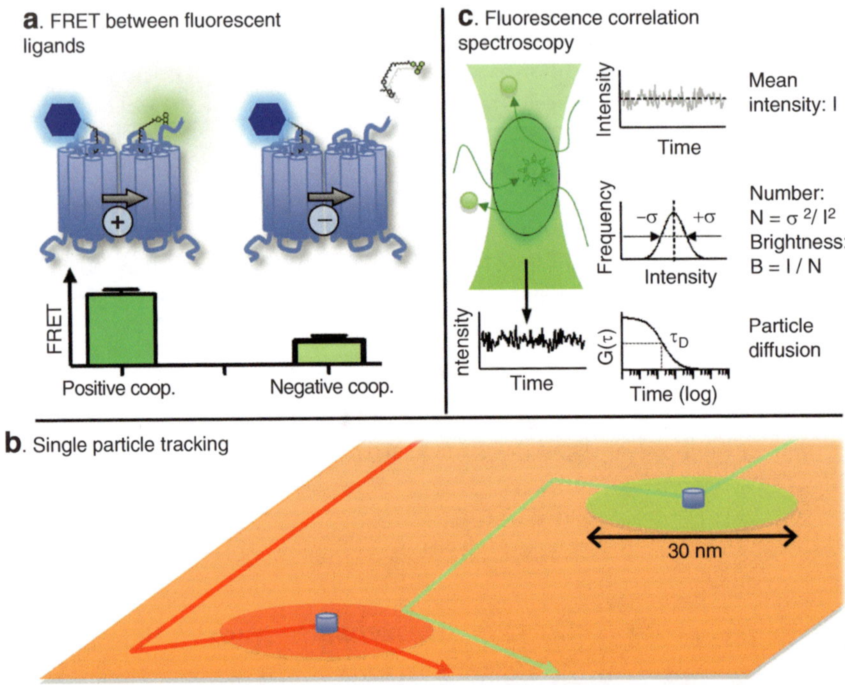

Fig. 3 *Fluorescent ligands bring new insight on GPCR oligomerization.* (**a**) Fluorescent ligands allow FRET measurement between occupied receptors in an oligomer, possibly influenced by cooperativity phenomena. (**b**) Single-particle tracking enables a tenfold improvement in receptor localization and thus illustrates the dynamics of receptors intermittently codiffusing in pairs in close proximity (30 nm). (**c**) Fluorescence correlation spectroscopy allows the monitoring of fluorescence variations originating from fluorescence particles entering and exiting the excitation volume. Analysis of these fluctuations provides information on the number of particles monitored, their brightness, and a coefficient corresponding to their ability to diffuse in the sample

developed by several groups [18, 20, 117], GPCR oligomers can present a wide variety of pharmacological profiles resulting from cooperativity between orthosteric binding sites [6, 8]. In fact, different cooperativity phenomena have been observed in the mammary gland, when using antagonists or agonists. This result has been related to the difference in FRET signals observed when labeling the mammary gland with fluorescent antagonists or agonists [116]. Furthermore, it was shown that ligand binding had no impact on the oligomeric state of oxytocin receptors. Cooperativity therefore appears to result exclusively from the conformational change induced by the first ligand binding, promoting or preventing the binding of a second ligand (Fig. 3a). Finally, although other hypotheses have been proposed to explain cooperativity, such as receptor/G-protein ratio [12], oligomerization explains both negative and positive cooperativity phenomena [18].

More recently, evidence has also been brought concerning the existence of heteromers in the rat hypothalamus. In their study, Kern and colleagues use a very similar TR-FRET approach with a fluorescent red ghrelin ligand in association with a Lumi4-Tb-conjugated secondary antibody labeling the dopamine D2 receptor [118]. Membrane preparations of wild-type rat hypothalamus show FRET, indicating the presence of ghrelin-dopamine heteromers, whereas striatum and membrane preparations from ghrelin knockout animals show no FRET. Taken together, both studies show that oligomerization can be characterized for receptors in their native environment and that TR-FRET and fluorescent ligands are an effective combination for these studies.

3.3 *Perspectives for Imaging Endogenous GPCR Oligomers with TR-FRET*

These preliminary results on the de facto existence of oligomeric GPCRs in their native context may however benefit from further development to bring additional data regarding the localization and functioning of these structures. Recently, Rajapakse et al. [119, 120] presented the premises of a TR-FRET microscopy technique compatible with the use of lanthanide cryptates. This could bring important improvements in both the signal-to-noise ratio and a better discrimination between fluorescence emissions by the different fluorophores. These two aspects are fundamental in order to detect oligomers in a context of low physiological expression levels and high background fluorescence originating from the biological environment.

Although this strategy provides a simpler way to image FRET, it also needs further development. This is especially true for the detection of fluorescence in a time-resolved manner. The use of intensified CCDs, as is the case in this study, provides a better sensitivity but poorer spatial resolution. Spatial resolution is essential when studying membrane proteins. It allows the verification that the labeled population of receptor is indeed addressed to the cell surface. It also offers the possibility to monitor receptors as they engage in different aspects of their signaling pattern. In their review, Grecco and Vereer aim at replacing FRET microscopy among other super-resolution microscopy techniques [121], concluding on "the unique capability of FRET to detect interactions at the nanometer scale." Indeed, if recent advances in fluorescence microscopy techniques have permitted to cross the diffraction barrier [122], they still lack the resolution to resolve possible GPCR oligomers. However, if time-resolved FRET microscopy can undeniably provide invaluable information regarding close-distance interactions, it must nonetheless take advantage of these novel microscopy technique developments to monitor as precisely as possible the activity of GPCRs.

Furthermore, fluorescent ligands show the promise of being adapted for direct, whole animal, in vivo imaging. Josan et al. synthetized fluorescent GPCR ligands that were successfully injected in mice to visualize specific labeling [123]. Such an approach could make possible the identification of GPCRs throughout the organism and eventually discriminate between receptors forming oligomeric structures.

4 Emerging Concepts of GPCR Oligomerization: Stability and Dynamics of Complexes

The field of GPCR oligomerization has long been focused on deciding whether GPCRs naturally formed oligomers or remained monomeric. After efforts from numerous groups to show that GPCR oligomers could exist in heterologous conditions, and after recent evidence showing that some oligomers do indeed exist in a native context, growing evidence shows that GPCRs could in fact coexist as both monomeric and oligomeric structures. Where classic FRET techniques were limited to bringing evidence of the possibility of oligomerization, original microscopy and spectroscopy techniques have allowed researchers to dwell deeper in the dynamics of GPCRs and bring evidence of even more complex phenomena. This section aims at illustrating how novel approaches have recently brought to light new paradigms of GPCR oligomerization and receptor dynamics at the cell surface.

4.1 Stability of Oligomers: Monomer-Oligomer Equilibriums?

Dorsch and colleagues have attempted to bring evidence of a possible equilibrium between oligomers for β-adrenergic receptors [124]. In this study, receptors were labeled with fluorescent proteins, the first receptor with an extracellular YFP and the second with an intracellular CFP. Immobilizing the first receptor with an antibody recognizing the YFP and after photobleaching a region of interest, they measured the rate at which receptors labeled with a CFP diffused to that region. The fluorescence recovery was rapid in the case of the β_1-adrenergic receptor, suggesting unstable oligomers that did not restrain free diffusion of receptors, whereas the β_2-adrenergic receptor did not allow fluorescence recovery in the photobleached region, suggesting a more stable oligomeric structure. These results however must be put in perspective with the known differences for these two receptors. Rybin et al. [125] previously showed that in cardiac myocytes, β_2-adrenergic receptors presented a tendency to segregate in caveolae compartments, explaining a more constrained diffusion compared to β_1-adrenergic receptors. In perspective with the above results, Fallahi-Sichani et al. [126] showed in a computational model that receptors targeting could be simply decided by their tendency to form stable or unstable quaternary structures. In this case, stable monomers or oligomers preferentially segregated in less fluid compartments such as caveolae microdomains, whereas unstable oligomers were free to diffuse throughout the membrane.

The above studies show that the unstable oligomer formed by β_1-adrenergic receptors can lead to a greater mobility, as seen by Dorsch et al., whereas the stable oligomer formed by β_2-adrenergic receptors can impair its diffusion and be constrained in caveolae microdomains. More importantly, these studies show that GPCR oligomerization cannot be considered as a universal law for all GPCRs and that the stability of these quaternary structures is a crucial point to better understand their function and interactions with the cell membrane.

4.2 Dynamics of GPCRs

GPCR dynamics and diffusion on the cell surface have been recently studied using fluorescent ligands and single-particle tracking (SPT) observed by total internal reflection microscopy (TIRF) [127, 128]. Briefly, TIRF microscopy allows a very fine excitation of the cell basal membrane, where only a few labeled receptors may be present, increasing the signal-to-noise ratio. By identifying fluorescence spots corresponding to single fluorophores, it is possible to pinpoint their localization with great precision (single-particle imaging) and to follow their diffusion in time (SPT). These studies have demonstrated that receptors diffused freely throughout the cell membrane, either single or in pairs. Hern and colleagues [127] identified that 10 % of the labeled receptors diffused in pairs, another 10 % dissociated and reassociated rapidly in pairs, and the remaining 80 % diffused as monomeric receptors. Again, these results bring a new dimension to GPCR oligomerization [129]. However, if SPT allows an increase in spatial resolution, as the single fluorophore can be pinpointed to the center of mass of its diffraction-limited spot [130], the resolution achieved in these experiments is not higher than 20–30 nm, a value depending on fluorescence intensity, receptor density, and temporal resolution (Fig. 3b). To overcome receptor density, new techniques have recently been adapted, such as PALM or STORM, using photoactivatable fluorophores [122]. However, resolution remains dependent on fluorescence intensity and, to our knowledge, has not been reported higher to 30 nm in live cells [122, 131–133]. At the scale of dimerization, this is still insufficient to definitively conclude to a transient dimerization of these receptors and to rule out the possibility that they could simply be codiffusing inside the same microdomains of approximately 50 nm [134]. In the age of super-resolution microscopy, RET techniques still remain essential to monitor close protein interactions and, combined with recently developed microscopy techniques, could bring the missing answers to oligomerization phenomena.

4.3 Identifying Oligomeric Populations

A critical question regarding GPCR oligomerization is the proportion of receptors forming oligomers and receptors remaining as monomers. The possibility to identify single molecules with single-particle tracking microscopy allows the identification and localization of single FRET pairs. Indeed, the conformation of immobilized and purified membrane proteins has been studied by TIRF microscopy, which allowed the measurement of rapid conformational changes occurring in a single receptor [135]. As single FRET pairs are observed, this allows the measurement of unique FRET events and therefore the calculations of this energy transfer efficiency. These variations in FRET efficiency illustrate the change in conformation of the receptor. Just as recently, Sakon and colleagues have been able to measure single-molecule FRET (smFRET) at the cell surface [136]. This shows how smFRET could be used to study GPCR oligomerization. Indeed, these techniques could bring insight into the different quaternary structures adopted by a GPCR. For example, in the case of an intermediate FRET signal, it could potentially allow the distinction between receptors adopting an intermediate configuration and receptors clearly segregating into two distinct configurations. Whereas classic FRET studies provide an average of FRET intensity over several events, smFRET is able to resolve the heterogeneity among a population of receptors. Thus, smFRET could bring answers concerning the proportion of GPCRs forming oligomers as well as their conformation at the cell surface.

4.4 Ligand-Induced Dimerization

The dynamics of individual fluorophores and individual receptors can also be studied by fluorescence correlation spectroscopy (FCS). By using a confocal or two-photon microscope to excite fluorescently labeled receptors in a reduced confocal volume, FCS allows the quantification of individual labeled receptors diffusing in that volume (Fig. 3c). By measuring the variations in the emitted fluorescence in this volume, it is possible to extract the diffusion parameters of particles entering and exiting the volume as well as their average brightness. Analyzing data from FCS measurements, Ilien and colleagues observed a doubling of the brightness of diffusing GFP-tagged muscarinic M_1 receptors in presence of pirenzepine. They concluded that pirenzepine binding to the muscarinic M_1 receptor induced its dimerization [137]. Interestingly, as mentioned by the authors, throughout the literature, contradictory results have been reported on ligand-induced dimerization. An absence of effect has generally been observed for antagonists or agonists on muscarinic oligomerization [138, 139], but ligand-induced association and dissociation has also been described for other class A GPCRs [140].

Recently, this spectroscopy method has been adapted to TIRF and confocal scanning microscopy, to provide additional localization information of oligomers with number and molecular brightness analysis [141, 142]. By simply accumulating several images and analyzing the variation of fluorescence for each individual pixel, it is possible to quantify the fluctuation in fluorescence and resolve the number and the brightness of the diffusing particles. Therefore, this could give an indication on the size, the diffusion, and the localization of receptor complexes [143].

5 Conclusion

The concept of GPCR oligomerization has always been tightly linked with the development of fluorescence methods. As the first techniques of energy transfer were limited to the use of fluorescent proteins, the study of GPCR oligomerization has long been confined to heterologous systems with chimeric receptors transfected into cell lines. These models are not able to completely mimic the physiological context of these receptors, because of variations depending on cell types or simply matters of overexpression well beyond physiological levels. Thus, the concept of oligomerization was challenging to extrapolate to physiology.

The use of extracellular epitopes and tags to label receptors at the cell surface has allowed biologists to more finely control the expression level of receptors. More importantly, the use of fluorescent ligands, requiring no modification in the sequence of receptors, eventually enabled the direct labeling of receptors in native tissues. Along with results confirming positive or negative cooperativity between binding sites, this provided long-awaited information on the oligomerization of GPCRs in vivo.

In a distinct direction, advances in labeling and microscopy techniques have continuously provided researchers with new methods to monitor GPCR dynamics. Indeed, with the development of original fluorophores presenting more stable and brighter emissions, as well as photoconvertible fluorophores compatible with super-resolution imaging, GPCRs can be monitored more closely than ever before. Combining such super-resolution techniques with single-molecule FRET would allow even further precision in discriminating direct interactions and monitoring oligomer dynamics. Obviously, the actual structure of GPCR oligomers would only be resolved by crystallography, as has recently been achieved for some receptors binding antagonists [144, 145]. Finally, this has led these past years to original concepts regarding the possible existence and dynamics of GPCR oligomeric structures.

Once again, as these works rely mostly on the use of chimeric receptors and engineered cell lines, they only intend to provide potential mechanisms and structures. The growing availability and use of fluorescent ligands provide the possibility to adapt these cutting-edge techniques to GPCRs in their native context. The combined use of these advanced fluorescence methods will eventually bring us closer to understanding the complex mechanisms underlying GPCR functioning and allow us to develop ever more specifically targeted and efficient drugs.

References

1. Marianayagam NJ, Sunde M, Matthews JM (2004) The power of two: protein dimerization in biology. Trends Biochem Sci 29(11):618–625
2. MacKinnon R (2003) Potassium channels. FEBS Lett 555(1):62–65

3. Li E, Hristova K (2010) Receptor tyrosine kinase transmembrane domains: function, dimer structure and dimerization energetics. Cell Adh Migr 4(2):249–254
4. Mattera R, Pitts BJ, Entman ML, Birnbaumer L (1985) Guanine nucleotide regulation of a mammalian myocardial muscarinic receptor system. Evidence for homo- and heterotropic cooperativity in ligand binding analyzed by computer-assisted curve fitting J Biol Chem 260(12):7410–7421
5. Wreggett KA, Wells JW (1995) Cooperativity manifest in the binding properties of purified cardiac muscarinic receptors. J Biol Chem 270(38):22488–22499
6. Armstrong D, Strange PG (2001) Dopamine D2 receptor dimer formation: evidence from ligand binding. J Biol Chem 276(25):22621–22629
7. Rios CD, Jordan BA, Gomes I, Devi LA (2001) G-protein-coupled receptor dimerization: modulation of receptor function. Pharmacol Ther 92(2–3):71–87
8. Albizu L, Balestre MN, Breton C, Pin JP, Manning M, Mouillac B, Barberis C, Durroux T (2006) Probing the existence of G protein-coupled receptor dimers by positive and negative ligand-dependent cooperative binding. Mol Pharmacol 70(5):1783–1791
9. Springael JY, Le Minh PN, Urizar E, Costagliola S, Vassart G, Parmentier M (2006) Allosteric modulation of binding properties between units of chemokine receptor homo- and hetero-oligomers. Mol Pharmacol 69(5):1652–1661
10. Birdsall NJ (2010) Class A GPCR heterodimers: evidence from binding studies. Trends Pharmacol Sci 31(11):499–508
11. Pin JP, Galvez T, Prezeau L (2003) Evolution, structure, and activation mechanism of family 3/C G-protein-coupled receptors. Pharmacol Ther 98(3):325–354
12. Chabre M, Deterre P, Antonny B (2009) The apparent cooperativity of some GPCRs does not necessarily imply dimerization. Trends Pharmacol Sci 30(4):182–187
13. Skrabanek L, Murcia M, Bouvier M, Devi L, George SR, Lohse MJ, Milligan G, Neubig R, Palczewski K, Parmentier M, Pin JP, Vriend G, Javitch JA, Campagne F, Filizola M (2007) Requirements and ontology for a G protein-coupled receptor oligomerization knowledge base. BMC Bioinformatics 8:177
14. Khelashvili G, Dorff K, Shan J, Camacho-Artacho M, Skrabanek L, Vroling B, Bouvier M, Devi LA, George SR, Javitch JA, Lohse MJ, Milligan G, Neubig RR, Palczewski K, Parmentier M, Pin JP, Vriend G, Campagne F, Filizola M (2010) GPCR-OKB: the G protein coupled receptor oligomer knowledge base. Bioinformatics 26(14):1804–1805
15. Galvez T, Parmentier ML, Joly C, Malitschek B, Kaupmann K, Kuhn R, Bittiger H, Froestl W, Bettler B, Pin JP (1999) Mutagenesis and modeling of the GABAB receptor extracellular domain support a venus flytrap mechanism for ligand binding. J Biol Chem 274(19):13362–13369
16. Galvez T, Prezeau L, Milioti G, Franek M, Joly C, Froestl W, Bettler B, Bertrand HO, Blahos J, Pin JP (2000) Mapping the agonist-binding site of GABAB type 1 subunit sheds light on the activation process of GABAB receptors. J Biol Chem 275(52):41166–41174
17. Duthey B, Caudron S, Perroy J, Bettler B, Fagni L, Pin JP, Prezeau L (2002) A single subunit (GB2) is required for G-protein activation by the heterodimeric GABA(B) receptor. J Biol Chem 277(5):3236–3241
18. Durroux T (2005) Principles: a model for the allosteric interactions between ligand binding sites within a dimeric GPCR. Trends Pharmacol Sci 26(7):376–384
19. Han Y, Moreira IS, Urizar E, Weinstein H, Javitch JA (2009) Allosteric communication between protomers of dopamine class A GPCR dimers modulates activation. Nat Chem Biol 5(9):688–695
20. Rovira X, Pin JP, Giraldo J (2010) The asymmetric/symmetric activation of GPCR dimers as a possible mechanistic rationale for multiple signalling pathways. Trends Pharmacol Sci 31(1):15–21
21. White JF, Grodnitzky J, Louis JM, Trinh LB, Shiloach J, Gutierrez J, Northup JK, Grisshammer R (2007) Dimerization of the class A G protein-coupled neurotensin receptor NTS1 alters G protein interaction. Proc Natl Acad Sci USA 104(29):12199–12204

22. Arcemisbehere L, Sen T, Boudier L, Balestre MN, Gaibelet G, Detouillon E, Orcel H, Mendre C, Rahmeh R, Granier S, Vives C, Fieschi F, Damian M, Durroux T, Baneres JL, Mouillac B (2010) Leukotriene BLT2 receptor monomers activate the G(i2) GTP-binding protein more efficiently than dimers. J Biol Chem 285(9):6337–6347
23. Comps-Agrar L, Kniazeff J, Norskov-Lauritsen L, Maurel D, Gassmann M, Gregor N, Prezeau L, Bettler B, Durroux T, Trinquet E, Pin JP (2011) The oligomeric state sets GABA(B) receptor signalling efficacy. EMBO J 30(12):2336–2349
24. Pellissier LP, Barthet G, Gaven F, Cassier E, Trinquet E, Pin J-P, Marin P, Dumuis A, Bockaert J, Baneres J-L, Claeysen S (2011) G protein activation by serotonin type 4 receptor dimers: evidence that turning on two protomers is more efficient. J Biol Chem 286(12): 9985–9997
25. Fribourg M, Moreno JL, Holloway T, Provasi D, Baki L, Mahajan R, Park G, Adney SK, Hatcher C, Eltit JM, Ruta JD, Albizu L, Li Z, Umali A, Shim J, Fabiato A, MacKerell AD Jr, Brezina V, Sealfon SC, Filizola M, Gonzalez-Maeso J, Logothetis DE (2011) Decoding the signaling of a GPCR heteromeric complex reveals a unifying mechanism of action of antipsychotic drugs. Cell 147(5):1011–1023
26. Forster T (1946) Energiewanderung Und Fluoreszenz. Naturwissenschaften 33(6):166–175
27. Clegg RM (1995) Fluorescence resonance energy transfer. Curr Opin Biotechnol 6(1):103–110
28. Lakowicz JR (2006) Principles of fluorescence spectroscopy, 3rd edn. Springer, New York
29. Shimomura O, Johnson FH, Saiga Y (1962) Extraction, purification and properties of aequorin, a bioluminescent protein from the luminous hydromedusan, Aequorea. J Cell Comp Physiol 59:223–239
30. Prasher DC, Eckenrode VK, Ward WW, Prendergast FG, Cormier MJ (1992) Primary structure of the *Aequorea victoria* green-fluorescent protein. Gene 111(2):229–233
31. Matz MV, Fradkov AF, Labas YA, Savitsky AP, Zaraisky AG, Markelov ML, Lukyanov SA (1999) Fluorescent proteins from nonbioluminescent *Anthozoa* species. Nat Biotechnol 17(10):969–973
32. Wiedenmann J, Oswald F, Nienhaus GU (2009) Fluorescent proteins for live cell imaging: opportunities, limitations, and challenges. IUBMB Life 61(11):1029–1042
33. Patterson GH, Lippincott-Schwartz J (2002) A photoactivatable GFP for selective photolabeling of proteins and cells. Science 297(5588):1873–1877
34. Chudakov DM, Verkhusha VV, Staroverov DB, Souslova EA, Lukyanov S, Lukyanov KA (2004) Photoswitchable cyan fluorescent protein for protein tracking. Nat Biotechnol 22(11):1435–1439
35. Matsuda T, Miyawaki A, Nagai T (2008) Direct measurement of protein dynamics inside cells using a rationally designed photoconvertible protein. Nat Methods 5(4):339–345
36. Maurel D, Banala S, Laroche T, Johnsson K (2010) Photoactivatable and photoconvertible fluorescent probes for protein labeling. ACS Chem Biol 5(5):507–516
37. Miyawaki A (2011) Proteins on the move: insights gained from fluorescent protein technologies. Nat Rev Mol Cell Biol 12(10):656–668
38. Vilardaga JP, Bunemann M, Krasel C, Castro M, Lohse MJ (2003) Measurement of the millisecond activation switch of G protein-coupled receptors in living cells. Nat Biotechnol 21(7):807–812
39. Valentin G, Verheggen C, Piolot T, Neel H, Coppey-Moisan M, Bertrand E (2005) Photoconversion of YFP into a CFP-like species during acceptor photobleaching FRET experiments. Nat Methods 2(11):801
40. Miyawaki A, Tsien RY (2000) Monitoring protein conformations and interactions by fluorescence resonance energy transfer between mutants of green fluorescent protein. Methods Enzymol 327:472–500
41. Zacharias DA, Violin JD, Newton AC, Tsien RY (2002) Partitioning of lipid-modified monomeric GFPs into membrane microdomains of live cells. Science 296(5569):913–916

42. Gaibelet G, Planchenault T, Mazeres S, Dumas F, Arenzana-Seisdedos F, Lopez A, Lagane B, Bachelerie F (2006) CD4 and CCR5 constitutively interact at the plasma membrane of living cells: a confocal fluorescence resonance energy transfer-based approach. J Biol Chem 281(49):37921–37929
43. Hollins B, Kuravi S, Digby GJ, Lambert NA (2009) The c-terminus of GRK3 indicates rapid dissociation of G protein heterotrimers. Cell Signal 21(6):1015–1021
44. Berney C, Danuser G (2003) FRET or no FRET: a quantitative comparison. Biophys J 84(6):3992–4010
45. Sun Y, Periasamy A (2010) Additional correction for energy transfer efficiency calculation in filter-based Forster resonance energy transfer microscopy for more accurate results. J Biomed Opt 15(2):020513
46. Zheng J, Trudeau MC, Zagotta WN (2002) Rod cyclic nucleotide-gated channels have a stoichiometry of three CNGA1 subunits and one CNGB1 subunit. Neuron 36(5):891–896
47. Vogel SS, Thaler C, Koushik SV (2006) Fanciful FRET. Sci STKE 2006 (331):re2 http://stke.sciencemag.org/cgi/content/abstract/sigtrans;2006/331/re2
48. Dinant C, van Royen ME, Vermeulen W, Houtsmuller AB (2008) Fluorescence resonance energy transfer of GFP and YFP by spectral imaging and quantitative acceptor photobleaching. J Microsc 231(Pt 1):97–104
49. Ayoub MA, Pfleger KD (2010) Recent advances in bioluminescence resonance energy transfer technologies to study GPCR heteromerization. Curr Opin Pharmacol 10(1):44–52
50. Achour L, Kamal M, Jockers R, Marullo S (2011) Using Quantitative BRET to Assess G Protein-Coupled Receptor Homo- and Heterodimerization. Methods Mol Biol 756:183–200
51. Pfleger KD, Eidne KA (2006) Illuminating insights into protein-protein interactions using bioluminescence resonance energy transfer (BRET). Nat Methods 3(3):165–174
52. Pfleger KD, Seeber RM, Eidne KA (2006) Bioluminescence resonance energy transfer (BRET) for the real-time detection of protein-protein interactions. Nat Protoc 1(1):337–345
53. Kocan M, See HB, Seeber RM, Eidne KA, Pfleger KD (2008) Demonstration of improvements to the bioluminescence resonance energy transfer (BRET) technology for the monitoring of G protein-coupled receptors in live cells. J Biomol Screen 13(9):888–898
54. Kocan M, Pfleger KD (2011) Study of GPCR-protein interactions by BRET. Methods Mol Biol 746:357–371
55. Breton B, Sauvageau E, Zhou J, Bonin H, Le Gouill C, Bouvier M (2010) Multiplexing of multicolor bioluminescence resonance energy transfer. Biophys J 99(12):4037–4046
56. De A, Ray P, Loening AM, Gambhir SS (2009) BRET3: a red-shifted bioluminescence resonance energy transfer (BRET)-based integrated platform for imaging protein-protein interactions from single live cells and living animals. FASEB J 23(8):2702–2709
57. Ayoub MA, Couturier C, Lucas-Meunier E, Angers S, Fossier P, Bouvier M, Jockers R (2002) Monitoring of ligand-independent dimerization and ligand-induced conformational changes of melatonin receptors in living cells by bioluminescence resonance energy transfer. J Biol Chem 277(24):21522–21528
58. Terrillon S, Durroux T, Mouillac B, Breit A, Ayoub MA, Taulan M, Jockers R, Barberis C, Bouvier M (2003) Oxytocin and vasopressin V1a and V2 receptors form constitutive homo- and heterodimers during biosynthesis. Mol Endocrinol 17(4):677–691
59. Angers S, Salahpour A, Bouvier M (2002) Dimerization: an emerging concept for G protein-coupled receptor ontogeny and function. Annu Rev Pharmacol Toxicol 42:409–435
60. James JR, Oliveira MI, Carmo AM, Iaboni A, Davis SJ (2006) A rigorous experimental framework for detecting protein oligomerization using bioluminescence resonance energy transfer. Nat Methods 3(12):1001–1006
61. Bouvier M, Heveker N, Jockers R, Marullo S, Milligan G (2007) BRET analysis of GPCR oligomerization: newer does not mean better. Nat Methods 4(1):3–4; author reply 4
62. Hu CD, Chinenov Y, Kerppola TK (2002) Visualization of interactions among bZIP and Rel family proteins in living cells using bimolecular fluorescence complementation. Mol Cell 9(4):789–798

63. Hu CD, Grinberg AV, Kerppola TK (2006) Visualization of protein interactions in living cells using bimolecular fluorescence complementation (BiFC) analysis. Curr Protoc Cell Biol Chapter 21:Unit 21. 23
64. Kerppola TK (2006) Design and implementation of bimolecular fluorescence complementation (BiFC) assays for the visualization of protein interactions in living cells. Nat Protoc 1(3):1278–1286
65. Kerppola TK (2006) Complementary methods for studies of protein interactions in living cells. Nat Methods 3(12):969–971
66. Kerppola TK (2006) Visualization of molecular interactions by fluorescence complementation. Nat Rev Mol Cell Biol 7(6):449–456
67. Vidi PA, Ejendal KF, Przybyla JA, Watts VJ (2011) Fluorescent protein complementation assays: new tools to study G protein-coupled receptor oligomerization and GPCR-mediated signaling. Mol Cell Endocrinol 331(2):185–193
68. Vidi PA, Chen J, Irudayaraj JM, Watts VJ (2008) Adenosine A(2A) receptors assemble into higher-order oligomers at the plasma membrane. FEBS Lett 582(29):3985–3990
69. Guo W, Urizar E, Kralikova M, Mobarec JC, Shi L, Filizola M, Javitch JA (2008) Dopamine D2 receptors form higher order oligomers at physiological expression levels. EMBO J 27 (17):2293–2304
70. Urizar E, Yano H, Kolster R, Gales C, Lambert N, Javitch JA (2011) CODA-RET reveals functional selectivity as a result of GPCR heteromerization. Nat Chem Biol 7(9):624–630
71. Heroux M, Hogue M, Lemieux S, Bouvier M (2007) Functional calcitonin gene-related peptide receptors are formed by the asymmetric assembly of a calcitonin receptor-like receptor homo-oligomer and a monomer of receptor activity-modifying protein-1. J Biol Chem 282(43):31610–31620
72. Pietraszewska-Bogiel A, Gadella TW (2011) FRET microscopy: from principle to routine technology in cell biology. J Microsc 241(2):111–118
73. Tadross MR, Park SA, Veeramani B, Yue DT (2009) Robust approaches to quantitative ratiometric FRET imaging of CFP/YFP fluorophores under confocal microscopy. J Microsc 233(1):192–204
74. Coulon V, Audet M, Homburger V, Bockaert J, Fagni L, Bouvier M, Perroy J (2008) Subcellular imaging of dynamic protein interactions by bioluminescence resonance energy transfer. Biophys J 94(3):1001–1009
75. Perroy J (2010) Subcellular dynamic imaging of protein-protein interactions in live cells by bioluminescence resonance energy transfer. Methods Mol Biol 591:325–333
76. Zhang J, Campbell RE, Ting AY, Tsien RY (2002) Creating new fluorescent probes for cell biology. Nat Rev Mol Cell Biol 3(12):906–918
77. Resch-Genger U, Grabolle M, Cavaliere-Jaricot S, Nitschke R, Nann T (2008) Quantum dots versus organic dyes as fluorescent labels. Nat Methods 5(9):763–775
78. Bates M, Huang B, Dempsey GT, Zhuang X (2007) Multicolor super-resolution imaging with photo-switchable fluorescent probes. Science 317(5845):1749–1753
79. Tzeng YK, Faklaris O, Chang BM, Kuo Y, Hsu JH, Chang HC (2011) Superresolution imaging of albumin-conjugated fluorescent nanodiamonds in cells by stimulated emission depletion. Angewandte Chemie 50(10):2262–2265
80. Mathis G (1995) Probing molecular interactions with homogeneous techniques based on rare earth cryptates and fluorescence energy transfer. Clin Chem 41(9):1391–1397
81. Selvin PR (2000) The renaissance of fluorescence resonance energy transfer. Nat Struct Biol 7(9):730–734
82. Selvin PR (2002) Principles and biophysical applications of lanthanide-based probes. Annu Rev Biophys Biomol Struct 31:275–302
83. Pin JP, Maurel D, Comps-Agrar L, Monnier C, Rives ML, Doumazane E, Rondard P, Durroux T, Prezeau L, Trinquet E (2010) Time-resolved FRET approaches to study GPCR complexes. In: Siehler S, Milligan G (eds) G protein-coupled receptors: structure signaling and physiology. Cambridge University Press, Cambridge, pp 67–89

84. Pradidarcheep W, Labruyere WT, Dabhoiwala NF, Lamers WH (2008) Lack of specificity of commercially available antisera: better specifications needed. J Histochem Cytochem 56 (12):1099–1111
85. Beermann S, Seifert R, Neumann D (2012) Commercially available antibodies against human and murine histamine H(4)-receptor lack specificity. Naunyn Schmiedebergs Arch Pharmacol 385(2):125–135
86. Bodei S, Arrighi N, Spano P, Sigala S (2009) Should we be cautious on the use of commercially available antibodies to dopamine receptors? Naunyn Schmiedebergs Arch Pharmacol 379(4):413–415
87. Jensen BC, Swigart PM, Simpson PC (2009) Ten commercial antibodies for alpha-1-adrenergic receptor subtypes are nonspecific. Naunyn Schmiedebergs Arch Pharmacol 379(4):409–412
88. Michel MC, Wieland T, Tsujimoto G (2009) How reliable are G-protein-coupled receptor antibodies? Naunyn Schmiedebergs Arch Pharmacol 379(4):385–388
89. Pradidarcheep W, Stallen J, Labruyere WT, Dabhoiwala NF, Michel MC, Lamers WH (2009) Lack of specificity of commercially available antisera against muscarinergic and adrenergic receptors. Naunyn Schmiedebergs Arch Pharmacol 379(4):397–402
90. Rocheville M, Lange DC, Kumar U, Sasi R, Patel RC, Patel YC (2000) Subtypes of the somatostatin receptor assemble as functional homo- and heterodimers. J Biol Chem 275 (11):7862–7869
91. McVey M, Ramsay D, Kellett E, Rees S, Wilson S, Pope AJ, Milligan G (2001) Monitoring receptor oligomerization using time-resolved fluorescence resonance energy transfer and bioluminescence resonance energy transfer. The human delta-opioid receptor displays constitutive oligomerization at the cell surface, which is not regulated by receptor occupancy. J Biol Chem 276(17):14092–14099
92. Maurel D, Kniazeff J, Mathis G, Trinquet E, Pin JP, Ansanay H (2004) Cell surface detection of membrane protein interaction with homogeneous time-resolved fluorescence resonance energy transfer technology. Anal Biochem 329(2):253–262
93. Urizar E, Montanelli L, Loy T, Bonomi M, Swillens S, Gales C, Bouvier M, Smits G, Vassart G, Costagliola S (2005) Glycoprotein hormone receptors: link between receptor homodimerization and negative cooperativity. EMBO J 24(11):1954–1964
94. O'Hare HM, Johnsson K, Gautier A (2007) Chemical probes shed light on protein function. Curr Opin Struct Biol 17(4):488–494
95. Keppler A, Gendreizig S, Gronemeyer T, Pick H, Vogel H, Johnsson K (2003) A general method for the covalent labeling of fusion proteins with small molecules in vivo. Nat Biotechnol 21(1):86–89
96. George N, Pick H, Vogel H, Johnsson N, Johnsson K (2004) Specific labeling of cell surface proteins with chemically diverse compounds. J Am Chem Soc 126(29):8896–8897
97. Griffin BA, Adams SR, Tsien RY (1998) Specific covalent labeling of recombinant protein molecules inside live cells. Science 281(5374):269–272
98. Gautier A, Juillerat A, Heinis C, Correa IR Jr, Kindermann M, Beaufils F, Johnsson K (2008) An engineered protein tag for multiprotein labeling in living cells. Chem Biol 15(2):128–136
99. Los GV, Wood K (2007) The HaloTag: a novel technology for cell imaging and protein analysis. Methods Mol Biol 356:195–208
100. Maurel D, Comps-Agrar L, Brock C, Rives ML, Bourrier E, Ayoub MA, Bazin H, Tinel N, Durroux T, Prezeau L, Trinquet E, Pin JP (2008) Cell-surface protein-protein interaction analysis with time-resolved FRET and snap-tag technologies: application to GPCR oligomerization. Nat Methods 5(6):561–567
101. Comps-Agrar L, Kniazeff J, Brock C, Trinquet E, Pin JP (2012) Stability of GABAB receptor oligomers revealed by dual TR-FRET and drug-induced cell surface targeting. FASEB J 26(8):3430–3439
102. Doumazane E, Scholler P, Zwier JM, Eric T, Rondard P, Pin JP (2011) A new approach to analyze cell surface protein complexes reveals specific heterodimeric metabotropic glutamate receptors. FASEB J 25(1):66–77

103. Monnier C, Tu H, Bourrier E, Vol C, Lamarque L, Trinquet E, Pin JP, Rondard P (2011) Trans-activation between 7TM domains: implication in heterodimeric GABA(B) receptor activation. EMBO J 30(1):32–42
104. Hoffmann C, Gaietta G, Bunemann M, Adams SR, Oberdorff-Maass S, Behr B, Vilardaga JP, Tsien RY, Ellisman MH, Lohse MJ (2005) A FlAsH-based FRET approach to determine G protein-coupled receptor activation in living cells. Nat Methods 2(3):171–176
105. Ambrosio M, Lohse MJ (2010) Microscopy: GPCR dimers moving closer. Nat Chem Biol 6(8):570–571
106. Scherrer G, Tryoen-Toth P, Filliol D, Matifas A, Laustriat D, Cao YQ, Basbaum AI, Dierich A, Vonesh JL, Gaveriaux-Ruff C, Kieffer BL (2006) Knockin mice expressing fluorescent delta-opioid receptors uncover G protein-coupled receptor dynamics in vivo. Proc Natl Acad Sci USA 103(25):9691–9696
107. Zwier JM, Roux T, Cottet M, Durroux T, Douzon S, Bdioui S, Gregor N, Bourrier E, Oueslati N, Nicolas L, Tinel N, Boisseau C, Yverneau P, Charrier-Savournin F, Fink M, Trinquet E (2010) A fluorescent ligand-binding alternative using Tag-lite(R) technology. J Biomol Screen 15(10):1248–1259
108. Briddon SJ, Kellam B, Hill SJ (2011) Design and use of fluorescent ligands to study ligand-receptor interactions in single living cells. Methods Mol Biol 746:211–236
109. Cottet M, Faklaris O, Zwier JM, Trinquet E, Pin J-P, Durroux T (2011) Original fluorescent ligand-based assays open new perspectives in G-protein coupled receptor drug screening. Pharmaceuticals 4(1):202–214
110. Lodowski DT, Palczewski K (2011) Chapter 1 The impact of G protein-coupled receptor (GPCR) structures on understanding signal transduction. In: Giraldo J, Pin JP (eds) G protein-coupled receptors: from structure to function. The Royal Society of Chemistry, Cambridge, UK, pp 1–27
111. Durroux T, Peter M, Turcatti G, Chollet A, Balestre MN, Barberis C, Seyer R (1999) Fluorescent pseudo-peptide linear vasopressin antagonists: design, synthesis, and applications. J Med Chem 42(7):1312–1319
112. Terrillon S, Cheng LL, Stoev S, Mouillac B, Barberis C, Manning M, Durroux T (2002) Synthesis and characterization of fluorescent antagonists and agonists for human oxytocin and vasopressin V(1)(a) receptors. J Med Chem 45(12):2579–2588
113. Mouillac B, Manning M, Durroux T (2008) Fluorescent agonists and antagonists for vasopressin/oxytocin G protein-coupled receptors: usefulness in ligand screening assays and receptor studies. Mini Rev Med Chem 8(10):996–1005
114. Kuder K, Kiec-Kononowicz K (2008) Fluorescent GPCR ligands as new tools in pharmacology. Curr Med Chem 15(21):2132–2143
115. Daly CJ, Ross RA, Whyte J, Henstridge CM, Irving AJ, McGrath JC (2010) Fluorescent ligand binding reveals heterogeneous distribution of adrenoceptors and 'cannabinoid-like' receptors in small arteries. Br J Pharmacol 159(4):787–796
116. Albizu L, Cottet M, Kralikova M, Stoev S, Seyer R, Brabet I, Roux T, Bazin H, Bourrier E, Lamarque L, Breton C, Rives ML, Newman A, Javitch J, Trinquet E, Manning M, Pin JP, Mouillac B, Durroux T (2010) Time-resolved FRET between GPCR ligands reveals oligomers in native tissues. Nat Chem Biol 6(8):587–594
117. Franco R, Casado V, Mallol J, Ferre S, Fuxe K, Cortes A, Ciruela F, Lluis C, Canela EI (2005) Dimer-based model for heptaspanning membrane receptors. Trends Biochem Sci 30(7):360–366
118. Kern A, Albarran-Zeckler R, Walsh HE, Smith RG (2012) Apo-ghrelin receptor forms heteromers with DRD2 in hypothalamic neurons and is essential for anorexigenic effects of DRD2 agonism. Neuron 73(2):317–332
119. Rajapakse HE, Gahlaut N, Mohandessi S, Yu D, Turner JR, Miller LW (2010) Time-resolved luminescence resonance energy transfer imaging of protein-protein interactions in living cells. Proc Natl Acad Sci USA 107(31):13582–13587

120. Rajapakse HE, Miller LW (2012) Time-resolved luminescence resonance energy transfer imaging of protein-protein interactions in living cells. Methods Enzymol 505:329–345
121. Grecco HE, Verveer PJ (2011) FRET in cell biology: still shining in the age of super-resolution? Chemphyschem: Eur J Chem Physics Phys Chem 12(3):484–490
122. Schermelleh L, Heintzmann R, Leonhardt H (2010) A guide to super-resolution fluorescence microscopy. J Cell Biol 190(2):165–175
123. Josan JS, Morse DL, Xu L, Trissal M, Baggett B, Davis P, Vagner J, Gillies RJ, Hruby VJ (2009) Solid-phase synthetic strategy and bioevaluation of a labeled delta-opioid receptor ligand Dmt-Tic-Lys for in vivo imaging. Org Lett 11(12):2479–2482
124. Dorsch S, Klotz KN, Engelhardt S, Lohse MJ, Bunemann M (2009) Analysis of receptor oligomerization by FRAP microscopy. Nat Methods 6(3):225–230
125. Rybin VO, Xu X, Lisanti MP, Steinberg SF (2000) Differential targeting of beta -adrenergic receptor subtypes and adenylyl cyclase to cardiomyocyte caveolae. A mechanism to functionally regulate the cAMP signaling pathway. J Biol Chem 275(52):41447–41457
126. Fallahi-Sichani M, Linderman JJ (2009) Lipid raft-mediated regulation of G-protein coupled receptor signaling by ligands which influence receptor dimerization: a computational study. PLoS ONE 4(8):e6604
127. Hern JA, Baig AH, Mashanov GI, Birdsall B, Corrie JE, Lazareno S, Molloy JE, Birdsall NJ (2010) Formation and dissociation of M1 muscarinic receptor dimers seen by total internal reflection fluorescence imaging of single molecules. Proc Natl Acad Sci USA 107 (6):2693–2698
128. Kasai RS, Suzuki KGN, Prossnitz ER, Koyama-Honda I, Nakada C, Fujiwara TK, Kusumi A (2011) Full characterization of GPCR monomer-dimer dynamic equilibrium by single molecule imaging. J Cell Biol 192(3):463–480
129. Lambert NA (2010) GPCR dimers fall apart. Sci Signal 3 (115):pe12
130. Mortensen KI, Churchman LS, Spudich JA, Flyvbjerg H (2010) Optimized localization analysis for single-molecule tracking and super-resolution microscopy. Nat Methods 7(5):377–381
131. Jones SA, Shim SH, He J, Zhuang X (2011) Fast, three-dimensional super-resolution imaging of live cells. Nat Methods 8(6):499–508
132. Hess ST, Gould TJ, Gudheti MV, Maas SA, Mills KD, Zimmerberg J (2007) Dynamic clustered distribution of hemagglutinin resolved at 40 nm in living cell membranes discriminates between raft theories. Proc Natl Acad Sci U S A 104(44):17370–17375
133. Pellett PA, Sun X, Gould TJ, Rothman JE, Xu MQ, Correa IR Jr, Bewersdorf J (2011) Two-color STED microscopy in living cells. Biomed Opt Express 2(8):2364–2371
134. Mayor S, Rao M (2004) Rafts: scale-dependent, active lipid organization at the cell surface. Traffic 5(4):231–240
135. Zhao Y, Terry D, Shi L, Weinstein H, Blanchard SC, Javitch JA (2010) Single-molecule dynamics of gating in a neurotransmitter transporter homologue. Nature 465(7295):188–193
136. Sakon JJ, Weninger KR (2010) Detecting the conformation of individual proteins in live cells. Nat Methods 7(3):203–205
137. Ilien B, Glasser N, Clamme JP, Didier P, Piemont E, Chinnappan R, Daval SB, Galzi JL, Mely Y (2009) Pirenzepine promotes the dimerization of muscarinic M1 receptors through a three-step binding process. J Biol Chem 284(29):19533–19543
138. Zeng FY, Wess J (1999) Identification and molecular characterization of m3 muscarinic receptor dimers. J Biol Chem 274(27):19487–19497
139. Goin JC, Nathanson NM (2006) Quantitative analysis of muscarinic acetylcholine receptor homo- and heterodimerization in live cells: regulation of receptor down-regulation by heterodimerization. J Biol Chem 281(9):5416–5425
140. Grant M, Collier B, Kumar U (2004) Agonist-dependent dissociation of human somatostatin receptor 2 dimers: a role in receptor trafficking. J Biol Chem 279(35):36179–36183

141. Ross JA, Digman MA, Wang L, Gratton E, Albanesi JP, Jameson DM (2011) Oligomerization state of dynamin 2 in cell membranes using TIRF and number and brightness analysis. Biophys J 100(3):L15–17
142. Digman MA, Dalal R, Horwitz AF, Gratton E (2008) Mapping the number of molecules and brightness in the laser scanning microscope. Biophys J 94(6):2320–2332
143. Sergeev M, Costantino S, Wiseman PW (2006) Measurement of monomer-oligomer distributions via fluorescence moment image analysis. Biophys J 91(10):3884–3896
144. Wu B, Chien EY, Mol CD, Fenalti G, Liu W, Katritch V, Abagyan R, Brooun A, Wells P, Bi FC, Hamel DJ, Kuhn P, Handel TM, Cherezov V, Stevens RC (2010) Structures of the CXCR4 chemokine GPCR with small-molecule and cyclic peptide antagonists. Science 330 (6007):1066–1071
145. Manglik A, Kruse AC, Kobilka TS, Thian FS, Mathiesen JM, Sunahara RK, Pardo L, Weis WI, Kobilka BK, Granier S (2012) Crystal structure of the micro-opioid receptor bound to a morphinan antagonist. Nature 485(7398):321–326

Application of Quantitative Fluorescence Microscopic Approaches to Monitor Organization and Dynamics of the Serotonin$_{1A}$ Receptor

Md. Jafurulla and Amitabha Chattopadhyay

Abstract G protein-coupled receptors (GPCRs) are the largest class of molecules involved in signal transduction across membranes and represent major targets in the development of novel drug candidates in all clinical areas. Recent advances in understanding of the nonrandom distribution of GPCRs, G proteins, and effector molecules have given rise to new challenges and complexities in cellular signaling by GPCRs. In this article, we provide specific examples on the application of quantitative fluorescence microscopic approaches to monitor organization and dynamics of the serotonin$_{1A}$ receptor (a GPCR) in live cells. This assumes broader relevance due to the emerging theme that GPCR function depends on its organization and dynamics. We envisage that with progress in understanding of receptor organization and dynamics, our knowledge of GPCR function would improve considerably, thereby enabling to design better therapeutic strategies to combat diseases related to malfunctioning of GPCRs.

Keywords Actin · GPCR · Oligomerization · Homo-FRET · Serotonin$_{1A}$ receptor · zFCS

Contents

M. Jafurulla and A. Chattopadhyay (✉)
Centre for Cellular and Molecular Biology, Council of Scientific and Industrial Research, Hyderabad 500 007, India
e-mail: amit@ccmb.res.in

Y. Mély and G. Duportail (eds.), *Fluorescent Methods to Study Biological Membranes*, Springer Ser Fluoresc (2013) 13: 417–438, DOI 10.1007/4243_2012_58,

Abbreviations

Bodipy-FL PC	2-(4,4-difluoro-5,7-dimethyl-4-bora-3a,4a-diaza-*s*-indacene-3-pentanoyl)-1-hexadecanoyl-*sn*-glycero-3-phosphocholine
CD	Cytochalasin D
DRM	Detergent-resistant membrane
EYFP	Enhanced yellow fluorescent protein
F-actin	Filamentous actin
FCS	Fluorescence correlation spectroscopy
FPR	*N*-formyl peptide receptor
FRET	Fluorescence resonance energy transfer
G-actin	Globular actin
GFP	Green fluorescent protein
GPCR	G protein-coupled receptor
Jas	Jasplakinolide
LatA	Latrunculin A
MβCD	Methyl-β-cyclodextrin
p-MPPI	4-(2′-Methoxy)-phenyl-1-[2′-(*N*-2″-pyridinyl)-*p*-iodobenzamido]ethyl-piperazine
zFCS	Z-Scanning FCS

1 Introduction

G protein-coupled receptors (GPCRs) represent the largest and most diverse protein family in mammals, involved in signal transduction across membranes [69, 80]. GPCRs are seven transmembrane domain proteins and include >800 members which are encoded by ~5 % of human genes [98]. In terms of evolution, they constitute the most versatile group of proteins in the human genome [68]. GPCRs dictate physiological responses to a diverse array of stimuli from biogenic amines, peptides, glycoproteins, lipids, and nucleotides to even photons. As a consequence, these receptors mediate multiple physiological processes such as neurotransmission, cellular metabolism, secretion, cellular differentiation, growth, and inflammatory and immune responses. GPCRs therefore have emerged as major targets for the development of novel drug candidates in all clinical areas [29]. It is estimated that ~50 % of clinically prescribed drugs act as either agonists or antagonists of GPCRs, and several ligands of GPCRs are listed among the top 100 globally selling drugs, which points out their immense therapeutic potential [84]. Importantly, GPCRs have been shown to be hijacked by pathogenic organisms for their entry and infection [5, 25].

The common structural features of G protein-coupled receptors include seven transmembrane domains with an extracellular N-terminus and a cytoplasmic C-terminus. Cellular signaling by GPCRs involves their activation by ligands present in the extracellular environment and the subsequent transduction of signals to the interior of the cell through concerted changes in their transmembrane domain structure [18]. The major paradigm in GPCR signaling is that ligand-mediated stimulation of these receptors leads to the activation of heterotrimeric GTP-binding proteins (G proteins) [62]. The multiple components of GPCR signal transduction such as different types of receptors and G protein subunits provide cells with the versatility to customize their responses to a diverse array of signals and stimuli.

The classical view of uniform distribution of GPCRs, G proteins, and effectors and their free diffusion on the cell surface is inadequate in explaining the specific and rapid signaling responses characteristic of GPCRs [61, 63]. This suggests that receptor-G protein interactions may be dependent on their spatiotemporal organization in the membrane. The organization and restricted mobility of signaling components such as receptors, G proteins, and effectors in the membrane, in addition to their specificity of interaction, are now believed to be important in spatiotemporal regulation of GPCR signaling [14, 33, 36, 63, 83]. In this overall scenario, the observation that GPCRs are not uniformly present on the plasma membrane but are concentrated in specific membrane domains enriched in cholesterol assumes relevance [63]. Understanding the role of cholesterol and cytoskeletal components (believed to be crucial in the organization of membrane domains) in GPCR function therefore represents a challenging aspect of GPCR signaling [49, 59].

Fluorescence-based approaches enjoy certain advantages over other approaches in monitoring GPCR functions such as receptor-receptor and receptor-ligand interactions, real-time assessment of signal transduction, receptor dynamics in the plasma membrane, and intracellular trafficking of receptors [46, 56, 74, 83]. This is due to their enhanced sensitivity, minimal perturbation, multiplicity of measurable parameters, and suitable time scales that allow the analysis of several biologically relevant molecular processes. In this review, we provide representative examples of the application of contemporary fluorescence-based microscopic approaches to explore receptor organization and dynamics in live cells using the serotonin$_{1A}$ receptor as a typical member of the GPCR family.

2 The Serotonin$_{1A}$ Receptor

The serotonin$_{1A}$ (5-HT$_{1A}$) receptor is an important neurotransmitter receptor and belongs to the GPCR superfamily. It is the most extensively studied among serotonin receptors for a variety of reasons [44, 75]. The serotonin$_{1A}$ receptor plays a key role in the generation and modulation of various cognitive, behavioral, and developmental functions such as sleep, mood, addiction, depression, anxiety, aggression, and learning [60]. The serotonin$_{1A}$ receptor serves as an important target in the development of therapeutic agents for neuropsychiatric disorders. Agonists [4] and antagonists [21] of the serotonin$_{1A}$ receptor represent major classes of molecules with potential therapeutic effects in anxiety- or stress-related disorders. Mutant (knockout) mice lacking the serotonin$_{1A}$ receptor exhibit enhanced anxiety-related behavior and represent an important animal model for the analysis of complex traits such as anxiety disorders and aggression in higher animals [17]. Work from our laboratory has comprehensively shown that membrane cholesterol [64, 67, 70, 71, 87] and sphingolipids [38, 65, 90] are necessary for the function of the serotonin$_{1A}$ receptor. In addition, the organization and dynamics of the receptor have been shown to depend on membrane cholesterol [72], sphingolipids [13], and the actin cytoskeleton [14].

3 A Novel Approach to Monitor Detergent Insolubility of GPCRs

The contemporary model of biological membranes involves the concept of lateral heterogeneities in the membrane, collectively termed as membrane domains. Although the concept of membrane domains is not new [47], understanding of the organization of such domains has considerably improved in the last two decades due to huge advancements in fluorescence spectroscopy and microscopy. In particular, the use of green fluorescent proteins (GFP) and its variants as reporter molecules in the cellular milieu has allowed direct visualization of signaling and real-time trafficking in live cells [19, 22, 23, 74, 93]. Such domains (sometimes termed as "lipid rafts" [54]) are believed to be important for the maintenance of membrane structure and function, although characterizing the spatiotemporal resolution of these domains has proven to be challenging [11, 37]. These specialized regions contribute to variable patchiness of the membrane and are enriched in specific lipids and proteins. They are believed to facilitate cellular trafficking and sorting [89], signal transduction [88], and pathogen entry [73, 79]. It is therefore crucial to appreciate the dynamic organization of membrane-bound components to understand cellular signaling [14, 37, 88].

Insolubility of membrane components in nonionic detergents such as Triton X-100 at low temperature has been widely used as a biochemical criterion to identify, isolate, and characterize membrane domains [6, 8, 31]. The tight acyl chain packing in cholesterol-sphingolipid-rich membrane regions is believed to provide detergent resistance to these regions (referred as detergent-resistant membranes (DRMs)) enriched in these lipids and to the proteins which reside in them [85]. Interestingly, resistance to detergent extraction continues to be a major tool to monitor membrane domains in spite of concerns of membrane perturbation due to the use of detergents [28].

Proteins in DRMs are generally detected either by immunoblotting or ligand binding. Unfortunately, these approaches are found to be not suitable when ligand

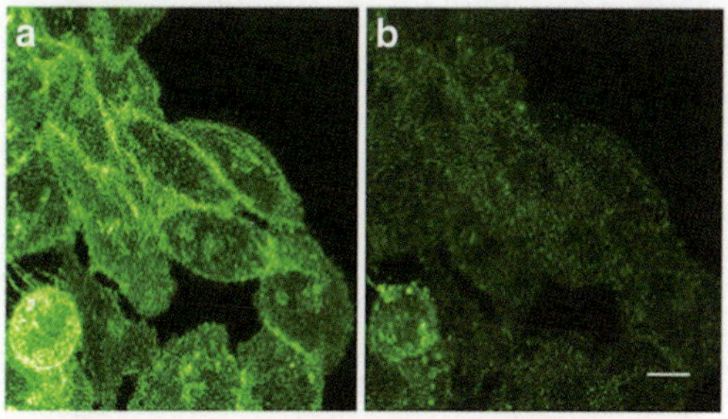

Fig. 1 Detergent insolubility of the serotonin$_{1A}$ receptor tagged to EYFP in CHO cells. Cells stably expressing serotonin$_{1A}$ receptors tagged to EYFP are shown (**a**) before and (**b**) after treatment with cold Triton X-100 (0.05 %, w/v). The images represent combined midplane confocal sections of the same group of cells before and after detergent extraction. The scale bar represents 10 μm. Reproduced from [42]. See [42] for details

binding of the protein is compromised in presence of detergent [2] or is limited by availability of antibodies with high specificity [99]. The use of GFP technology offers an alternative approach that can overcome these limitations. Using GFP fluorescence, detergent insolubility of the serotonin$_{1A}$ receptor has been directly determined [42, 43]. This approach relies on quantitation of integrated fluorescence intensity of membrane proteins in live cells (using the linear range, avoiding saturation of the detection system (photomultiplier tube)) before and after detergent treatment (see Fig. 1). Utilizing this GFP-based approach, it was found that ~26 % of serotonin$_{1A}$-EYFP receptors were localized in DRMs [42]. These results imply that the domain organization of serotonin$_{1A}$ receptors is heterogeneous and only a small (~26 %) population of receptors resides in DRMs. The validation of this approach came from later studies utilizing Western blot and copatching [78, 82] and a detergent-free approach [45].

Importantly, this method can be used in exploring membrane localization of any membrane protein, suitably tagged with GFP or its variants. Interestingly, this method has the potential to be used in large-scale screening of detergent insolubility of membrane proteins and receptors by monitoring insolubility of the fluorescently tagged protein by automated fluorescence imaging system capable of handling multiple samples.

4 Z-Scanning FCS: Exploring Confined Receptor Dynamics

Fluorescence correlation spectroscopy (FCS) represents a powerful and sensitive technique for measuring molecular diffusion [26]. In this approach, the small spontaneous fluctuations in fluorescence intensity due to diffusion of fluorophores into and out of an open confocal volume are monitored. The resultant autocorrelation function contains information on diffusion coefficient and number of particles in the sampling volume. However, single point FCS measurements could result in inaccurate estimation of diffusion parameters in case of membrane-bound molecules. This is due to the fact that the dimension of the membrane bilayer (~5 nm thick) is about three orders of magnitude smaller than that of the diffraction-limited spot size along the optic axis (~1 μm), which is typical axial length in FCS measurements [30]. This problem can be avoided in a variation of the FCS measurement, termed as z-scanning FCS (zFCS). In zFCS, the uncertainty in the positioning of the focused beam is avoided by a scan along the z-axis (z-scan) in which the diffusion times are determined in steps as the z-axis is scanned in small increments ([3, 11, 27, 35]; see left panel, Fig. 2a). Since the time taken to diffuse through a circular area (beam spot) scales with the square of the radius, the characteristic diffusion time (τ_D) for zFCS measurements is related to the projected area illuminated by the diffraction-limited spot on the plane of the plasma membrane (see right panel, Fig. 2a). A plot of diffusion time versus transverse area of the confocal volume yields the "FCS diffusion laws" that give information on the organization at submicron level (such as confinement and/or partitioning) of the diffusing species ([51, 95]; see Fig. 2b). According to FCS diffusion laws, the linear fit of data when extrapolated to zero spot width provides information on the nature of confinement experienced by the diffusing molecule. In case of free diffusion, the intercept is close to zero. In cases where the diffusing species experiences confinement, a negative intercept is obtained (Fig. 2b).

In a recent work, we combined the principle of zFCS and FCS diffusion laws to address the organization and confined dynamics of the serotonin$_{1A}$ receptor in live cell membranes [11]. Our results showed that a fluorescently labeled phospholipid (Bodipy-FL PC) exhibited free, Brownian diffusion in the membrane, as evident from zero intercept (left panel, Fig. 2b). In contrast, the serotonin$_{1A}$ receptor displayed confined diffusion, as apparent from the negative intercept of the plot of τ_D versus Δz^2 for control cells (right panel, Fig. 2b). The typical dependence of diffusion time for the serotonin$_{1A}$ receptor on the position of the focus on the z-axis under conditions of increasing actin destabilization is shown in Fig. 3a and c. Based on our previous work in which we observed an increase in receptor mobile fraction upon actin destabilization [14], the apparent confinement exhibited by the serotonin$_{1A}$ receptor in zFCS measurements could be due to the actin cytoskeleton network. This was confirmed when the receptor confinement was found to be progressively reduced (evident from a reduction in negative intercept, see Fig. 3b and d) upon treatment with increasing concentrations of cytochalasin D, which acts as a potent inhibitor of actin polymerization. We envisage that combined application of zFCS and the FCS diffusion laws could be a powerful tool to explore membrane heterogeneity at the submicron level.

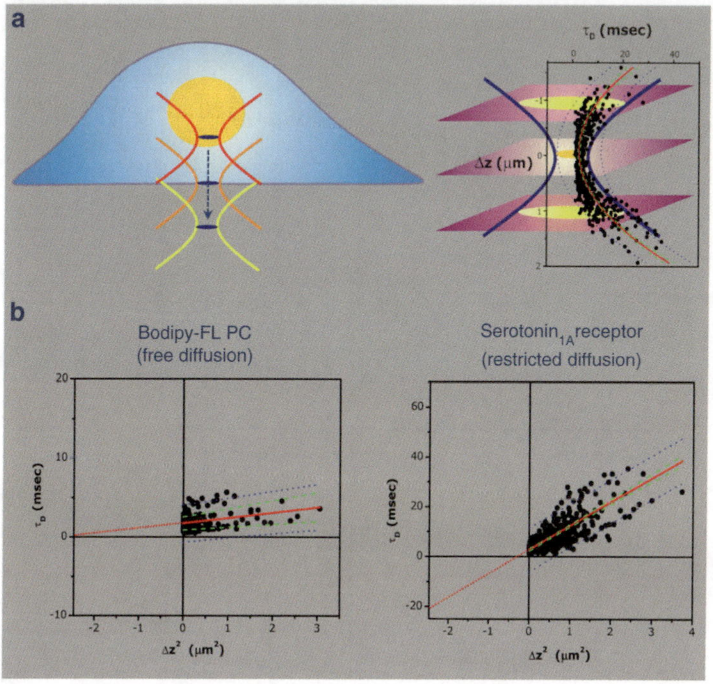

Fig. 2 Principle of z-scanning FCS (zFCS) and the application of FCS diffusion laws. (**a**) A schematic representation of the illumination profile in zFCS and the dependence of lateral diffusion time on the z position of the focus. The *z*-axis depicted in the figure (*blue dotted arrow*) is the optic axis (perpendicular to the image plane). The focus of the diffraction-limited illumination profile of the laser spot is moved in incremental steps (marked by arbitrary *z* positions) along the *z*-axis resulting in a parabolic scaling of the projected area. A region of interest can therefore be probed at increasing length scales. This allows probing the plasma membrane conveniently when fluorescence is predominantly originating from this region. Because the time taken to diffuse through a circular area should scale with the square of the radius, the characteristic diffusion time (τ_D) for zFCS measurements is related to the projected area illuminated by the diffraction-limited spot on the plane of the plasma membrane (shown in *panel on right*). (**b**) Representative plots of lateral diffusion time versus Δz^2 (FCS diffusion laws) for a freely diffusing phospholipid (Bodipy-FL PC) and the serotonin$_{1A}$ receptor with restricted diffusion in the plasma membrane are shown. The 95 % confidence interval (*green dashed line*) and 95 % prediction band (*blue dotted line*) for the fitted data are also shown. Note that Bodipy-FL PC exhibits free, Brownian diffusion in the membrane, as apparent from zero intercept (*left panel*). On the other hand, the serotonin$_{1A}$ receptor displays confined diffusion, evident from the negative intercept of the plot (*right panel*). Adapted and modified from [11]. See text and [11] for more details

5 Quantitation of Actin Reorganization

The actin cytoskeleton is involved in a multitude of cellular responses besides providing structural support. We have previously reported that the mobility and oligomerization of serotonin$_{1A}$ receptors are regulated by the integrity of actin

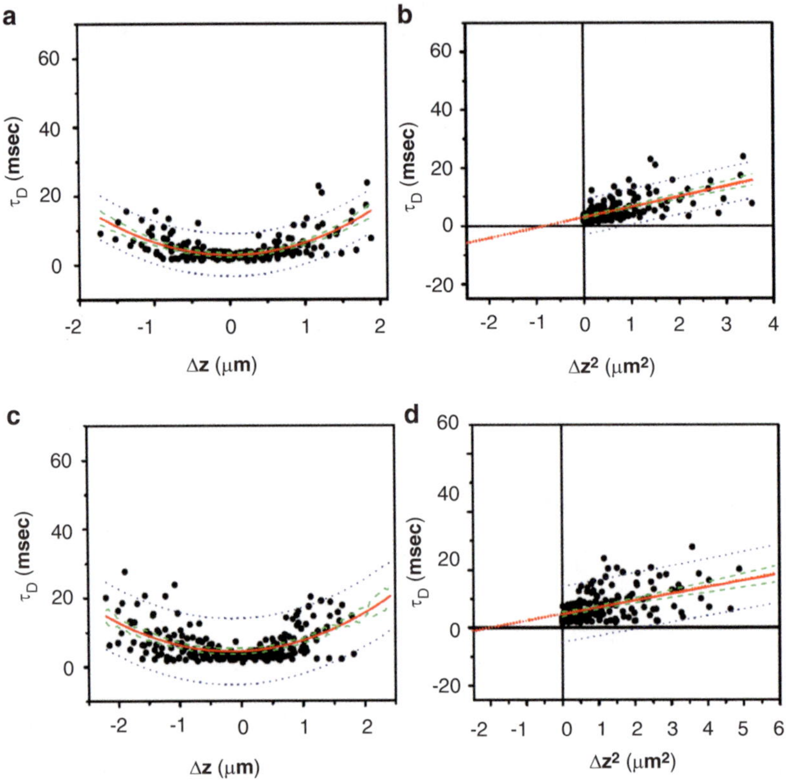

Fig. 3 Monitoring the role of actin cytoskeleton on the dynamics of serotonin$_{1A}$-EYFP receptors using zFCS. The actin cytoskeleton is destabilized utilizing cytochalasin D. Panels (**a**) and (**c**) show the dependence of lateral diffusion time for serotonin$_{1A}$-EYFP receptors on the z position of the focus, upon treatment with 5 and 10 μM cytochalasin D, respectively. Panels (**b**) and (**d**) show the corresponding plots of lateral diffusion time versus Δz^2 for data shown in panels (**a**) and (**c**). Data points were fitted to a straight line (shown as *red solid line*) and extrapolated to determine the intercept. The green dashed line shows 95 % confidence interval, and the blue dotted line shows 95 % prediction band for the fitted data. Reproduced from [11]. See text and [11] for more details

cytoskeleton meshwork [11, 12, 14, 16]. Our results showed an enhanced signaling efficiency of the serotonin$_{1A}$ receptor upon actin cytoskeleton destabilization. In addition, the mobility of the serotonin$_{1A}$ receptor in the plasma membrane was found to be strongly correlated with its signaling efficiency, thereby implying that the actin cytoskeleton could play a regulatory role in receptor signaling [14]. Interestingly, while the role of the actin cytoskeleton in cellular processes such as trafficking and motility has been studied [40], the reorganization of the actin cytoskeleton upon GPCR signaling has been rarely addressed. It is therefore important to develop approaches to estimate cellular actin and its modulation upon signaling.

Actin is one of the most abundant cytosolic proteins in eukaryotic cells and exists in both monomeric (globular or G-actin) and polymeric (filamentous or F-actin) forms. F-actin is maintained in cells in dynamic equilibrium with soluble G-actin. The extent of actin polymerization and depolymerization (actin reorganization) is orchestrated by a number of actin-binding proteins in response to a variety of stimuli [9]. Such reorganization offers a mechanism in which dynamic changes in the actin cytoskeleton could act as a transducer in communicating signaling transients. However, study of actin reorganization in cells is limited due to lack of methods to monitor reorganization in a quantitative fashion. Toward this effect, we recently developed an advanced microscopy-based image reconstruction technique to quantitatively monitor changes in the actin cytoskeletal network. This technique is based on high-magnification imaging, followed by image reconstruction [15]. Multiple sections (z-scans) of images acquired at high (63x) magnification were selected from the base of the cells, and an iso-surface (i.e., a contour made upon joining voxels of equal fluorescence intensity) of selected sections is generated. The best estimate of total volume enclosed by the iso-surface was obtained by normalizing it to the projected area of the cells in the given field. The major advantage of this approach using high-magnification imaging, over other methods using low-magnification imaging, is its dynamic range and overall reproducibility.

We utilized this technique in order to quantitatively estimate actin reorganization. To validate this approach, the reorganization of actin in cells treated with various actin-reorganizing agents such as cytochalasin D, latrunculin A, and jasplakinolide was quantitated (see Fig. 4). Alexa Fluor 546-phalloidin was used to stain the actin cytoskeleton. To obtain a quantitative measure of actin reorganization, iso-surfaces of the same selection of sections were generated and the corresponding iso-surface images are shown in Fig. 5a–d. Figure 5e shows the total volume enclosed by the iso-surface, normalized to the projected area of the cells, which provides the best estimate of changes in F-actin content under different treatment conditions. As evident from the figure, latrunculin A and cytochalasin D, known to induce F-actin depolymerization, show reduction in the normalized volume of F-actin over control levels. Jasplakinolide, on the other hand, known to induce F-actin formation, shows an increase in F-actin over control [15]. These results confirm the reliability of this method for quantitation of actin reorganization.

As mentioned earlier, we have previously reported that upon actin cytoskeleton destabilization, the signaling efficiency of the serotonin$_{1A}$ receptor is enhanced along with a correlated increase in its lateral mobility (i.e., mobile fraction) [14]. Interestingly, we also observed an increase in receptor mobility upon increasing cellular cAMP levels by directly activating adenylyl cyclase, implying a possible reorganization of the actin cytoskeleton with changes in cAMP level. In addition, cAMP/PKA signaling is known to induce significant changes in cellular architecture [32]. However, the role of cAMP/PKA-mediated reorganization of the actin cytoskeleton upon GPCR signaling has not been explored. To address such possibility of cytoskeletal reorganization upon GPCR signaling, we monitored the effect of serotonin$_{1A}$ receptor signaling on the actin cytoskeleton organization [15]. The serotonin$_{1A}$ receptor signals via Gα_i-mediated inhibition of adenylyl cyclase,

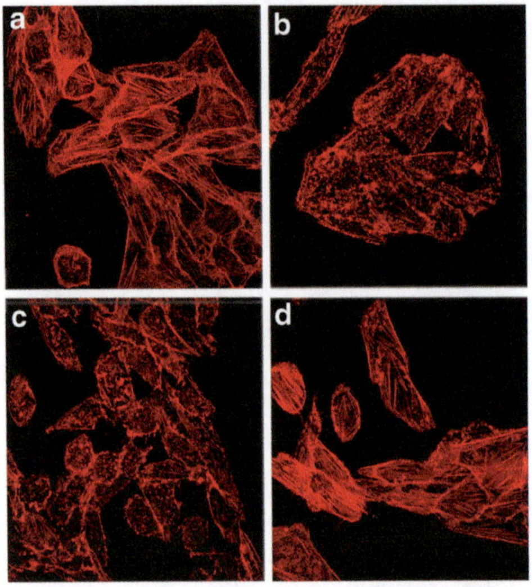

Fig. 4 Visualization of the organization of actin cytoskeleton upon treatment with actin-reorganizing agents in CHO cells. Alexa Fluor 546-phalloidin was used to stain the actin cytoskeleton, and a projection of 10–12 sections from the base (~3 μm from the coverslip into the cell) is shown. Projected images for cells treated with (**a**) DMSO, (**b**) cytochalasin D, (**c**) latrunculin A, and (**d**) jasplakinolide are shown. The scale bar represents 20 μm. Reproduced from [15]. See [15] for more details

resulting in the lowering of cAMP level and consequent downstream signaling [75, 77]. Our results showed that upon treatment of cells with forskolin, a direct activator of adenylyl cyclase resulted in elevated levels of cAMP, which in turn led to a reduction of F-actin. Interestingly, treatment with H-89, an inhibitor of PKA, rescued F-actin depolymerization to a large extent. These observations suggested that elevated cAMP levels result in actin depolymerization, mediated by activation of PKA. Our results further showed that stimulation of cells with serotonin led to an increase in F-actin compared to control cells. Importantly, the reduction of F-actin brought about by stimulation of cells with forskolin was recovered upon treatment of cells with forskolin in the presence of serotonin. Overall, our results showed that the serotonin$_{1A}$ receptor signaling led to increase in F-actin content (actin reorganization) via PKA-mediated pathway. Importantly, these results highlight the possibility of a vice versa control mechanism of receptor signaling and actin reorganization upon each other. Our approach could be utilized to quantitate actin reorganization upon signaling by other membrane receptors, which could give better insight into involvement of these receptors in several cellular processes in which actin reorganization has been predicted and/or reported.

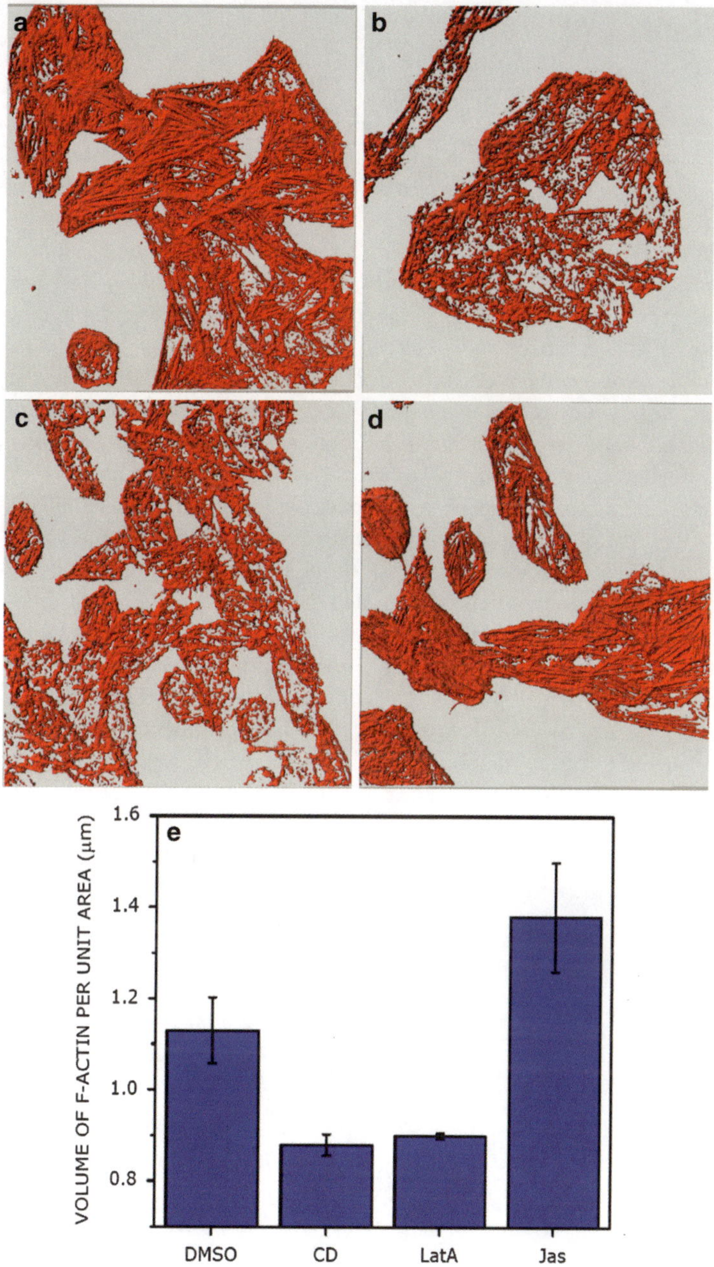

Fig. 5 Quantitation of F-actin from the iso-surfaces generated using the IsoSurface tool in Imaris. Iso-surfaces of the same selection of sections as shown in Fig. 4 were generated. The volume formed by the generated iso-surface to the projected area of cells in the field is then normalized using software provided with LSM 510 Meta confocal microscope. The iso-surface images of cells treated with (**a**) DMSO, (**b**) cytochalasin D, (**c**) latrunculin A, and (**d**) jasplakinolide are shown. Panel (**e**) shows the quantitation of F-actin. Reproduced from [15]. See [15] for more details

6 GPCR Oligomerization

GPCR oligomerization is an interesting and exciting aspect of contemporary receptor biology since it is believed to be an important determinant for cellular signaling [58, 86]. Such oligomerization is implicated in proper folding of receptors, thereby providing the framework for efficient and controlled signal transduction. The potential implications of such oligomerization are far reaching, specially keeping in mind the role of GPCRs as major drug targets [20]. Evidence of GPCR dimers or higher-order oligomers has been recently reported [1, 12, 48, 66, 96] and implicated in receptor trafficking, signaling, and pharmacology. Interestingly, oligomerization of certain GPCRs has been shown to be constitutive [12, 24].

Oligomerization of GPCRs in live cell membranes has been studied mostly by utilizing fluorescence-based resonance energy transfer methods such as hetero-FRET (FRET between two different fluorophores) and bioluminescence resonance energy transfer [55]. The major inherent complications of hetero-FRET measurements arise from the use of receptors conjugated to two different probes and a lack of control in their relative expression levels [39, 57]. Homo-FRET (FRET between two identical fluorophores) is a simpler variant of energy transfer because it takes place between like fluorophores and therefore requires only a single type of fluorophore. Fluorophores with a relatively small Stokes' shift will have a greater probability of undergoing homo-FRET. In addition, homo-FRET measurements can provide an estimate of higher-order oligomerization [81, 97], which is a serious limitation with hetero-FRET measurements. This becomes crucial particularly in the microheterogeneous membrane environment where multiple types of oligomeric clusters can coexist. Importantly, homo-FRET is manifested by a reduction in fluorescence anisotropy, a parameter that is largely independent of the concentration of fluorophores [92]. Homo-FRET leads to depolarization of the emission because of the lack of correlation between the orientation of the initially photoselected donor and the secondarily excited molecules [52].

6.1 Monitoring Receptor Oligomerization Utilizing Homo-FRET

In view of the advantages of homo-FRET, we utilized this approach to explore the oligomerization state of the serotonin$_{1A}$ receptor [12]. Homo-FRET was monitored by the increase in fluorescence anisotropy upon progressive photobleaching of the receptor, in which fluorescence depolarization due to energy transfer is prevented by photobleaching of FRET acceptors [94]. Our results showed that the initial anisotropy of serotonin$_{1A}$-EYFP receptors in control cells is significantly low (~0.22) compared to the fundamental anisotropy (r_o) of EYFP (0.38) (see Fig. 6a). The observed depolarization of emission has been attributed to energy

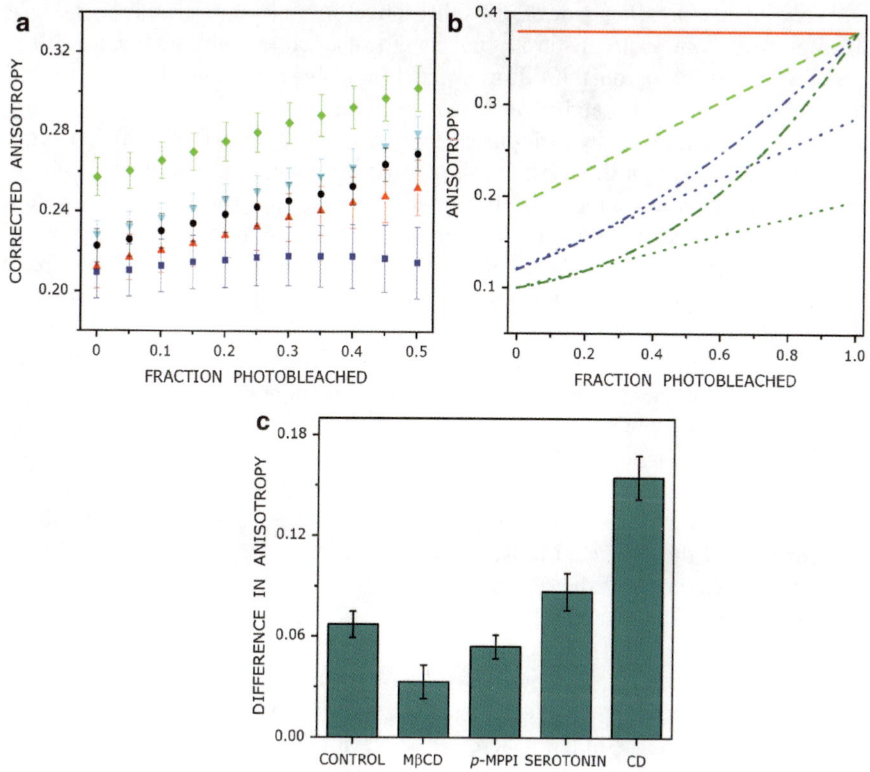

Fig. 6 Exploring receptor oligomerization utilizing homo-FRET. (**a**) Receptor oligomerization is monitored by increase in fluorescence anisotropy upon photobleaching of serotonin$_{1A}$-EYFP receptors under various conditions. Anisotropies corrected for microscopic aperture-induced depolarization upon photobleaching are shown for untreated cells (●, *black*) and cells treated with the agonist serotonin (▲, *red*), the antagonist *p*-MPPI (▼, *cyan*), cytochalasin D (■, *blue*), and MβCD (◆, *green*). Membrane cholesterol was depleted using methyl-β-cyclodextrin (MβCD) and actin cytoskeleton destabilization is achieved by cytochalasin D treatment. (**b**) Simulation of the enhancement in fluorescence anisotropy upon progressive photobleaching for a population of homogeneous oligomers containing *N* subunits. *N*-values correspond to 1 (monomer, *red* ——), 2 (dimer, *light green* ------), 3 (trimer, *blue* — - - —), and 4 (tetramer, *deep green* — - —). The extrapolated anisotropies derived from initial points corresponding to fractional bleaching (<0.3) in the case of trimer and tetramer are also shown (. . .) (**c**) Difference between the extrapolated anisotropy (extrapolated to complete photobleaching) and the predicted anisotropy of serotonin$_{1A}$-EYFP receptors under various treatment conditions. The extrapolated anisotropy is estimated from a linear fit of the photobleaching data shown in panel (**a**). The magnitude of deviation provides a measure of the extent of oligomerization. Reproduced from [12]. See text and [12] for more details

transfer (homo-FRET) between oligomers of the receptor molecules. Figure 6a shows that there is a steady increase in fluorescence anisotropy of serotonin$_{1A}$-EYFP receptors with progressive photobleaching, which is expected for a system undergoing homo-FRET.

We utilized a previously developed theoretical formalism for deducing the type of oligomers from such anisotropy enhancement upon photobleaching data [97]. This formalism relies on the difference between the extrapolated and predicted (0.38) anisotropy values at 100 % photobleaching limit for predicting oligomeric state, with the larger the difference, the greater the fraction of higher-order oligomers. In other words, with increasing oligomerization, the extrapolated anisotropy shows higher deviation from the predicted anisotropy. The predicted variation of fluorescence anisotropy with increased photobleaching for a homogeneous distribution of monomers, dimers, trimers, and tetramers (assuming an anisotropy of 0.38 for monomers) is shown in Fig. 6b. Due to experimental limitation of achieving very high degree of fractional bleaching, we compared the linearly extrapolated anisotropy with the predicted anisotropy to infer the presence of higher-order oligomers (see Fig. 6b). Therefore, the oligomeric state and inherent mobility of the serotonin$_{1A}$-EYFP receptors in principle can be obtained by fitting the anisotropy bleaching data to the theoretical formalism. On the basis of the observed increase in fluorescence anisotropy upon progressive photobleaching (Fig. 6a), and the analysis of data based on the difference between the extrapolated anisotropy and the predicted anisotropy (Fig. 6c), we proposed the presence of constitutive oligomers of the serotonin$_{1A}$ receptor [12].

Utilizing the same method, we further monitored the effect of membrane cholesterol, actin cytoskeleton, and ligand stimulation on the oligomerization state of the serotonin$_{1A}$ receptor. Our results showed that agonist (serotonin) stimulation and actin cytoskeletal destabilization led to reduction in the initial anisotropy and increase in the difference between the extrapolated anisotropy and the predicted anisotropy compared to control conditions (Fig. 6c). This suggested increased contribution from higher-order oligomers under these conditions. In contrast, antagonist (*p*-MPPI) treatment and cholesterol depletion led to an increase in initial anisotropy and reduction in the difference between the extrapolated anisotropy and the predicted anisotropy than in control conditions (Fig. 6c). These results indicate that antagonist treatment and cholesterol depletion effectively reduced the population of higher-order oligomers.

Taken together, our results show the presence of constitutive oligomers of the serotonin$_{1A}$ receptor and reorganization of higher-order oligomers in response to ligand activation, membrane cholesterol depletion, and actin cytoskeleton destabilization. This method could be applied to other membrane receptors to determine their oligomeric state in native and in various conditions. Such knowledge of receptor oligomerization state and reorganization under various pathophysiological conditions assume greater significance in the pharmacology of GPCRs since oligomerization gives rise to pharmacological diversity [91], opening new avenues for therapeutics.

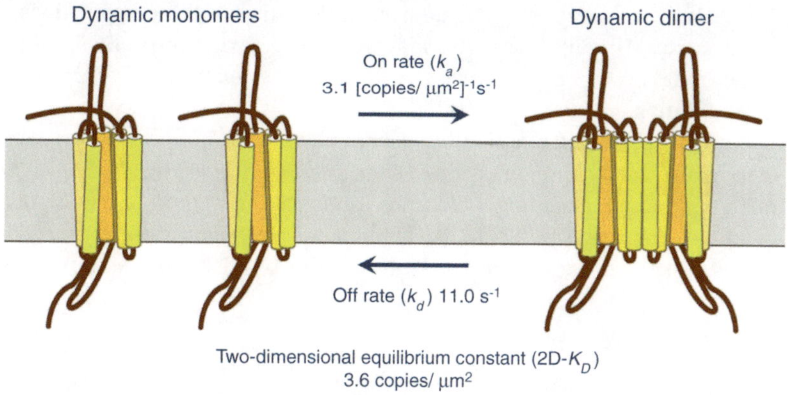

Fig. 7 Kinetics of GPCR oligomerization: dynamic monomer-dimer equilibrium of *N*-formyl peptide receptor (FPR) along with kinetic parameters for two-dimensional equilibrium dissociation constant for the dimer (2D-K_D), the rate constant for monomer association (k_a), and the rate constant for dimer dissociation (k_d). Adapted and modified from [48]

6.2 *Quantitation of GPCR Oligomerization: Kinetics and Equilibrium*

In spite of significant enhancement of our understanding of receptor organization and signaling, our present knowledge of cellular signal transduction events is mostly qualitative in nature. Our lack of ability to predict time-dependent changes in concentrations of important signaling molecules (such as receptors and effectors) could be attributed to a lack of approaches for quantitative evaluation of such parameters. A comprehensive understanding of signal transduction could be accomplished by development of approaches for quantitation of time-dependent number densities (concentrations) of key molecules involved in such processes.

Toward this effect, a recent work by Kasai et al. [48] addresses the issue of clear differentiation and quantitation of monomer and dimer population of receptors in living cells. In this study, the monomer and dimer receptor densities and the dynamic equilibrium between receptor monomers and dimers were monitored in case of a chemoattractant GPCR, the *N*-formyl peptide receptor (FPR). This was achieved by use of a single fluorescent molecule imaging method taking care of the fraction of molecules that actually fluoresce (labeling efficiency). In addition, molecular level interactions of the receptors were detected by employing bimolecular fluorescence complementation [23] at single-molecule level. Such robust analysis allows determination of the actual number of monomers and dimers at the plasma membrane of live cells and therefore the two-dimensional equilibrium dissociation constant (2D-K_D) for the dimers dissociating to monomers and the rate constant for monomer association (k_a) and the rate constant for dimer dissociation (k_d) (see Fig. 7). The analysis revealed that at physiological levels of receptor expression, ~41 % of FPRs exist as dynamic (transient) dimers, and the remaining

~59 % as dynamic monomers. The transience (lifetime) of these populations were estimated to be ~91 and 150 ms for dimers and monomers, respectively. Interestingly, the dynamic monomer-dimer equilibrium was found to be independent of ligand stimulation.

The knowledge of receptor number density in various oligomeric states and the parameters of receptor monomer-dimer dynamic equilibrium would help in better and quantitative understanding of receptor organization at the plasma membrane. This method, in principle, could be applied to other cell surface receptors to determine the monomer and dimer concentrations and kinetic and equilibrium parameters.

7 Conclusions and Future Perspectives

As mentioned earlier, GPCRs have emerged as major targets for the development of novel drug candidates in all clinical areas [29, 84]. Interestingly, although GPCRs represent ~50 % of present drug targets, only a small fraction of all GPCRs are presently targeted by drugs [53]. This points out the exciting possibility that GPCRs that are not recognized yet could be potential drug targets for diseases that are difficult to treat by currently available drugs. Importantly, GPCR function and signaling has been shown to depend on receptor organization [50, 63] and dynamics [7, 14, 34, 41, 76]. In this scenario, it is crucial to understand the organization and dynamics of GPCRs with respect to their signaling function. With progress in knowledge on receptor organization and dynamics at increasing spatiotemporal resolution [10], our overall understanding of GPCR function would improve significantly. This would enhance our ability to design better therapeutic strategies to combat diseases related to malfunctioning of GPCRs.

Acknowledgments Work in A.C.'s laboratory was supported by the Council of Scientific and Industrial Research and Department of Science and Technology, Govt. of India. We thank Sourav Ganguly for help in making Fig. 2. A.C. is an adjunct professor at the Special Centre for Molecular Medicine of Jawaharlal Nehru University (New Delhi, India) and Indian Institute of Science Education and Research (Mohali, India) and an honorary professor of the Jawaharlal Nehru Centre for Advanced Scientific Research (Bangalore, India). A.C. gratefully acknowledges J.C. Bose Fellowship (Department of Science and Technology, Govt. of India). Some of the work described in this article was carried out by former members of A.C.'s research group whose contributions are gratefully acknowledged. We thank members of our laboratory for critically reading the manuscript.

References

1. Albizu L, Cottet M, Kralikova M, Stoev S, Seyer R, Brabet I, Roux T, Bazin H, Bourrier E, Lamarque L, Breton C, Rives ML, Newman A, Javitch J, Trinquet E, Manning M, Pin JP, Mouillac B, Durroux T (2010) Time-resolved FRET between GPCR ligands reveals oligomers in native tissues. Nat Chem Biol 6:587–594

2. Banerjee P, Joo JB, Buse JT, Dawson G (1995) Differential solubilization of lipids along with membrane proteins by different classes of detergents. Chem Phys Lipids 77:65–78
3. Benda A, Benes M, Marecek V, Lhotsky A, Hermens WT, Hof M (2003) How to determine diffusion coefficients in planar phospholipid systems by confocal fluorescence correlation spectroscopy. Langmuir 19:4120–4126
4. Blier P, Ward NM (2003) Is there a role for 5-HT$_{1A}$ agonists in the treatment of depression? Biol Psychiatry 53:193–203
5. Broder CC, Dimitrov DS (1996) HIV and the 7-transmembrane domain receptors. Pathobiology 64:171–179
6. Brown DA, Rose JK (1992) Sorting of GPI-anchored proteins to glycolipid-enriched membrane subdomains during transport to the apical cell surface. Cell 68:533–544
7. Calvert PD, Govardovskii VI, Krasnoperova N, Anderson RE, Lem J, Makino CL (2001) Membrane protein diffusion sets the speed of rod phototransduction. Nature 411:90–94
8. Chamberlain LH (2004) Detergents as tools for the purification and classification of lipid rafts. FEBS Lett 559:1–5
9. dos Remedios CG, Chhabra D, Kekic M, Dedova IV, Tsubakihara M, Berry DA, Nosworthy NJ (2003) Actin binding proteins: regulation of cytoskeletal microfilaments. Physiol Rev 83:433–473
10. Eggeling C, Ringemann C, Medda R, Schwarzmann G, Sandhoff K, Polyakova S, Belov VN, Hein B, von Middendorff C, Schönle A, Hell SW (2009) Direct observation of the nanoscale dynamics of membrane lipids in a living cell. Nature 457:1159–1163
11. Ganguly S, Chattopadhyay A (2010) Cholesterol depletion mimics the effect of cytoskeletal destabilization on membrane dynamics of the serotonin$_{1A}$ receptor: a zFCS study. Biophys J 99:1397–1407
12. Ganguly S, Clayton AHA, Chattopadhyay A (2011) Organization of higher-order oligomers of the serotonin$_{1A}$ receptor explored utilizing homo-FRET in live cells. Biophys J 100:361–368
13. Ganguly S, Paila YD, Chattopadhyay A (2011) Metabolic depletion of sphingolipids enhances the mobility of the human serotonin$_{1A}$ receptor. Biochem Biophys Res Commun 411:180–184
14. Ganguly S, Pucadyil TJ, Chattopadhyay A (2008) Actin cytoskeleton-dependent dynamics of the human serotonin$_{1A}$ receptor correlates with receptor signaling. Biophys J 95:451–463
15. Ganguly S, Saxena R, Chattopadhyay A (2011) Reorganization of the actin cytoskeleton upon G-protein coupled receptor signaling. Biochim Biophys Acta 1808:1921–1929
16. Ganguly S, Singh P, Manoharlal R, Prasad R, Chattopadhyay A (2009) Differential dynamics of membrane proteins in yeast. Biochem Biophys Res Commun 387:661–665
17. Gardier AM (2009) Mutant mouse models and antidepressant drug research: focus on serotonin and brain-derived neurotrophic factor. Behav Pharmacol 20:18–32
18. Gether U (2000) Uncovering molecular mechanisms involved in activation of G protein-coupled receptors. Endocr Rev 21:90–113
19. Giepmans BNG, Adams SR, Ellisman MH, Tsien RY (2006) The fluorescent toolbox for assessing protein location and function. Science 312:217–224
20. González-Maeso J, Ang RL, Yuen T, Chan P, Weisstaub NV, López-Giménez JF, Zhou M, Okawa Y, Callodo LF, Milligan G, Gingrich JA, Filizola M, Meana JJ, Sealfon SC (2008) Identification of a serotonin/glutamate receptor complex implicated in psychosis. Nature 452:93–97
21. Griebel G (1999) 5-HT$_{1A}$ receptor blockers as potential drug candidates for the treatment of anxiety disorders. Drug News Perspect 12:484–490
22. Haldar S, Chattopadhyay A (2009) The green journey. J Fluoresc 19:1–2
23. Haldar S, Chattopadhyay A (2009) Green fluorescent protein: a molecular lantern that illuminates the cellular interior. J Biosci 34:169–172
24. Harding PJ, Attrill H, Boehringer J, Ross S, Wadhams GH, Smith E, Armitage JP, Watts A (2009) Constitutive dimerization of the G-protein coupled receptor, neurotensin receptor 1, reconstituted into phospholipid bilayers. Biophys J 96:964–973

25. Harrison T, Samuel BU, Akompong T, Hamm H, Mohandas N, Lomasney JW, Haldar K (2003) Erythrocyte G protein-coupled receptor signaling in malarial infection. Science 301:1734–1736
26. Haustein E, Schwille P (2007) Fluorescence correlation spectroscopy: novel variations of an established technique. Annu Rev Biophys Biomol Struct 36:151–169
27. He H-T, Marguet D (2011) Detecting nanodomains in living cell membrane by fluorescence correlation spectroscopy. Annu Rev Phys Chem 62:417–436
28. Heerklotz H (2002) Triton promotes domain formation in lipid raft mixtures. Biophys J 83:2693–2701
29. Heilker R, Wolff M, Tautermann CS, Bieler M (2009) G-protein coupled receptor-focused drug discovery using a target class platform approach. Drug Discov Today 14:231–240
30. Hess ST, Huang S, Heikal AA, Webb WW (2002) Biological and chemical applications of fluorescence correlation spectroscopy: a review. Biochemistry 41:697–705
31. Hooper NM (1999) Detergent-insoluble glycosphingolipid/cholesterol-rich membrane domains, lipid rafts and caveolae. Mol Membr Biol 16:145–156
32. Howe AK (2004) Regulation of actin-based cell migration by cAMP/PKA. Biochim Biophys Acta 1692:159–174
33. Huang C, Hepler JR, Chen LT, Gilman AG, Anderson RG, Mumby SM (1997) Organization of G proteins and adenylyl cyclase at the plasma membrane. Mol Biol Cell 8:2365–2378
34. Hui KL, Wang C, Grooman B, Wayt J, Upadhyaya A (2012) Membrane dynamics correlate with formation of signaling clusters during cell spreading. Biophys J 102:1524–1533
35. Humpolícková J, Gielen E, Benda A, Fagulova V, Vercammen J, vandeVen M, Hof M, Ameloot M, Engelborghs Y (2006) Probing diffusion laws within cellular membranes by Z-scan fluorescence correlation spectroscopy. Biophys J 91:L23–L25
36. Hur E-M, Kim KT (2002) G protein-coupled receptor signalling and cross-talk: achieving rapidity and specificity. Cell Signal 14:397–405
37. Jacobson K, Mouritsen OG, Anderson RGW (2007) Lipid rafts: at a crossroad between cell biology and physics. Nat Cell Biol 9:7–14
38. Jafurulla M, Pucadyil TJ, Chattopadhyay A (2008) Effect of sphingomylinase treatment on ligand binding activity of human serotonin$_{1A}$ receptors. Biochim Biophys Acta 1778:2022–2025
39. James JR, Oliveira MI, Carmo AM, Iaboni A, Davis SJ (2006) A rigorous experimental framework for detecting protein oligomerization using bioluminescence resonance energy transfer. Nat Methods 3:1001–1006
40. Janmey PA (1998) The cytoskeleton and cell signaling: component localization and mechanical coupling. Physiol Rev 78:763–781
41. Jans DA, Peters R, Jans P, Fahrenholz F (1991) Vasopressin V_2 receptor mobile fraction and ligand-dependent adenylate cyclase activity are directly correlated in LLC-PK1 renal epithelial cells. J Cell Biol 114:53–60
42. Kalipatnapu S, Chattopadhyay A (2004) A GFP fluorescence based approach to determine detergent insolubility of the human serotonin$_{1A}$ receptor. FEBS Lett 576:455–460
43. Kalipatnapu S, Chattopadhyay A (2005) Membrane organization of the human serotonin$_{1A}$ receptor monitored by detergent insolubility using GFP fluorescence. Mol Membr Biol 22:539–547
44. Kalipatnapu S, Chattopadhyay A (2007) Membrane organization and function of the serotonin$_{1A}$ receptor. Cell Mol Neurobiol 27:1097–1116
45. Kalipatnapu S, Chattopadhyay A (2007) Membrane organization of the serotonin$_{1A}$ receptor monitored by a detergent-free approach. Cell Mol Neurobiol 27:463–474
46. Kalipatnapu S, Pucadyil TJ, Chattopadhyay A (2007) Membrane organization and dynamics of the serotonin$_{1A}$ receptor monitored using fluorescence microscopic approaches. In: Chattopadhyay A (ed) Serotonin receptors in neurobiology, new frontiers in neuroscience series. CRC Press, Boca Raton, pp 41–60
47. Karnovsky MJ, Kleinfeld AM, Hoover RL, Klausner RD (1982) The concept of lipid domains in membranes. J Cell Biol 94:1–6

48. Kasai RS, Suzuki KGN, Prossnitz ER, Koyama-Honda I, Nakada C, Fujiwara TK, Kusumi A (2011) Full characterization of GPCR monomer-dimer dynamic equilibrium by single molecule imaging. J Cell Biol 192:463–480
49. Kusumi A, Nakada C, Ritchie K, Murase K, Suzuki K, Murakoshi H, Kasai RS, Kondo J, Fujiwara T (2005) Paradigm shift of the plasma membrane concept from the two-dimensional continuum fluid to the partitioned fluid: high-speed single-molecule tracking of membrane molecules. Annu Rev Biophys Biomol Struct 34:351–378
50. Lajoie P, Goetz JG, Dennis JW, Nabi IR (2009) Lattices, rafts and scaffolds: domain regulation of receptor signaling at the plasma membrane. J Cell Biol 185:381–385
51. Lenne P-F, Wawrezinieck L, Conchonaud F, Wurtz O, Boned A, Guo X-J, Rigneault H, He H-T, Marguet D (2006) Dynamic molecular confinement in the plasma membrane by microdomains and the cytoskeleton meshwork. EMBO J 25:3245–3256
52. Lidke DS, Nagy P, Barisas BG, Heintzmann R, Post JN, Lidke KA, Clayton AH, Arndt-Jovin DJ, Jovin TM (2003) Imaging molecular interactions in cells by dynamic and static fluorescence anisotropy (rFLIM and emFRET). Biochem Soc Trans 31:1020–1027
53. Lin SH, Civelli O (2004) Orphan G protein-coupled receptors: targets for new therapeutic interventions. Ann Med 36:204–214
54. Lingwood D, Simons K (2010) Lipid rafts as a membrane-organizing principle. Science 327:46–50
55. Lohse MJ (2010) Dimerization in GPCR mobility and signaling. Curr Opin Pharmacol 10:53–58
56. Lohse MJ, Nikolaev VO, Hein P, Hoffmann C, Vilardaga J-P, Bünemann M (2008) Optical techniques to analyze real-time activation and signaling of G-protein-coupled receptors. Trends Pharmacol Sci 29:159–165
57. Meyer BH, Segura JM, Martinez KL, Hovius R, George N, Johnsson K, Vogel H (2006) FRET imaging reveals that functional neurokinin-1 receptors are monomeric and reside in membrane microdomains of live cells. Proc Natl Acad Sci USA 103:2138–2143
58. Milligan G (2007) G protein-coupled receptor dimerisation: molecular basis and relevance to function. Biochim Biophys Acta 1768:825–835
59. Mukherjee S, Maxfield FR (2004) Membrane domains. Annu Rev Cell Dev Biol 20:839–866
60. Müller CP, Carey RJ, Huston JP, De Souza Silva MA (2007) Serotonin and psychostimulant addiction: focus on 5-HT$_{1A}$-receptors. Prog Neurobiol 81:133–178
61. Neubig RR (1994) Membrane organization in G-protein mechanisms. FASEB J 8:939–946
62. Oldham WM, Hamm HE (2006) Structural basis of function in heterotrimeric G proteins. Q Rev Biophys 39:117–166
63. Ostrom RS, Insel PA (2004) The evolving role of lipid rafts and caveolae in G protein-coupled receptor signaling: implications for molecular pharmacology. Br J Pharmacol 143:235–245
64. Paila YD, Chattopadhyay A (2010) Membrane cholesterol in the function and organization of G-protein coupled receptors. Subcell Biochem 51:439–466
65. Paila YD, Ganguly S, Chattopadhyay A (2010) Metabolic depletion of sphingolipids impairs ligand binding and signaling of human serotonin$_{1A}$ receptors. Biochemistry 49:2389–2397
66. Paila YD, Kombrabail M, Krishnamoorthy G, Chattopadhyay A (2011) Oligomerization of the serotonin$_{1A}$ receptor in live cells: a time-resolved fluorescence anisotropy approach. J Phys Chem B 115:11439–11447
67. Paila YD, Murty MRVS, Vairamani M, Chattopadhyay A (2008) Signaling by the human serotonin$_{1A}$ receptor is impaired in cellular model of Smith-Lemli-Opitz Syndrome. Biochim Biophys Acta 1778:1508–1516
68. Perez DM (2003) The evolutionarily triumphant G-protein-coupled receptor. Mol Pharmacol 63:1202–1205
69. Pierce KL, Premont RT, Lefkowitz RJ (2002) Seven-transmembrane receptors. Nat Rev Mol Cell Biol 3:639–650

70. Pucadyil TJ, Chattopadhyay A (2004) Cholesterol modulates ligand binding and G-protein coupling to serotonin$_{1A}$ receptors from bovine hippocampus. Biochim Biophys Acta 1663:188–200
71. Pucadyil TJ, Chattopadhyay A (2006) Role of cholesterol in the function and organization of G-protein coupled receptors. Prog Lipid Res 45:295–333
72. Pucadyil TJ, Chattopadhyay A (2007) Cholesterol depletion induces dynamic confinement of the G-protein coupled serotonin$_{1A}$ receptor in the plasma membrane of living cells. Biochim Biophys Acta 1768:655–668
73. Pucadyil TJ, Chattopadhyay A (2007) Cholesterol: a potential therapeutic target in *Leishmania* infection? Trends Parasitol 23:49–53
74. Pucadyil TJ, Kalipatnapu S, Chattopadhyay A (2005) Membrane organization and dynamics of the G-protein-coupled serotonin$_{1A}$ receptor monitored using fluorescence-based approaches. J Fluoresc 15:785–796
75. Pucadyil TJ, Kalipatnapu S, Chattopadhyay A (2005) The serotonin$_{1A}$ receptor: A representative member of the serotonin receptor family. Cell Mol Neurobiol 25:553–580
76. Pucadyil TJ, Kalipatnapu S, Harikumar KG, Rangaraj N, Karnik SS, Chattopadhyay A (2004) G-protein-dependent cell surface dynamics of the human serotonin$_{1A}$ receptor tagged to yellow fluorescent protein. Biochemistry 43:15852–15862
77. Raymond JR, Mukhin YV, Gettys TW, Garnovskaya MN (1999) The recombinant 5-HT$_{1A}$ receptor: G protein coupling and signalling pathways. Br J Pharmacol 127:1751–1764
78. Renner U, Glebov K, Lang T, Papusheva E, Balakrishnan S, Keller B, Richter DW, Jahn R, Ponimaskin E (2007) Localization of the mouse 5-hydroxytryptamine$_{1A}$ receptor in lipid microdomains depends on its palmitoylation and is involved in receptor-mediated signaling. Mol Pharmacol 72:502–513
79. Riethmüller J, Riehle A, Grassmé H, Gulbins E (2006) Membrane rafts in host-pathogen interactions. Biochim Biophys Acta 1758:2139–2147
80. Rosenbaum DM, Rasmussen SGF, Kobilka BK (2009) The structure and function of G-protein-coupled receptors. Nature 459:356–363
81. Runnels LW, Scarlata SF (1995) Theory and application of fluorescence homotransfer to melittin oligomerization. Biophys J 69:1569–1583
82. Sahu S, Saxena R, Chattopadhyay A (2012) Cholesterol depletion modulates detergent resistant fraction of human serotonin$_{1A}$ receptors. Mol Membr Biol (in press)
83. Saxena R, Chattopadhyay A (2011) Membrane organization and dynamics of the serotonin$_{1A}$ receptor in live cells. J Neurochem 116:726–733
84. Schlyer S, Horuk R (2006) I want a new drug: G-protein-coupled receptors in drug development. Drug Discov Today 11:481–493
85. Schroeder RJ, Ahmed SN, Zhu Y, London E, Brown DA (1998) Cholesterol and sphingolipid enhance the Triton X-100 insolubility of glycosylphosphatidylinositol-anchored proteins by promoting the formation of detergent-insoluble ordered membrane domains. J Biol Chem 273:1150–1157
86. Shanti K, Chattopadhyay A (2000) A new paradigm in the functioning of G-protein-coupled receptors. Curr Sci 79:402–403
87. Shrivastava S, Pucadyil TJ, Paila YD, Ganguly S, Chattopadhyay A (2010) Chronic cholesterol depletion using statin impairs the function and dynamics of human serotonin$_{1A}$ receptors. Biochemistry 49:5426–5435
88. Simons K, Toomre D (2000) Lipid rafts and signal transduction. Nat Rev Mol Cell Biol 1:31–39
89. Simons K, van Meer G (1988) Lipid sorting in epithelial cells. Biochemistry 27:6197–6202
90. Singh P, Chattopadhyay A (2012) Removal of sphingomyelin headgroup inhibits the ligand binding function of hippocampal serotonin$_{1A}$ receptors. Biochem Biophys Res Commun 419:321–325
91. Terrillon S, Bouvier M (2004) Roles of G-protein-coupled receptor dimerization. EMBO Rep 5:30–34

92. Tramier M, Piolot T, Gautier I, Mignotte V, Coppey J, Kemnitz K, Durieux C, Coppey-Moisan M (2003) Homo-FRET versus hetero-FRET to probe homodimers in living cells. Methods Enzymol 360:580–597
93. Tsien RY (1998) The green fluorescent protein. Annu Rev Biochem 67:509–544
94. Varma R, Mayor S (1998) GPI-anchored proteins are organized in submicron domains at the cell surface. Nature 394:798–801
95. Wawrezinieck L, Rigneault H, Marguet D, Lenne P-F (2005) Fluorescence correlation spectroscopy diffusion laws to probe the submicron cell membrane organization. Biophys J 89:4029–4042
96. Woehler A, Wlodarczyk J, Ponimaskin EG (2009) Specific oligomerization of the 5-HT1A receptor in the plasma membrane. Glycoconj J 26:749–756
97. Yeow EKL, Clayton AHA (2007) Enumeration of oligomerization states of membrane proteins in living cells by homo-FRET spectroscopy and microscopy: theory and application. Biophys J 92:3098–3104
98. Zhang Y, DeVries ME, Skolnick J (2006) Structure modeling of all identified G protein-coupled receptors in the human genome. PLoS Comput Biol 2:88–99
99. Zhou FC, Patel TD, Swartz D, Xu Y, Kelley MR (1999) Production and characterization of an anti-serotonin 1A receptor antibody which detects functional 5-HT$_{1A}$ binding sites. Brain Res Mol Brain Res 69:186–201

TNF Receptor Membrane Dynamics Studied with Fluorescence Microscopy and Spectroscopy

Felix Neugart, Darius Widera, Barbara Kaltschmidt, Christian Kaltschmidt, and Mike Heilemann

Abstract Sensitive fluorescence techniques opened novel opportunities to study the function and interaction of proteins in living cells. Here, we review the contribution of fluorescence correlation spectroscopy (FCS) and single-molecule tracking to study the dynamics of TNF receptor 1 and 2 (TNFR1/2). Although these techniques greatly differ, both report a similar behavior of TNF receptor (TNFR) species in Hela cells under different experimental conditions. FCS as well as single-molecule tracking revealed an increase of the diffusion coefficient of TNFR1 after treating cells with methyl-cyclodextrin.

In addition, FCS studies of the activation of TNFR1 showed that ligand binding hardly affects its diffusion coefficient.

In contrast, unstimulated TNFR2 was observed to diffuse faster than TNFR1, whereas ligand stimulation of TNFR2 decreases the diffusion coefficient.

In conclusion, the results indicate that the two TNFRs compartmentalize in distinct domains of the plasma membrane most likely determined by the respective transmembrane domains and/or transmembrane domain near regions.

F. Neugart (✉)
Institute of Cell Biology and Immunology, University of Stuttgart, Allmandring 31, D-70569 Stuttgart, Germany

Leica Microsystems CMS GmbH, Am Friedensplatz 3, D-68165 Mannheim, Germany
e-mail: felix.neugart@leica-microsystems.com

D. Widera (✉) and C. Kaltschmidt
Cell Biology, University of Bielefeld, Universitätsstr. 25, D-33501 Bielefeld, Germany
e-mail: darius.widera@uni.bielefeld.de

B. Kaltschmidt
Molecular Neurobiology, University of Bielefeld, Universitätsstr. 25, D-33501 Bielefeld, Germany

M. Heilemann (✉)
Department of Biotechnology & Biophysics, Julius-Maximilians-Universität, Am Hubland, D-97074 Würzburg, Germany
e-mail: mike.heilemann@uni-wuerzburg.de

Y. Mély and G. Duportail (eds.), *Fluorescent Methods to Study Biological Membranes*, Springer Ser Fluoresc (2013) 13: 439–456, DOI 10.1007/4243_2012_60,

Keywords FCS · sptPALM · TNFR1 · TNFR2

Contents

1 Introduction

Signaling via tumor necrosis factor (TNF) plays an important role in the immune system, for instance, regarding the regulation of innate immune responses and inflammation. As known for many membrane receptors, the function of TNF receptors (TNFRs) is closely related to their state of multimerization and interactions with membrane micro-domains. Thus, investigating the molecular mechanism of TNFR-activation, which in turn initiates further signaling cascades, requires studies in living cells.

For many decades fluorescence microscopy has been a state-of-the-art method to study such biomolecular dynamics in living cells. In particular, the availability of fluorescent proteins, which can be genetically fused to any target protein and expressed in a living cell, has paved the way for in vivo studies. The development of advanced fluorescence spectroscopy and microscopy techniques allowed studying even single, fluorescently labeled molecules [1], such that subpopulations of biomolecules could be observed and heterogeneities can be resolved. These developments made it possible to address the complexity of biological processes such as signal transduction at the plasma membrane in much more detail.

In this article, we review recent work that used advanced fluorescence spectroscopy and microscopy methods to study the dynamics of TNFR1 and TNFR2 in the plasma membrane of living cells. Furthermore, underlining novel insights gained by these methods, we summarize their potential impact on biology of TNFRs.

2 TNFRs 1 and 2 and Their Dynamics, Association, and Activation

Beside its very well-known role as a crucial mediator of inflammation, TNF signaling has been described to have an impact on various pathological disease symptoms like diabetes (type 2), heart failure, artheriosclerosis, tumorigenesis, and tumor metastasis [2–4].

The most prominent ligand of TNFRs, TNF-α has firstly been cloned, purified, and characterized by Aggerwal's group in 1984 [5, 6]. Since then, additional cytokines of the TNF-superfamily have been identified, for example, CD95L, TRAIL, or RANKL. Most of the ligands of the TNF-superfamily are type II transmembrane proteins. They harbor the so-called TNF-homology domain in its c-terminal extracellular domain. Importantly, membrane-bound homotrimeric TNF-superfamily members can be cleaved by metalloproteinases of the ADAM-family resulting in the formation of soluble TNFs. Remarkably, membrane-bound and soluble TNF do have nonoverlapping capabilities to activate the two distinct TNFRs. The two ligand forms bind both receptors, but soluble TNF can fully activate TNFR1 only, whereas the membrane-bound form can activate both receptors (reviewed in [7]).

Most TNFR-superfamily members are transmembrane receptors containing cytoplasmic, transmembrane, and extracellular domains and can be classified into three groups based on their signaling properties and sequences.

The members of the first group, comprising Fas/CD95/ Apo1/APT1 and TNFR1 (p55/60; CD120a), contain a death domain (DD) in the cytoplasmic tail. The second group, including TNFR2 (p75/80; CD120b), CD40, and CD27, is characterized by one or more TRAF interacting motifs (TIM). Activation of these receptors results in a recruitment of TRAF proteins and subsequent activation of multiple signal transduction pathways such as Nuclear Factor-kappaB (NF-κB) [8, 9], p38 mitogen-activated protein kinases (p38 MAPK), Jun N-terminal kinase (JNK, reviewed in [10]), and Rho GTPases [11]. In contrast to the first and second group, members of the third group consisting of human TRAIL-R3, TRAIL-R4 and OPG, have no functional intracellular signaling domains or motifs or exist as soluble proteins. Remarkably, none of the mammalian TNFRs and further TNFR-superfamily members have enzymatic activity itself.

The classic activation model proposes that binding of the ligand induces trimerization of TNFRs (Fig. 1). However, in their extracellular part, TNFR1 and TNFR2 contain both the so-called pre–ligand binding assembly domain (PLAD) which has been shown to be necessary for signal initiation after binding of the ligand without being directly in contact with the latter.

Recent evidence suggests that TNFRs may pre-assemble via their PLADs and form oligomers at the plasma membrane.

Importantly, most of the models for assembly of TNFRs are based on experimental data obtained using classic biochemical methods like Western blotting in cell lysates or fixed cells. These methods naturally drastically impact the natural environment of the membrane receptors and cannot avoid producing artifacts, such that until now, the question of pre- or post-ligand oligomerization remains unsolved.

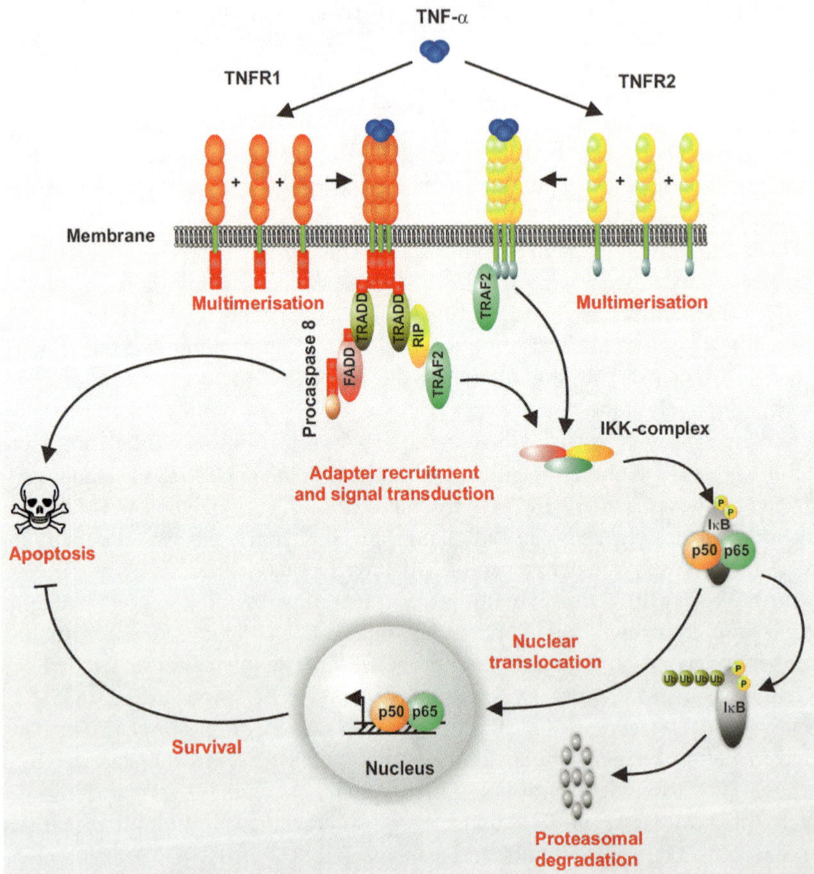

Fig. 1 Classic model for TNF-α induced signaling. Activation of the TNFR by the ligand binding leads to multimerization of the monomeric TNFRs 1 and/or 2 and subsequent recruitment of adapter proteins such as TRADD, FADD, RIP, and TRAF2. Depending on the adapter, the binding of TNF leads either to activation of Procaspase 8, resulting in apoptosis, or to signal transmission to the IKK complex which mediates NF-κB signaling. IKKs phosphorylate IκB family inhibitory molecules targeting IκB for degradation within the 26 S proteasome and frees the nuclear localization signal on the DNA-binding NF-κB subunits p65/p50. After nuclear import, target gene transcription is initiated leading to survival of the cell

Besides the molecular interaction-driven trimerization, several reports suggested that membrane micro-domains and, therefore, the spatial distribution of TNFRs at the plasma membrane may have a tremendous impact on downstream signaling pathways. Consequently, it has been reported that after ligand binding, TNFR1/TNF complexes can translocate into membrane micro-domains and that an inhibition of this process can switch the signaling from anti-apoptotic NF-κB activation to the pro-apoptotic signaling [12]. A contrary mechanism was reported by Doan et al. in a myeloid cell line [13]. In this study, membrane micro-domains were crucial for the activation of p42 MAPK, but not for the NF-κB signaling, whereas in human airway

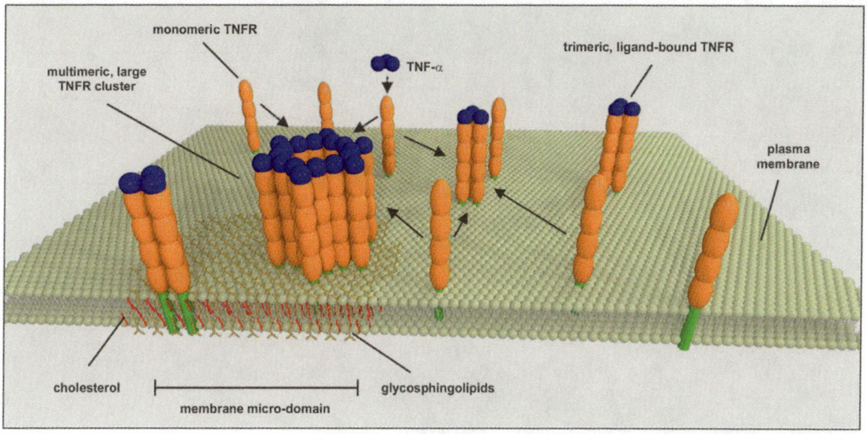

Fig. 2 Potential influence of ligand binding on multimerization and dynamics of TNFRs. In the classic view, the binding of TNF-α leads to trimerization of monomeric TNFRs. However, ligand binding may also lead to recruitment of TNFRs into plasma membrane micro-domains and formation of large, multimeric TNFR clusters. Such receptor-trapping within micro-domains may have tremendous impact on the signaling. Since these clustering events occur at spatial scales which are below the diffraction limit of classic light microscopy, super-resolution microscopy may provide helpful insights into this phenomenon. Furthermore, multimerization affects also the dynamics of receptors within the membrane which can be sophisticatedly investigated by experimental techniques such as FCS and sptPALM

smooth muscle cells, activation of RhoA, but neither of MAPK nor NF-κB, was micro-domain dependant [14]. Importantly, recruitment of receptors to micro-domains may lead to the formation of large receptor clusters and favor multimerization. However, the size of micro-domains (~5–200 nm) is in general below the diffraction limit of classic light microscopy making them only visible using modern super-resolution techniques [15]. A further degree of complexity is provided by the fact that plasma membrane–anchored proteins can be transiently trapped in cholesterol-mediated molecular complexes resulting in a dynamic steady-state system including assembly and disassembly of receptor clusters [16].

Thus the imaging of receptor clustering should be performed ideally in living cells, since not only the lateral organization of the TNFRs (TNFR clusters and their size), but also its long- and short-term dynamics may be vital for the signaling process. Furthermore, the association of TNFRs on the plasma membrane can change rapidly in response to ligand binding, i.e., the oligomerization and potential interactions with plasma membrane micro-domains.

Modern fluorescence spectroscopy and super-resolution microscopy with single-molecule resolution provide for the first time the possibility to address the important questions of the biological impact of the assembly of the TNFRs (oligomerization), their association with membrane micro-domains, and the receptor dynamics before and after ligand binding (Fig. 2). Each of the methods has its particular strengths, which are discussed in the following and may provide data potentially extending existing models concerning the complex TNF signaling network.

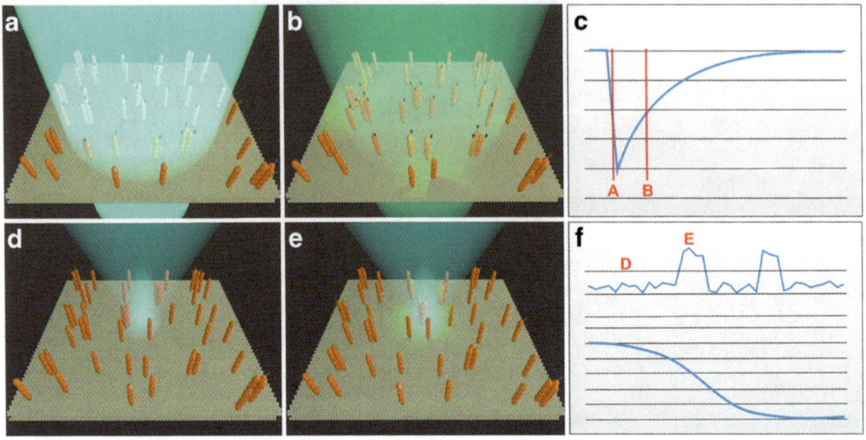

Fig. 3 (**a**) High-intensity illumination causes photo-bleaching of fluorophores in a region of interest (ROI). (**b**) After photoleaching, fluorophores outside the ROI will translocate into the ROI by diffusion or transport processes. Moderate illumination shows the recovery of fluorescence in the ROI. (**c**) The evaluation of the recovery of the mean fluorescence intensity of the ROI gives model-dependent information of the diffusion behavior of the fluorescent proteins. (**d**) A confocal spot illuminates a small region of the plasma membrane. (**e**) When a fluorescent molecule diffuses through the confocal volume, observed fluorescence signal can be detected. (**f**) A second-order autocorrelation function is calculated from the fluorescence time trace and provides the mean diffusion coefficient and concentration of fluorophores

3 Investigation of Membrane Protein Dynamics with Fluorescence microscopy

Among all the applicable methods to study the dynamics of membrane proteins, fluorescence recovery after photo bleaching (FRAP) is the most commonly used one (Fig. 3a–c). Using FRAP, not only the distribution of the observed parameter (e.g., the diffusion coefficient) is averaged but also its spatial distribution within the size of the region of interest. Moreover, FRAP is useful to determine the diffusion coefficient of fluorescent particles but requires homogenous environment with a unique diffusion behavior. However, the evaluation of proteins with an inhomogeneous diffusion behavior is only accessible by detailed modeling and impossible if there is no detailed knowledge about the underlying diffusion model [17].

Fluorescence correlation spectroscopy (FCS) (Fig. 3d–f) is a powerful method to measure the diffusion of a target molecule using confocal microscopy within a small volume of about 1 fl. [18]. In contrast to FRAP, FCS strongly reduces spatial averaging and gives access to the stoichiometry of biomolecular complexes via an analysis of the photon counting histogram (PCH) [19, 20]. The interpretation of experimental data obtained by FCS is less model dependent compared to FRAP and allows a direct extraction of the diffusion coefficient. Among different modifications and extensions of the original principle of FCS, fluorescence cross correlation spectroscopy (FCCS) provides additional information on binding characteristics of differently labeled proteins or on the stoichiometry of heteromolecular complexes [21, 22].

A third approach – single-molecule tracking – enables to follow the fate of individual biomolecules. While bulk measurements only provide average values based on many events, single-molecule measurements can resolve subpopulations and heterogeneities in populations. This is of particular advantage when analyzing biological observables, which are in general inhomogeneously distributed. Concomitantly, single-molecule experiments bypass the need to synchronize biological processes, which is necessary to extract mechanistic subpopulations otherwise averaged over time [23, 24].

In summary, the analysis of single-molecule distributions often provides a deeper insight into mechanisms of membrane protein dynamics. In this regard, we outline in the following the current research on the dynamics of TNFR1 and TNFR2 via FCS and sptPALM.

4 Fluorescence Correlation Spectroscopy (FCS) Reveals Topological Segregation of TNFR1 and TNFR2

Fluorescence correlation spectroscopy (FCS) analyzes information of the statistical fluctuation of the fluorescence intensity of a system in equilibrium. Remarkably, a high numerical aperture confocal microscope can reach a focal volume of less than 1 fl. Using FCS, the modulation of the number of fluorescent molecules diffusing through such small confocal volume can be analyzed via a second-order autocorrelation function (Fig. 3d–f) [18].

Observables from FCS measurements are the time-averaged diffusion coefficient and the time-averaged number of diffusing particles with respect to the observation volume. Additional information can be obtained from photon counting histogram (PCH) analysis, which provides information on the distribution of particle brightness, i.e., the number of fluorophores (e.g., proteins) within a diffusing particle (e.g., cluster). Moreover, a series of FCS measurements access not only the evolution of the dynamical behavior but also changes of concentration and stoichiometry of the particle of interest [19, 20].

4.1 Diffusion of TNFR1 and TNFR2

In a recent study, we investigated the dynamics of TNFR1- and TNFR2-GFP fusion-proteins in the plasma membrane via FCS measurements [25]. While TNFR2 shows a narrow, homogenous, and relatively fast distribution of diffusion coefficient (average value = 0.31 $\mu m^2/s$), TNFR1 was significantly slower and the distribution of its diffusion coefficient spread over a significantly broader range

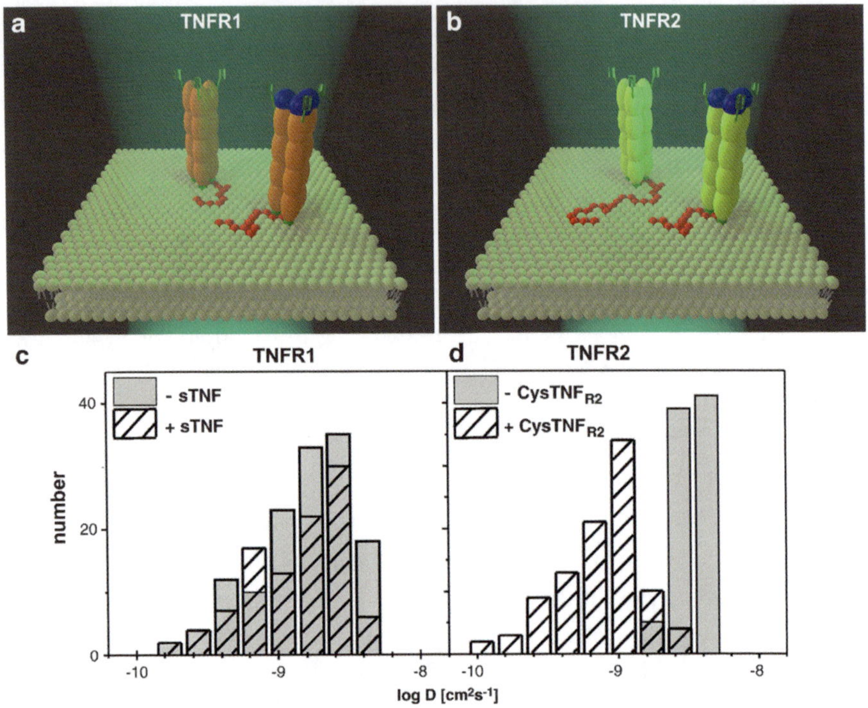

Fig. 4 HeLa cells were transiently transfected with pTNFR1Δ42-EGFP and pTNFR2-EGFP, respectively. Cells were stimulated by the addition of 100 ng/ml sTNF (**a** and **c**) or TNFR2-selective CysTNF$_{R2}$ (**b** and **d**). The diffusion coefficients were calculated from the autocorrelation curves generated from the FCS measurements. (**a**) Sketch of TNFR1 diffusion (relatively slow) through the plasma membrane, (**b**) Sketch of unstimulated TNFR2 diffuses (relatively fast) and stimulated TNFR2 (relatively slow). (**c**) Distribution of diffusion coefficients for TNFR1Δ42-EGFP before (*gray bars*) and 60 min after stimulation (*hatched bars*). (**d**) Distribution of diffusion coefficients for TNFR2-EGFP before (*gray bars*) and 30 min after stimulation (*hatched bars*) (Adapted from [25])

(average value = 0.12 μm^2/s) (data taken from [25]). Stimulation of TNFR2 with recombinant human soluble TNF (sTNF) prolonged the diffusion of the receptor and led to a diffusion behavior similar to that of TNFR1. In contrast, the diffusion of TNFR1 shows no significant changes due to stimulation with TNF (see Fig. 4, modified after [25]).

4.2 Investigation of the Underlying Mechanisms

Importantly, interactions with intracellular compartments such as the cytoskeleton may cause a hindering of the diffusion of transmembrane proteins. To address the impact of this fact on the diffusion of TNFRs, further experiments were performed

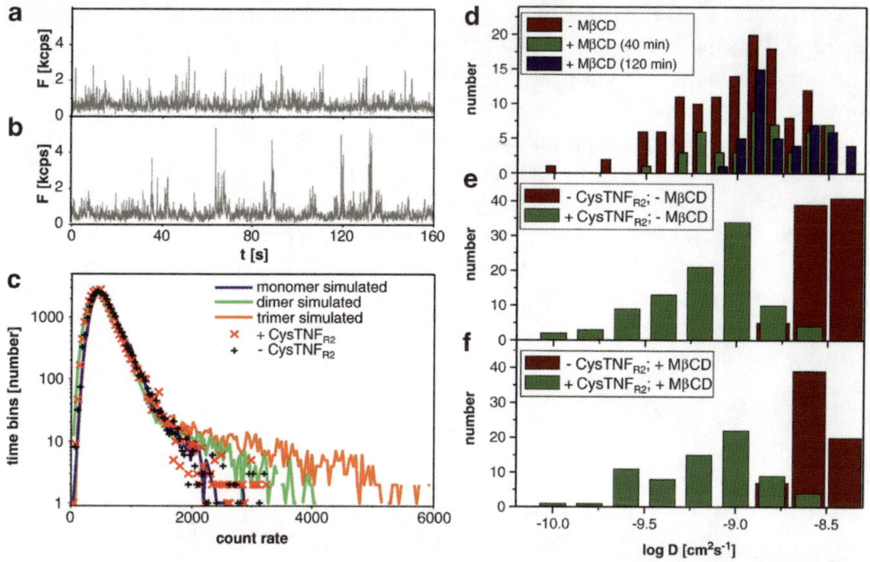

Fig. 5 (**a**) Evaluation of TNFR2-EGFP fluorescence intensity after stimulation. (**a** and **b**) Typical examples of fluorescence fluctuations of TNFR2-EGFP before (**a**) and 20 min after stimulation (**b**) with $CysTNF_{R2}$ (100 ng/ml) in HeLa cells transiently transfected with pTNFR2-EGFP. Every spike in the trace represents a single entity. (**c**) Photon counting histograms of TNFR2-EGFP before (*black crosses*), after stimulation (*red crosses*), simulated monomer (*blue*), simulated dimer (*green*), and trimer (*orange*). D-F. Effects of cholesterol depletion on the dynamics of TNFR diffusion. Distribution of the diffusion coefficients of TNFR1Δ42-EGFP (**d**) and TNFR2-EGFP (**e** and **f**), transiently expressed in HeLa cells. (**d**) Cells were left untreated (*red bars*) or had been pre-treated with 1 mM MβCD for 40 min (*green bars*) or 120 min (*blue bars*). About 100 measurements for each data set were recorded at different positions in the plasma membrane of more than 25 cells. (**e**) Cells were left unstimulated (*red bars*) or were stimulated with $CysTNF_{R2}$ (100 ng/ml) for 30 min. (**f**) HeLa cells transiently expressing TNFR2-EGFP had been pretreated with 1 mM MβCD for 30 min and FCS measurements were performed immediately after (*red bars*) or upon stimulation with $CysTNF_{R2}$ (100 ng/ml) for 30 min (Adapted from [25])

using mutants of TNFR1 and TNFR2 lacking the intracellular domain [25]. Interestingly, neither this approach nor the disruption of actin filaments by cytochalasin D treatment led to a significant difference in diffusion behavior of TNFR1/TNFR2.

One possible reason for the slow diffusion of TNFR1 may be an aggregation of membrane proteins. To address this question, PCH-analysis was used to evaluate the number of fluorescent proteins in a single diffusing compartment. However, the relatively low intensity maximum in bursts of fluorescence did not indicate for clustering. Moreover, PCH-analysis confirmed this observation quantitatively (Fig. 5a–c, modified after [25]).

To further investigate the association of TNFR1 and TNFR2 to cholesterol-rich membrane micro-domains, cells were treated with methyl-β-cyclodextrine (MβCD), which is known to deplete cholesterol from the plasma membrane. FCS measurements of MβCD-treated cells showed an increased diffusion coefficient of

TNFR1, whereas further treatment with TNF resulted in no significant changes. In contrast, FCS measurements of TNFR2 after MβCD-treatment showed no significant difference to untreated cells, neither before nor after stimulation with TNF. This observation provides strong evidence for localization of TNFR1 within cholesterol-rich micro-domains. As these domains are expected to be smaller than the diffraction limit, FCS measurements are expected to measure the slow diffusion of the entire domain with respect to the confocal volume. TNFR2, contrariwise, seems not to be associated with membrane micro-domains (Fig. 5d–f, data from [25]).

We conclude from these measurements that the two TNFRs are compartmentalized in distinct domains of the plasma membrane [25]. This compartmentalization might be determined by the respective transmembrane domains and/or transmembrane domain near regions.

5 Single-Particle Tracking PALM of TNFR1

5.1 Single-Particle Tracking

The idea of tracking single particles or molecules through scattering or fluorescence is to extract information on the diffusion properties from single trajectories. In the "classic" approach of single-molecule tracking of fluorescent species, a target molecule is either co-expressed with a fluorescent protein or labeled with a synthetic fluorophore or quantum dot [26]. Single fluorophores and fluorescent proteins are of small size but limited in photostability, which limits the observation time. Quantum dots allow long observation times, but the size of the label can influence the mobility of the tagged biomolecule and also prevent its access to specific confined cellular areas. The integration of a fluorescent protein requires co-expression and transfection protocols which need to be adjusted to yield very low concentrations of molecules (single-molecule density). Synthetic fluorophores or quantum dots require a chemical labeling procedure that might induce artifacts, for example, by unspecific labeling.

From the trajectory of a single-molecule, the type of diffusion as well as the diffusion coefficient can be extracted from mean square displacement (MSD) analysis of single trajectories. However, for some experiments, the fact that both the observation time and the number of events (single-molecules trajectories available per cell) are limited might be of a disadvantage. This, in particular, is the case if changes in dynamics of biomolecules are the subject of a study: this typically requires that (1) the observation time is long enough to cover the whole temporal feedback of a cellular response as well as (2) a sufficient number of events is observed in order to generate a statistical robustness and to extract information on subpopulations and heterogeneities [1].

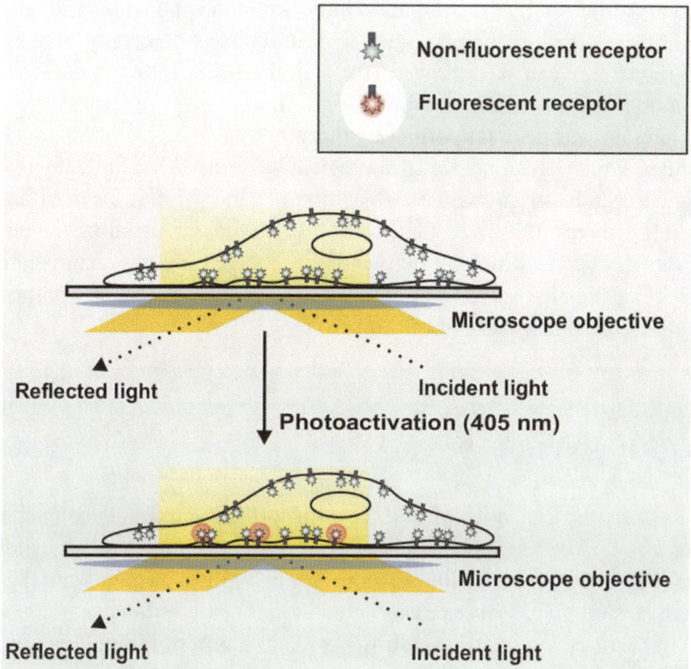

Fig. 6 Schematic representation of single-particle tracking with photoactivatable fluorescent proteins. A target protein is genetically fused to a fluorescent protein which in the beginning of the experiment is dark (*top*), and upon activation with light becomes fluorescent (*bottom*). Through selective activation of only a subset of the fluorophores, single target proteins can be detected and tracked. The large pool of "silent" (i.e., initially non-fluorescent) fluorophores allows continuous activation and observation of target proteins, over extended experimental time and in response to external stimuli

5.2 *Single-Particle Tracking of Photoactivatable Fluorescent Probes*

To increase the number of events observed for a single living cell, the combination of single-particle tracking and photoactivatable fluorescent proteins was introduced (single-particle tracking photoactivated-localization microscopy, sptPALM) [27]. In the original concept, a target molecule (e.g., a membrane receptor) is co-expressed with a fluorescent protein which becomes fluorescent only after activation with light (typically UV) (Fig. 6). Novel suitable fluorescent proteins for this purpose are constantly developed and also used in related methods of super-resolution fluorescence microscopy [28]. The advantages of this approach are threefold. Firstly, the co-expression of a fluorescent protein to a target molecule represents a stoichiometric labeling, and can be combined with the generation of a stable (and eventually inducible) cell line. Secondly, a large pool of non-activated fluorescent proteins are available in a cell, and the read-out of a subset of molecules (and trajectories) can be controlled and adjusted by the irradiation intensity of the photoactivation light source (Fig. 6). Different to classic single-molecule tracking where only a small number of single molecules can be followed (for a limited time), this approach allows to obtain a much larger number and provides a significantly better statistics (Fig. 6). This becomes in particular important to study heterogeneities or subpopulations of biomolecules which are inherent to biological systems [1]. Third, this technology paves way to study the dynamics of a biomolecule with respect to an external stimulus. This makes it, in particular, useful for studies of signaling molecules such as membrane receptors which in many cases change their dynamical behavior upon initiation by a ligand. Observation times of 30 min or more are possible on microscopes appropriate for live cell imaging.

5.3 sptPALM of TNFR1

In order to study the dynamics of TNFR1 with sptPALM in living cells, the receptor needs to be conjugated to a photoactivatable fluorescent protein such as tdEos [29]. tdEos belongs to the class of fluorescent proteins which change their fluorescence emission spectrum upon photoactivation (sometimes referred to as photoconversion) with UV light. In the native form, tdEos has a maximum of fluorescence emission at 516 nm, and in the photoconverted form, the maximum is at 582 nm.

The advantage of tdEos is that cells that were successfully transfected with TNFR1-tdEos can be identified in a separate detection channel prior to the experiment.

In the actual experiment, a subset of TNFR1-tdEos is activated using low-intensity irradiation with a UV light source. The fluorescence signal of activated TNFR1-tdEos is read out using 568 nm laser light. Individual single-molecule localizations are grouped into single-molecule trajectories using appropriate spatiotemporal filtering. The trajectories are further analyzed by calculating the mean square displacement (MSD) for each trajectory and plotting MSD over time (Fig. 7). The initial slope of this plot is approximated with a linear fit, and the diffusion coefficient is extracted.

In a recent study, we applied sptPALM to investigate the lateral dynamics of TNFR1 in living cells under different experimental conditions [30]. In particular, sptPALM measurement of TNFR1-tdEos in living Hela cells yielded an average diffusion coefficient of 0.14 $\mu m^2/s$ [30], which is in good accordance with the results obtained using FCS [25]. As a single-molecule approach, sptPALM provides information on the dynamic properties (velocity, diffusion coefficient) at the single-receptor level. Subpopulations of receptors, for example, slower receptor species that are either engaged to membrane micro-domains or receptor multimers, can readily be resolved from single-molecule histogram data (Fig. 8, modified after [30]). Noticeably, the introduction of a photoactivatable probe significantly increases the statistical level: whereas "classic" single-molecule tracking experiments are limited in the number of proteins that can be observed and tracked

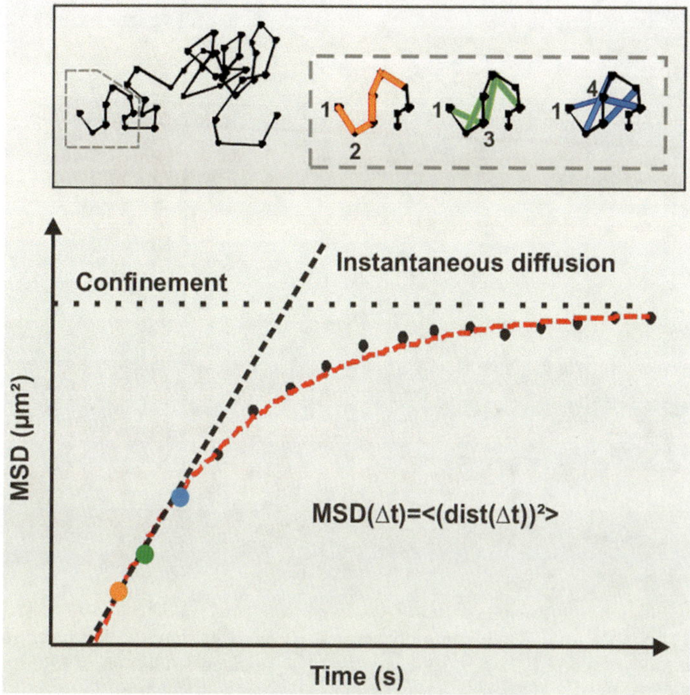

Fig. 7 Mean square displacement analysis of single trajectories. For each single trajectory (*top*), the mean square displacement is calculated and plotted against time (*bottom*). The instantaneous diffusion coefficient can be extracted from the slope of a linear fit to the first part of the plot

at the same time (by the diffraction limit) [31], sptPALM bypasses this limit by separating the observation in time. Typically, thousands of single-molecule trajectories can be recorded (subject to expression level) with this strategy. In addition, fast motion of proteins can be measured by adjusting a low density of activated fluorophores and allowing for larger step sizes in the trajectories; these events are typically not accessible with classic single-molecule tracking, and if so, at the expense of the number of events that can be observed. As a whole-cell approach with single-molecule resolution, sptPALM further offers the opportunity to generate diffusion coefficient maps of an entire cell [32, 33].

Dynamic properties of membrane receptors depend on their membrane environment, for example, membrane micro-domains, and local heterogeneities can be observed. As discussed earlier, FCS measurements have revealed a higher diffusion coefficient if cholesterol was depleted from the membrane using MβCD. This finding could be corroborated with sptPALM. In addition, the single-molecule statistics can be used to extract additional information on subpopulations and heterogeneities. For "native" TNFR1-tdEos, we have observed that the distribution of diffusion coefficients is rather broad and exhibits a shoulder of very slow receptor molecules, ranging from 0.0001 to 0.01 $\mu m^2/s$ (Fig. 8, data from [30]). Upon addition of MβCD,

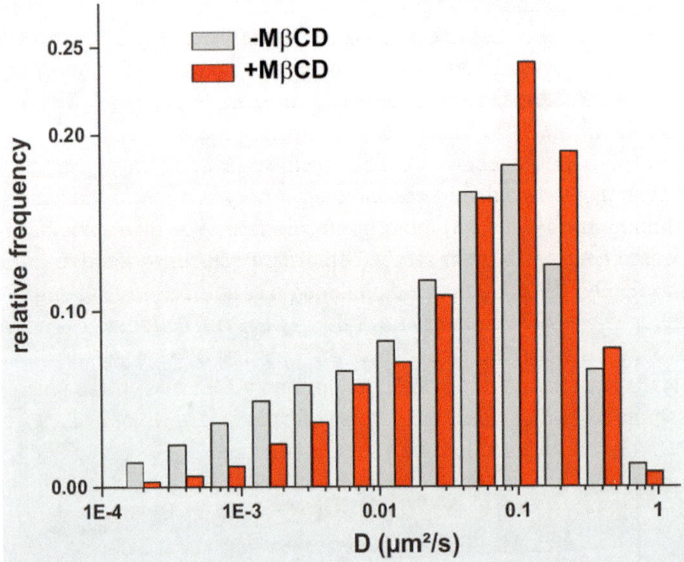

Fig. 8 Distribution of diffusion coefficients derived from single-molecule data of TNFR1-tdEos and TNFR1-tdEos + MβCD (cholesterol depletion). The distribution is rather broad for TNFR1-tdEos, and a population of slow receptor aggregates is apparent. Addition of MβCD leads to a disappearance of this slow population (Data from [30])

the distribution of diffusion coefficients is less broad, and in addition, we observe a decrease in this slow population (Fig. 8, data from [30]). This finding can be interpreted as large receptor clusters that are nearly immobile and associated with lipid micro-domains. Depletion of cholesterol changes the micro-architecture of the membrane, with the consequence that (1) large and immobile clusters of receptors disappear and (2) a population of fast receptor molecules appears. The average diffusion coefficient, which notably cannot reveal subpopulations of receptor molecules, increases from 0.14 (TNFR1-tdEos) to 0.20 $\mu m^2/s$ (TNFR1-tdEos + MβCD) (data from [30]). Importantly, the diffusion coefficient observed under cholesterol depletion stays similarly constant over time as observed for native TNFR1-tdEos (data from [30]).

6 Outlook

There has been considerable new information on association and activation of TNFRs in fixed and living cells, generated from the toolbox of advanced fluorescence methods. The technologies we introduce can similarly be used to investigate the dynamics of other cell membrane receptors. Overall, we aimed to instruct how a biological problem can be addressed by a careful choice of appropriate experimental methods.

The approaches described in this chapter can be further extended. Fluorescence cross-correlation spectroscopy (FCCS) [21, 22] is an additionally valuable tool that can be used to study the association and dissociation, and has been used for other biological systems [34]. Using this method, the endogenous receptor can be probed in a similar way by a fluorophore-labeled ligand (TNF-α) [33] and binding events of TNF-α with TNFRs can be detected. Importantly, nonspecific binding can be separated from specific binding events profiting from the knowledge of TNFRs mobility on the cell membrane that has been generated by FCS and sptPALM. Finally, a clever combination of both approaches can probe activated and non-activated receptors. Single-molecule photobleaching is a third possible route which has been successfully used to study the dimerization of EGF receptor [35]. Such approach can also be used to study the oligomerization of TNFRs and may help to understand its dynamics.

In summary, modern fluorescence spectroscopy and super-resolution microscopy with single-molecule resolution are of great value for investigation of receptor dynamics and oligomerization. Nevertheless, a number of biological questions on the function of TNF receptors still remain unsolved. For example, it is still not elucidated how many of the endogenous TNFRs are found as monomers, dimers, trimers, or multimers on a cell membrane. For the future, we can also expect this question to be addressed, as advanced fluorescence microscopy techniques may overcome current limitations and provide more valuable insights into TNFR-biology.

References

1. Hohlbein J, Gryte K, Heilemann M, Kapanidis AN (2010) Surfing on a new wave of single-molecule fluorescence methods. Phys Biol 7(3):031001. doi:S1478-3975(10)53547-7 [pii] 10.1088/1478-3975/7/3/031001
2. Hotamisligil GS, Murray DL, Choy LN, Spiegelman BM (1994) Tumor necrosis factor alpha inhibits signaling from the insulin receptor. Proc Natl Acad Sci USA 91(11):4854–4858
3. Bright JJ (2007) Curcumin and autoimmune disease. Adv Exp Med Biol 595:425–451
4. Pacifico F, Leonardi A (2006) Nf-kappab in solid tumors. Biochem Pharmacol 72(9):1142–1152
5. Gray PW, Aggarwal BB, Benton CV, Bringman TS, Henzel WJ, Jarrett JA, Leung DW, Moffat B, Ng P, Svedersky LP et al (1984) Cloning and expression of cdna for human lymphotoxin, a lymphokine with tumour necrosis activity. Nature 312(5996):721–724
6. Pennica D, Nedwin GE, Hayflick JS, Seeburg PH, Derynck R, Palladino MA, Kohr WJ, Aggarwal BB, Goeddel DV (1984) Human tumour necrosis factor: precursor structure, expression and homology to lymphotoxin. Nature 312(5996):724–729
7. Aggarwal BB (2003) Signalling pathways of the tnf superfamily: a double-edged sword. Nat Rev Immunol 3(9):745–756. doi:10.1038/nri1184 nri1184 [pii]
8. Schmitz ML, Mattioli I, Buss H, Kracht M (2004) Nf-kappab: a multifaceted transcription factor regulated at several levels. Chembiochem 5(10):1348–1358
9. Bonizzi G, Karin M (2004) The two nf-kappab activation pathways and their role in innate and adaptive immunity. Trends Immunol 25(6):280–288
10. Wajant H, Pfizenmaier K, Scheurich P (2003) Tumor necrosis factor signaling. Cell Death Differ 10(1):45–65. doi:10.1038/sj.cdd.4401189 4401189 [pii]

11. Mathew SJ, Haubert D, Kronke M, Leptin M (2009) Looking beyond death: a morphogenetic role for the tnf signalling pathway. J Cell Sci 122(Pt 12):1939–1946. doi:122/12/1939 [pii] 10.1242/jcs.044487
12. Legler DF, Micheau O, Doucey MA, Tschopp J, Bron C (2003) Recruitment of tnf receptor 1 to lipid rafts is essential for tnfalpha-mediated nf-kappab activation. Immunity 18(5):655–664
13. Doan JE, Windmiller DA, Riches DW (2004) Differential regulation of tnf-r1 signaling: lipid raft dependency of p42mapk/erk2 activation, but not nf-kappab activation. J Immunol 172(12):7654–7660. doi:172/12/7654 [pii]
14. Hunter I, Nixon GF (2006) Spatial compartmentalization of tumor necrosis factor (tnf) receptor 1-dependent signaling pathways in human airway smooth muscle cells. Lipid rafts are essential for tnf-alpha-mediated activation of rhoa but dispensable for the activation of the nf-kappab and mapk pathways. J Biol Chem 281(45):34705–34715. doi:M605738200 [pii] 10.1074/jbc.M605738200
15. van de Linde S, Wolter S, Heilemann M, Sauer M (2010) The effect of photoswitching kinetics and labeling densities on super-resolution fluorescence imaging. J Biotechnol 149(4):260–266. doi:10.1016/j.jbiotec.2010.02.010
16. Eggeling C, Ringemann C, Medda R, Schwarzmann G, Sandhoff K, Polyakova S, Belov VN, Hein B, von Middendorff C, Schonle A, Hell SW (2009) Direct observation of the nanoscale dynamics of membrane lipids in a living cell. Nature 457(7233):1159–1162. doi:10.1038/nature07596
17. Sprague BL, McNally JG (2005) Frap analysis of binding: proper and fitting. Trends Cell Biol 15(2):84–91. doi:10.1016/j.tcb.2004.12.001
18. Garcia-Saez AJ, Schwille P (2008) Fluorescence correlation spectroscopy for the study of membrane dynamics and protein/lipid interactions. Methods 46(2):116–122. doi:10.1016/j.ymeth.2008.06.011
19. Anikovsky M, Wiltshire ZD, Weisshart K, Petersen NO (2011) Photon counting histogram analysis for two-dimensional systems. Chemphyschem: European J Chemical Physics Physical Chem 12(13):2439–2448. doi:10.1002/cphc.201100414
20. Huang B, Perroud TD, Zare RN (2004) Photon counting histogram: One-photon excitation. Chemphyschem: European J Chemical Physics Physical Chem 5(10):1523–1531. doi:10.1002/cphc.200400176
21. Bacia K, Kim SA, Schwille P (2006) Fluorescence cross-correlation spectroscopy in living cells. Nat Methods 3(2):83–89. doi:10.1038/nmeth822
22. Hwang LC, Wohland T (2007) Recent advances in fluorescence cross-correlation spectroscopy. Cell Biochem Biophys 49(1):1–13
23. Joo C, Balci H, Ishitsuka Y, Buranachai C, Ha T (2008) Advances in single-molecule fluorescence methods for molecular biology. Annu Rev Biochem 77:51–76. doi:10.1146/annurev.biochem.77.070606.101543
24. Garcia-Saez AJ, Schwille P (2007) Single molecule techniques for the study of membrane proteins. Appl Microbiol Biotechnol 76(2):257–266. doi:10.1007/s00253-007-1007-8
25. Gerken M, Krippner-Heidenreich A, Steinert S, Willi S, Neugart F, Zappe A, Wrachtrup J, Tietz C, Scheurich P (2010) Fluorescence correlation spectroscopy reveals topological segregation of the two tumor necrosis factor membrane receptors. Biochim Biophys Acta 1798 (6):1081–1089. doi:10.1016/j.bbamem.2010.02.021
26. Groc L, Lafourcade M, Heine M, Renner M, Racine V, Sibarita JB, Lounis B, Choquet D, Cognet L (2007) Surface trafficking of neurotransmitter receptor: comparison between single-molecule/quantum dot strategies. J Neurosci: Official J Soc Neurosci 27(46):12433–12437. doi:10.1523/JNEUROSCI.3349-07.2007
27. Manley S, Gillette JM, Patterson GH, Shroff H, Hess HF, Betzig E, Lippincott-Schwartz J (2008) High-density mapping of single-molecule trajectories with photoactivated localization microscopy. Nat Methods 5(2):155–157. doi:10.1038/nmeth.1176

28. Betzig E, Patterson GH, Sougrat R, Lindwasser OW, Olenych S, Bonifacino JS, Davidson MW, Lippincott-Schwartz J, Hess HF (2006) Imaging intracellular fluorescent proteins at nanometer resolution. Science 313(5793):1642–1645. doi:10.1126/science.1127344
29. Wiedenmann J, Ivanchenko S, Oswald F, Schmitt F, Rocker C, Salih A, Spindler KD, Nienhaus GU (2004) Eosfp, a fluorescent marker protein with UV-inducible green-to-red fluorescence conversion. Proc Natl Acad Sci USA 101(45):15905–15910. doi:10.1073/pnas.0403668101
30. Heidbreder M, Zander C, Malkusch S, Widera D, Kaltschmidt B, Kaltschmidt C, Nair D, Choquet D, Sibarita JB, Heilemann M (2012) Tnf-alpha influences the lateral dynamics of tnf receptor i in living cells. Biochim Biophys Acta. doi:10.1016/j.bbamcr.2012.06.026
31. Ma X, Wang Q, Jiang Y, Xiao Z, Fang X, Chen YG (2007) Lateral diffusion of tgf-beta type i receptor studied by single-molecule imaging. Biochem Biophys Res Commun 356(1):67–71. doi:S0006-291X(07)00372-5 [pii] 10.1016/j.bbrc.2007.02.080
32. Franklin KJ (1948) Circulatory concepts. St Bartholomew's Hospital J 52(1):2, passim
33. Giannone G, Hosy E, Levet F, Constals A, Schulze K, Sobolevsky AI, Rosconi MP, Gouaux E, Tampe R, Choquet D, Cognet L (2010) Dynamic superresolution imaging of endogenous proteins on living cells at ultra-high density. Biophys J 99(4):1303–1310. doi:10.1016/j.bpj.2010.06.005
34. Neugart F, Zappe A, Buk DM, Ziegler I, Steinert S, Schumacher M, Schopf E, Bessey R, Wurster K, Tietz C, Borsch M, Wrachtrup J, Graeve L (2009) Detection of ligand-induced cntf receptor dimers in living cells by fluorescence cross correlation spectroscopy. Biochim Biophys Acta 1788(9):1890–1900. doi:10.1016/j.bbamem.2009.05.013
35. Teramura Y, Ichinose J, Takagi H, Nishida K, Yanagida T, Sako Y (2006) Single-molecule analysis of epidermal growth factor binding on the surface of living cells. EMBO J 25(18):4215–4222. doi:10.1038/sj.emboj.7601308

HIV-1 Gag Directed Assembly of Retroviral Particles Investigated by Quantitative Fluorescence Imaging

Hugues de Rocquigny, Hocine Gacem, Pascal Didier, Jean-Luc Darlix, and Yves Mély

Abstract HIV-1 particle assembly is driven by oligomerization of the Gag polyprotein precursor in infected cells. Translation of the full-length viral RNA results in the synthesis of large amounts of Gag molecules which oligomerize upon the combined interactions of its C-terminal NC domain with the genomic RNA and of its N-terminal myristate and matrix basic residues with cellular membranes. HIV assembly has been studied during the past two decades mostly by means of biochemical techniques on transfected or infected cells as well as *in vitro* using purified Gag molecules. More recently, understanding the mechanisms of viral assembly moved a big step forward due to the utilization of fluorescently labeled viral proteins, notably Gag, and of fluorescent microscopy techniques able to track single viral particles. In this chapter, we will summarize recent imaging data on HIV-1 Gag assembly at the level of the plasma membrane where viral particles bud in the form of cell-free viruses or can be transmitted to adjacent naïve cells through virological synapses.

Keywords AIDS · FLIM · Fluorescence Microscopy · FRET · HIV-1 · Membrane

Contents

H. de Rocquigny (✉), H. Gacem, P. Didier, J.-L. Darlix and Y. Mély
Laboratoire de Biophotonique et Pharmacologie, UMR 7213 CNRS, Faculté de Pharmacie, Université de Strasbourg, 74, Route du Rhin, 67401 ILLKIRCH Cedex, France
e-mail: hderocquigny@unistra.fr

Y. Mély and G. Duportail (eds.), *Fluorescent Methods to Study Biological Membranes*, Springer Ser Fluoresc (2013) 13: 457–478, DOI 10.1007/4243_2012_55,

Abbreviations

CA	Capsid
ESCRT	Endosomal sorting complex required for transport
FCS	Fluorescence correlation spectroscopy
FFS	Fluorescence fluctuation spectroscopy
FlAsH	4′,5′-Bis(arsenoso)fluorescein
FLIM	Fluorescence lifetime imaging microscopy
FRAP	Fluorescence recovery after photobleaching
Gag	Group-specific antigen
HIV	Human immunodeficiency virus
IN	Integrase
MA	Matrix
MuLV	Murine leukemia virus
NC	Nucleocapsid
ORF	Open reading frame
PM	Plasma membrane
Pol	Polymerase
PR	Protease
ReAsH	4′,5′-Bis(arsenoso)resorufin
RNA	Ribonucleic acid
RSV	Rous sarcoma virus
RT	Reverse transcriptase
TIRFM	Total internal reflection fluorescence microscopy
TMgp41	Transmembrane glycoprotein 41
UTR	Untranslated region
VLP	Virus-like particles
Vpr	Viral protein R
vRNP	Viral ribonucleoparticle
VSV	Vesicular stomatitis virus

1 Introduction

The human immunodeficiency virus type 1 (HIV-1) is the causative agent of AIDS (acquired immunodeficiency syndrome) and belongs to the Lentivirus genus of the retrovirus family. The spherical viral particle has a diameter of 100–120 nm with an outer envelope originating from the lipid bilayer of the host cell membrane (Fig. 1a). The two viral envelope glycoprotein subunits, namely, the surface SUgp120 and the transmembrane TMgp41 proteins in the form of trimers, are anchored in the virion envelope. SUgp120 interacts with the target cell receptors while TMgp41 mediates the fusion of the infectious particle with the cell plasma membrane. The inner virion structure is filled by the matrix protein (MAp17) and

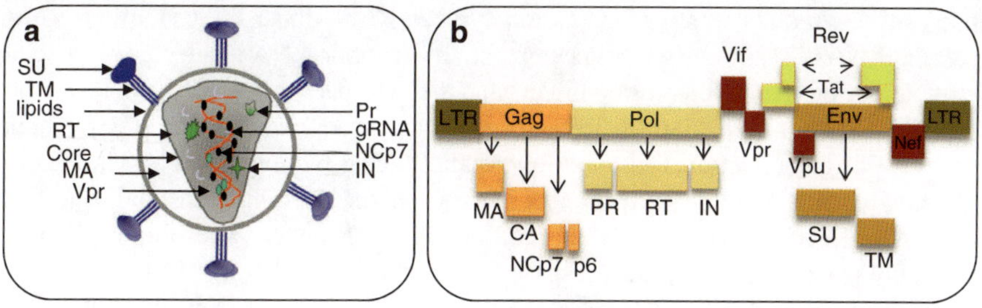

Fig. 1 Schematic representation of HIV (**a**) and genetic organization of the viral DNA (**b**). (**a**) Scheme of the mature HIV-1 particle. From outside to inside: trimeric viral glycoproteins SU/TM (*blue*) anchored to the lipid envelope from cellular origin (*gray circle*), matrix (MA, *white*), and core composed of capsid proteins and the viral ribonucleoprotein (vRNP) containing the dimeric RNA genome (*red*) coated by nucleocapsid proteins (NCp7, black) together with molecules of reverse transcriptase (RT), Vpr, integrase (IN), and protease (Pr) (*green*). (**b**) Genetic organization of HIV-1 viral DNA. The long terminal repeats (LTR, *brown*) flank the ORF sequences for Gag (*orange*), Pol (*yellow*), and Env (*brown*) coding for the structural proteins, enzymes, and glycoproteins, respectively. In addition, the sequences encoding for the viral cofactors Tat and Rev (bright *yellow*) and the helper factors Vif, Vpr, Vpu, and Nef (dark *red*) are shown. Not shown are the *cis*-acting sequences required for genomic RNA replication located in and next to the LTRs

contains the capsid protein (CAp24) organized as a conical core that contains the viral ribonucleoparticle (vRNP). This vRNP is composed of the positive sense genomic RNA of 9.6 kbases, in a dimeric form, coated by about 1,400 to 2,000 nucleocapsid protein (NCp7) molecules together with viral protein R (Vpr) molecules and the viral enzymes reverse transcriptase (RT), integrase (IN), and protease (PR) [1].

As a matter of fact, viral particle assembly results from the initial oligomerization of the newly made Gag and Gag-Pol polyprotein precursor molecules (Fig. 1b, orange), resulting from the translation of the full-length viral RNA by the host cell ribosome machinery [2]. As a consequence of their accumulation in the cytoplasm, Gag and Gag-Pol polyproteins tend to self-assemble [3]. Actually, the accepted scheme for HIV-1 Gag assembly stipulates that two platforms are driving Gag oligomerization, namely, an RNA platform recruiting Gag via NC-RNA interactions and a second one corresponding to a lipid membrane targeted by the N-terminal myristate and basic residues of Gag-MA [4, 5]. In this context, a major element that orchestrates the correct Gag oligomerization is the capsid (CA) domain as it can form series of hexameric rings connected to each other [6] (and ref in [7]). Another driving Gag assembly element is the nucleocapsid domain (NC) that contributes to Gag oligomerization [8]. During the assembly process, small nascent Gag oligomers are first formed in the cytoplasm [9, 10], which causes a conformational change of MA exposing the N-terminal hydrophobic myristate (myristyl switch). This in turn targets the Gag oligomers to the plasma membrane (PM) [11–14]. The PM anchorage of Gag is stabilized by specific interactions of the MA basic region HBR with PI(4,5)P2 [15, 16], a component of the PM intracytoplasmic leaflet [17–20]. As stated above, one of the two platforms for Gag assembly is an RNA, either the genomic RNA or cellular RNAs [21]. In the wild-type context, the genomic RNA is selected by specific interactions between the NC zinc fingers and the viral packaging Psi signal located in the 5'UTR [4, 21–27]. Thus, three major molecular interactions are driving the bona fide Gag oligomerization and assembly, namely, Gag binding to RNA via the NC domain, Gag interaction with phospholipids via the MA domain, and Gag-Gag protein interactions via the CA and NC domains. However, where and how Gag assembly is taking place in the HIV-1 infected cells is poorly understood.

In the course of assembly, Gag oligomers are trafficking from the translating ribosomes to the PM, where virion budding is taking place [28]. In order to complete virion assembly, trafficking and ultimately budding, Gag was found to hijack cellular proteins of the cytoskeleton and the ESCRT (endosomal sorting complex required for transport) and to recruit viral proteins such as Vpr and Vif and the viral envelope proteins [1, 29–31].

Understanding the Gag assembly process was rendered possible through the use of a panel of biochemical analyses on *in vitro* model systems and model cell lines such as HeLa cells. While these technical approaches provided important insights into HIV-1 (and other retrovirus) assembly, their low resolution hampered monitoring assembly and trafficking of single viral particles, or at least of a small number of them. Therefore, in an attempt to investigate Gag assembly and trafficking in cells at the molecular level, recombinant Gag was labeled with fluorescent tags, thus allowing its visualization by fluorescent microscopy techniques on either fixed or living cells [32].

This chapter focuses on the use of fluorescent microscopy techniques that are currently used to investigate Gag assembly and trafficking toward the PM, where newly made virions bud from the cell in the form of cell-free viruses or else are transmitted to adjacent naïve cells through virological synapses.

2 Fluorescently Labeled Retroviral Gag Proteins

Fluorescent Gag fusion proteins are now becoming popular tools to visualize retrovirus (HIV-1, MoMuLV, RSV) assembly in living cells as well as newly made particles. In early studies, the green fluorescent protein (GFP) of *Aequorea victoria* or its derivatives was fused to the Gag C terminus and expressed via a strong promoter (Fig. 2). After DNA transfection in various cell types, Gag-GFP was found throughout the cytoplasm [33] and at the cell PM [34]. However, virus-like particles (VLP) containing Gag-GFP were shown by electron microscopy (EM) to exhibit aberrant morphology, indicating that addition of a 27 kDa protein to the Gag C terminus has a strong impact on virus morphology [35, 36]. Moreover, this Gag chimera could not be used in the viral context, since the 3′ region of the HIV-1 gag open reading frame overlaps with that of Pol. More recently, the group of B Muller [37] engineered the Gag-iGFP mutant where eGFP is inserted close to the C terminus of MA. Interestingly, fluorescently labeled viruses prepared by DNA

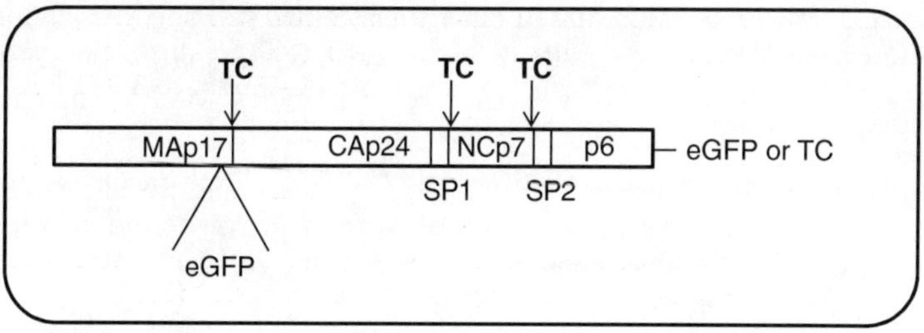

Fig. 2 Insertion of a fluorescent reporter in the HIV-1 Gag polyprotein. The HIV-1 Gag polyprotein precursor, composed of 550 amino acids, is processed by the viral protease, generating the MA, CA, NCp7, and p6 viral proteins. The Gag polyprotein was labeled by eGFP, and also mCherry, CFP, or Venus, at the indicated positions. TC refers to the tetracysteine tag

transfection of equimolar amounts of plasmids expressing Gag wt and Gag-iGFP were shown to resemble wild-type viruses with respect to particle production, particle morphology, and virus infectivity [37].

However, as the size of the fluorescent protein moiety might alter the biological activity of the Gag polyprotein, this prompted the search for smaller potent tags for cell imaging. To that end, a 12-amino-acid tetracysteine sequence was inserted at the C terminus of the HIV-1 Gag protein (Fig. 2) [38–40]. This tag interacts with high affinity with a membrane-permeable biarsenical fluorescein derivative, FlAsH (or its red analog ReAsH) [41]. This interaction strongly enhances the fluorescence emission of the chromophore, but in the context of a rather strong fluorescence background, due to the free chromophore and to the chromophore bound to cellular proteins containing cystein-rich domains [40, 42]. Addition of this tag at the C terminus of Gag did not interfere with virion assembly [40, 42] and packaging of the viral Vpr proteins [43]. Viruses produced with TC inserted downstream of MA were as infectious as the wild-type counterpart [38, 44]. However, TC insertion before or next to NC was found to cause a twofold reduction of viral infectivity [39]. The TC tag has also been inserted next to the integrase (IN) domain, at the C terminus of pol. This resulted in the production of infectious viruses and, upon cell infection, allowed visualization of the microtubule-assisted movements of the HIV-1 preintegration complex (PIC) from the cell periphery to the nucleus and within the nuclear compartment [45, 46].

3 Gag-Gag Interaction Visualized by FRET and Correlation Spectroscopy

Where and when are Gag oligomerization and assembly taking place in infected cells? Although this has been the subject of a large number of studies, it is still a matter of debate. At first, HIV-1 was considered to be a C-type retrovirus, and thus, Gag assembly was considered to occur at the level of the PM. In this model which appears to occur in HeLa, Cos, and Jurkat T transformed cell lines, newly made Gag is directly targeted to the PM [47]. In a second model, Gag assembles in intracellular vesicles such as, for instance, multivesicular bodies (MVB) in human macrophages, and subsequently, newly formed virions are delivered in the extracellular medium by means of vesicular trafficking [42, 48–50]. Both mechanisms of Gag assembly and trafficking were shown to take place in several cell lines, while either one was predominant in other cells, indicating that the cellular context has a major influence on the mechanism of Gag assembly and trafficking (reviewed in [28]).

Major contributions on retroviral Gag assembly and trafficking were obtained by molecular imaging. Gag-Gag interactions at the level of the PM were first visualized by two-photon imaging, using Rous sarcoma virus (RSV), a type C alpha retrovirus [51]. In this study, DF-1 cells (chicken fibroblast cells) were transfected by plasmid DNAs expressing RSV Gag-CFP and Gag-YFP, and a z-stack series was carried out to record the FRET between these two reporters throughout the cell. FRET corresponds to a nonradiative energy transfer between the fluorescent donor and an acceptor when they are less than 10 nm apart, a distance corresponding to intermolecular protein-protein interactions [52–54]. The FRET was larger at the interface between the cell and the cover slip and also at the dorsal cell surface in agreement with the expected localization of the virus budding sites. Gag-Gag interaction at the level of the PM was confirmed using an approach similar to FRET with HIV-1 Gag fluorescent chimeras expressed in Mel JuSo cells (human melanoma cell line) [55], HeLa cells [56, 57], and epithelial 293 T cells [58]. All these studies used the fluorescence intensity changes of the donor and acceptor after photobleaching of the acceptor. Alternatively, another strategy is to monitor by FLIM the fluorescence lifetime of FlAsH linked to HIV-1 Gag-TC. Indeed, Gag polymerization clusters the FlAsH chromophores, resulting in a decrease of their fluorescence lifetime due to self-quenching and/or exciton coupling [59, 60]. This self-quenching of fluorescein derivatives was already successfully used to study protein-protein and protein-lipid interaction [61–63]. To carry out these experiments, cells were stained with FlAsH and scanned continuously for about 30 s to achieve appropriate photon statistics to analyze the fluorescence decays. Data were analyzed with a commercial software package (SPCImage V2.8, Becker & Hickl, Germany), and FLIM images were constructed using an arbitrary color scale, ranging from blue (short lifetime) to red (long lifetime) corresponding to the different lifetimes of the donor (see scale on Fig. 3, panel 1). As a control, naive HeLa cells show a diffuse fluorescence characterized by a fluorescence lifetime of 3.52 ns (Fig. 3, panel 1, green pseudocolor). When Hela cells were transfected by a plasmid DNA encoding for Gag-TC, Gag accumulates at the level of the PM, and the FlAsH lifetime drops to 2.57 ns (Fig. 3, panel 2, blue pseudocolor).

FRET was also used to study the relationship between Gag oligomerization and its localization at the PM and to investigate the impact of Gag mutations on these processes. For example, substituting methionine 369 in the SP1 spacer for alanine decreased the fluorescence lifetime value of the chromophore to 3.08 ns (Fig. 3, panel 3), confirming the importance of SP1 in HIV-1 $Pr55^{Gag}$ assembly [64, 65]. Mutation in the capsid C terminus, which is not required for Gag-PM interaction, has a severe impact on Gag-Gag FRET [56] in agreement with the role of this

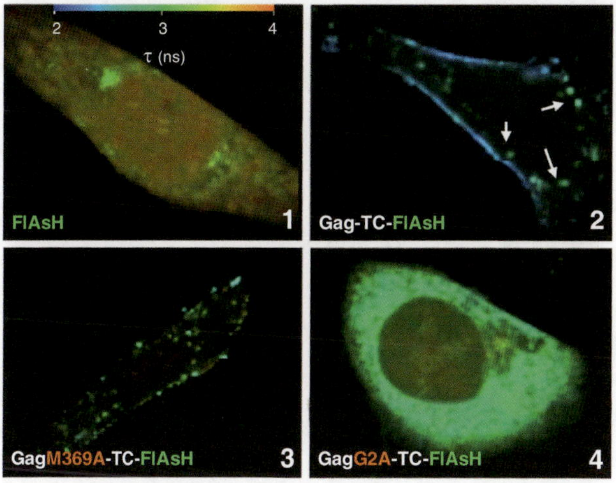

Fig. 3 Gag-Gag interaction monitored by FLIM. HeLa cells were transfected by a plasmid DNA encoding for Gag-TC [40]. Twenty-four hours post-transfection, cells were labeled with the FlAsH derivative [43]. Then, two-photon FLIM images were recorded, and measured lifetimes were imaged using an arbitrary color scale (*top* of panel **1**). Panel **1**: control naive HeLa cells incubated with FlAsH. The chromophore is homogeneously distributed with an average lifetime of 3.52 ns. Panel **2**: HeLa cell expressing Gag-TC. The close proximity of Gag-TC-FlAsH molecules decreases their fluorescence lifetime to 2.52 ns, reflecting Gag multimerization. Gag-containing dots with an intermediate lifetime (*green color*) might correspond to Gag oligomers. Panel **3**: Substitution of methionine 369 in the SP1 spacer for alanine ($Pr55^{GagM369A\text{-}TC}$) severely impairs $Pr55^{Gag}$ assembly, as shown by the 3.08 ns fluorescence lifetime value obtained for this mutant. Panel **4**: HeLa cell transfected with the GagG2A-TC mutant. The substitution of Gly for Ala hampers membrane anchoring of Gag. The fluorescence lifetime homogenously distributed throughout the cytoplasm is equal to 2.87 ns indicating that myristylation-defective $Pr55^{Gag}$ can oligomerize but does not form high molecular weight structures. This figure is adapted from Fig. 7 of [43]

domain in Gag oligomerization [66, 67]. When the membrane-binding domain of RSV Gag [68] or HIV-1 Gag was modified [12, 13, 55], the accumulation of Gag at the PM was impaired, but FRET was observed in the cytoplasm either by intensity measurements [51, 56] or by FLIM (Fig. 3, panel 4) with an intermediate fluorescence lifetime of 2.87 ns, indicating that membrane-defective Gag can oligomerize in the cytoplasm [55, 69–71].

As already stated, one domain important for HIV-1 Gag assembly is NC [27], since it drives assembly through its ability to bind RNA, preferentially the genomic RNA [72–74]. NC may also drive Gag-Gag interaction through an RNA-independent mechanism since replacement of NC by polypeptides capable of forming oligomers can rescue Gag assembly [8]. In agreement with the key role of NC in Gag assembly, monitoring by fluorescence microscopy of Gag-delta-NC assembly revealed the absence of FRET [51, 56] and a diffusion coefficient equal to that of eGFP [51]. In conclusion, these fluorescence microscopy studies correlate with the biochemical analyses and indicate that Gag most probably starts assembling in the cytoplasm and subsequently Gag oligomers traffic to the PM, where anchoring takes place.

It remains to be established where Gag oligomer formation takes place and with which stoichiometry. Using FLIM, complexes with a lifetime of 2.82 ns were observed in the cytoplasm (Fig. 3, panel 2, white arrows), in line with the presence of low molecular weight HIV-1 $Pr55^{Gag}$ complexes in subcellular compartments [10, 70]. Recently, cytoplasmic Gag-Gag interactions were characterized by fluorescence fluctuation spectroscopy (FFS) coupled to a z-scan approach [75]. This method can determine the number of fluorescent molecules present in complexes diffusing in a small volume. The analysis relies on the measurement of the brightness of individual complexes compared to that of individual fluorescent proteins. It was found that the Gag stoichiometry in the viral-like particles (VLPs) formed in living cells depended on the quantity of transfected plasmid DNA, ranging from 750 to 2500 Gag molecules per VLP [76]. Moreover, by measuring the average diffusion time, it was possible to determine a hydrodynamic diameter of about 130 nm for the VLPs, independent from the amount of transfected plasmid DNA. This observation suggests that VLPs form in the cytoplasm and that a full coverage of the VLP by the Gag protein is not required (for a recent review [77]).

Assembly of Gag at the PM was also investigated using recombinant Gag labeled with *eos* derivatives downstream from MA [78]. The EOS Monomeric photoactivable protein from the stony coral *Lobophyllia hemprichii* changes its spectral properties, in response to UV irradiation with high photon count and high contrast ratio between the converted and nonconverted forms [79]. After protein excitation at 488 nm, a green emission is monitored at 516 nm. In addition, the protein emission is red shifted to 581 nm after irradiation at 405 nm and excitation at 516 nm. The group of Lamb [78] combined total internal reflection fluorescence (TIRF) and wide-field (WF) microscopy to address the question of whether VLPs detected at the surface of HeLa cells resulted from Gag clusters arriving from the cytoplasm or from Gag molecules assembling at the PM [80]. The TIRF technique uses an evanescent wave to selectively illuminate and excite fluorophores in a restricted area of the specimen immediately adjacent to the glass-water interface. The evanescent wave is generated only when the incident light is totally internally reflected at the glass-water interface. The evanescent electromagnetic field decays exponentially from the interface and penetrates to a depth of only 75–200 nm into the sample medium [81]. Thus, the TIRF microscopy preferentially photoconverts Gag-EOS localized at the cell surface. To combine WF and TIRF approaches, WF images were recorded in the green channel with an excitation of the Gag-EOS at 488 nm, while TIRF images were recorded in the red channel with an excitation at 516 nm after a pulse at 405 nm (Fig. 4a).

As depicted in Fig. 4b, before irradiation, Gag-EOS either in the cytoplasm or at the PM emits only in the green channel. In contrast, after 90-s illumination at 405 nm (Fig. 4, C), almost 30 % of Gag-EOS molecules were photoconverted as shown by the numerous fluorescent dots observed in the red channel. Next, this experiment was repeated 15 min after illumination. If newly formed VLPs would result from Gag clusters migrating from the cytoplasm, then the number of green

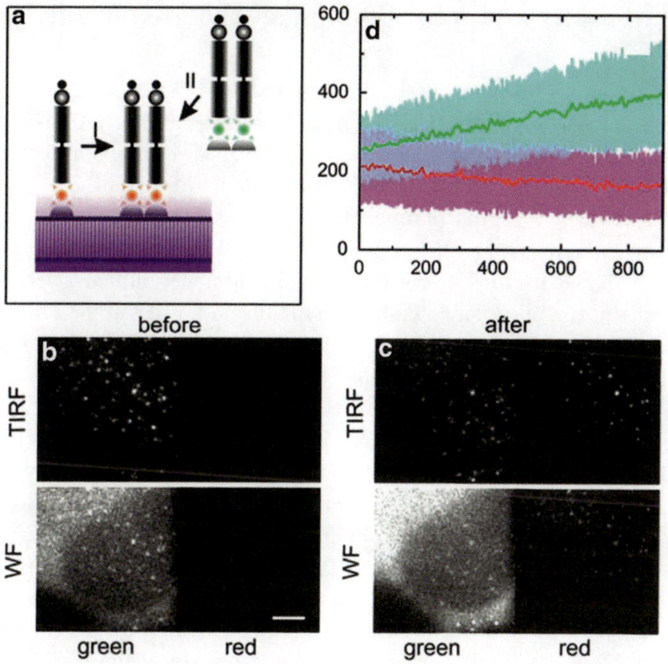

Fig. 4 VLP assembly results from the recruitment of cytosolic Gag clusters. (**a**) Photoconversion scheme and intracellular localization of Gag-EOS chimera proteins. The *green spheres* correspond to the nonphotoconverted form of Gag-EOS ($\lambda_{exc} = 488$ nm and $\lambda_{em} = 516$ nm) observable in the epifluorescence mode. The *red spheres* correspond to the photoconverted form of Gag-EOS ($\lambda_{exc} = 516$ nm and $\lambda_{em} = 581$ nm) after a photoconversion pulse at 405 nm in the TIRF mode. Gag coming from the cytoplasm (II) will be detected in the *green channel* while Gag already anchored at the plasma membrane (I) prior to the 405-nm pulse will be detected in the *red channel*. (**b–c**) Live-cell images of HeLa cells transfected with a mixture of pCHIV and pCHIVmEosFP. Before photoconversion (**b**), WF and TIRF images in the *green channel* allow the detection of Gag dots within the cytoplasm and at the level of the plasma membrane. After photoconversion (**c**), dots are detected in the *green channel* (Gag-EOS coming from the cytoplasm) and in the *red channel* (photoconverted Gag-EOS at the PM) of the TIRF mode. The scale bar corresponds to 5 µm. (**d**) Fluorescence intensity with 488-nm excitation (*green line*, nonphotoconverted species) and with 561-nm excitation (*red line*, photoconverted) of the plasma membrane measured in the vicinity of 23 assembly sites as a function of time after photoconversion. The standard deviation of the individual traces is given as a halo (Adapted from Fig. 5 of [78])

dots at the PM level should increase. On the contrary, if newly formed VLPs would result from the assembly of Gag clusters diffusing at the cell surface, then the emission in the red channel should be constant. As shown in Fig. 4d, the fluorescence intensity of the green channel increases as a function of time, suggesting that the quantity of nonactivated Gag-EOS increases. Thus, VLP formation at the surface of HeLa cells likely results from the accumulation of Gag oligomers coming from the cytoplasm.

4 Gag Mobility and VLP Budding at the Membrane Level

Intracellular Gag mobility at the level of the PM was first studied by means of an elegant method developed by Hope et al. [82]. HeLa cells were transfected with a plasmid DNA expressing a photoactivable GFP (PA-GFP) that corresponds to GFP with threonine 203 changed for isoleucine [83] and is used as EOS. To determine the motion of Gag, a region of interest (ROI) was photoactivated by a pulse at 405 nm, and the emission of the activated PA-GFP was measured at 516 nm in a second ROI. Gag-PA-GFP was considered as being unassembled when the fluorescence was diffuse, while Gag assembled in the form of VLPs at the PM appeared as bright spots. Diffuse Gag was found to move very fast within the cytoplasm, in agreement with fluorescence correlation spectroscopy (FCS) data on Gag RSV [51] and with the fact that Gag interacts with molecular motors [84]. An identical diffusion time was also found with G2AGag-PA-GFP, suggesting that intracellular Gag oligomers are not anchored to membranes. This high mobility might well facilitate Gag recruitment by growing Gag oligomers during virus assembly.

To monitor individual particles in live cells using TIRF, HeLa cells were transfected with plasmids expressing both Gag and Gag-GFP proteins. Using this strategy, Jouvenet et al. found that budding particles could be divided in two populations [57]. The first one appeared as dots 8–9 min after the initial detection of fluorescence in the cytoplasm and stayed at the level of the PM, showing slow lateral movements (lateral velocity $= 0.007\ \mu m.s^{-1}$). Interestingly, the steady-state fluorescence brightness of these dots remains constant indicating completeness of VLP assembly. Meanwhile, a second population of Gag puncta emerged and disappeared rapidly along the z-axis and colocalized with the endocytic marker CD63 [82]. These two distinct behaviors of Gag VLPs appeared independent from the presence of GFP or mCherry. Nevertheless, the first population was found to appear earlier (5–6 min) when higher quantities of transfected plasmids were used.

This TIRF-based quantitative analysis of individual particles at the cell periphery was also used in combination with epifluorescence microscopy to characterize the Gag-containing complexes both at the PM and in the cytoplasm [78]. To be close to virological-like conditions, HeLa cells were transfected with a mixture of plasmids expressing all the viral proteins. Viral budding could be divided into three phases. During the first phase, most (up to 80 %) of the Gag-containing complexes trafficking toward the PM display a rapid increase in their fluorescence intensity. This phase corresponds to the progressive assembly of the initial Gag oligomers to form VLPs. An average VLP assembly rate of $4.3 \times 10^{-3}\ s^{-1}$ was measured, in good agreement with that obtained by Jouvenet et al. [57]. Moreover, variations in these kinetic rates, as measured in individual cells, suggest that additional events such as packaging of nonstructural viral proteins or protein processing could delay the assembly process. After the fast increase in the fluorescence intensity of the Gag clusters, a stationary phase was observed where the fluorescence signal fluctuates about a constant value. This phase could reflect the interaction between the Gag-containing VLPs and membrane components to initiate the budding process. Part of the fluorescence signal in this phase could also correspond to VLPs

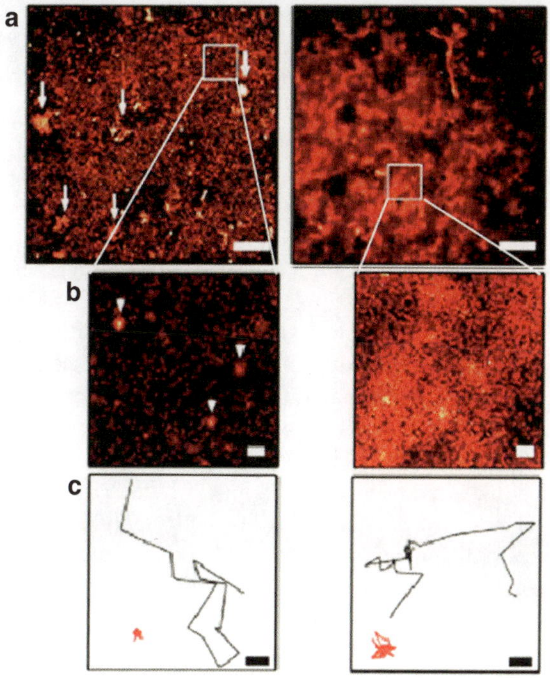

Fig. 5 PALM images of Gag and VSV-G. Cos cells expressing either Gag-EOS (images **a** and **b**, *left*) or VSV-G-EOS (images **a** and **b**, *right*) proteins were imaged by PALM (500 s of integration, 10,000 images). Various sizes of bright Gag-containing dots are visible and highlighted by white arrows while VSV-G-EOS is homogeneously distributed. Scale bars, 2 μm. (**c**) Representative Gag or VSV-G single molecule trajectories representing diffuse (*black line*) or confined (*red line*) movements. Scale bars correspond to 100 nm (Adapted from Fig. 1 of [86])

immobilized between the coverslip and the cell surface. In the third phase, the fluorescence intensity was seen to decay with a rate of $2.7 \times 10^{-3}\ s^{-1}$. This phase is thought to correspond either to virus release, photobleaching, or endocytosis.

Virus release as cell-free particles was also investigated through real-time observation of single particle motions using highly sensitive fluorescence microscopy techniques. Experiments were performed at 37 °C using a thermostated chamber or a temperature-controlled stage. Individual particles were identified based on the brightness of the centroid of individual spots [80]. To describe the dynamics of the particle motion, x and y coordinates were used to reconstruct trajectories based on the nearest-neighbor algorithm [85]. Then, by calculating the time-dependent mean square displacements for each trajectory, different diffusion regimes (free and anomalous) were evidenced. Interestingly, this approach of single particle tracking was also combined with high-resolution spectroscopy using PALM with a TIRF objective [86]. To perform PALM, Cos7 cells were transfected with plasmids expressing both Gag-EOS and VSV-G-EOS, where VSV-G is the pantropic envelope G protein from the vesicular stomatitis virus, used to pseudotype HIV-1 particles and HIV-1-based vectors [86]. As shown in Fig. 5, dots of Gag-EOS with a size of 100–200 nm were observed (Fig. 5a and b, left panels) in good agreement with data from electron microscopy [9, 87]. Calculation of trajectories (Fig. 5c) showed various populations of Gag with Brownian and anomalous diffusion as well as much larger clusters that were immobile. The average diffusion coefficient of Gag in live cells was $0.11 \pm 0.08\ \mu m^2/s$, in good agreement with the value reported by [78]. It was surprising that no active transport was evidenced in this study, though Gag was reported to interact with several molecular motors [84, 88–91]. In contrast to Gag, VSV-G-EOS is homogeneously distributed across the plasma membrane and does not form bright dots. In addition to a population of immobile VSV-G-Eos molecules, a population of mobile molecules with an average diffusion coefficient of $0.14 \pm 0.08\ \mu m^2/s$ was observed, in line with fluorescence recovery after photobleaching data [92]. The high density of dynamic information available through this approach allowed the authors to draw a spatially resolved map of single molecule diffusion coefficients. This map highlighted large clusters of immobile Gag-Eos molecules, which may correspond to regions where VLPs were accumulating. Although VSV-G-Eos is characterized as a highly mobile molecule, less mobile regions were detected and may represent filipodial structures where molecular mobility was slowed down due to membrane folding.

5 Cell-to-Cell Transmission of Retroviral Gag Followed by Live Imaging

Single particle tracking can be used to follow cell-to-cell transmission of retroviral particles, from infected to noninfected naïve cells (reviewed in [93–96]). This transmission necessitates extensive cell-cell contacts [97] and was associated to higher retrovirus infectivity [98–102]. More recently, these cell-to-cell contacts were described as subcellular structures referred to as virological synapses (VS) [93, 103–106] that are involved in virus transmission from monocyte-derived macrophages (MDM) to T CD4$^+$ cells [38], from T cells to T cells [107] [108], and from dendritic to T cells [109, 110]. This mode of transfer is thought to hide the virus from the immune system even though VS were recently shown to be sensitive to neutralizing antibodies [111].

Gag trafficking toward the VS was monitored using Gag-TC proteins labeled with FlAsH in live MDMs [38]. Labeled Gag proteins were found to accumulate both at the PM and within internal compartments of the MDMs. These compartments are rich in tetraspanin proteins [112, 113] and might form a complex labyrinth of interconnected membranes [114]. Interestingly, real-time visualization of HIV-1 infected MDMs after addition of naïve T cells showed that intracellular Gag-containing compartments accumulate in less than 30 min at the level of MDM-MDM or MDM-T cell contacts with a slight preference for the latter [102]. These internal compartments differ from endosomes since mutations of basic residues that maintained Gag in internal vesicles block the translocation of mutated Gag to sites of cell-cell contacts [115]. In contrast, the envelope protein was not involved in this trafficking in agreement with the fact that the two proteins follow distinct routes before budding takes place [19].

To further understand the role of these VS, Gag trafficking during T cell transmission of HIV-1 was followed by quantitative video microscopy [107]. These experiments were carried out using Jurkat cells expressing Gag-iGFP and primary T $CD4^+$ cells. Images obtained from Jurkat diffusing as free cells are characterized by a homogeneous fluorescent staining of Gag-iGFP (Fig. 6a, panels a and b). In sharp contrast, in 80 % of the infected Jurkat cells contacting naïve T $CD4^+$ cells, Gag-iGFP molecules accumulated at the contact sites in an envelope-dependent manner (Fig. 6a, panels c and d). Interestingly, when this experiment was carried out with Cerulean and Venus Gag chimera molecules, a FRET signal was obtained at this cell-cell contact demonstrating a link between Gag assembly and Gag accumulation at the level of the VS. Moreover, spatiotemporal monitoring of Gag trafficking showed that formation of a new synaptic button was associated with a rapid and directional motion of membrane-associated Gag molecules toward the cell-cell contact site (Fig. 6b). As a consequence of this rapid motion, a zone depleted of Gag molecules was observed in the periphery of the newly formed synaptic button (Fig. 6c) in agreement with the local active transport [116]. Once the synaptic button was completed, the preformed Gag clusters were seen to rapidly diffuse with an average velocity of 0.1 to 0.25 μm/s in the direction of the synapse to fuse with the button structure (Fig. 6d). Using high-resolution confocal microscopy and 3D image reconstruction, the synapse appeared as a button with a diameter of 1 to 3 μm, resembling the previously reported morphology [117]. Concomitant with VS formation, patches of Gag molecules were observed to enter the target T cell within vesicles, suggesting that the VS favors an endocytic route for viral infection (Fig. 6e). In fact, HIV-1 antigens were found colocalized with endocytic route markers such as dynamin and clathrin emphasizing the role of this pathway for T cell to T cell HIV-1 transfer and maturation [108, 118].

An alternative model of virus transmission through filopodia was proposed by the group of Mothes for the Moloney leukemia virus (MLV) [94]. Filopodia, also referred to as cytoneme, correspond to plasma membrane projections that bridge cells and serve as conduits for signal exchanges between two cells [119]. In the context of virus-producing cells, filopodia are emitted by noninfected cells to avoid virus delivery into the cortical actin filament-dense areas [90]. Cos-1 cells expressing Gag-CFP and Env-YFP from MLV were cocultured with naïve XC cells expressing the MLV receptor mCAT1-CFP [120, 121]. As shown in Fig. 7a, superimposed images revealed the production of a huge quantity of virus at the surface of the Cos-1 cells (yellow carpet) as well as the presence of filopodia full of viruses between infected and naïve cells (Fig. 7a, arrows a–e). Interestingly, particles appeared as yellow dots close to the infected cells and as green dots close to the naïve cells, probably as a consequence of a progressive increase in the concentrations of the receptor protein. Moreover, a colocalization of the viral envelope and CAT proteins was observed, suggesting that these two proteins act to maintain a filopodia linkage between the two cells (Fig. 7b) [122]. Time lap experiments revealed that viruses migrate along these filopodia with an average

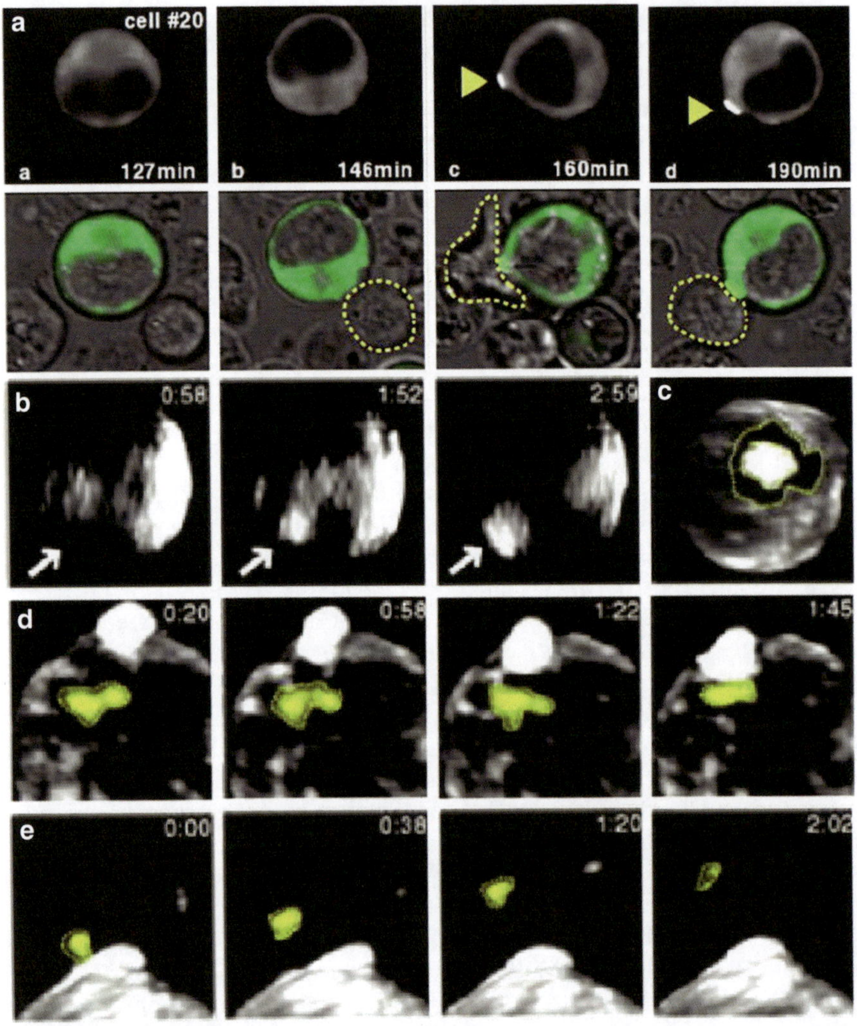

Fig. 6 Formation of Gag-containing synaptic buttons. (**a**) Jurkat cells were transfected with plasmid DNA expressing Gag-iGFP and cultured with T $CD4^+$ cells. Time lap experiments were carried out for 4 h. The upper line corresponds to fluorescence images, while the lower one corresponds to an overlay between these fluorescent images and phase contrast images. Panels *a* and *b* show that Gag-iGFP is diffusely distributed in the cytoplasm. Panels *c* and *d* show the distribution of Gag-iGFP after a contact between an infected Jurkat cell and a naive T cell and the formation of a synaptic button. (**b**) Dynamics of Gag-iGFP motion during formation of a synaptic button at the site of adhesion. (**c**) Ring-shaped zone around the synaptic button which is depleted from surrounding Gag molecules, suggesting that Gag is recruited in the synaptic button. (**d**) Dynamics of Gag clusters in the vicinity of the synaptic button. (**e**) Transmission of Gag-containing complexes from a Jurkat cell to a T cell through a synaptic button. During this transmission, Gag-containing patches (*yellow*) were seen to emerge from the synaptic button, next to leave it, and then migrate inside the T cell. For **b**, **d**, and **e**, time unit indicated on the panels is seconds (Adapted from Figs. 1 and 2 of [107])

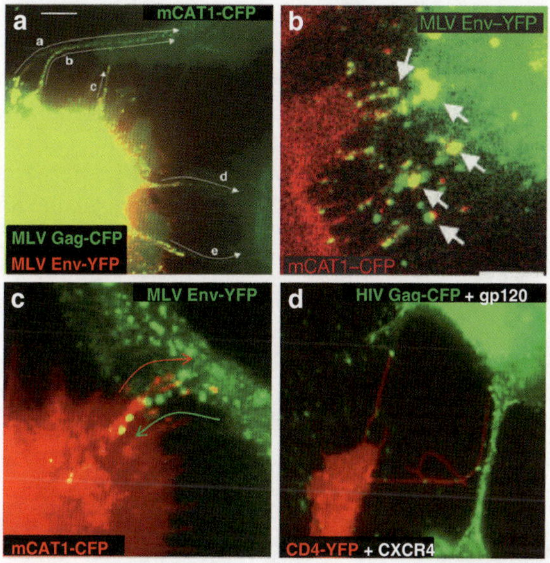

Fig. 7 Filopodia-based model for virus transfer. (**a**) Cos cells were transfected with plasmid DNAs expressing MLV Gag-CFP and Env-YFP and cocultured with naïve XC cells expressing mCAT1-CFP. YFP intensity was arbitrary colored in *red* and CFP in *green*. Arrows *a*, *b*, *c*, *d*, and *e* indicate the directions of the virus traffic in the filopodia. (**b**) Colocalization of MLV receptor (mCTA1) in *red* with MLV Env in *green*. The *yellow dots* suggest a colocalization of the two partners at contact points between cells and filopodia. (**c**) Superimposition of time lap images showing the elongation of the filopodia (mCAT1-YFP, *red arrow*) and the movement of MLV particles (Gag-GFP). (**d**) Thin filopodia emitted from XC cells containing HIV particles in *green* (Gag-CFP). The red color corresponds to the expression of CD4-YFP/CXCR4 (Adapted from Figs. 1 and 5 of [122])

speed of 0.7 μm/min and thus required an average of 18 min to migrate from one cell to the other [122]. Interestingly, while the cytoneme growth was directed from noninfected to infected cells (Fig. 7c, red arrow), the virus moves toward the naïve cell through an actin- and myosin-mediated retrograde transport of the envelope receptor [90, 122] (Fig. 7c, green arrow). This model of filopodia-based transmission was also found with other cell types and retroviruses. Indeed, experiments were also carried out with Cos cells expressing HIV-1 Gag fused to GFP, cocultured with XC cells expressing CXCR4 and CD4-YFP. As seen in Fig. 7d, particles of HIV-1 are transferred from Cos cells (green) toward naïve cells (red). However, only a few particles were observed, suggesting that this transmission mode is less efficient for HIV than for MLV or that the cells used are not relevant for HIV-1 infection.

6 Concluding Remarks

The Gag polyprotein precursor codes for the major structural proteins of the retroviral particle and as such drives retrovirus assembly in infected cells. In an attempt to gain a better understanding of retrovirus assembly, notably that of

HIV-1, MLV, and RSV, recombinant fluorescent Gag polyproteins were expressed alone or in the virus context in different cell lines and visualized using microscopy approaches with high spatial and temporal resolutions. As summarized in this chapter, interesting new data have been obtained indicating that Gag assembly most probably starts in the cytoplasm by the formation of small oligomeric complexes that rapidly traffic toward the PM or intracellular membranes (i.e., MVB), ultimately leading to virion formation and dissemination. Else, when Gag-producing cells are engaged in cell-cell interactions, Gag particles accumulate at cell-cell contact points, and viral particles are transferred to noninfected cells via cell-cell junctions, called the virological synapses. Similar approaches have also been used to monitor Gag-RNA interaction [123–126], recruitment of cellular proteins that promote viral budding [127, 128], and interaction between TRIM5α and incoming HIV-1 virions [129] (for a recent review, see [130]). In a close future, since assembly, budding, and infection are closely interconnected [118], emerging high-resolution techniques will allow addressing some unresolved questions such as the following: (1) where does Gag specifically recognize the genomic RNA which represents one of the two assembly platforms? (2) where does the interaction of Gag with its different partners to incorporate viral and cellular proteins takes place and how this influences Gag/genomic RNA interaction, assembly, and trafficking? (3) what are the mechanisms that regulate the stoichiometry of all the different partners that are incorporated in the nascent particles?

Acknowledgments We apologize to our colleagues whose work could not be cited here since this chapter was focused mainly on Gag-Gag interaction. Thanks are due to INSERM, CNRS, ANRS, and Sidaction for their continuous support.

References

1. Chertova E, Chertov O, Coren LV, Roser JD, Trubey CM, Bess JW Jr, Sowder RC 2nd, Barsov E, Hood BL, Fisher RJ, Nagashima K, Conrads TP, Veenstra TD, Lifson JD, Ott DE (2006) Proteomic and biochemical analysis of purified human immunodeficiency virus type 1 produced from infected monocyte-derived macrophages. J Virol 80:9039–9052
2. Balvay L, Lopez Lastra M, Sargueil B, Darlix JL, Ohlmann T (2007) Translational control of retroviruses. Nature Rev 5:128–140
3. Ganser-Pornillos BK, Yeager M, Sundquist WI (2008) The structural biology of HIV assembly. Curr Opin Struct Biol 18:203–217
4. Cimarelli A, Darlix JL (2002) Assembling the human immunodeficiency virus type 1. Cell Mol Life Sci 59:1166–1184
5. Datta SA, Temeselew LG, Crist RM, Soheilian F, Kamata A, Mirro J, Harvin D, Nagashima K, Cachau RE, Rein A (2011) On the role of the SP1 domain in HIV-1 particle assembly: a molecular switch? J Virol 85:4111–4121
6. Li S, Hill CP, Sundquist WI, Finch JT (2000) Image reconstructions of helical assemblies of the HIV-1 CA protein. Nature 407:409–413
7. Ako-Adjei D, Johnson MC, Vogt VM (2005) The retroviral capsid domain dictates virion size, morphology, and coassembly of gag into virus-like particles. J Virol 79:13463–13472

8. Zhang Y, Qian H, Love Z, Barklis E (1998) Analysis of the assembly function of the human immunodeficiency virus type 1 gag protein nucleocapsid domain. J Virol 72:1782–1789
9. Adamson CS, Freed EO (2007) Human immunodeficiency virus type 1 assembly, release, and maturation. Adv Pharmacol (San Diego Calif) 55:347–387
10. Kutluay SB, Bieniasz PD (2010) Analysis of the initiating events in HIV-1 particle assembly and genome packaging. PLoS Pathog 6:e1001200
11. Datta SA, Zhao Z, Clark PK, Tarasov S, Alexandratos JN, Campbell SJ, Kvaratskhelia M, Lebowitz J, Rein A (2007) Interactions between HIV-1 Gag molecules in solution: an inositol phosphate-mediated switch. J Mol Biol 365:799–811
12. Bryant M, Ratner L (1990) Myristoylation-dependent replication and assembly of human immunodeficiency virus 1. Proc Natl Acad Sci USA 87:523–527
13. Gottlinger HG, Sodroski JG, Haseltine WA (1989) Role of capsid precursor processing and myristoylation in morphogenesis and infectivity of human immunodeficiency virus type 1. Proc Natl Acad Sci USA 86:5781–5785
14. Tang C, Loeliger E, Luncsford P, Kinde I, Beckett D, Summers MF (2004) Entropic switch regulates myristate exposure in the HIV-1 matrix protein. Proc Natl Acad Sci USA 101:517–522
15. Ono A, Ablan SD, Lockett SJ, Nagashima K, Freed EO (2004) Phosphatidylinositol (4,5) bisphosphate regulates HIV-1 Gag targeting to the plasma membrane. Proc Natl Acad Sci USA 101:14889–14894
16. Saad JS, Miller J, Tai J, Kim A, Ghanam RH, Summers MF (2006) Structural basis for targeting HIV-1 Gag proteins to the plasma membrane for virus assembly. Proc Natl Acad Sci USA 103:11364–11369
17. Chukkapalli V, Oh SJ, Ono A (2010) Opposing mechanisms involving RNA and lipids regulate HIV-1 Gag membrane binding through the highly basic region of the matrix domain. Proc Natl Acad Sci USA 107:1600–1605
18. Chukkapalli V, Hogue IB, Boyko V, Hu WS, Ono A (2008) Interaction between the human immunodeficiency virus type 1 Gag matrix domain and phosphatidylinositol-(4,5)-bisphosphate is essential for efficient gag membrane binding. J Virol 82:2405–2417
19. Checkley MA, Luttge BG, Freed EO (2011) HIV-1 envelope glycoprotein biosynthesis, trafficking, and incorporation. J Mol Biol 410:582–608
20. Hamard-Peron E, Muriaux D (2011) Retroviral matrix and lipids, the intimate interaction. Retrovirology 8:15
21. Muriaux D, Mirro J, Harvin D, Rein A (2001) RNA is a structural element in retrovirus particles. Proc Natl Acad Sci USA 98:5246–5251
22. Parent LJ, Gudleski N (2011) Beyond plasma membrane targeting: role of the MA domain of Gag in retroviral genome encapsidation. J Mol Biol 410:553–564
23. Campbell S, Fisher RJ, Towler EM, Fox S, Issaq HJ, Wolfe T, Phillips LR, Rein A (2001) Modulation of HIV-like particle assembly in vitro by inositol phosphates. Proc Natl Acad Sci USA 98:10875–10879
24. Campbell S, Vogt VM (1997) In vitro assembly of virus-like particles with Rous sarcoma virus Gag deletion mutants: identification of the p10 domain as a morphological determinant in the formation of spherical particles. J Virol 71:4425–4435
25. Darlix JL, Godet J, Ivanyi-Nagy R, Fosse P, Mauffret O, Mely Y (2011) Flexible nature and specific functions of the HIV-1 nucleocapsid protein. J Mol Biol 410:565–581
26. Rein A, Datta SA, Jones CP, Musier-Forsyth K (2011) Diverse interactions of retroviral Gag proteins with RNAs. Trends Biochem Sci 36:373–380
27. Muriaux D, Darlix JL (2010) Properties and functions of the nucleocapsid protein in virus assembly. RNA Biol 7:744–753
28. Balasubramaniam M, Freed EO (2011) New insights into HIV assembly and trafficking. Physiology (Bethesda) 26:236–251
29. Bieniasz PD (2009) The cell biology of HIV-1 virion genesis. Cell Host Microbe 5:550–558
30. Weiss ER, Gottlinger H (2011) The role of cellular factors in promoting HIV budding. J Mol Biol 410:525–533
31. Ott DE (2008) Cellular proteins detected in HIV-1. Rev Med Virol 18:159–175

32. Jouvenet N, Simon SM, Bieniasz PD (2011) Visualizing HIV-1 assembly. J Mol Biol 410:501–511
33. Perrin-Tricaud C, Davoust J, Jones IM (1999) Tagging the human immunodeficiency virus gag protein with green fluorescent protein. Minimal evidence for colocalisation with actin. Virology 255:20–25
34. Hermida-Matsumoto L, Resh MD (2000) Localization of human immunodeficiency virus type 1 Gag and Env at the plasma membrane by confocal imaging. J Virol 74:8670–8679
35. Larson DR, Johnson MC, Webb WW, Vogt VM (2005) Visualization of retrovirus budding with correlated light and electron microscopy. Proc Natl Acad Sci USA 102:15453–15458
36. Pornillos O, Higginson DS, Stray KM, Fisher RD, Garrus JE, Payne M, He GP, Wang HE, Morham SG, Sundquist WI (2003) HIV Gag mimics the Tsg101-recruiting activity of the human Hrs protein. J Cell Biol 162:425–434
37. Muller B, Daecke J, Fackler OT, Dittmar MT, Zentgraf H, Krausslich HG (2004) Construction and characterization of a fluorescently labeled infectious human immunodeficiency virus type 1 derivative. J Virol 78:10803–10813
38. Gousset K, Ablan SD, Coren LV, Ono A, Soheilian F, Nagashima K, Ott DE, Freed EO (2008) Real-time visualization of HIV-1 GAG trafficking in infected macrophages. PLoS Pathog 4:e1000015
39. Pereira CF, Ellenberg PC, Jones KL, Fernandez TL, Smyth RP, Hawkes DJ, Hijnen M, Vivet-Boudou V, Marquet R, Johnson I, Mak J (2011) Labeling of multiple HIV-1 proteins with the biarsenical-tetracysteine system. PLoS One 6:e17016
40. Rudner L, Nydegger S, Coren LV, Nagashima K, Thali M, Ott DE (2005) Dynamic fluorescent imaging of human immunodeficiency virus type 1 gag in live cells by biarsenical labeling. J Virol 79:4055–4065
41. Adams SR, Campbell RE, Gross LA, Martin BR, Walkup GK, Yao Y, Llopis J, Tsien RY (2002) New biarsenical ligands and tetracysteine motifs for protein labeling in vitro and in vivo: synthesis and biological applications. J Am Chem Soc 124:6063–6076
42. Perlman M, Resh MD (2006) Identification of an intracellular trafficking and assembly pathway for HIV-1 gag. Traffic (Copenhagen, Denmark) 7:731–745
43. Fritz JV, Dujardin D, Godet J, Didier P, De Mey J, Darlix JL, Mely Y, de Rocquigny H (2010) HIV-1 Vpr oligomerization but not that of Gag directs the interaction between Vpr and Gag. J Virol 84:1585–1596
44. Turville SG, Aravantinou M, Stossel H, Romani N, Robbiani M (2008) Resolution of de novo HIV production and trafficking in immature dendritic cells. Nat Methods 5:75–85
45. Arhel N, Genovesio A, Kim KA, Miko S, Perret E, Olivo-Marin JC, Shorte S, Charneau P (2006) Quantitative four-dimensional tracking of cytoplasmic and nuclear HIV-1 complexes. Nat Methods 3:817–824
46. Arhel NJ, Charneau P (2009) Bisarsenical labeling of HIV-1 for real-time fluorescence microscopy. Methods Mol Biol 485:151–159
47. Ono A (2009) HIV-1 assembly at the plasma membrane: Gag trafficking and localization. Future Virol 4:241–257
48. Dong X, Li H, Derdowski A, Ding L, Burnett A, Chen X, Peters TR, Dermody TS, Woodruff E, Wang JJ, Spearman P (2005) AP-3 directs the intracellular trafficking of HIV-1 Gag and plays a key role in particle assembly. Cell 120:663–674
49. Grigorov B, Arcanger F, Roingeard P, Darlix JL, Muriaux D (2006) Assembly of infectious HIV-1 in human epithelial and T-lymphoblastic cell lines. J Mol Biol 359:848–862
50. Nydegger S, Foti M, Derdowski A, Spearman P, Thali M (2003) HIV-1 egress is gated through late endosomal membranes. Traffic (Copenhagen, Denmark) 4:902–910
51. Larson DR, Ma YM, Vogt VM, Webb WW (2003) Direct measurement of Gag-Gag interaction during retrovirus assembly with FRET and fluorescence correlation spectroscopy. J Cell Biol 162:1233–1244
52. Bastiaens PI, Squire A (1999) Fluorescence lifetime imaging microscopy: spatial resolution of biochemical processes in the cell. Trends Cell Biol 9:48–52
53. Day RN, Periasamy A, Schaufele F (2001) Fluorescence resonance energy transfer microscopy of localized protein interactions in the living cell nucleus. Methods 25:4–18

54. Voss TC, Demarco IA, Day RN (2005) Quantitative imaging of protein interactions in the cell nucleus. Biotechniques 38:413–424
55. Derdowski A, Ding L, Spearman P (2004) A novel fluorescence resonance energy transfer assay demonstrates that the human immunodeficiency virus type 1 Pr55Gag I domain mediates Gag-Gag interactions. J Virol 78:1230–1242
56. Hogue IB, Hoppe A, Ono A (2009) Quantitative fluorescence resonance energy transfer microscopy analysis of the human immunodeficiency virus type 1 Gag-Gag interaction: relative contributions of the CA and NC domains and membrane binding. J Virol 83:7322–7336
57. Jouvenet N, Bieniasz PD, Simon SM (2008) Imaging the biogenesis of individual HIV-1 virions in live cells. Nature 454:236–240
58. Hubner W, Chen P, Del Portillo A, Liu Y, Gordon RE, Chen BK (2007) Sequence of human immunodeficiency virus type 1 (HIV-1) Gag localization and oligomerization monitored with live confocal imaging of a replication-competent, fluorescently tagged HIV-1. J Virol 81:12596–12607
59. Krishnamoorthy G, Duportail G, Mely Y (2002) Structure and dynamics of condensed DNA probed by 1,1'-(4,4,8,8-tetramethyl-4,8-diazaundecamethylene)bis[4-[[3- methylbenz-1,3-oxazol-2-yl]methylidine]-1,4-dihydroquinolinium] tetraiodide fluorescence. Biochemistry 41:15277–15287
60. Lakowicz JR (1980) Fluorescence spectroscopic investigations of the dynamic properties of proteins, membranes and nucleic acids. J Biochem Biophys Methods 2:91–119
61. Deka C, Lehnert BE, Lehnert NM, Jones GM, Sklar LA, Steinkamp JA (1996) Analysis of fluorescence lifetime and quenching of FITC-conjugated antibodies on cells by phase-sensitive flow cytometry. Cytometry 25:271–279
62. Runnels LW, Scarlata SF (1995) Theory and application of fluorescence homotransfer to melittin oligomerization. Biophys J 69:1569–1583
63. Lakowicz JR (2001) Radiative decay engineering: biophysical and biomedical applications. Anal Biochem 298:1–24
64. Accola MA, Ohagen A, Gottlinger HG (2000) Isolation of human immunodeficiency virus type 1 cores: retention of Vpr in the absence of p6(gag). J Virol 74:6198–6202
65. Krausslich HG, Facke M, Heuser AM, Konvalinka J, Zentgraf H (1995) The spacer peptide between human immunodeficiency virus capsid and nucleocapsid proteins is essential for ordered assembly and viral infectivity. J Virol 69:3407–3419
66. Liang C, Hu J, Whitney JB, Kleiman L, Wainberg MA (2003) A structurally disordered region at the C terminus of capsid plays essential roles in multimerization and membrane binding of the gag protein of human immunodeficiency virus type 1. J Virol 77:1772–1783
67. von Schwedler UK, Stray KM, Garrus JE, Sundquist WI (2003) Functional surfaces of the human immunodeficiency virus type 1 capsid protein. J Virol 77:5439–5450
68. Callahan EM, Wills JW (2000) Repositioning basic residues in the M domain of the *Rous sarcoma* virus gag protein. J Virol 74:11222–11229
69. Li H, Dou J, Ding L, Spearman P (2007) Myristoylation is required for human immunodeficiency virus type 1 Gag-Gag multimerization in mammalian cells. J Virol 81:12899–12910
70. Lee YM, Liu B, Yu XF (1999) Formation of virus assembly intermediate complexes in the cytoplasm by wild-type and assembly-defective mutant human immunodeficiency virus type 1 and their association with membranes. J Virol 73:5654–5662
71. Morikawa Y, Hockley DJ, Nermut MV, Jones IM (2000) Roles of matrix, p2, and N-terminal myristoylation in human immunodeficiency virus type 1 Gag assembly. J Virol 74:16–23
72. Burniston MT, Cimarelli A, Colgan J, Curtis SP, Luban J (1999) Human immunodeficiency virus type 1 Gag polyprotein multimerization requires the nucleocapsid domain and RNA and is promoted by the capsid-dimer interface and the basic region of matrix protein. J Virol 73:8527–8540

73. Bowzard JB, Bennett RP, Krishna NK, Ernst SM, Rein A, Wills JW (1998) Importance of basic residues in the nucleocapsid sequence for retrovirus Gag assembly and complementation rescue. J Virol 72:9034–9044
74. Ott DE, Coren LV, Shatzer T (2009) The nucleocapsid region of human immunodeficiency virus type 1 Gag assists in the coordination of assembly and Gag processing: role for RNA-Gag binding in the early stages of assembly. J Virol 83:7718–7727
75. Chen Y, Wu B, Musier-Forsyth K, Mansky LM, Mueller JD (2009) Fluorescence fluctuation spectroscopy on viral-like particles reveals variable gag stoichiometry. Biophys J 96:1961–1969
76. Wright ER, Schooler JB, Ding HJ, Kieffer C, Fillmore C, Sundquist WI, Jensen GJ (2007) Electron cryotomography of immature HIV-1 virions reveals the structure of the CA and SP1 Gag shells. EMBO J 26:2218–2226
77. Briggs JA, Krausslich HG (2011) The molecular architecture of HIV. J Mol Biol 410:491–500
78. Ivanchenko S, Godinez WJ, Lampe M, Krausslich HG, Eils R, Rohr K, Brauchle C, Muller B, Lamb DC (2009) Dynamics of HIV-1 assembly and release. PLoS Pathog 5:e1000652
79. Wiedenmann J, Ivanchenko S, Oswald F, Schmitt F, Rocker C, Salih A, Spindler KD, Nienhaus GU (2004) EosFP, a fluorescent marker protein with UV-inducible green-to-red fluorescence conversion. Proc Natl Acad Sci USA 101:15905–15910
80. Ruthardt N, Lamb DC, Brauchle C (2011) Single-particle tracking as a quantitative microscopy-based approach to unravel cell entry mechanisms of viruses and pharmaceutical nanoparticles. Mol Ther 19:1199–1211
81. Axelrod D (2008) Chapter 7: Total internal reflection fluorescence microscopy. Methods Cell Biol 89:169–221
82. Gomez CY, Hope TJ (2006) Mobility of human immunodeficiency virus type 1 Pr55Gag in living cells. J Virol 80:8796–8806
83. Patterson GH, Lippincott-Schwartz J (2004) Selective photolabeling of proteins using photoactivatable GFP. Methods 32:445–450
84. Martinez NW, Xue X, Berro RG, Kreitzer G, Resh MD (2008) Kinesin KIF4 regulates intracellular trafficking and stability of the human immunodeficiency virus type 1 Gag polyprotein. J Virol 82:9937–9950
85. Sbalzarini IF, Koumoutsakos P (2005) Feature point tracking and trajectory analysis for video imaging in cell biology. J Struct Biol 151:182–195
86. Manley S, Gillette JM, Patterson GH, Shroff H, Hess HF, Betzig E, Lippincott-Schwartz J (2008) High-density mapping of single-molecule trajectories with photoactivated localization microscopy. Nat Methods 5:155–157
87. Ottmann M, Gabus C, Darlix JL (1995) The central globular domain of the nucleocapsid protein of human immunodeficiency virus type 1 is critical for virion structure and infectivity. J Virol 69:1778–1784
88. Kim W, Tang Y, Okada Y, Torrey TA, Chattopadhyay SK, Pfleiderer M, Falkner FG, Dorner F, Choi W, Hirokawa N, Morse HC 3rd (1998) Binding of murine leukemia virus Gag polyproteins to KIF4, a microtubule-based motor protein. J Virol 72:6898–6901
89. Lehmann M, Milev MP, Abrahamyan L, Yao XJ, Pante N, Mouland AJ (2009) Intracellular transport of human immunodeficiency virus type 1 genomic RNA and viral production are dependent on dynein motor function and late endosome positioning. J Biol Chem 284:14572–14585
90. Lehmann MJ, Sherer NM, Marks CB, Pypaert M, Mothes W (2005) Actin- and myosin-driven movement of viruses along filopodia precedes their entry into cells. J Cell Biol 170:317–325
91. Tang Y, Winkler U, Freed EO, Torrey TA, Kim W, Li H, Goff SP, Morse HC 3rd (1999) Cellular motor protein KIF-4 associates with retroviral Gag. J Virol 73:10508–10513

92. Kenworthy AK, Nichols BJ, Remmert CL, Hendrix GM, Kumar M, Zimmerberg J, Lippincott-Schwartz J (2004) Dynamics of putative raft-associated proteins at the cell surface. J Cell Biol 165:735–746
93. Feldmann J, Schwartz O (2010) HIV-1 virological synapse: live imaging of transmission. Viruses 2:1666–1680
94. Mothes W, Sherer NM, Jin J, Zhong P (2010) Virus cell-to-cell transmission. J Virol 84:8360–8368
95. Waki K, Freed EO (2010) Macrophages and cell-cell spread of HIV-1. Viruses 2:1603–1620
96. Sattentau QJ (2010) Cell-to-cell spread of retroviruses. Viruses 2:1306–1321
97. Sourisseau M, Sol-Foulon N, Porrot F, Blanchet F, Schwartz O (2007) Inefficient human immunodeficiency virus replication in mobile lymphocytes. J Virol 81:1000–1012
98. Phillips DM (1994) The role of cell-to-cell transmission in HIV infection. AIDS (London, England) 8:719–731
99. Cameron PU, Freudenthal PS, Barker JM, Gezelter S, Inaba K, Steinman RM (1992) Dendritic cells exposed to human immunodeficiency virus type-1 transmit a vigorous cytopathic infection to CD4+ T cells. Science 257:383–387
100. Ayehunie S, Groves RW, Bruzzese AM, Ruprecht RM, Kupper TS, Langhoff E (1995) Acutely infected Langerhans cells are more efficient than T cells in disseminating HIV type 1 to activated T cells following a short cell-cell contact. AIDS Res Hum Retroviruses 11:877–884
101. Pope M, Betjes MG, Romani N, Hirmand H, Cameron PU, Hoffman L, Gezelter S, Schuler G, Steinman RM (1994) Conjugates of dendritic cells and memory T lymphocytes from skin facilitate productive infection with HIV-1. Cell 78:389–398
102. Carr JM, Hocking H, Li P, Burrell CJ (1999) Rapid and efficient cell-to-cell transmission of human immunodeficiency virus infection from monocyte-derived macrophages to peripheral blood lymphocytes. Virology 265:319–329
103. Jolly C, Kashefi K, Hollinshead M, Sattentau QJ (2004) HIV-1 cell to cell transfer across an Env-induced, actin-dependent synapse. J Exp Med 199:283–293
104. Martin N, Sattentau Q (2009) Cell-to-cell HIV-1 spread and its implications for immune evasion. Curr Opin HIV AIDS 4:143–149
105. McDonald D, Wu L, Bohks SM, KewalRamani VN, Unutmaz D, Hope TJ (2003) Recruitment of HIV and its receptors to dendritic cell-T cell junctions. Science 300:1295–1297
106. Jolly C, Sattentau QJ (2004) Retroviral spread by induction of virological synapses. Traffic (Copenhagen, Denmark) 5:643–650
107. Hubner W, McNerney GP, Chen P, Dale BM, Gordon RE, Chuang FY, Li XD, Asmuth DM, Huser T, Chen BK (2009) Quantitative 3D video microscopy of HIV transfer across T cell virological synapses. Science 323:1743–1747
108. Bosch B, Grigorov B, Senserrich J, Clotet B, Darlix JL, Muriaux D, Este JA (2008) A clathrin-dynamin-dependent endocytic pathway for the uptake of HIV-1 by direct T cell-T cell transmission. Antiviral Res 80:185–193
109. Wang JH, Janas AM, Olson WJ, KewalRamani VN, Wu L (2007) CD4 coexpression regulates DC-SIGN-mediated transmission of human immunodeficiency virus type 1. J Virol 81:2497–2507
110. Jouve M, Sol-Foulon N, Watson S, Schwartz O, Benaroch P (2007) HIV-1 buds and accumulates in "nonacidic" endosomes of macrophages. Cell Host Microbe 2:85–95
111. Martin N, Welsch S, Jolly C, Briggs JA, Vaux D, Sattentau QJ (2010) Virological synapse-mediated spread of human immunodeficiency virus type 1 between T cells is sensitive to entry inhibition. J Virol 84:3516–3527
112. Deneka M, Pelchen-Matthews A, Byland R, Ruiz-Mateos E, Marsh M (2007) In macrophages, HIV-1 assembles into an intracellular plasma membrane domain containing the tetraspanins CD81, CD9, and CD53. J Cell Biol 177:329–341

113. Krementsov DN, Rassam P, Margeat E, Roy NH, Schneider-Schaulies J, Milhiet PE, Thali M (2010) HIV-1 assembly differentially alters dynamics and partitioning of tetraspanins and raft components. Traffic (Copenhagen, Denmark) 11:1401–1414
114. Welsch S, Groot F, Krausslich HG, Keppler OT, Sattentau QJ (2011) Architecture and regulation of the HIV-1 assembly and holding compartment in macrophages. J Virol 85:7922–7927
115. Ono A, Freed EO (2004) Cell-type-dependent targeting of human immunodeficiency virus type 1 assembly to the plasma membrane and the multivesicular body. J Virol 78: 1552–1563
116. Sol-Foulon N, Sourisseau M, Porrot F, Thoulouze MI, Trouillet C, Nobile C, Blanchet F, di Bartolo V, Noraz N, Taylor N, Alcover A, Hivroz C, Schwartz O (2007) ZAP-70 kinase regulates HIV cell-to-cell spread and virological synapse formation. EMBO J 26:516–526
117. Rudnicka D, Feldmann J, Porrot F, Wietgrefe S, Guadagnini S, Prevost MC, Estaquier J, Haase AT, Sol-Foulon N, Schwartz O (2009) Simultaneous cell-to-cell transmission of human immunodeficiency virus to multiple targets through polysynapses. J Virol 83:6234–6246
118. Dale BM, McNerney GP, Thompson DL, Hubner W, de Los Reyes K, Chuang FY, Huser T, Chen BK (2011) Cell-to-cell transfer of HIV-1 via virological synapses leads to endosomal virion maturation that activates viral membrane fusion. Cell Host Microbe 10:551–562
119. Roy S, Hsiung F, Kornberg TB (2011) Specificity of Drosophila cytonemes for distinct signaling pathways. Science 332:354–358
120. Sherer NM, Lehmann MJ, Jimenez-Soto LF, Ingmundson A, Horner SM, Cicchetti G, Allen PG, Pypaert M, Cunningham JM, Mothes W (2003) Visualization of retroviral replication in living cells reveals budding into multivesicular bodies. Traffic (Copenhagen, Denmark) 4:785–801
121. Yan Y, Jung YT, Wu T, Kozak CA (2008) Role of receptor polymorphism and glycosylation in syncytium induction and host range variation of ecotropic mouse gammaretroviruses. Retrovirology 5:2
122. Sherer NM, Lehmann MJ, Jimenez-Soto LF, Horensavitz C, Pypaert M, Mothes W (2007) Retroviruses can establish filopodial bridges for efficient cell-to-cell transmission. Nat Cell Biol 9:310–315
123. Abrahamyan LG, Chatel-Chaix L, Ajamian L, Milev MP, Monette A, Clement JF, Song R, Lehmann M, DesGroseillers L, Laughrea M, Boccaccio G, Mouland AJ (2010) Novel Staufen1 ribonucleoproteins prevent formation of stress granules but favour encapsidation of HIV-1 genomic RNA. J Cell Sci 123:369–383
124. Molle D, Segura-Morales C, Camus G, Berlioz-Torrent C, Kjems J, Basyuk E, Bertrand E (2009) Endosomal trafficking of HIV-1 gag and genomic RNAs regulates viral egress. J Biol Chem 284:19727–19743
125. Jouvenet N, Simon SM, Bieniasz PD (2009) Imaging the interaction of HIV-1 genomes and Gag during assembly of individual viral particles. Proc Natl Acad Sci USA 106:19114–19119
126. Kemler I, Meehan A, Poeschla EM (2010) Live-cell coimaging of the genomic RNAs and Gag proteins of two lentiviruses. J Virol 84:6352–6366
127. Baumgartel V, Ivanchenko S, Dupont A, Sergeev M, Wiseman PW, Krausslich HG, Brauchle C, Muller B, Lamb DC (2011) Live-cell visualization of dynamics of HIV budding site interactions with an ESCRT component. Nat Cell Biol 13:469–474
128. Jouvenet N, Zhadina M, Bieniasz PD, Simon SM (2011) Dynamics of ESCRT protein recruitment during retroviral assembly. Nat Cell Biol 13:394–401
129. Campbell EM, Perez O, Anderson JL, Hope TJ (2008) Visualization of a proteasome-independent intermediate during restriction of HIV-1 by rhesus TRIM5alpha. J Cell Biol 180:549–561
130. Campbell EM, Hope TJ (2008) Live cell imaging of the HIV-1 life cycle. Trends Microbiol 16:580–587

Index

Y. Mély and G. Duportail (eds.), *Fluorescent Methods to Study Biological Membranes*, Springer Ser Fluoresc (2013) 13: 479–486, DOI 10.1007/4243_2012,

Printed by Printforce, the Netherlands